86.00
68I

/138

Nitroazoles

The C-Nitro Derivatives of Five-Membered N– and N,O– Heterocycles

VOLUME 1
Organic Nitro Chemistry

Managing Editor:
Dr. Henry Feuer
Purdue University
West Lafayette, Indiana 47907 (U.S.A.)

EDITORIAL BOARD

Hans H. Baer
Ottawa, Canada
Robert G. Coombes
London, England
Leonid T. Eremenko
Chernogolovka, USSR
Milton B. Frankel
Canoga Park, CA, USA
Mortimer J. Kamlet
Silver Springs, MD, USA
Nathan Kornblum
West Lafayette, IN, USA
Philip C. Myhre
Clairmont, CA, USA
Arnold T. Nielsen
China Lake, CA, USA
Wayland E. Noland
Minneapolis, MN, USA

George A. Olah
Los Angeles, CA, USA
Noboru Ono
Kyoto, Japan
C.N.R. Rao
Bangalore, India
John H. Ridd
London, England
Glen A. Russell
Ames, IA, USA
Dieter Seebach
Zurich, Switzerland
Francois Terrier
Rouen, France
Heinz G. Viehe
Louvain-la-Neuve, Belgium
Zhou Fa-qi
Peking, China

© 1986 VCH Publishers, Inc. Deerfield Beach, Florida

Distribution: VCH Verlagsgesellschaft mbH, P.O. Box 1260/1280, D-6940 Weinheim,
Federal Republic of Germany

USA and Canada: VCH Publishers, Inc., 303 N.W. 12th Avenue, Deerfield Beach, FL 33442-1705, USA

Nitroazoles

The C-Nitro Derivatives of Five-Membered N– and N,O– Heterocycles

Joseph H. Boyer

Joseph H. Boyer
Department of Chemistry
University of New Orleans
Lakefront
New Orleans, Louisiana 70148

Library of Congress Cataloging-in-Publication Data

Boyer, Joseph H., 1922–
 Nitroazoles: the C-nitro derivatives of N– and N,O– five-membered heterocycles.

 (Organic nitro chemistry series; 1)
 Bibliography: p.
 Includes index.
 1. Nitro compounds. 2. Pyrylium compounds. I. Title. II. Series.
 QD401.B76 1986   547'.593   86-15667
 ISBN 0-89573-148-7

© 1986 VCH Publishers, Inc.

This work is subject to copyright.

All rights are reserved, whether the whole or part of the material is concerned, specifically those of translation, reprinting, re-use of illustrations, broadcasting, reproduction by photocopying machine or similar means, and storage in data banks.

Registered names, trademarks, etc. used in this book, even when not specifically marked as such, are not to be considered unprotected by law.

Printed in the United States of America.

ISBN 0-89573-148-7 VCH Publishers
ISBN 3-527-26513-9 VCH Verlagsgesellschaft

Series Foreword

In the organic nitro chemistry era of the fifties and early sixties, a great emphasis of the research was directed towards the synthesis of new compounds that would be useful as potential ingredients in explosives and propellants.

In recent years, the emphasis of research has been directed more and more toward utilizing nitro compounds as reactive intermediates in organic synthesis. The activating effect of the nitro group is exploited in carrying out many organic reactions, and its facile transformation into various functional groups has broadened the importance of nitro compounds in the synthesis of complex molecules.

It is the purpose of the Series to review the field of organic nitro chemistry in its broadest sense by including structurally related classes of compounds such as nitroamines, nitrates, nitrones, and nitrile oxides. It is intended that the contributors who are active investigators in the various facets of the field will provide a concise presentation of recent advances that have generated a renaissance in nitro chemistry research.

Henry Feuer
Purdue University

Preface

The advisability of reviewing the chemistry of nitroheterocycles was stimulated by many discussions. It was generally believed that significant contributions from the development of the area needed to be more readily accessible. After an invitation from Professor Henry Feuer to contribute to his Series on Organic Nitro Chemistry, a formulation of the present treatment was determined.

In this volume the chemistry of the nitro derivatives of the monocyclic and polycyclic five-membered N– and N,O– heterocycles is presented. It covers the azoles — pyrroles, imidazoles, pyrazoles, triazoles, tetrazoles, isoxazoles, oxazoles, and oxadiazoles — and the corresponding azolines and azolidines. Discussions of polycyclic compounds have been restricted by the exclusion of examples in which a nitro substituent is not located on one of these rings of immediate interest. Attention to the chemistry of side-chain functional groups diminishes as their remoteness from a heterocyclic nitro substituent increases.

Throughout the discussions attention is given to the illustration of chemical principles and to the collection of data to provide the reader with the latest information and a critical evaluation. References to reviews of related material are given to provide background information. A liberal selection of examples of preparations, reactions, and properties is offered; however, an exhaustive coverage of the literature cannot be claimed.

Most of the information has been provided by investigations initiated to satisfy requirements for useful high-energy materials, desirable biological properties, and compounds required for preparative work. A rapid development of the chemistry of the nitroazoles in the Soviet Union began about 1960 and has provided more journal publications of research in the area than were found for any other country. The Russian emphasis on investigating nitrotriazoles, nitroisoxazoles, and nitrooxadiazoles has been outstanding.

Cyclic formulas are shown in accordance with the general practice of omitting hydrogen atoms attached to ring carbon atoms (hydrogen attached to a ring nitrogen atom is shown) except where their presence for clarity is indicated. To avoid the possibility of confusion, rings are shown with the appropriate number of double bonds. In general, nomenclature follows the style of *Chemical Abstracts*. However, less systematic and trivial nomenclature has been used on occasion where no sacrifice of clarity ensued.

Financial support from the Office of Naval Research is gratefully acknowledged. It provided not only partial salary for the author but also stenographic, drafting, and clerical services otherwise unavailable at the University of Illinois at Chicago, where most of the work was carried out. I am indebted to Walter Berkowicz, Diane Bily, Patricia Cambell, Myrna Egdorf, Regina Gierlowski, and Frank Sticha for typing the manuscript; to Catherin Ho, Deborah Kallick, May Szeto, Arvydas Zygas, and Christopher Wieczorek for drafting formulas and equations; and to Stanley Merchant, Silvio Emery, Chulani Karunatilake, Linda Rosen, and May Szeto for clerical and library assistance. To Govindarajulu Kumar, Thanikavelu Manimaran, and Philip Pagoria I give special thanks for their careful proofreading. Finally I wish to thank the Lawrence Livermore Laboratory for financial assistance in the initiation of the project, and to the University of Illinois at Chicago for one quarter sabbatical leave to devote to writing the manuscript.

Contents

PREFACE	
CHAPTER 1. NITROPYRROLES	1
INTRODUCTION	1
NITRATION OF SIMPLE PYRROLES	1

Nitric Acid in Acetic Anhydride
Pyrroles and Alkylpyrroles
Acetamidopyrroles
Acylpyrroles
Halogenopyrroles
Miscellaneous Pyrroles
Dinitration
Nitric Acid Alone and with Other Acids
Nitrous Acid, Nitrite or Nitrate Esters

RING CLOSURE AND RING TRANSFORMATION	22
NITRATION OF CONDENSED RING SYSTEMS	27

Indoles
Azaindoles
Indolizines
Pyrrolo[2,1-*b*]thiazoles
Thiazolo[2,3,4-*c,d*]pyrrolizines
5-*H*-Pyrrolo[2,1-*a*]isoindoles

CHEMICAL PROPERTIES	40

pKa Values for Nitropyrrolecarboxylic Acids
Reduction
Deoxygenation
Oxidation
Decarboxylation
Nucleophilic Reactions
Mobility of NH and CH Bonds

Electrophilic Reactions
Free Radical Reactions
Unclassified Reactions
Thermal and Photochemistry

PHYSICAL PROPERTIES 62

Ultraviolet Spectroscopy
Nuclear Magnetic Resonance Spectroscopy
Mass Spectroscopy
Infrared Spectroscopy
Correlation Methods

BIOLOGICAL PROPERTIES 70

REFERENCES 72

CHAPTER 2. NITROIMIDAZOLES 79

INTRODUCTION 79

PREPARATIONS OF SIMPLE NITROIMIDAZOLES 80

Nitration and Nitrolysis
Imidazole and alkylimidazoles
α-Hydroxyalkyl and acylimidazoles
Arylimidazoles
Imidazole halides
Imidazole amines, ethers, and sulfides
Nitronium Tetrafluoroborate
Dinitrogen Tetroxide
Acetyl Nitrate
Nitrolysis of a Diazonium Group
Nitration by Nitrites
Nitration of 2-lithioimidazoles
Polynitrations

NITROIMIDAZOLES BY RING FORMATION 93

OXIDATION OF 2-AMINOIMIDAZOLES 94

BIOCHEMICAL CONVERSION OF 2-AMINOIMIDAZOLES 95

CONDENSED RING SYSTEMS 95

Nitration and Nitrolysis
Imidazo[2,1-*b*]thiazoles
Imidazo[1,2-*a*]imidazoles
Imidazo[1,2-*a*]pyridines
Imidazo[1,5-*a*]pyridines

Imidazo[1,2-b]pyridazines
Imidazo[1,2-a]pyrimidines
Imidazo[4,5-d]pyrimidines (purines)
Imidazo[1,2-a]benzimidazoles
Imidazo[1,2-f]xanthines
Ring Formation
Imidazo[5,1-b]thiazoles
Imidazo[1,2-a]pyrimidines
2,3-Dihydroimidazo[2,1-b]oxazoles
Di- and tetrahydroimidazo[1,5-c]-1,2,4-triazines
Imidazo[2,1-d]-1,4-oxazepines
Imidazo[2,1-a]isoquinolines
9H-Imidazo[1,2-a]indoles

REACTIONS OF NITROIMIDAZOLES 107
Halogenation
C-Alkylation
C-Acylation
N-Alkylation, Amination, and Arylation
Alkylations
Aminations
Arylations
Arylalkylations
Alkylations with functionalized alkyl groups
Glycosylations
N-Acylations
Condensations
Methylene groups
Amines
Carboximidates
Aldehyde Derivatives
Oximes and nitrones
Hydrazones
Schiff bases
Vinylimidazoles
α-Hydroxybenzylimidazoles
Replacements of Ring Halogens and Pseudohalogens
Halogen interchange
Cyanide
Amines and amides
Alkoxides and Phenoxides
Mercaptides
Replacement of Ring Nitro Groups
Rearrangements
Replacements in Side Chains

2-Chloromethylimidazoles
Imidazolecarbonyl chlorides
Esterifications and ester reactions
1-ω-Halo(pseudohalo)alkylimidazoles
        Oxidations in Side Chains
        Chemical Reductions
Radical anions
Hydroxylamines
Amines
        Catalytic Reduction
        Electrochemical and Radiolysis Reduction
        Imidazoles as Acids
NH Acids
CH Acids

SPECTROSCOPY	154

        Nuclear Magnetic Resonance Spectroscopy

Hydrogen
Carbon
Nitrogen

        Infrared Spectroscopy
        Ultraviolet Spectroscopy
        Mass Spectroscopy
        Photoelectron Spectroscopy
        X-Ray Spectroscopy
        Electron Spin Resonance Spectroscopy

OTHER PHYSICAL PROPERTIES	163
BIOCHEMICAL PROPERTIES	165
REFERENCES	166
CHAPTER 3. NITROPYRAZOLES	186
INTRODUCTION	186
PREPARATIONS OF MONOCYCLIC PYRAZOLES	186

        Nitration and Nitrolysis
Pyrazole, alkyl, and arylpyrazoles
Pyrazolyl halides
Pyrazole carboxylic acid derivatives
Hydroxy and alkoxypyrazoles
Amino and amidopyrazoles
Pyrazole-N-oxides
Silylpyrazoles

Rearrangements of 1-Nitropyrazoles
Pyrazole diazonium compounds
Oxidation of amino and Nitrosopyrazoles
Aromatization of Nitropyrazolines and Nitropyrazolidines

RING FORMATIONS 207

Nitrilimines and Nitroethenes
Nitrocarbonyl compounds and Hydrazines
Nitroketenaminals and Hydrazines
α-Nitro-β-hydrazinoacrylates

RING TRANSFORMATIONS 211

Pyrimidines
Pyrrolidines
Isoxazoles
Pyridines
Pyridazines

POLYCYCLIC NITROPYRAZOLES 216

Indazoles
Pyrazolo[1,5-a]pyridines
Pyrazolo[4,3-b]pyridines
Pyrazolo[3,4-b]pyridines
Pyrazolo[3,4-d]pyrimidines
Pyrazolo[1,5-a]pyrimidines
4H-Pyrazolo[1,5-a]benzimidazoles
Pyrazolo[1,5-c]quinazolines
Pyrazolo[1,5-a]-s-triazines
2H,5H-Pyrazolo[4,3-c]pyrazoles

REACTIONS OF NITROPYRAZOLES 223

N-Alkylation and N-Arylation
Acylation
Halogenation
Halogen Replacement
Replacement of Nitro Groups
Catalytic Reduction
Chemical Reduction
Conversions of Aminonitropyrazoles
Oxidation
Metal Salts of Nitropyrazoles
Acids and Bases
Thermolysis
Photolysis
Picrolonic acid
Miscellaneous

PROPERTIES 248

Nuclear Magnetic Resonance Spectroscopy
Infrared Spectroscopy
Ultraviolet Spectroscopy
Mass Spectroscopy
X-Ray Spectroscopy
Other Physical Properties
Biological Properties

REFERENCES 258

CHAPTER 4. NITROTRIAZOLES AND NITROTETRAZOLES 268

INTRODUCTION 268

NITROTRIAZOLE PREPARATIONS 268

Nitration
Rearrangement of Nitramines
Conversion of Diazonium Salts
Ring Formations
Nucleophilic Substitutions
Oxidation of Aminotriazoles
Fused Ring Triazoles

NITROTETRAZOLE PREPARATIONS 276

REACTIONS 277

Alkylations
Replacement of Nitro and Halo Groups
Reduction
Miscellaneous

PROPERTIES 291

Acid-Base
Spectroscopy
Miscellaneous

REFERENCES 296

CHAPTER 5. NITRODERIVATIVES OF ISOXAZOLES, OXAZOLES, AND OXADIAZOLES 301

INTRODUCTION 301

ISOXAZOLES 301

Preparations
Nitration and Nitrolysis
Nitrones, Nitronates, and Tetranitromethane Reactions
Nitrile Oxide Reactions
Other Acyclic Nitro Compound Reactions
Ring Conversions
Diazoalkane Reactions
Reactions
Isoxazolines to Isoxazoles
Reduction
Displacement of the Nitro Group
Side Chain Reactions
Ring-Opening

PROPERTIES 335

Acid-Base
Infrared Spectroscopy
Ultraviolet Spectroscopy
Nuclear Magnetic Resonance Spectroscopy
Mass Spectroscopy
Other Physical-Chemical Properties
Biological Properties

OXAZOLES 340

OXADIAZOLES 341

Preparations
Reactions
Chemical and Physical Properties
Biological Properties

REFERENCES 344

Chapter 1

Nitropyrroles

INTRODUCTION

Brief discussions of the nitropyrrole, pyrroline, and pyrrolidine derivatives of mono- and polycyclic systems are found in books[1-3] and other treatises[4,5] that offer a broad coverage of the chemistry of the heterocycles. As recently as 1950 nitropyrroles were conspicuously absent in a chapter, "The Chemistry of Pyrrole and its Derivatives."[6] Over 200 publications since that time have brought them out of obscurity, but there are only a few reports on nitropyrrolines and nitropyrrolidines. The literature has been covered through *Chem. Abstr.* **1983**, *98*; in an addendum to this work selected publications from *Chem. Abstr.* **1983**, *99* and **1984**, *100* are listed.

NITRATION OF SIMPLE PYRROLES

Nitric Acid in Acetic Anhydride

Pyrroles and Alkylpyrroles

At $-10°$ for 90 min pyrrole and nitric acid (70%) in a slight molar excess in acetic anhydride produced a mixture (70%) of 2- and 3-nitropyrroles (**1-1, 1-2**).

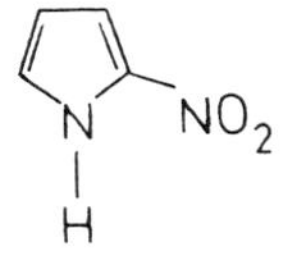

1-1 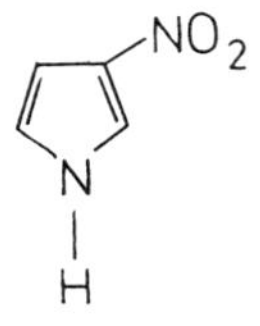   **1-2**

This is a preparative reaction for 2-nitropyrrole (80% of the product mixture) that can be isolated by chromatography.[7-9]

Acetyl nitrate was spectroscopically detected in the reaction mixture, Eq. (1-1). Its role as an effective nitrating species was supported by kinetic results, by the recovery of unchanged pyrrole when acetic anhydride was replaced by acetic acid, acetonitrile, ether, nitromethane, and tetrahydrofuran at 0°C, and by investigations with acetyl nitrate (from the distillation of a mixture of acetic anhydride and dinitrogen pentoxide) and with dinitrogen pentoxide.

Acetyl nitrate (separately prepared) and pyrrole gave considerable amounts of black tar and lower yields of nitropyrroles difficult to reproduce. Since nitrite (3%) was detected in samples of acetyl nitrate the probability of a competitive nitrosation was recognized. That nitrosopyrroles were not precursors to nitropyrroles was shown by investigations on mixtures of pyrrole with dinitrogen tetroxide in acetic anhydride at 20° and with nitric acid and dinitrogen tetroxide (0.03 mol) in acetic anhydride at 20°. Nitropyrroles were not detected in the opaque mixture from the former and were detected in trace amount ($<0.1\%$) from the latter experiment.[7]

Dinitrogen pentoxide was moderately effective in nitrating pyrrole at $-10°$ in acetonitrile and nitromethane; its failure to give the reaction in acetic anhydride, acetic acid, or tetrahydrofuran was apparently the result of competitive oxidative degradation.[7]

$$(CH_3CO)_2O + HNO_3 \rightleftarrows CH_3COONO_2 + CH_3CO_2H$$
$$CH_3COONO_2 + HNO_3 \rightleftarrows CH_3CO_2H + N_2O_5$$
$$2\,N_2O_5 \rightleftarrows 2\,N_2O_4 + O_2 \qquad\qquad (1\text{-}1)$$

At 25° for 18 h or at 50° for 1 h, a mixture of nitric and acetic acids transformed pyrrole into a black tar. Neither pyrrole nor a nitropyrrole could be detected in the mixture. Similar "polymerization" of pyrrole by strong acid has long been known. Over a temperature range from $-70°$ to 20° pyrrole and a slight molar excess of nitric acid (98%) in acetic anhydride gave 2- and 3-nitropyrroles in a fairly constant ratio of 4:1.[7]

Competitive nitrations in acetic anhydride at 0° established the order of reactivity: benzene < toluene < p-xylene < thiophene < mesitylene < pyrrole. Reactivities at the 2- and 3-positions in pyrrole were estimated to be 130,000 and 30,000 times the reactivity of benzene, respectively.[7] The nitration of a pyrrole to an N-nitropyrrole has not been reported.

The α-directing effect in 1-methylpyrrole (**1-3a**) is apparently weaker than it is in pyrrole since one equivalent of nitric acid in acetic anhydride at $-10°$ to 5° gave 1-methyl-2-nitropyrrole (**1-4a**) (37%) and 1-methyl-3-nitropyrrole (**1-5a**) (21%) after purification of a crude product mixture (80%).[8] A similar ratio of about 2:1 for the 2- and 3-nitro isomers was obtained in the nitration of 1-phenylpyrrole (**1-3f**).[10] Somewhat different results were obtained in a recent study of the nitration of pyrroles (**1-3a–d**) in acetic anhydride. The following yields for mixtures of 1-alkyl-2- (**1-4a–e**) and 3-nitropyrroles (**1-5a–e**) and the

2-nitro/3-nitro product ratio were obtained: R = CH$_3$, 34%, 3.15; R = (CH$_3$)$_2$CH, 62%, 0.56; R = (CH$_3$)$_3$C, 55%, 0.25; R = p-O$_2$NC$_6$H$_5$, 34%, 1.43; R = o-O$_2$NC$_6$H$_5$, 53%, 1.07. Larger 1-substituents permitted less nitration at the 2-position:[11] 1-benzylpyrrole (**1-3g**) gave the 2-nitro derivative (**1-4g**) (28%) and the 3-nitro derivative (**1-5g**) (42%).[12] When the initial molar ratio of nitric acid to the pyrrole (**1-3g**) was 2:1 the yield of derivative (**1-5g**) was diminished but 1-benzyl-2,4-dinitropyrrole (**1-6**) was also obtained (15%). Nitration of either mononitro compound (**1-4g**) or (**1-5g**) also gave the dinitro compound (**1-6**), but attempts to further nitrate by nitric acid in either acetic anhydride or sulfuric acid at temperatures 0°–100° were unsuccessful.[12] In a recently reported procedure cupric nitrate in acetic anhydride at 25° for 2 h converted 1-triisopropylsilylpyrrole to the 3-nitro derivative; desilylation by treatment with tetra-*n*-butylammonium fluoride in tetrahydrofuran at 25° for 5 min gave 3-nitropyrrole in an overall yield of 80%.[13]

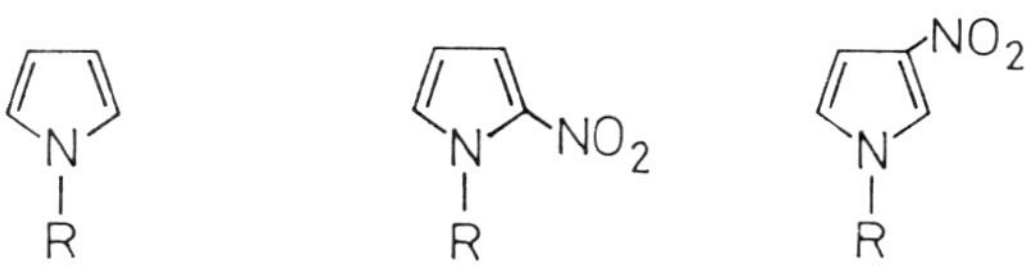

1-3   **1-4**   **1-5**

a, R = CH$_3$; **b**, R = (CH$_3$)$_2$CH; **c**, R = (CH$_3$)$_3$C;
d, R = p-O$_2$NC$_6$H$_5$; **e**, R = o-O$_2$NC$_6$H$_5$;
f, R = C$_6$H$_5$; **g**, R = C$_6$H$_5$CH$_2$

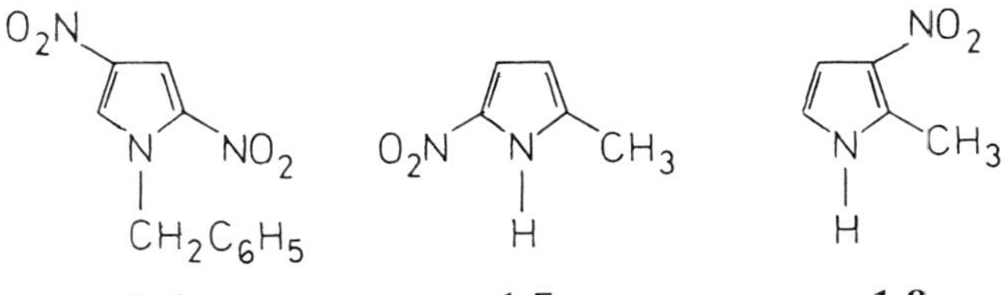

1-6   **1-7**   **1-8**

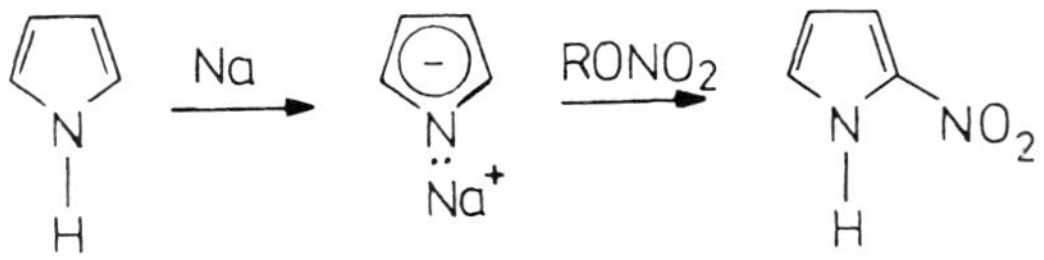

(1-2)

Identification of the only product isolated from the treatment of the sodium salt of pyrrole and isoamyl nitrate as 2-nitropyrrole (very low yield) rather than its 3-isomer (an earlier assignment), Eq. (1-2), was not inconsistent with these results.[9]

A red solid mixture obtained from the nitration of the 2-methylpyrrole in acetic anhydride at −50° gave yellow 2-methyl-5-nitropyrrole (**1-7**) (12%) and yellow 2-methyl-3-nitropyrrole (**1-8**) (2%) after chromatographic purification.[14] Other products which may represent further nitration remain to be identified. That 2-methyl-4-nitropyrrole was not detected was compatible with the small amount of 1,2-dimethyl-4-nitropyrrole (**1-11**) in the inefficient nitration of 1,2-dimethylpyrrole (**1-9**).[15] Nitration of 3-methylpyrrole has not been investigated.

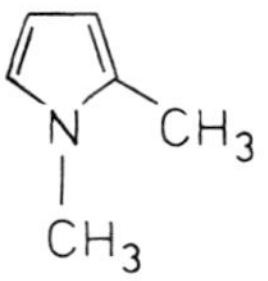

1-9 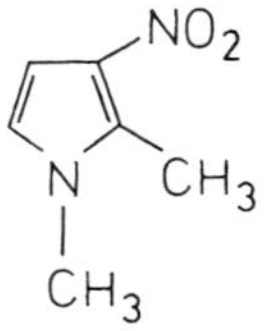 **1-10** 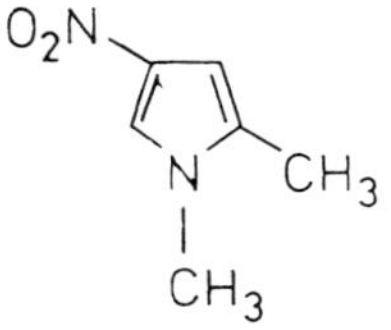 **1-11** 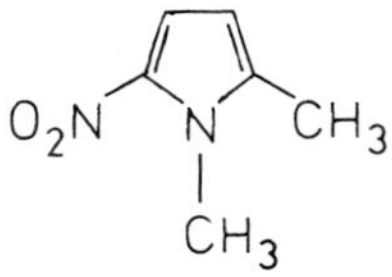 **1-12**

Nitration of the pyrrole (**1-9**) at $-5°$ in acetic anhydride gave the three expected mononitroderivatives, **1-10** (8%), **1-11** (5%), and **1-12** (13%). A dark sticky resin was obtained from 1,2,5-trimethylpyrrole when a similar nitration was tempted. After extensive chromatographic separation of the resin mixture 1,2,5-trimethyl-3-nitropyrrole (**1-13**) (2%) was isolated.[15] Dihydropyrrolizines gave 5- (**1-14**), 6- (**1-15**), and 7-nitro derivatives (**1-16**). The product ratio was controlled by the size of R'.[16]

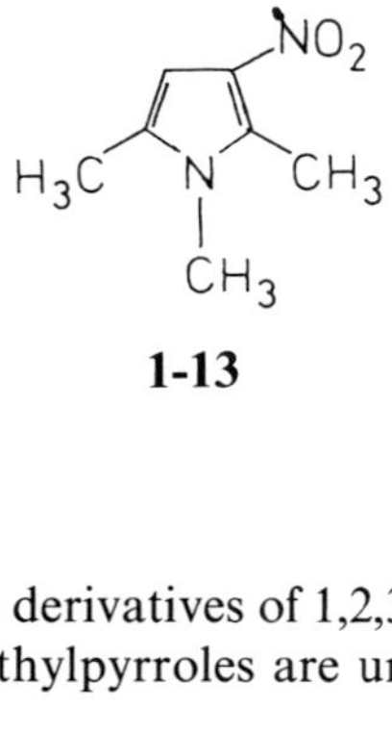

1-13

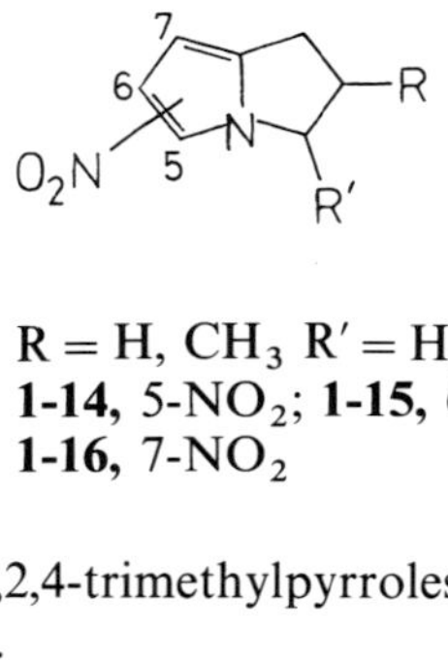

R = H, CH$_3$ R' = H, CH$_3$, (CH$_3$)$_3$C
1-14, 5-NO$_2$; **1-15**, 6-NO$_2$;
1-16, 7-NO$_2$

Nitro derivatives of 1,2,3- and 1,2,4-trimethylpyrroles and 1,2,3,4- and 1,2,3,5-tetramethylpyrroles are unknown.

Acetamidopyrroles

In the absence of a stabilizing electron-withdrawing substituent, aminopyrroles are generally sensitive to oxidation and to acids. Although the corresponding acetamidopyrroles are also susceptible to oxidation, they have been successfully nitrated under mild conditions. In acetic anhydride at $-15°$ 1-methyl-3-acetamidopyrrole (**1-17**) reacted with nitric acid (65%) to produce a mixture of its yellow mononitroderivatives: 2- (39%), 4- (7%), and 5- (2%). A similar nitration of the more sensitive 1,2-dimethyl-3-acetamidopyrrole (**1-18**) gave a low yield of an equimolar mixture of its two yellow mononitro derivatives.[17] Nitramide formation was apparently noncompetitive.

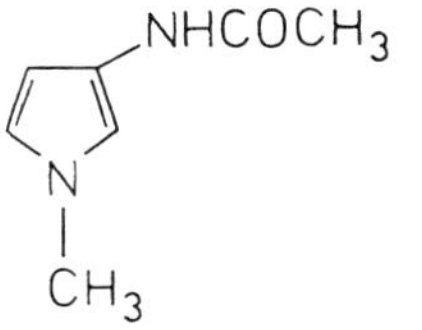

1-17 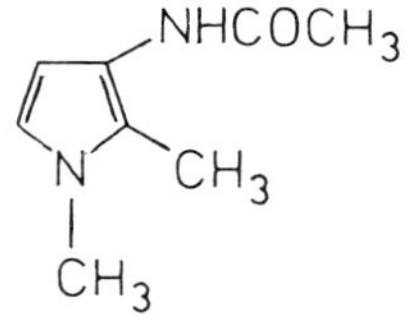   **1-18**

Nitric acid (65%) in acetic anhydride at 10° converted ethyl 1-methyl-4-acetamidopyrrole-2-carboxylate to a mixture (65%) of its 5-nitro (3 parts) and 5-nitroso (1 part) derivatives, Eq. (1-3).[17] A generation of the nitrosating agent can be attributed to a redox reaction between nitric acid and the pyrrole.

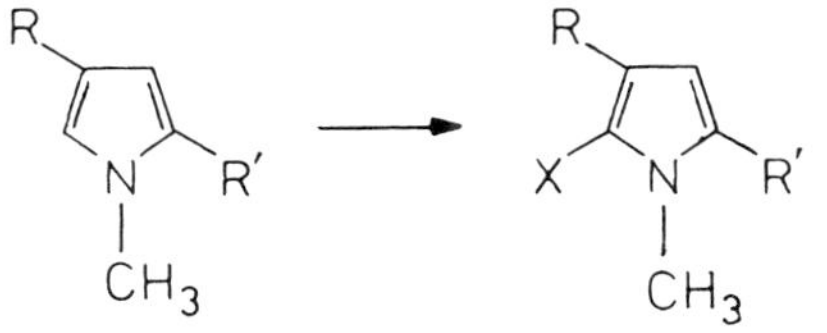

$$R = NHCOCH_3 \quad X = NO_2, NO$$
$$R' = CO_2C_2H_5 \tag{1-3}$$

Acylpyrroles

Mononitration efficiency ranged from 25% to 45% for 1-acyl-pyrrole (**1-19**) and to 60% for 2-acyl- (**1-20**) and 3-acyl- (**1-21**) isomers when carried out in acetic anhydride at −15° for 30 min. Equimolar amounts of 2- and 3-nitro derivatives of the pyrrole (**1-19**) ($R = CH_3$, OCH_3) were obtained along with trace amounts of 2- and 3-nitropyrroles apparently produced by hydrolysis. Alkaline hydrolysis of the crude product mixture, followed by a simple chromatographic separation, offered a more acceptable preparation for 3-nitropyrrole than was available by the direct nitration of pyrrole.[18] Nitration of 1-benzenesulfonylpyrrole followed by hydrolysis was recommended as a particularly useful route to 3-nitropyrrole free from contamination with the isomeric 2-nitropyrrole.[19]

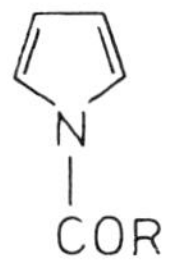

1-19 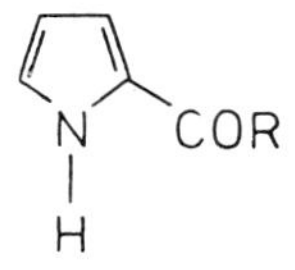   **1-20** 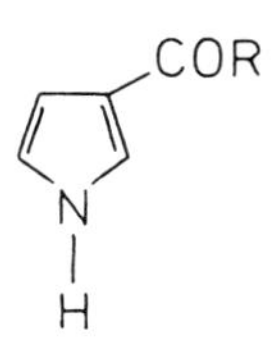   **1-21**

Nearly equimolar amounts of the 4- and 5-nitro derivatives were obtained from a similar nitration of 2-acylpyrroles (**1-20**) ($R = H$, CH_3, OH, OCH_3). There was no evidence for nitration at the 3-position.[18–21,23,24] The sensitive aldehyde (**1-20**) (R = H) gave each of the two mononitro derivatives in 20% yields.[20,21] A 5-nitro derivative was obtained from 3-acylpyrroles (**1-21**) (R = OH, OCH$_3$) with no evidence for the formation of 2- and 4-nitro derivatives.[22]

A recent report claimed that the nitration of 2-formylpyrrole (**1-20**) (R = H) in concentrated aqueous nitric acid at −2° gave a higher yield of the mixture of 4-nitro- (64 parts) and 5-nitro-2-formylpyrrole (36 parts). A similar reaction at −20° was claimed. In an unspecified yield 4-nitro- (37 parts) and 5-nitro-2-formylpyrrole (63 parts) were claimed for the same reaction by acetyl nitrate at −2°. The higher yield of the 4-isomer in concentrated nitric acid was attributed to nitration of a protonated form of the aldehyde (**1-20**) (R = H). An attempted nitration of 2-formylpyrrole in oleum resulted in a fire.[23]

Other 2-acylpyrroles (**1-20**) also gave 4- and 5-nitro derivatives by nitration in acetic anhydride: RCO = COCH$_2$Cl: 4-NO$_2$ (35%) and 5-NO$_2$ (25%); RCO = COCHCl$_2$: 4-NO$_2$ (22%) and 5-NO$_2$ (20%); RCO = COCCl$_3$: 4-NO$_2$ (40%) and 5-NO$_2$ (25%); RCO = CO(CH$_2$)$_3$Cl: 4-NO$_2$ (22%) and 5-NO$_2$ (20%); RCO = COCON(CH$_2$CH$_2$)$_2$O: 4-NO$_2$ (25%) and 5-NO$_2$ (12%).[24] A claim for a nearly doubled yield of 77% for the 4-nitro derivative of the pyrrole (**1-20**) (RCO = COCCl$_3$) with the comment that the 5-nitro isomer (yield not given) was also obtained[25] has not been confirmed.

A similar nitration of 1-methyl-2-formylpyrrole (**1-22**) at 0° for 30 min gave a mixture (41%) of the 3-nitro, 4-nitro, and 5-nitro derivatives in a ratio of 2:35:11.[26] In another report[27] equimolar amounts of the latter two isomers were claimed. Although nitration has sometimes replaced the formyl group,[28] neither replacement nor oxidation was noted in this instance.[26]

1-22   **1-23**   **1-24**

The recommended nitration of ethyl 1-methylpyrrole-2-carboxylate (**1-23**) called for strict temperature control at 30 ± 1° in the otherwise normal procedure. Lower temperature retarded the reaction and slightly higher temperatures lowered the yield by promoting the formation of by-products. The product mixture (42%) was nearly pure ethyl 1-methyl-4-nitropyrrole-2-carboxylate (**1-24**) with a trace amount of a coupled product (**1-25**). With no temperature control the reaction became brisk and produced a mixture in lowered yield of nearly equimolar amounts of the 4-nitro-**1-24** and the 5-nitro-**1-26** derivatives along with the dimeric compound **1-25**. Since nitric acid at a higher temperature did not convert a mixture of the compounds **1-24** and **1-26** into the dimeric compound **1-25**, it was believed that the oxidative dimerization occurred before both mononitrations had occurred.[15]

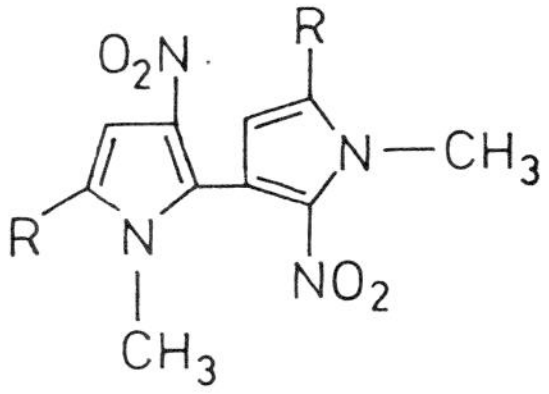

1-25
R = CO$_2$C$_2$H$_5$

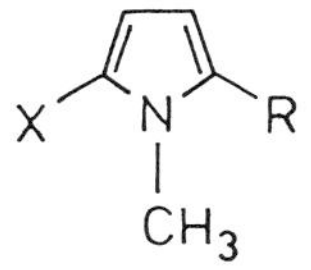

1-26 X = NO$_2$
R = CO$_2$C$_2$H$_5$
1-27 X = CH$_3$
R = CO$_2$C$_2$H$_5$

An attempted nitration of ethyl 1,5-dimethylpyrrole-2-carboxylate (**1-27**) in acetic anhydride gave a dark resin from which a trace of product could be extracted;[15] under similar conditions nitration of 1-methylpyrrole-2-carboxylic acid (**1-28**) gave its 4-nitro-derivative (**1-29**) (11%) in the first step of a preparation of distamycin A, an antiviral agent.[29]

In the nitration of 2,5-diphenyl-3-pyrryl 3,5-dimethyl/diphenyl-4-isoxazoyl ketone (**1-30**) the only reaction detected gave the 4-nitro derivatives (**1-31, 1-32**) (75%, 80%).[30] A slightly lower efficiency was found in the similar transformation of 2,5-diphenyl-3-phenacylsulfonylpyrrole into its 4-nitro derivative (**1-33**) (65%).[31] Since nitromethane was successful as a solvent for acetic anhydride, nitric acid, and the pyrrole (**1-30**) in the formation of **1-31** and **1-32**, its further investigation in this role is warranted.

There is added interest in the nitration of 1-hydroxy-2-methyl-3-acetyl-5-(5-methylisoxazol-3-yl)pyrrole in acetic anhydride to its 4-nitro derivative (**1-34**) (80%)[32] insofar as it is a rare example of the nitration of an N-hydroxypyrrole. Oxidation was apparently not competitive.

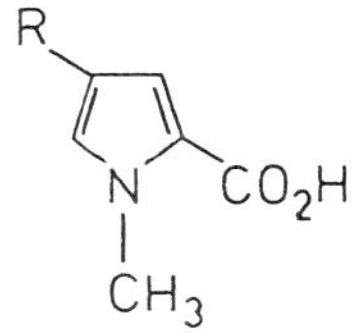

1-28 R = H
1-29 R = NO$_2$

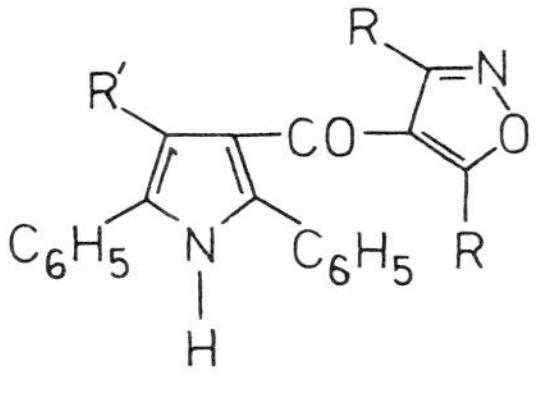

1-30 R = CH$_3$/C$_6$H$_5$, R′ = H
1-31 R = CH$_3$, R′ = NO$_2$
1-32 R = C$_6$H$_5$, R′ = NO$_2$

The 4-nitro- and 5-nitro- derivatives of 2-acetylpyrrole gave respectively the 4,5-dinitro (**1-35**) and 3,5-dinitro (**1-36**) derivatives in low yield; a trace of 2,4-dinitropyrrole (by replacement of an acetyl group) from the former and a trace of the 4,5-dinitro derivative (**1-35**) from the latter were also detected. The single product, methyl 3,5-dinitropyrrole-2-carboxylate (**1-38**) in good yield was obtained from the ester (**1-37**); but the ester group was replaced by the nitro group in the nitration of the triiodoester (**1-39**) into 2-nitro-3,4,5-triiodopyrrole (**1-40**).[33]

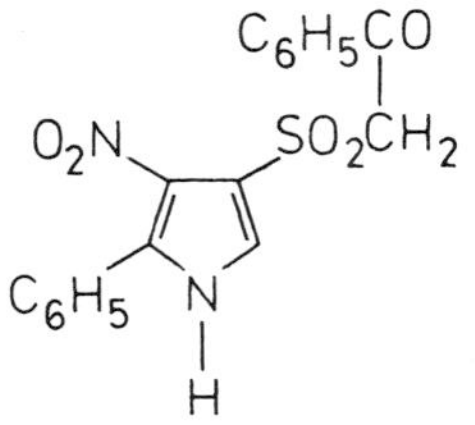

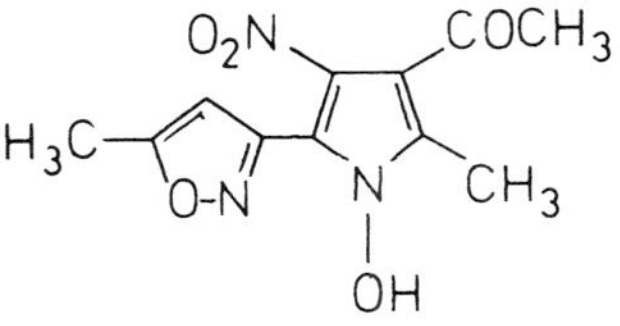

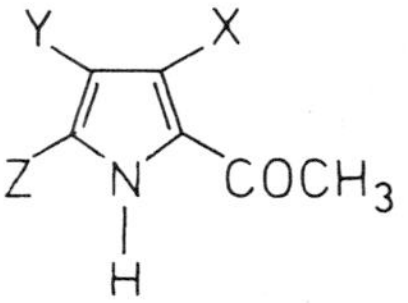

1-35 X = H, Y = Z = NO$_2$
1-36 Y = H, X = Z = NO$_2$

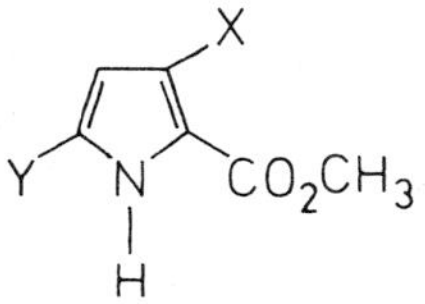

1-37 X = H, Y = NO$_2$
1-38 X = Y = NO$_2$

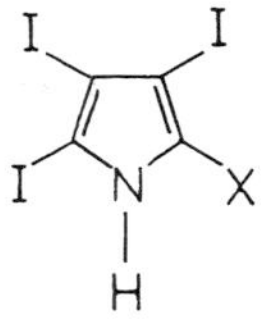

1-39 X = CO$_2$CH$_3$
1-40 X = NO$_2$

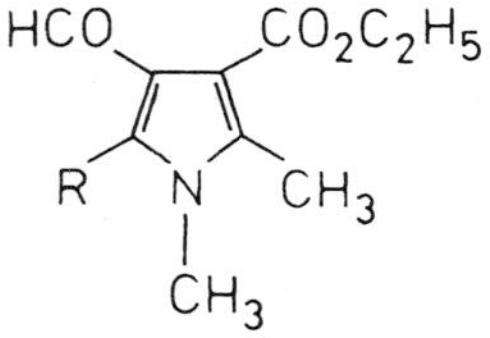

1-41 R = H
1-42 R = NO$_2$

Nitration of ethyl 1,2-dimethyl-4-formylpyrrole-3-carboxylate (**1-41**) in acetic anhydride to its 5-nitro derivative (**1-42**) (96%) was highly efficient.[34] It is of interest that neither acyl group was replaced. In contrast, nitration of ethyl 5-formylpyrrole-2-carboxylate (**1-43a**) replaced the formyl group to give the 5-nitropyrrole ester (**1-43b**) (16%), and also gave ethyl 4-nitro-5-formylpyrrole-2-carboxylate (**1-43c**) (18%) and ethyl 3-nitro-5-formylpyrrole-2-carboxylate (**1-43d**) (14%).[28] Similarly, ethyl 4-formylpyrrole-2-carboxylate (**1-44a**) gave ethyl 5-nitro-4-formylpyrrole-2-carboxylate (**1-44b**) (29%), the 4-nitropyrrole ester (**1-44c**) (2%), and ethyl 3-nitro-4-formylpyrrole-2-carboxylate (**1-44d**) (3%).[28]

Five 4-substituted derivatives (**1-45a–e**) of ethyl 5-methylpyrrole-2-carboxylate were selected for a nitration study. Nitric acid in acetic anhydride at −20° for 16 h converted the 3-bromo (**1-45a**) to the 3-nitro derivative (**1-45f**) (20%), but failed to nitrate the 3-formyl derivative (**1-45b**). Aqueous concen-

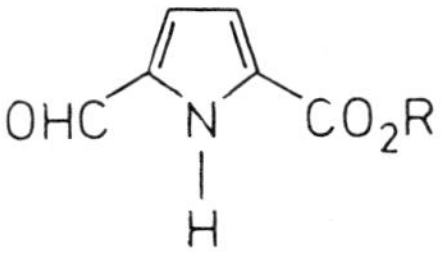

1-43a (OHC–pyrrole–CO₂R, 2,5)

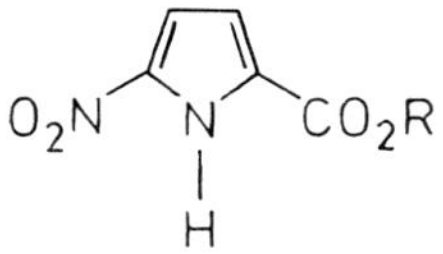

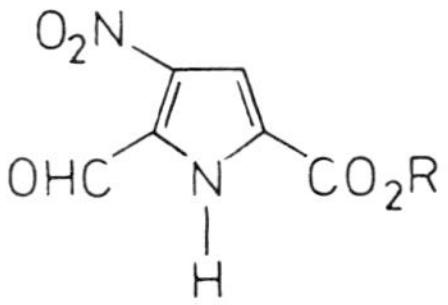

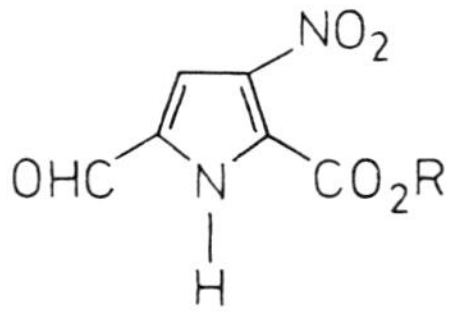

1-43b (O₂N–pyrrole–CO₂R, 2,5)

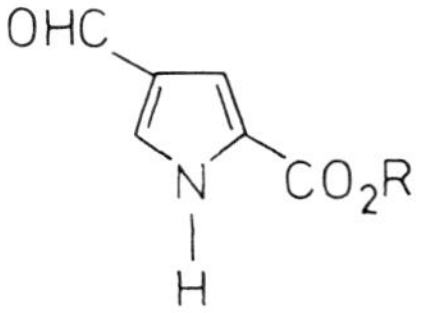

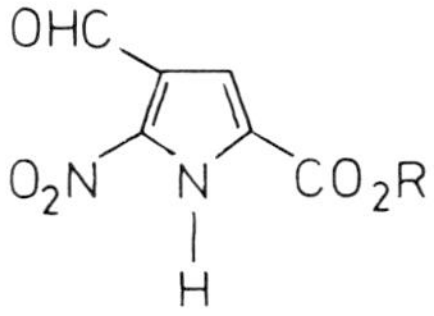

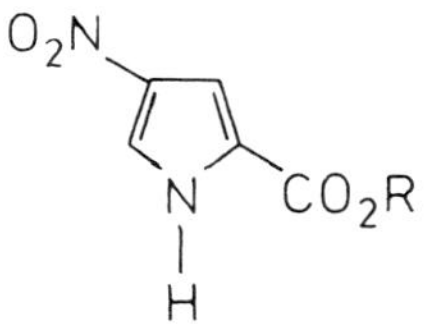

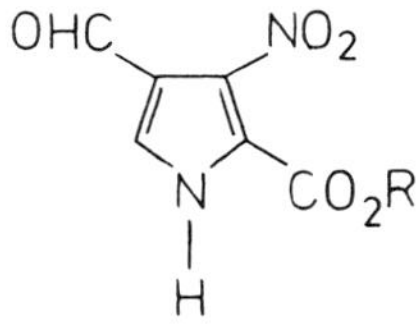

1-43c (3-O₂N, 2-OHC, 5-CO₂R)

1-43d (3-NO₂, 2-OHC, 5-CO₂R)

1-44a (3-OHC, 5-CO₂R)

1-44b (3-OHC, 2-O₂N, 5-CO₂R)

1-44c (3-O₂N, 5-CO₂R)

1-44d (3-OHC, 4-NO₂, 2-CO₂R)

trated nitric acid converted **1-45a** to a complex mixture, failed to nitrate **1-45b** at 0° for 45 min, but converted it to **1-45f** (34%) in an exothermic reaction at ambient temperatures, converted the 3-acetyl (**1-45c**) and 3-benzoyl (**1-45d**) derivatives to traces of **1-45f**, and the 3-methyl derivative (**1-45e**) by oxidation to the ethyl ester (**1-45g**) of 3-methylpyrrole-2,5-dicarboxylic acid.[35]

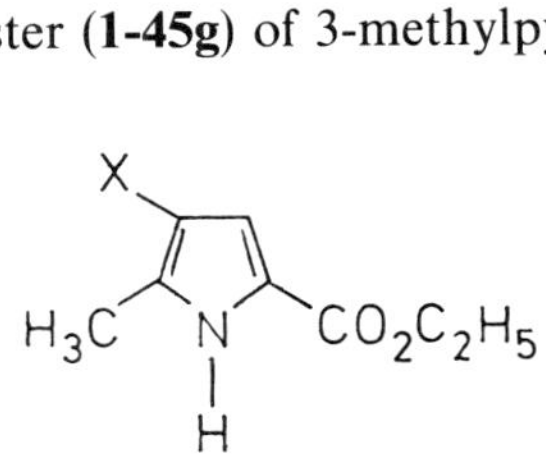

1-45a X = Br
1-45b X = CHO
1-45c X = COCH₃
1-45d X = COC₆H₅

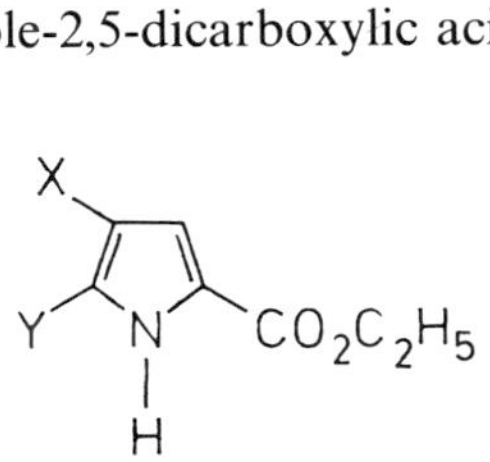

1-45e X = CH₃, Y = CH₃
1-45f X = NO₂, Y = CH₃
1-45g X = CH₃, Y = CO₂H

A compound, $C_8H_8N_2O_8Br_2$, produced from dimethyl pyrrole-2,5-dicarboxylate (**1-46a**), bromine, and nitric acid and from the 3,4-dibromo derivative of **1-46a** and nitric acid was thought to be the nitrate ester (**1-46b**) of dimethyl 2,5-dihydro-2,5-dihydroxy-3,4-dibromopyrrole-2,5-dicarboxylate. Heating an aqueous solution of the nitrate (**1-46b**) produced dimethyl 2,5-dihydroxy-2,5-dihydro-3,4-dibromofuran-2,5-dicarboxylate (**1-46c**), an assignment partially supported by an efficient conversion to dibromomaleic acid on treatment with nitric acid. Nitrous oxide probably resulted from the known breakdown of nitramine, an intermediate to be expected from a sequential conversion of the O-nitro compound (**1-46b**) to an N-nitro isomer (**1-46d**) and a diketone (**1-46e**), an expected precursor to the dihydrofuran (**1-46c**), Eq. (1-4).[36]

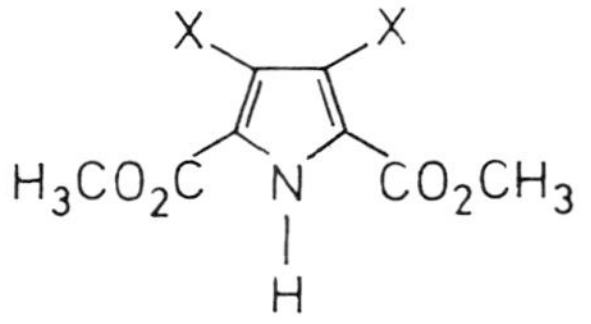

1-46a X = H

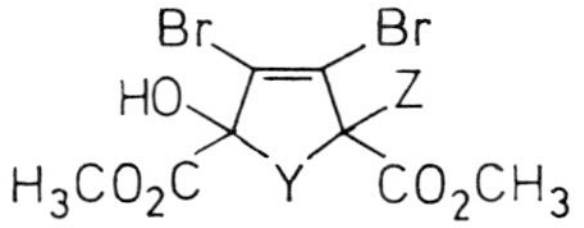

1-46b Y = NH, Z = ONO_2
1-46c Y = O, Z = OH

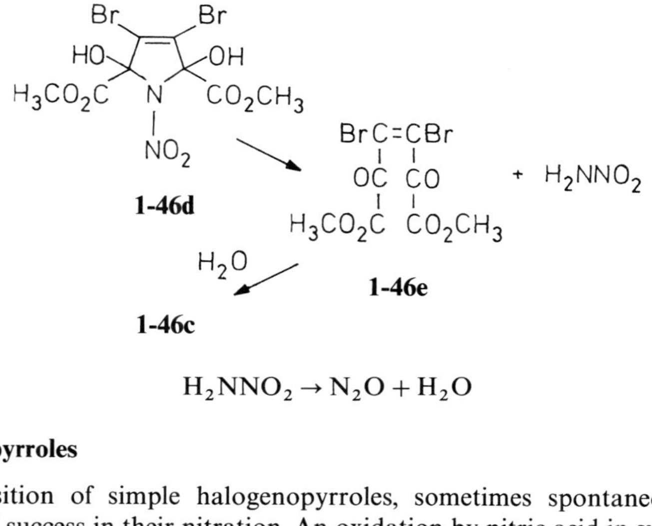

1-46d

1-46e

1-46c

$$H_2NNO_2 \rightarrow N_2O + H_2O \quad (1\text{-}4)$$

Halogenopyrroles

Decomposition of simple halogenopyrroles, sometimes spontaneous,[37] has minimized success in their nitration. An oxidation by nitric acid in sulfuric acid at 10° for 30 min accounted for the formation of 3-bromo-pyrrolo[2,3-b]pyridin-2-one (**1-48**) (70%) from the bromopyrrolopyridine (**1-47**). This interesting product resisted purification; its identification was based on IR and NMR spectroscopic analyses and mass spectroscopic determination of the molecular weight.[38]

Where nitric acid in acetic anhydride failed, concentrated aqueous nitric acid at 0° transformed both the 4-iodo- (**1-49a**) and 3,4-diiodo-**1-49b** derivatives of ethyl 5-methylpyrrole-2-carboxylate (**1-49c**) ($R = C_2H_5$) to the 4-nitro-3-iodo-derivative (**1-49d**), Eq. (1-5). Although the reaction may remain unaccounted for, two rationales were suggested. In one, a proton-initiated disproportionation of the iodo derivative (**1-49a**) to the pyrrole (**1-49c**) and its diiodide (**1-49b**) was followed by nitrolysis to the product (**1-49d**), Eq. (1-5). In the other, a rearrangement and deprotonation of an intermediate cation (**1-51**) followed its formation from the iodide (**1-49a**) and the nitronium cation, Eq. (1-6).[35] The formation of ethyl 5-methyl-4-iodo-3-nitropyrrole-2-carboxylate by the migration of the nitro group apparently did not occur. The proposed intermediacy of ethyl 5-methyl-4-nitropyrrole-2-carboxylate (**1-50**) was erroneous since compound **1-50**[39] failed to undergo iodination either by alkaline iodination or by iodine in nitric acid.[35]

1-47

1-48

1-49a
$X = CO_2R$

1-49b
$X = CO_2R$

1-49c
$X = CO_2R$

1-49d
$X = CO_2R$

(1-5)

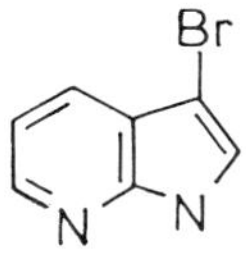

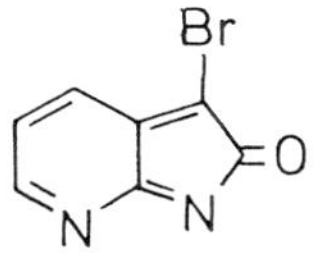

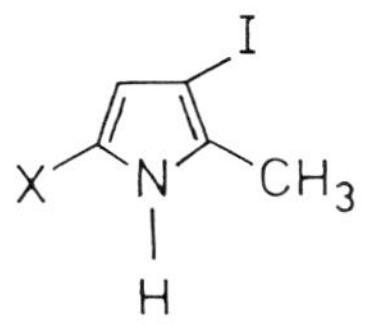

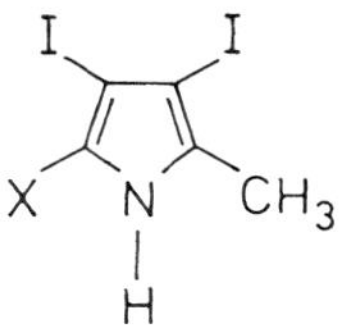

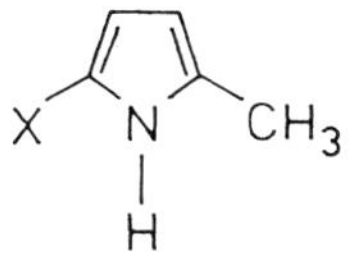

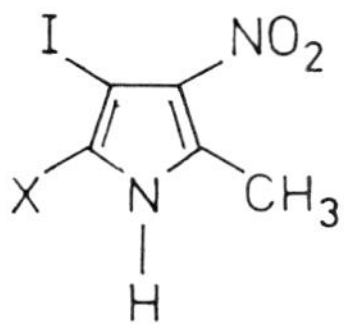

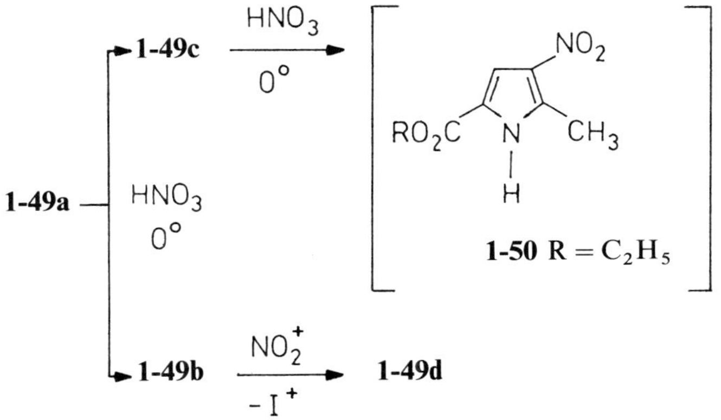

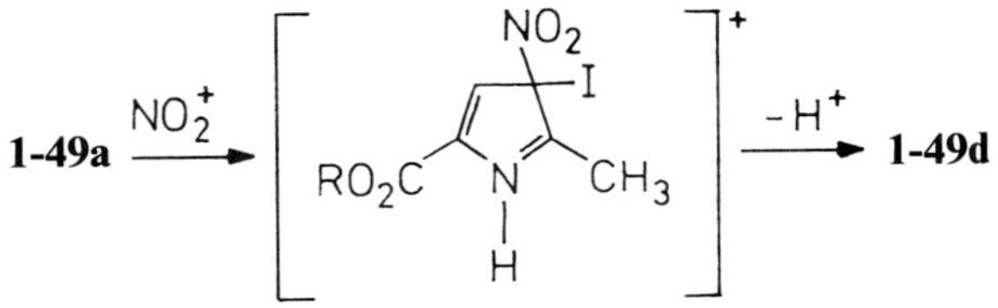

$$\textbf{1-49a} \xrightarrow{NO_2^+} \left[\begin{array}{c} \text{RO}_2\text{C}\overset{NO_2}{\underset{\underset{H}{N}}{\diagdown}}\overset{I}{\diagup}\text{CH}_3 \end{array} \right]^+ \xrightarrow{-H^+} \textbf{1-49d}$$

1-51 R = C$_2$H$_5$ (1-6)

The proposed electrophilic displacement of an iodonium cation from the diiodide (**1-49b**) resembled an older conversion of 2-acetyl-4,5-diiodopyrrole (**1-52a**) by nitric acid in acetic acid to 2-acetyl-3,4,5-triiodopyrrole (**1-52c**), Eq. (1-7).[40] Apparently the nitronium cation displaced an α-iodo group (as expected) since longer treatment with nitric acid produced diiodomaleimide.

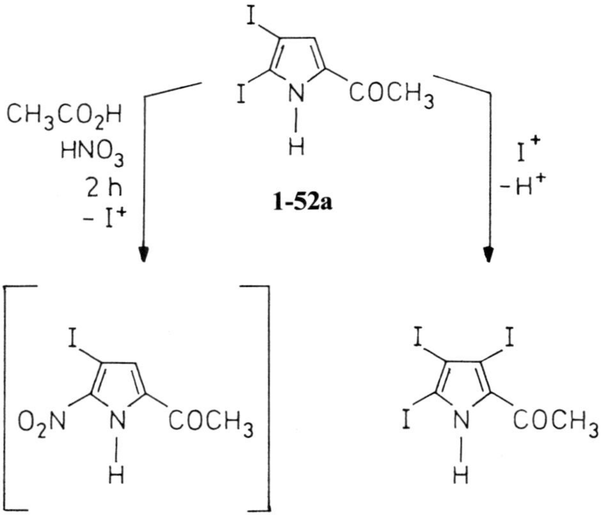

(1-7)

Both ethyl 3,5-dimethylpyrrole-2-carboxylate (**1-54**) and its 4-iodo derivative (**1-55**) gave the 4-nitro derivative (**1-53**) when treated with aqueous concentrated nitric acid, Eqs. (1-8, 1-9). In related reactions the 5-iodo substituent in pyrroles **1-56**, **1-57**, and **1-60** was replaced by the nitro group, Eqs. (1-10, 1-11).[41]

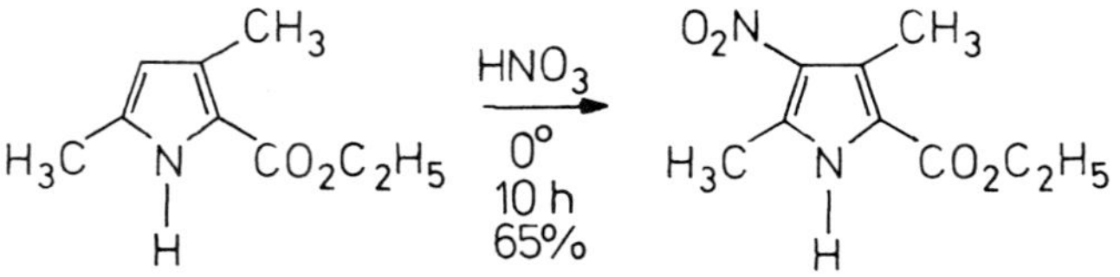

(1-8)

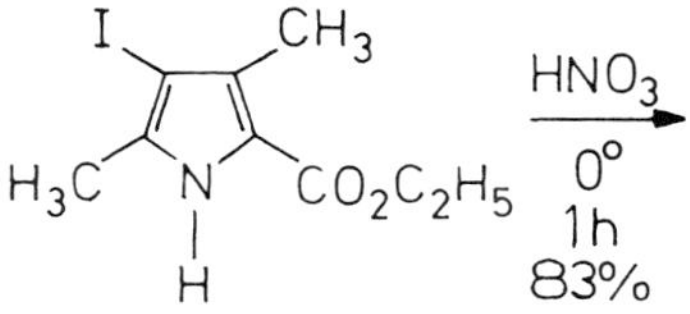

Nitration of Simple Pyrroles

(structure **1-55**: N-H pyrrole with I, CH₃, H₃C, CO₂C₂H₅ substituents) → HNO₃, 0°, 1h, 83% → **1-53** (1-9)

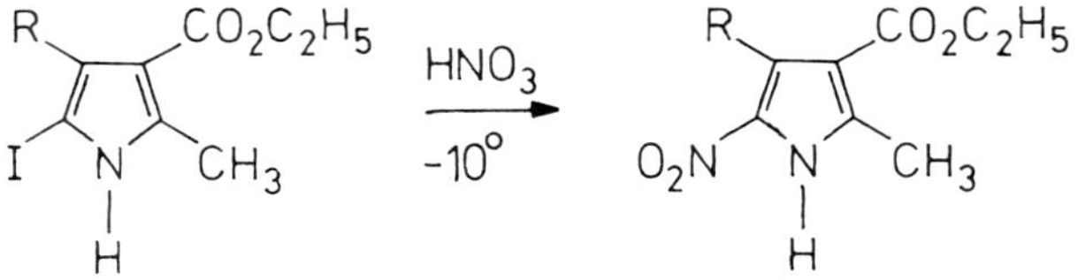

1-56 R = CH₃
1-57 R = I

HNO₃, −10° →

1-58 R = CH₃
1-59 R = I, 75%   (1-10)

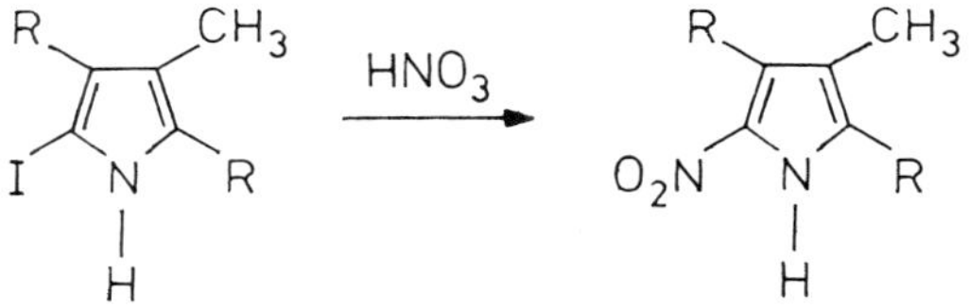

1-60
R = CO₂C₂H₅

HNO₃ →   (1-11)

The replaceability of pyrrole α-substituents by the nitro group in an electrophilic attack appears to follow the order: benzoyl < iodo < acetyl < carboxyl.[42] For example, nitration with nitric acid (d 1.5) at 15°–20° in acetic acid replaced the acetyl group in 2-acetyl-3,4,5-triiodopyrrole (**1-61a**) but 2-benzoyl-3,4,5-triiodopyrrole (**1-61b**) under the same conditions gave 2-benzoyl-3,4-diiodo-5-nitropyrrole by replacement of the 5-iodo group along with competitive oxidation into diiodomaleimide.[43] An apparent replacement of the formyl substituent may actually involve replacement of the carboxyl group of an intermediate acid. When these groups are β-pyrrole substituents, replacement by the nitro group may be less facile,[28] eg, compare the formations of **1-43b** (16%) and **1-44c** (2%); **1-78a** (36%) and **1-78b** (18%).

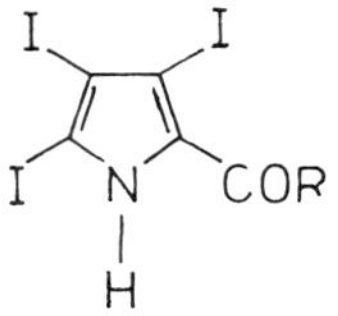

1-61a R = CH₃
1-61b R = C₆H₅

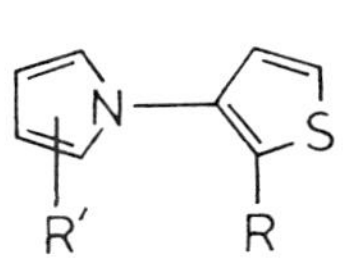

1-62 R = H
R′ = 2- and 3-NO₂

163 R = CO₂CH₃
R′ = 2- and 3-NO₂

Miscellaneous Pyrroles

Attempts (unspecified) to nitrate 1-(3-thienyl)pyrrole and 1-(2-methoxycarbonyl-3-thienyl) pyrrole gave mixtures of the corresponding 2- and 3-nitro derivatives (**1-62, 1-63**) in poor yields.[44]

Nitric acid and 2-cyano- or 1-methyl-2-cyanopyrrole in acetic anhydride at 0° gave mixtures (68% and 76%) of the 4-nitro- (**1-64**) and 5-nitro- (**1-65**) isomers.[45] Apparently the cyano substituent at the 2-position in pyrrole is less "*meta*-directing" than it is in the nitration of benzene.[46]

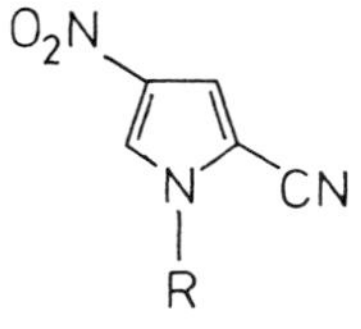

R	1-64/1-65
H	0.7
CH$_3$	2.4

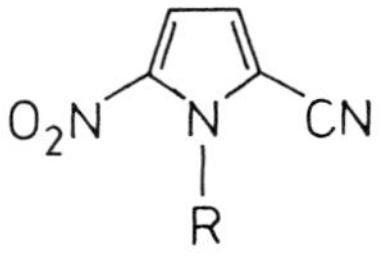

1-64   **1-65**

Dinitration

In acetic anhydride at −10° for 90 min pyrrole gave 2,4-dinitropyrrole (**1-66**) (48%) and 2,5-dinitropyrrole (**1-67**) (15%).[47]

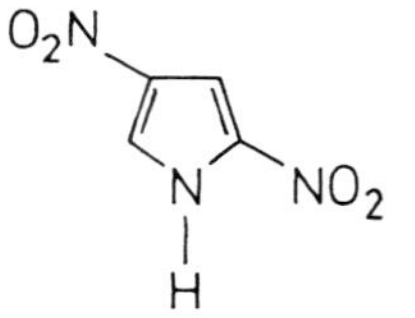

1-66 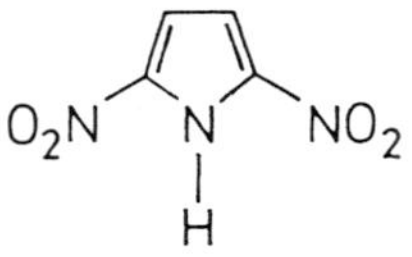   **1-67**

Dinitration of 1-methylpyrrole was most efficient when carried out in two steps. A mixture of mononitro derivatives carefully maintained at 20° was treated with fuming nitric acid in acetic anhydride at 20 ± 2° to give a mixture (78%) of the four possible dinitro derivatives: 2,3-dinitro- (5 parts), 2,4-dinitro- (74 parts), 2,5-dinitro- (10 parts), and 3,4-dinitro- (6 parts).[15,48]

Similar nitration directly transformed each 1,2-dimethylnitropyrrole at 0° into mixtures of dimethyldinitropyrroles (**1-68a–c**), Eqs. (1-12–1-14). A lower

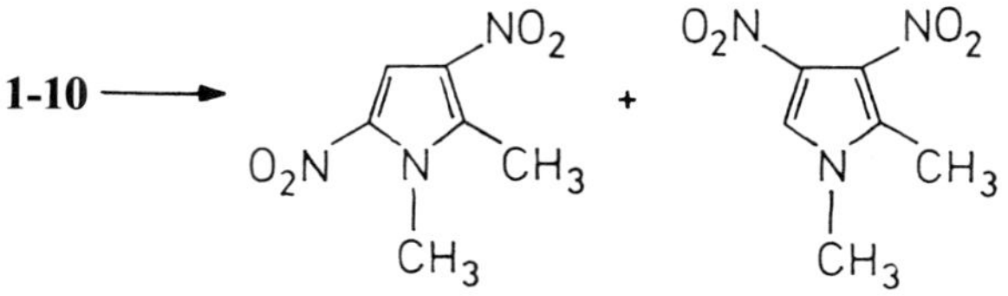

1-10 ⟶

1-68a 56%   **1-68b** 8%   (1-12)

1-11 → 1-68b 22% +   [structure of 1-68c]

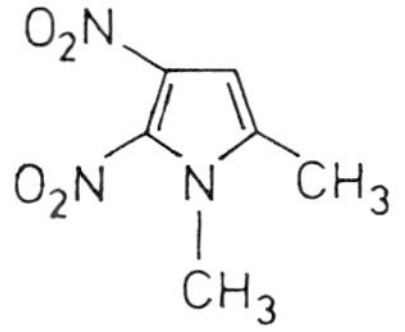

1-68c 10% (1-13)

1-12 → 1-68a 82% + 1-68c 2% (1-14)

yield of **1-68a** and **1-68b** from the pyrrole (**1-10**) was obtained when 100% nitric acid replaced 70% nitric acid in acetic anhydride. The facility of these nitrations was attributed to activation from the 2-methyl substituent.[15]

Nitric Acid Alone and with Other Acids

Both 2,4- and 2,5-dinitropyrrole were obtained from 2-nitropyrrole and a 4 M excess of nitric acid.[49] Kinetic studies of the nitrations of 2- and 3-nitropyrrole and of their 1-methyl derivatives by nitric acid in sulfuric acid over a range of concentrations at 25° revealed a similarity with the nitration of benzene.[50]

The first trinitropyrrole was prepared from 1-methyl-3,4-dinitropyrrole by treatment with a mixture of nitric acid (65%) in concentrated sulfuric acid at ambient temperature (< 50°). The product, 1-methyl-2,3,4-trinitropyrrole (**1-69**) (70%), was a colorless crystalline solid.[15]

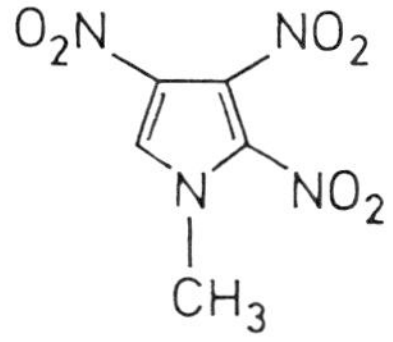

[structures of 1-69, 1-70, 1-71]

1-69 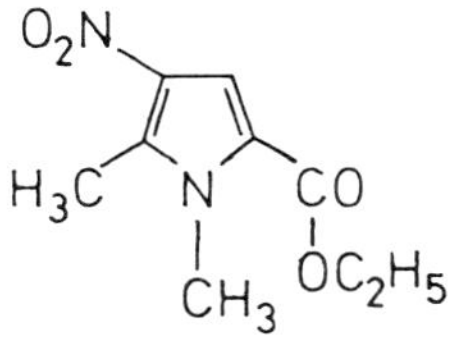   1-70 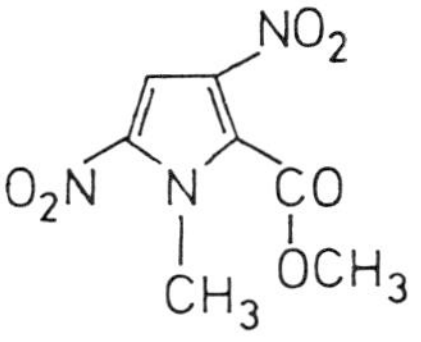   1-71

Nitric acid (70%) and ethyl 1,5-dimethylpyrrole-2-carboxylate (**1-27**) at 30 ± 1° for 1 h gave the 4-nitro derivative (**1-70**) (85%). A lower temperature retarded the reaction and increased by-product formation; and a slightly higher temperature (35°) decreased the yield by almost 50%. For comparison the 4-nitropyrrole (**1-24**) was recovered without change upon exposure to nitric acid (70%) at 90° for 1 h. Under the same conditions the isomeric 5-nitropyrrole (**1-26**) unexpectedly gave its 3-nitro derivative (**1-71**); and a markedly improved yield (61%) was obtained when sulfuric acid was also present.[15]

Mono- and dinitro derivatives were obtained from dimethylpyrrole-3,4-dicarboxylate (**1-72**), Eqs. (1-15, 1-16).[51] A mononitro derivative was slowly produced from the diethyl ester in an unspecified yield after 24 h.

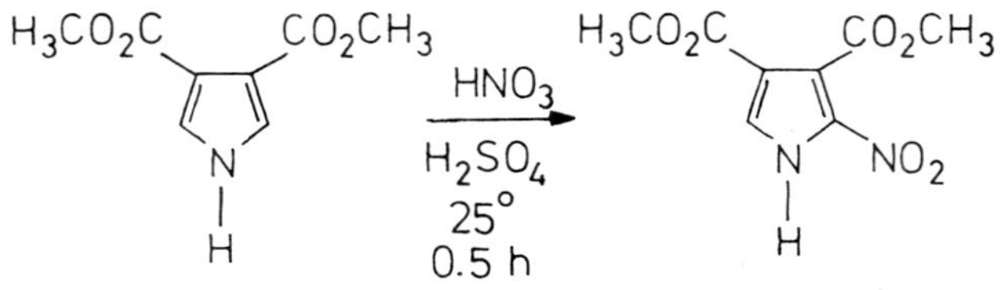

$$\text{1-72} \xrightarrow[\substack{H_2SO_4 \\ 25° \\ 0.5\ h}]{HNO_3} \text{[dimethyl 2-nitropyrrole-3,4-dicarboxylate]} \qquad (1\text{-}15)$$

$$\text{1-72} \xrightarrow[\substack{H_2SO_4 \\ 40\ h}]{HNO_3} \text{[dimethyl 2,5-dinitropyrrole-3,4-dicarboxylate]} \quad 59\% \qquad (1\text{-}16)$$

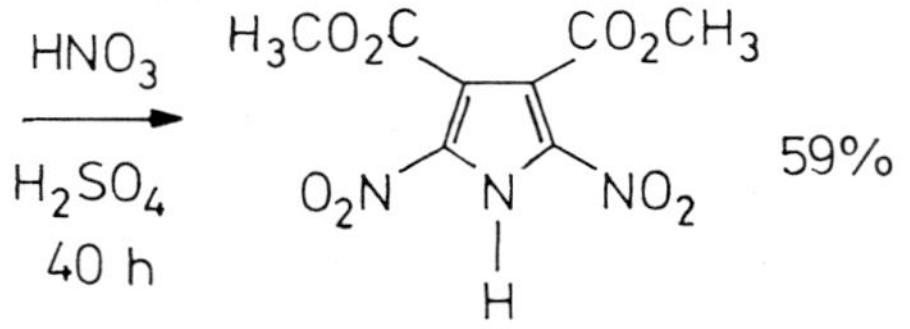

Acetic acid has served as solvent for the nitration of arylpyrroles (**1-73**) by nitric acid (65%–70%) at ambient temperatures to the 3-nitro and some 3,4-dinitro derivatives.[52-56] Similar treatments converted 1,2,3,5-tetraphenylpyrrole (**1-74a**) for 10 min to the 4-nitro derivative[53] and 1-methyl-2,5-diphenyl-3-α-pyridylpyrrole (**1-74b**) to its 4-nitro derivative.[54] Fuming nitric acid at 0° for several minutes transformed ethyl 1-methyl-4-α-pyridyl-5-phenylpyrrole-2-carboxylate (**1-74c**) to its 3-nitro derivative (26%);[54] similarly ethyl 3,5-dimethylpyrrole-2-carboxylate (**1-74d**) as well as its 4-acetyl derivative gave the 4-nitro derivative by treatment with nitric acid in acetic acid at 45°.[55,57] In other circumstances a 1-phenyl substituent was nitrated. For example, 1-phenylpyrrole in a mixture of nitric acid and sulfuric acids gave 1-*p*-nitrophenylpyrrole (25%).[10]

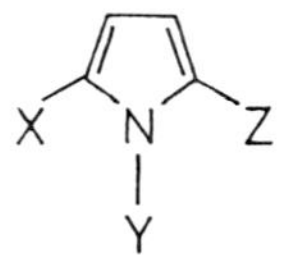

1-73
$X = Y = Z = C_6H_5$
$X = Z = C_6H_5,\ Y = p\text{-}CH_3C_6H_4$
$X = Z = C_6H_5,\ Y = C_6H_5CH_2$
$X = Z = C_6H_5,\ Y = CH_3$

1-74
a $W = X = Y = Z = C_6H_5$
b $W = \alpha\text{-}C_5H_4N,\ X = Z = C_6H_5,\ Y = CH_3$
c $W = \alpha\text{-}C_5H_4N,\ X = C_6H_5$
  $Y = CH_3,\ Z = CO_2C_2H_5$
d $W = Z = CH_3,\ X = CO_2C_2H_5,\ Y = H$

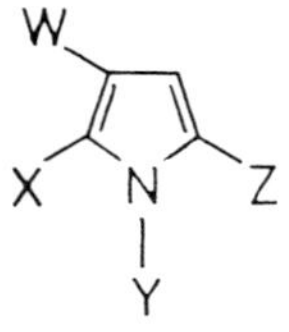

Potassium nitrate and 2-methyl-3-acetyl-5-(2-nitrophenyl) pyrrole in concentrated sulfuric acid at 0° for ½ h gave the 4-nitro derivative (**1-75**) (R = H) (40%). The related pyrroles (**1-76**) (R = CH_3, C_6H_5) were obtained in comparable yields by similar reactions. When the nitrations were carried out with nitric acid and acetic anhydride in nitromethane the pyrroles (**1-75**) (R = H, CH$_3$, C$_6$H$_5$) were obtained in poor yields (7%–11%) as 2-methyl-3-nitro-5(2-nitrophenyl)pyrrole (**1-76**) (R = H), and its 1-methyl- **1-76** (R = CH$_3$) and 1-phenyl- **1-76** (R = C$_6$H$_5$) derivatives (20%–25%) were produced by the replacement of an acetyl group with a nitro group.[58]

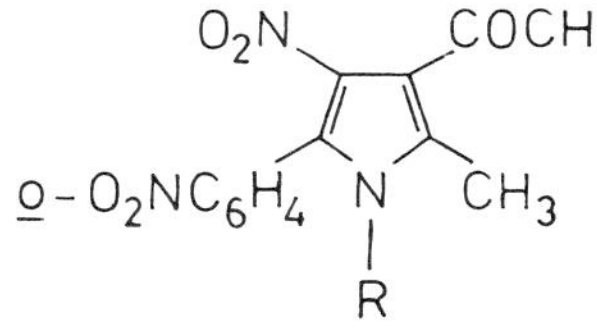

1-75
R = H, CH$_3$, C$_6$H$_5$

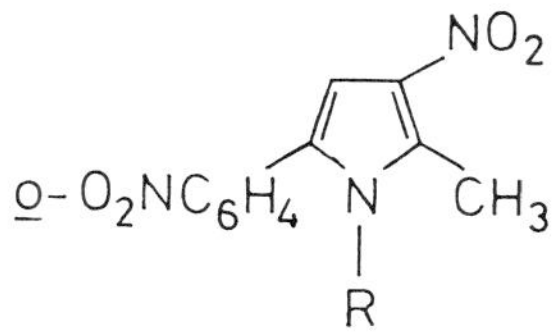

1-76
R = H, CH$_3$, C$_6$H$_5$

Replacement of an acetyl group also occurred when each of the pyrrole isomers (**1-77a** and **1-77b**) was treated with concentrated nitric acid at 10°–20° for 5–15 min to give ethyl 5-nitro-2,4-dimethylpyrrole-3-carboxylate (**1-78a**) (36%) and the isomeric ethyl 4-nitro-3,5-dimethylpyrrole-2-carboxylate (**1-78b**) (18%). A similar replacement gave methyl 5-nitro-2,4-dimethylpyrrole-3-carboxylate (31%) from the pyrrole (**1-77c**). Nitric acid also induced an oxidative rearrangement in which a 2-acyl- or a 2-acyloxy group migrated to the 3-position as a pyrrolinone (**1-79**) was produced, Eq. (1-17) and Table 1-1.[59]

In another oxidation nitric or nitrous acid converted tetraphenylpyrrole (**1-80**) to a hydroxytetraphenylpyrrolenine (**1-81**).[60]

Each of the three 1-methyldinitropyrroles was easily nitrated by one equivalent of nitric acid (100%) in concentrated sulfuric acid at 0°. The reactions were complete within 15 minutes, the time required for the addition, except for the

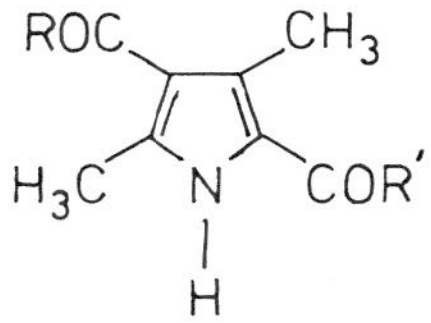

1-77

a R = OC$_2$H$_5$, R' = CH$_3$
b R = CH$_3$, R' = OC$_2$H$_5$
c R = OCH$_3$, R' = CH$_3$

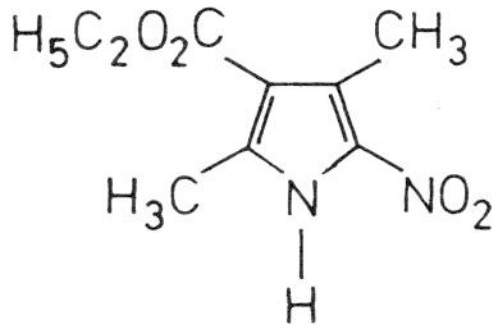

1-78a

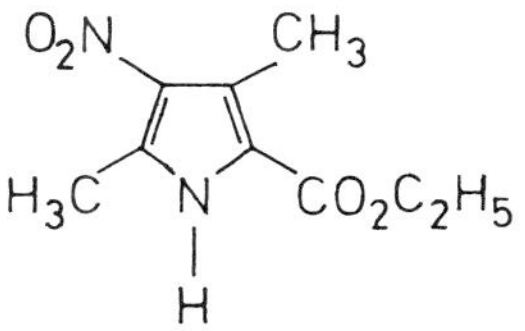

1-78b

1-77 $\xrightarrow{HNO_3}$ [structure **1-79**] (1-17)

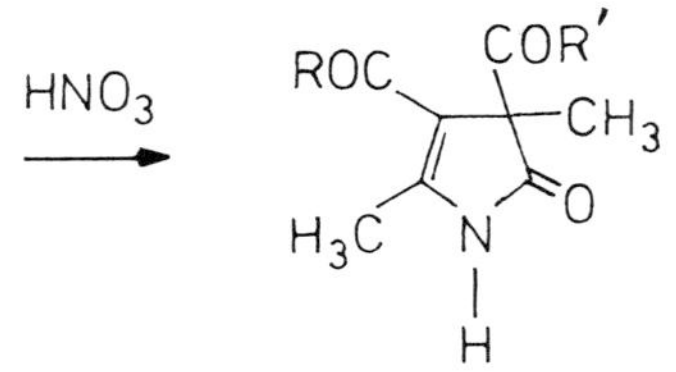

Table 1-1. Oxidative Rearrangements: **1-77** → **1-79**

R	OC_2H_5	OCH_3	CH_3	OCH_3
R'	CH_3	CH_3	OC_2H_5	OCH_3
Yield—(%)	26	10	32	13

R	OCH_3	OC_2H_5	OCH_3	OCH_3	OCH_3
R'	OC_2H_5	OC_2H_5	NCH_3 $\mid$ CH_3	NCH_3 $\mid$ C_2H_5	NC_2H_5 $\mid$ C_2H_5
Yield(%)	42	45	44	42	80

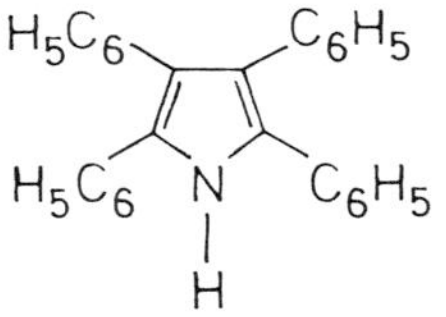

1-80 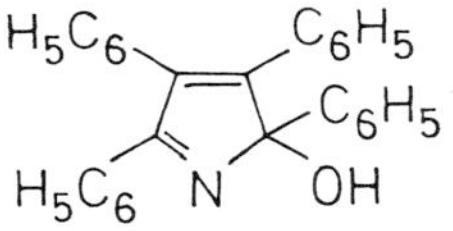   **1-81**

less reactive 1-methyl-2,4-dinitropyrrole (**1-84**), which had only partially reacted after 1 h, Eqs. (1-18–1-20).[11]

Control by a 1-alkyl (aryl) substituent in the nitration of the pyrrole ring was further revealed in the investigation of a series of 1-alkyl(aryl)-3-nitropyrroles (**1-5**). Each pyrrole (**1-5**), obtained from a 1-alkyl(aryl)pyrrole and one equivalent of nitric acid (100%) in acetic anhydride at −20° (0° when R = aryl), was nitrated in concentrated sulfuric acid at 0° (25° for R = o-nitrophenyl) with one

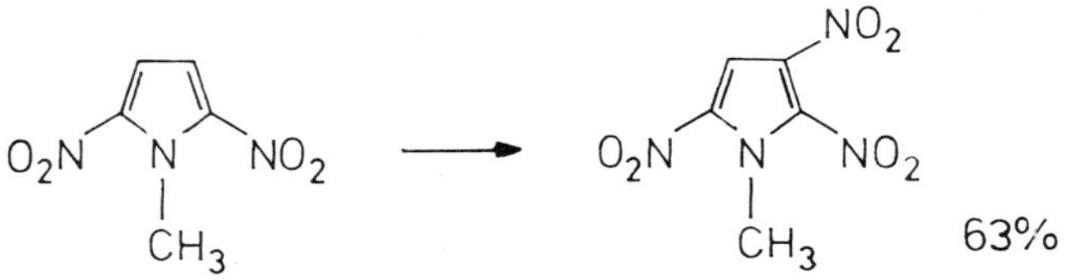

1-82 → **1-85**   63%   (1-18)

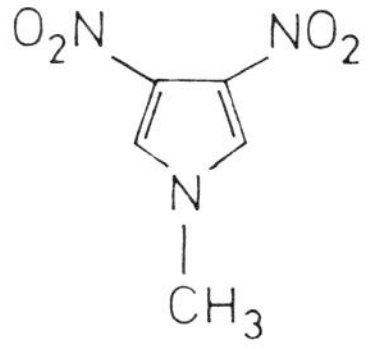

Nitration of Simple Pyrroles

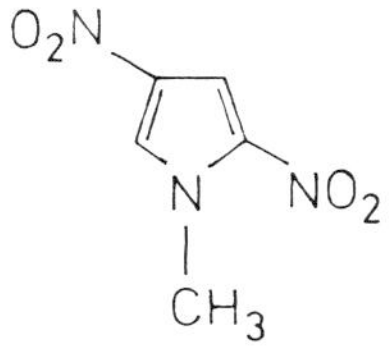

[Structure 1-83: 1-methyl-3,4-dinitropyrrole] → **1-69**, 75%   (1-19)

[Structure 1-84: 1-methyl-2,4-dinitropyrrole] → **1-85** + **1-69**   (1-20)

equivalent of nitric acid (100%). The product ratios (Table 1-2) showed an increase in the nitration at the 4-position as the 1-alkyl(aryl)substituent became bulkier; however, the nitration was always predominant at the 5-position to give 2,4-dinitro-1-R-pyrroles (**1-86**). To account for the smaller amounts of 2,3,4- and 2,3,5-trinitro derivatives some decomposition of the starting material (**1-5**) (none was recovered) was assumed. These formations of 1-alkyl(aryl)-3,4-dinitropyrroles (**1-87**) offer the first example of 3,4-disubstituted pyrroles by direct electrophilic substitution.[11]

Table 1-2. Nitration of 3-Nitro-1-R-pyrroles (**1-5**)

1-5	R	Derivatives of 1-R-pyrroles, Product Ratios			
		2,4-Dinitro (**1-86**)	3,4-Dinitro (**1-87**)	2,3,4-Trinitro (**1-88**)	2,3,5-Trinitro (**1-89**)
a	CH_3	74	13	9	4
b	$(CH_3)_2CH$	66	23	11	1
c	$(CH_3)_3C$	43	41	16	
d	$p\text{-}O_2NC_6H_5$	75	15	10	
e	$o\text{-}O_2NC_6H_5$	72	28		

Dinitration of the pyrrole (**1-83**) in fuming sulfuric acid (20% sulfur trioxide) by an excess of nitric acid (100%) at 65° for 15 min gave 1-methyl-2,3,4,5-tetranitropyrrole (**1-90**) (45%), the first example of a five-membered heterocycle with four C-nitro groups.[11] A calculated density of 1.9[61] added to its potential value as a high-energy polynitro aromatic compound.

Concentrated nitric acid converted the sodium salts (**1-91**, **1-93**) of the pyrrole-3- and the pyrrole-2-sulfonic acids to the corresponding 4- and 5-nitropyrroles (**1-92**, **1-94**).[62]

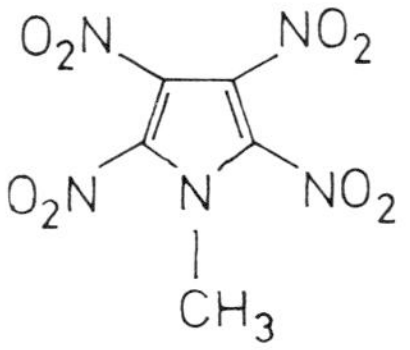

1-90

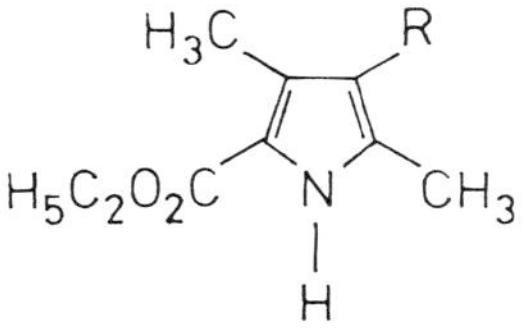

1-91 R = SO$_3$Na
1-92 R = NO$_2$

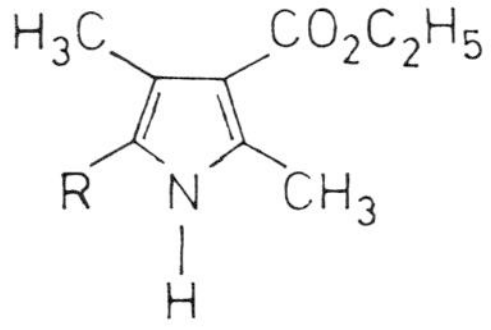

1-93 R = SO$_3$Na
1-94 R = NO$_2$

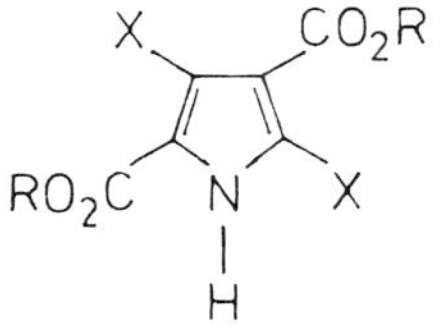

1-95a R = X = CH$_3$
1-95b R = CH$_3$, X = CO$_2$H
1-95c R = CH$_3$, X = NO$_2$

Replacement of methyl groups by nitration occurred in the conversion of the 3,5-dimethylpyrrole (**1-95a**) to the corresponding 3,5-dinitropyrrole (**1-95c**). It was assumed that initial oxidation to the pyrroledicarboxylic acid (**1-95b**) was followed by replacement of the carboxyl groups.[63] Since replacement of a methyl group by an aryl diazo group was recently reported,[64] Eq. (1-21), the direct replacement of a methyl group in the pyrrole (**1-95a**) by a nitro group must also be considered.

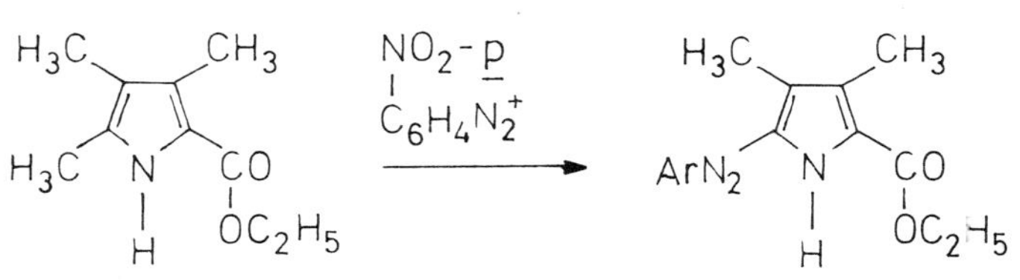

(1-21)

In the conversion of ethyl 4,5-dimethylpyrrole-2-carboxylate to the symmetrical dipyrrylmethane (**1-96**) (28%) by treatment with a mixture of nitric acid and acetic anhydride a methyl group was lost.[65]

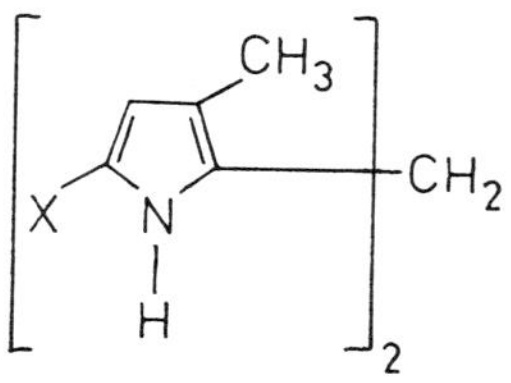

X = CO$_2$C$_2$H$_5$
1-96

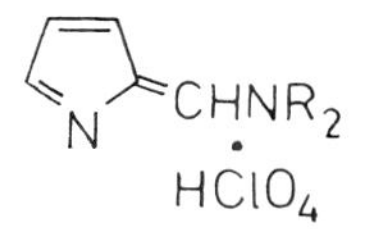

1-97a R = CH$_3$
1-97b R$_2$N = N(CH$_2$)$_4$

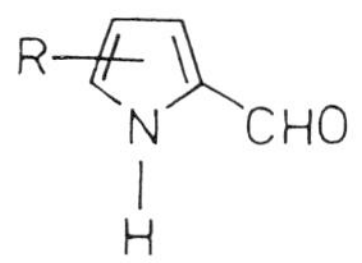

1-98a R = 4-NO$_2$
1-98b R = 5-NO$_2$

Nitration by nitric acid in acetic anhydride at 5°–8° transformed the perchlorate (**1-97a**) of 2-dimethylaminomethylene-2H-pyrrole (from 2-formylpyrrole and dimethylamine) to 4-nitro-**1-98a** (57%) and 5-nitro-2-formylpyrrole (**1-98b**) (25%).[66] In other experiments the perchlorate (**1-97b**) of 2-pyrrolidinomethylene-2H-pyrrole gave the pyrroles (**1-98a,b**) in a product ratio of 67:33 when treated with concentrated nitric acid at 0°.[23] Similar experiments with the corresponding derivative from morpholine gave a product ratio of 64:36. Nitration of the 2H-pyrrole (**1-97b**) in oleum <0° under nitrogen gave a dinitro-2-formylpyrrole (unisolated) that was oxidized by silver nitrate to the corresponding 4,5-dinitropyrrole-2-carboxylic acid (50%), a new compound characterized only by IR and NMR spectroscopy.[23]

Nitrous Acid, Nitrite or Nitrate Esters

In 1939 nitrations of 2,5-diphenyl- and 2,4,5-triphenylpyrrole by amyl nitrite in ether were reported. Conversions of intermediate isonitrosopyrroles by amyl nitrite to the corresponding nitropyrroles (**1-99a,b**), (Eq. (1-22)) was independently established.[67] Methylation by alkaline dimethyl sulfate was reported to give the 1-methyl derivatives; however, it was later suggested that the product from the triphenylnitropyrrole was a methyl nitronate (**1-101**).[68] Each nitropyrrole (**1-99a,b**)[67] was reduced by aluminum in alcoholic potassium hydroxide to a primary aminopyrrole (**1-100a,b**), the latter known by an independent preparation. Each amine was also produced from the corresponding methyl derivative by reduction with zinc and acetic acid, a reaction to be expected from a nitronate ester (**1-101**) but not from an N-methyl compound.

These preparations of the nitropyrrole (**1-99b**) (40%) and the amine (**1-100b**) have been confirmed, but in aqueous acetone nitrous acid converted 2,4,5-

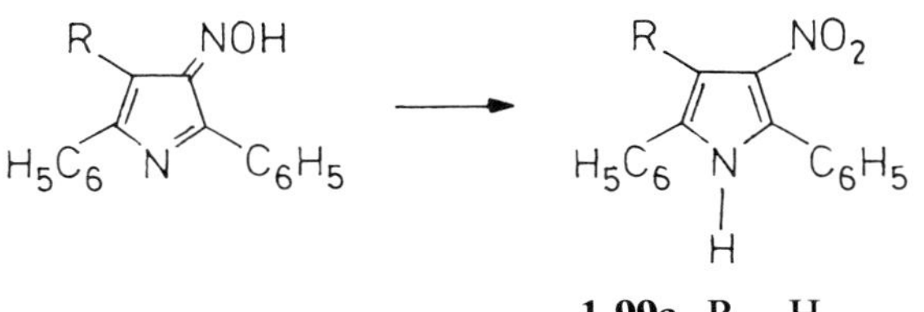

1-99a, R = H
1-99b, R = C$_6$H$_5$ (1-22)

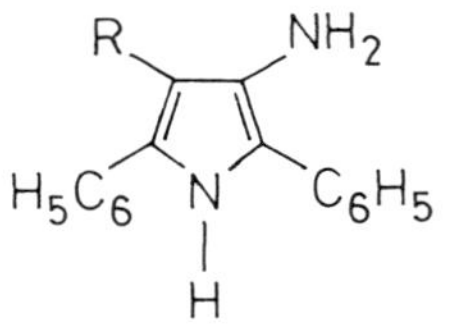

1-100a R = H
1-100b R = C₆H₅

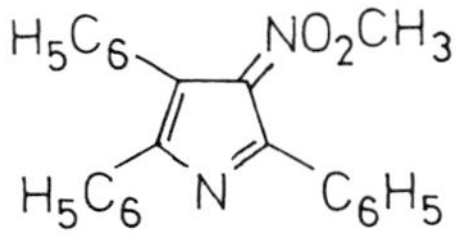

1-101

triphenylpyrrole to 3-diazo-2,4,5-triphenylpyrrole. Similar treatment converted 2,5-diphenylpyrrole to 3-diazo-4-nitro-2,5-diphenylpyrrole (**1-102**) (23%).[69]

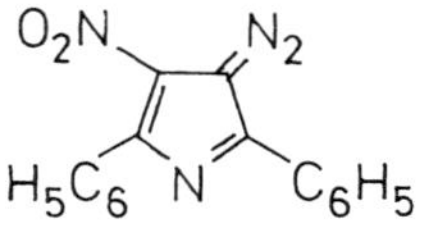

1-102

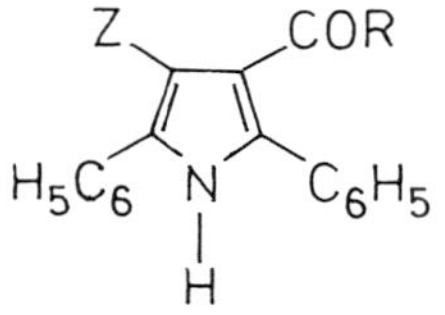

1-103 R = CH₃CO, C₆H₅CO
  a Z = NO₂
  b Z = NO

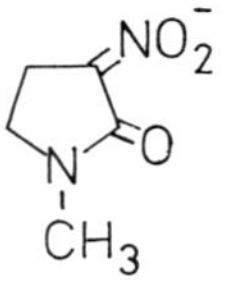

1-103c

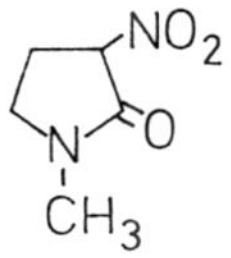

1-103d

Important reaction differences were established for amyl nitrite with and without sodium ethoxide in the conversions of 3-acetyl or 3-benzoyl-2,5-diphenylpyrrole. In the absence of sodium ethoxide the 4-nitro derivative (**1-103a**) was produced, whereas the 4-nitroso derivative (**1-103b**) was the product obtained when the base was present.[70, 71]

Both potassium *t*-butoxide and lithium diisopropylamide catalyzed the low temperature nitration of 1-methyl-2-pyrrolidinone by an alkyl nitrate to the 3-nitro derivative as the nitronate salt (**1-103c**). The nitro isomer (**1-103d**) (53%) was obtained on acidification.[72, 73]

RING CLOSURE AND RING TRANSFORMATION

Several ring closures with the retention of a nitro group as a pyrrole substituent are known. They include acid- and base-catalyzed reactions and dipolar additions.

An intramolecular base-catalyzed α-acylation converted ethyl N-(2-nitro-1-phenylethyl)oxamate (**1-104**) to 3-hydroxy-4-nitro-5-phenyl-3-pyrrolin-2-one (**1-105**) (64%) after acidification of its sodium enolate, Eq. (1-23). This cyclization provided an example of the rare α-acylation of a nitroparaffin derivative. The oxamate was obtained from ethoxalyl chloride and an adduct from ammonia and β-nitrostyrene.[74]

$$\underset{\textbf{1-104}}{\begin{array}{c}NO_2 \quad OC_2H_5 \\ | \qquad | \\ CH_2 \quad CO \\ | \qquad | \\ H_5C_6CHNHCO\end{array}} \xrightarrow[2.\ H_2SO_4]{1.\ NaOC_2H_5} \underset{\textbf{1-105}}{\begin{array}{c}O_2N \qquad OH \\ \diagdown\diagup \\ H_5C_6-N-O \\ | \\ H\end{array}} \qquad (1\text{-}23)$$

To prepare a 3,4-dinitropyrrole by cyclization has become attractive. Ingenious in design and simplicity, a reaction between the dipotassium salt of 2,3,3-trinitropropanal (from mucobromic acid and sodium nitrite in 70% yield),[75] aqueous formaldehyde, benzylamine, and a small amount of ammonium hydroxide was exothermic after an induction period (15 min) and gave 3,4-dinitro-1-benzylpyrrole (**1-106**) (18%), Eq. (1-24). Appropriate primary amines provided 1-β-hydroxyethyl (**1-107**) (17%), 1,2-dimethyl (**1-108**) (4%), and 1-methyl-2-ethyl (**1-109**) (13%) derivatives of 3,4-dinitropyrrole.[76]

$$\begin{array}{c}(NO_2)_2\underset{=}{C}^- \\ | \\ O_2N\underset{=}{C}CHO\end{array} \xrightarrow[\begin{array}{c}C_6H_5CH_2NH_2 \\ NH_4OH\end{array}]{CH_2O} \underset{\textbf{1-106}}{\begin{array}{c}O_2N \qquad NO_2 \\ \diagdown\diagup \\ N \\ | \\ CH_2C_6H_5\end{array}} \qquad (1\text{-}24)$$

$$\underset{\textbf{1-107}}{\begin{array}{c}O_2N \qquad NO_2 \\ \diagdown\diagup \\ N \\ | \\ CH_2CH_2OH\end{array}} \qquad \underset{\begin{array}{c}\textbf{1-108},\ X = CH_3 \\ \textbf{1-109},\ X = C_2H_5\end{array}}{\begin{array}{c}O_2N \qquad NO_2 \\ \diagdown\diagup \\ N \quad X \\ | \\ CH_3\end{array}}$$

Aqueous formaldehyde and the aniline salt of the anil of trinitropropanal produced a red solid, mp 130°–131°, with an unknown structure for the composition $C_{16}H_{13}N_3O_2$. A similar reaction but with acetone also present gave an orange solid $C_{13}H_{11}N_3O_2$, mp 202°–203°,[76] with unknown structure.

An adduct from aminoacetaldehyde ketal and nitroketene dithioketal cyclized on treatment with anhydrous hydrogen chloride to 2-methylthio-3-nitropyrrole (**1-110**) (54%).[77]

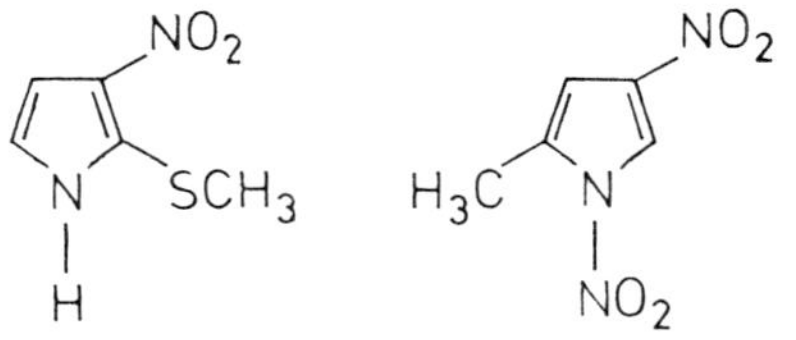

1-110

1-111

A cyclization to 1,3-dinitro-5-methylpyrrole (**1-111**) (4%) resulted from the treatment of sorbic acid (**1-112**) with sodium nitrite in water (pH 3.5) at 60° for 2 h, Eq. (1-25).[78] Ethyl nitrolic acid and a furoxan (**1-113**) were also produced.[79] The structure of the unique N-nitropyrrole (**1-111**) was established by elemental analysis, M^+ at m/e 171.0284, ^{1}H and ^{13}C NMR spectra compatible with a pyrrole derivative, absence of the pyrrole NH function as shown by no appreciable change in ^{1}H NMR signals at 25° by the addition of D_2O or a trace of piperidine and the absence of absorption around 3400–3500 cm^{-1} in the ir spectrum, the presence of an unsubstituted α-position by δ_C 124.3 and δ_H 8.88 and of a β-position by δ_C 113.2 and δ_H 7.91, C-NO$_2$ ν_{as} 1530 (NO$_2$) and ν_s 1345 (NO$_2$) cm^{-1}, and N-NO$_2$ ν_{as} 1590 (NO$_2$) cm^{-1}. Treatment with diazomethane gave 1,3-dinitro-2,5-dimethylpyrrole with a ^{13}C NMR spectrum that showed a long range coupling (d, J = 3.1 Hz), also present in the pyrrole (**1-111**), between 5-CH$_3$ and 4-H, but no coupling for the new 2-CH$_3$. Both pyrrole (**1-111**) and its 2-methyl derivative gave a mass spectrum base peak of m/e 39 (probably the cyclopropenyl cation), a general characteristic of pyrroles.[78]

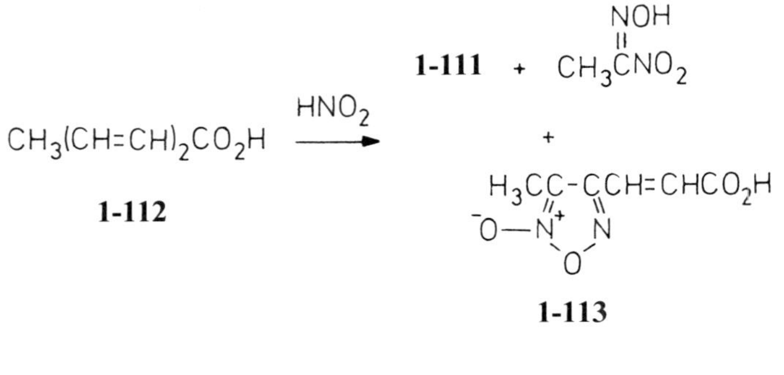

$$CH_3(CH=CH)_2CO_2H \xrightarrow{HNO_2} \textbf{1-111} + CH_3\overset{NOH}{\underset{\|}{C}}NO_2 + \textbf{1-113}$$

1-112

(1-25)

An alternative preparation for ethyl 4-nitropyrrole-2-carboxylate (**1-114**) (83%), needed for a preparation of distamycin, an antiviral agent, consisted in the condensation between nitromalonaldehyde and ethyl glycine, Eq. (1-26),[80] a reaction first reported in 1915.[81]

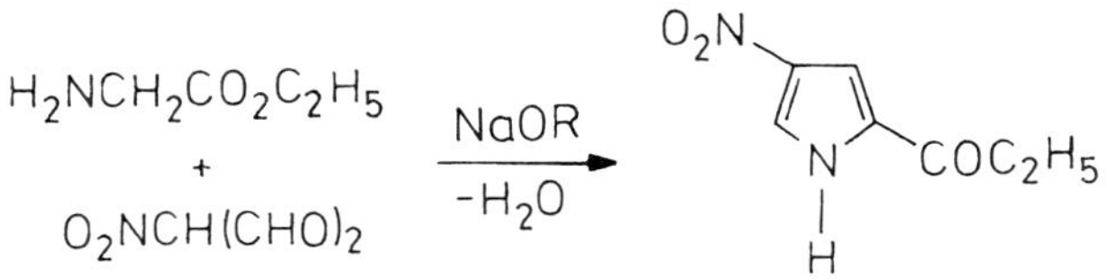

$$H_2NCH_2CO_2C_2H_5 + O_2NCH(CHO)_2 \xrightarrow[-H_2O]{NaOR}$$

[Structure **1-114**: pyrrole with O_2N at 4-position, COC_2H_5 at 2-position, NH]

1-114   (1-26)

Trichloronitroethylene and aniline gave 3-nitro-2-anilinoindole (**1-115**) (R = H) (89%) in a straightforward combination of replacement and ring closure reactions, Eq. (1-27).[82] In a similar reaction with p-nitroaniline, the 4′,5-dinitro derivative (76%) of the indole (**1-115**) was obtained.

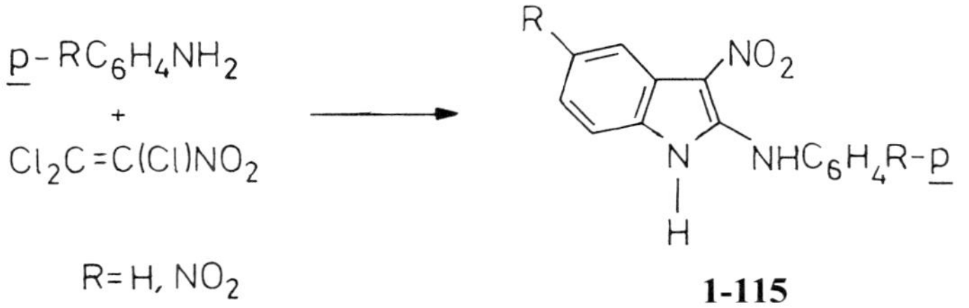

$$\underline{p}\text{-}RC_6H_4NH_2 + Cl_2C=C(Cl)NO_2 \longrightarrow$$

[Structure **1-115**: indole with R at 5-position, NO_2 at 3-position, $NHC_6H_4R\text{-}\underline{p}$ at 2-position]

R = H, NO_2 **1-115**   (1-27)

Nitroethylene and 4-methyl-5-propoxyoxazole (**1-116**) presumably gave a Diels-Alder adduct (**1-117**) that underwent ring-opening and reclosure to 2-acetyl-3-nitropyrrole (**1-118**) (57%), Eq. (1-28).[83] In a similar reaction, the oxazole (**1-116**) and nitropropene in benzene containing hydroquinone gave 2-acetyl-3-nitro-4-methylpyrrole (**1-119**) (44%) and 2,5-dimethyl-3-hydroxy-4-nitropyridine (**1-120**) (0.8%).[84] The latter (3%) was also obtained from the oxazole (**1-116**) and 1-nitro-2-propene.

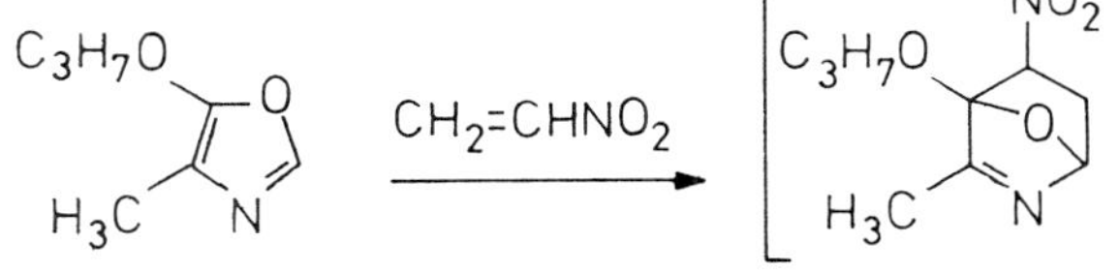

[Structure **1-116**: oxazole with C_3H_7O at 5-position, H_3C at 4-position]

$$\xrightarrow{CH_2=CHNO_2}$$

[Structure **1-117**: bicyclic Diels-Alder adduct with C_3H_7O, NO_2, H_3C]

1-116   **1-117**

$$\xrightarrow{H_2O} \begin{bmatrix} CH_3COCOCH(NO_2)CH_2NHCHOH \end{bmatrix} \longrightarrow$$

[Structure **1-118**: pyrrole with NO_2 at 3-position, $COCH_3$ at 2-position, NH]

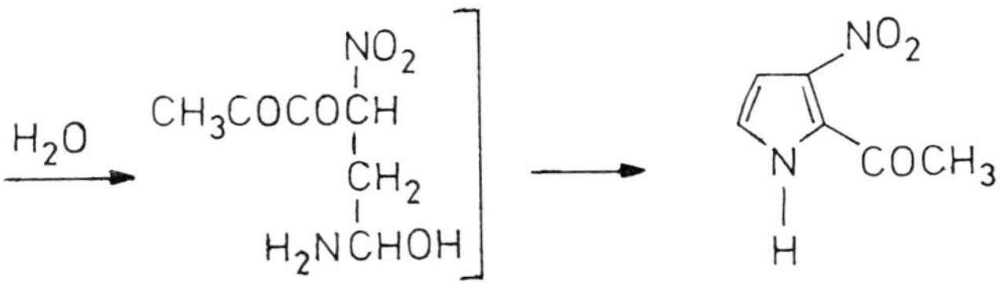

1-118   (1-28)

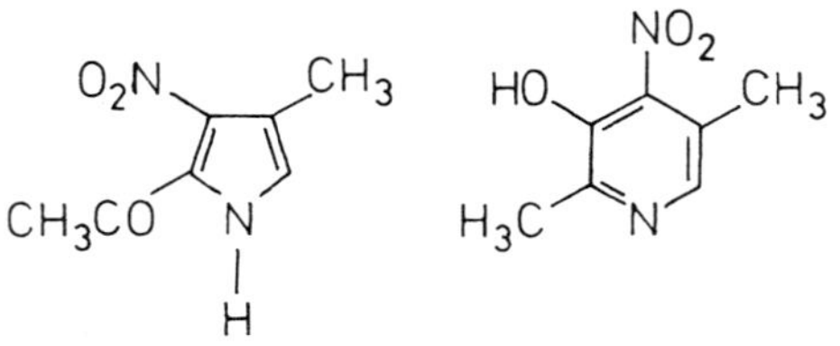

1-119 **1-120**

The combination of 2-chloro-1-nitroethylene and 2,4-diphenyl-3-methyloxazolium-5-oxide (**1-121a**) gave 1-methyl-2,5-diphenyl-3-nitropyrrole (**1-121b**) (76%), a product accounted for by a 1,3-dipolar reaction between the intermediate nitroacetylene and **1-121a**, followed by decarboxylation, Eq. (1-29a).[85]

Thermolysis of 2-azido-5-nitropyridine-1-oxide in benzene in a sealed tube at 100°for 11 h gave 2-cyano-4-nitropyrrole (**1-122a**) (19%); initial ring-opening with loss of nitrogen followed by a ring closure was proposed, Eq. (1-29b).[86]

Sodium cyanide in ethanol at 5° for 10 min converted 2-methyl-3methylamino-4-nitro-5-phenylimino-2,5-dihydroisothiazole to 2,5-dimethylamino-3-nitro-4-phenyliminopyrrole (**1-122b**) or a tautomer, Eq. (1-29c).[87]

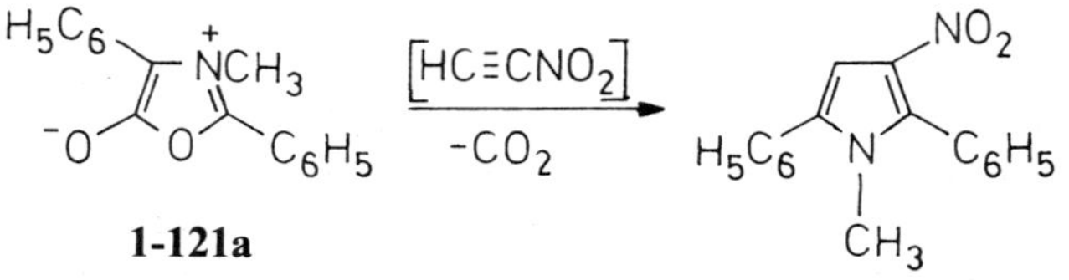

1-121a **1-121b** (1-29a)

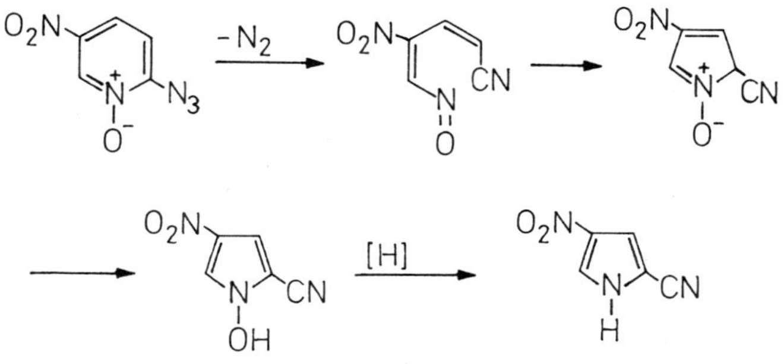

1-122a (1-29b)

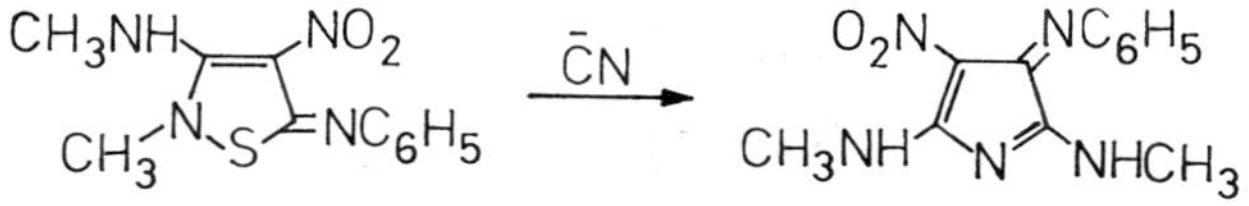

1-122b (1-29c)

Condensation reactions have brought about the formation of nitroindolizines; bromoacetaldehyde and 2-nitromethylpyridine gave 1-nitroindolizine, Eq. (1-29d);[88] acetonyl-3-cyanopyridinium bromide and nitromethane gave 1-nitro-2-methyl-6-cyanoindolizine, Eq. (1-29e).[89]

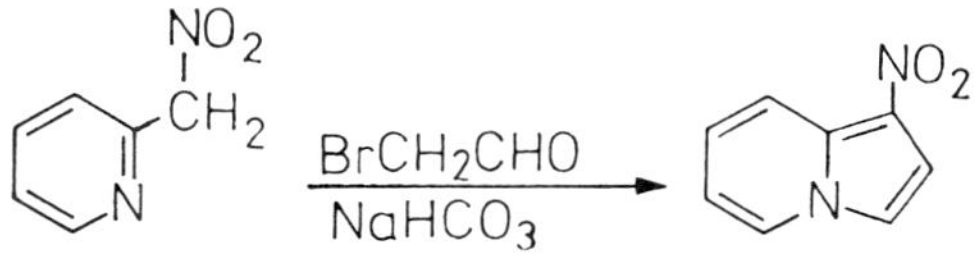

(1-29d)

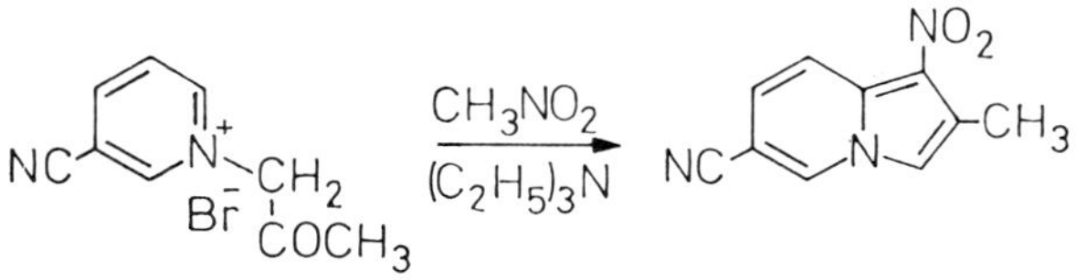

(1-29e)

NITRATION OF CONDENSED RING SYSTEMS

Indoles

Indole and 2-substituted indoles (**1-123**) have given 3-nitro derivatives (**1-124**) in reactions with an alkyl nitrate and sodium alkoxide, Eq. (1-30). Oxidation of oximino-3H-indoles also gave 3-nitroindoles, Eq. (1-31).[90]

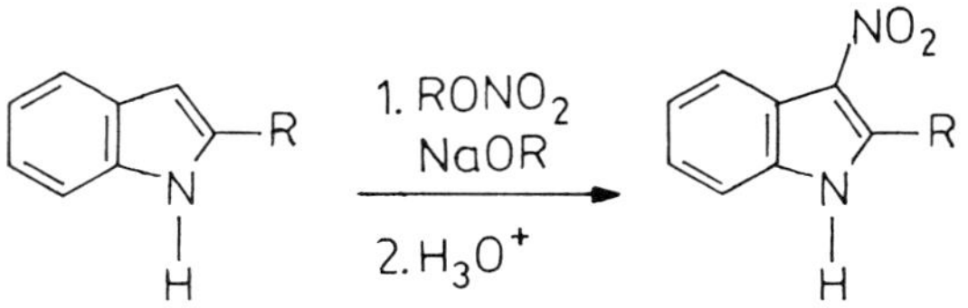

(1-30)

(1-31)

Amorphous dark brown solids were obtained from indole and its 1- and 3-methyl derivatives when treated with nitrate salts in concentrated sulfuric acid. Similar treatment of 2-methyl and 1,2-dimethylindole gave the corresponding 5-nitro derivatives (84%, 82%).[91] To obtain nitration in the pyrrole ring, 2-methylindole was nitrated by concentrated nitric acid alone or in acetic acid to give its 3,4-dinitro, 3,5-dinitro, 3,6-dinitro (39%), and 3,4,6-trinitro derivatives;

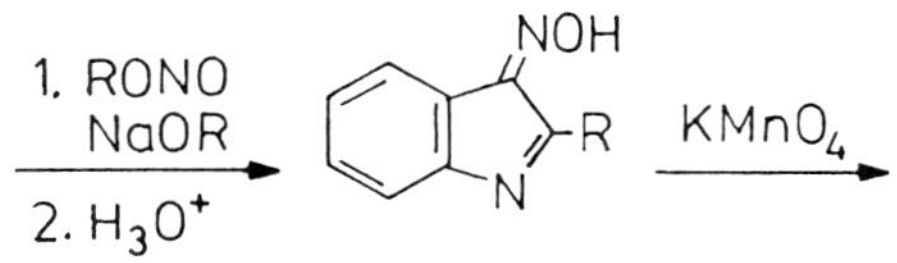

1,2-dimethylindole gave its 3,5-dinitro, 3,6-dinitro (28%), 3,4,6-trinitro, and 3,5,6-trinitro derivatives together with products of nitration restricted to the benzene ring;[92] and 2-phenylindole gave 3,6-dinitro-2-phenylindole (40%).[93]

An adduct from 9-benzoyltetrahydrocarbazole and nitric acid in acetic acid was assigned the structure **1-125**.[94] If correct, then a similar addition of nitric acid in reactions of other pyrrole derivatives (see **1-46a** → **1-46b** and **1-47** → **1-48**) may have occurred.

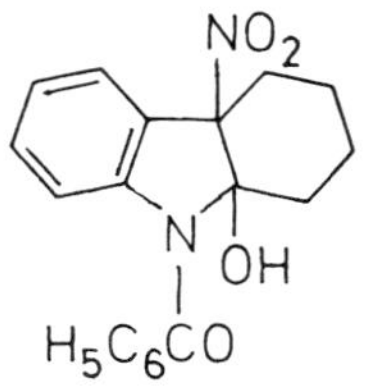

1-125

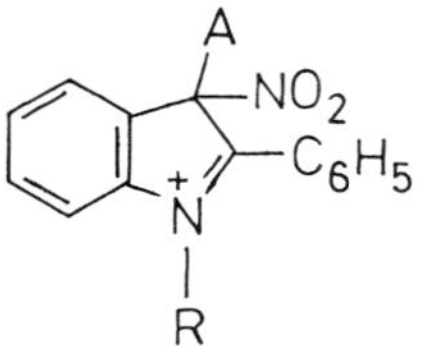

1-128

Recently, 3-nitroindoles (**1-127**) were obtained by electrophilic *ipso*- substitution of arylazo, hydroxymethyl, and acyl groups in derivatives (**1-126**) of 2-phenylindole treated with a 2 M ratio of nitric acid (70%) in acetic acid at 25°, Eq. (1-32), (Table 1-3).[95,96]

An intermediate σ-complex (**1-128**) may have been detected by a transient deep color which developed upon mixing each azoindole (**1-126a,b**) with nitric acid and was persistent for 70 h when derived from **1-126c**. When the blue solution was poured into water the starting materials were completely recovered, but when it was warmed for 30 min the nitro derivative (**1-127c**) was isolated, Eq. (1-32), (Table 1-3). Attempts to isolate the intermediate (**1-128**) were unsuccessful.

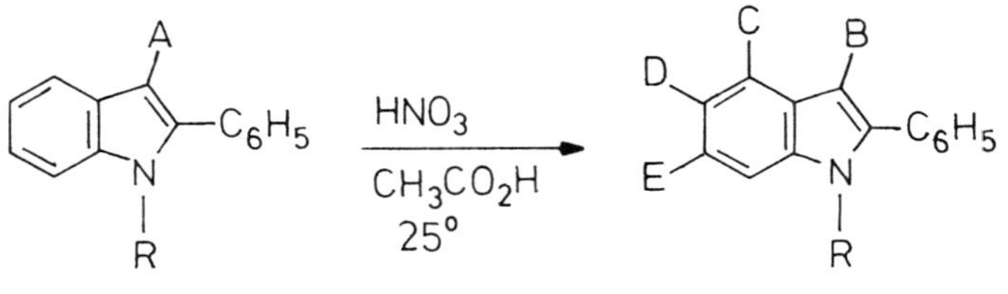

1-126   **1-127**   (1-32)

Sodium nitrite (one molar excess) and the 3-arylazoindoles (**1-126a,b,c**) in acetic acid at 25° also gave the 3-nitroindoles (**1-127a,b,c**) via the 3-nitrosoindoles (**1-129a,b,c**), as demonstrated by independent oxidation of the latter to the nitroindoles by sodium nitrite in acetic acid at 25°, Eq. (1-33), (Table 1-4). A competitive reaction gave 3-*p*-nitrophenylazoindoles (**1-130a,b,c**) also via intermediate nitroso compounds, Eq. (1-34), (Table 1-4).[95,96]

Nitric acid (70%) converted the nitrosoindole (**1-129b**) to the 3-nitroindole

Table 1-3. 3-Nitroindoles (1-127) from 3-Substituted Indoles (1-126)[a]

1-126	R	A	Time(h)	1-127	R	B	C	D	E	Yield(%)
a	H	$N_2C_6H_5$	24	a	H	NO_2	H	H'	H	86
				a'	H	NO_2	H	H	NO_2	14
b	CH_3	$N_2C_6H_5$	50	b	CH_3	NO_2	H	H	H	80
				b'	CH_3	NO_2	NO_2	H	H	20
c	C_2H_5	$N_2C_6H_5$	50	c	C_2H_5	NO_2	H	H	H	90
d	CH_3	$N_2C_6H_4NO_2$-p	96	b'	CH_3	NO_2	H	H	H	17
				d	CH_3	NO_2	H	NO_2	H	6
e	CH_3	C_2H_5	2	b'	CH_3	NO_2	H	H	H	35
f	CH_3	$COCH_3$	72	b'	CH_3	NO_2	H	H	H	40
				f	CH_3	$COCH_3$	H	NO_2	H	30
				f'	CH_3	$COCH_3$	H	H	NO_2	24
g	CH_3	CHO	1.5	b'	CH_3	NO_2	H	H	H	15
				g	CH_3	CHO	H	NO_2	H	60
h	CH_3	NO_2	0.5	h	CH_3	NO_2	H	H	NO_2	81

[a] Ref. 86.

$$\text{1-126a,b,c} \quad A = C_6H_5N_2 \xrightarrow[-C_6H_5N_2^+]{NO^+} \text{1-129a,b,c}$$

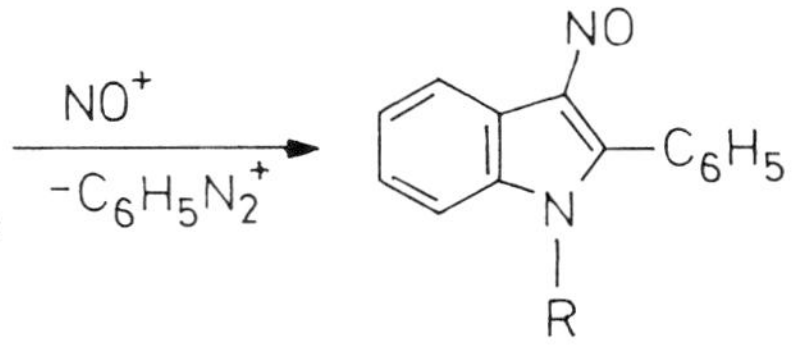

$$\text{1-129a,b,c} \xrightarrow{HONO} \text{1-127a,b,c} \qquad (1\text{-}33)$$

$$\text{1-126a,b,c} \quad 1 = C_6H_5N_2 \xrightarrow{HONO} \text{1-130a,b,c} \qquad (1\text{-}34)$$

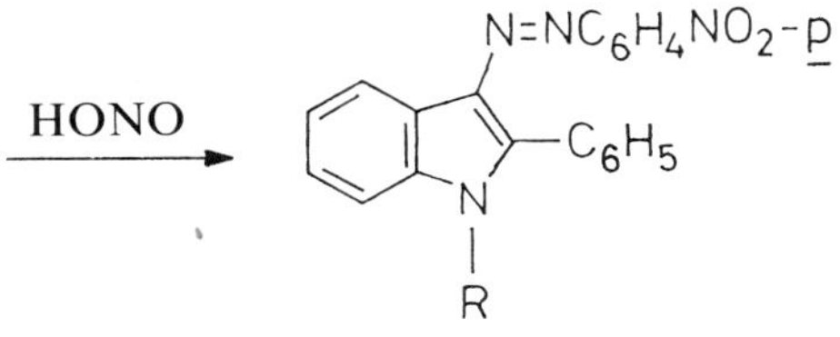

Table 1-4. Nitrosation of Indoles 1-126a,b,c[a]

1-126	R	Time(h)	1-127	Yield(%)	1-130	R	Yield(%)
a	H	24	a	82	a	H	18
b	CH_3	48	b	74	b	CH_3	26
c	CH_3	72	c	78	c	CH_3	22

[a] Ref. 86.

(**1-127b**) and 1-methyl-2-phenyl-3,5-dinitroindole (**1-127d**). These compounds were accounted for by an initial *ipso*-attack by the nitronium cation followed by ejection of a nitrosonium cation to produce the nitroindole (**1-127b**) and by nitration of the intermediate α-complex (**1-131**) to produce the dinitro compound (**1-127d**), Eq. (1-35a).[95,96] An oxidation of the nitroso compound (**1-129b**) by nitric acid was recognized as an alternative explanation but discarded since the nitration of the 3-nitroindole (**1-127b**) should give the 3,6-dinitro derivative which was not found (Tables 1-3, 1-4).

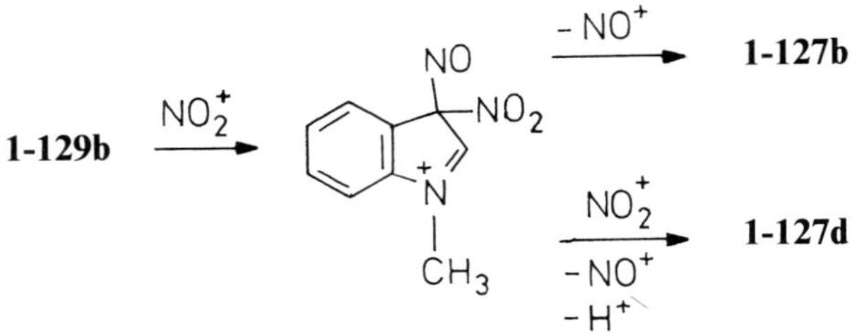

$$\textbf{1-129b} \xrightarrow{NO_2^+} \textbf{1-131} \begin{array}{c} \xrightarrow{-NO^+} \textbf{1-127b} \\ \xrightarrow[-H^+]{NO_2^+ \\ -NO^+} \textbf{1-127d} \end{array} \tag{1-35a}$$

Nitration of the σ-complex (**1-132**) and of the free base (**1-126f**) accounted for the formations of the 5-nitro- (**1-127f**) and the 6-nitro derivative (**1-127f'**) (Table 1-3), Eq. (1-35b), Eq. (1-36).[95,96]

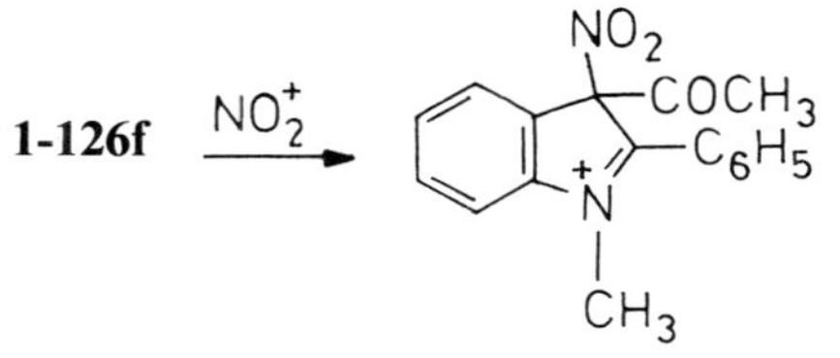

$$\textbf{1-126f} \xrightarrow{NO_2^+} \textbf{1-132} \tag{1-35b}$$

$$\textbf{1-127b} \xleftarrow{-CH_3\overset{+}{C}O} \textbf{1-132} \xrightarrow{-H^+} \textbf{1-127f'}$$

$$\textbf{1-132} \xrightarrow[-H^+]{NO_2^+ \quad -NO_2^+} \textbf{1-127f} \tag{1-36}$$

When the indole (**1-126e**) was nitrated at 25°, an e.s.r. signal of three bands of equal intensity was assigned to the nitro radical ($a^N = 29.2$ G). This was taken as evidence that an electron transfer preceded σ-complex formation for those indoles in which a 3-substituent increased the basicity, Eq. (1-37).[95,96] An e.s.r. signal for the nitro radical was not observed from the indoles (**1-126f,g**).

The *bis*-indoylmethanes (**1-133a**) and nitric acid (70%) (1:4) gave mononitroindole (**1-127b**) (15%), 1-methyl-2-phenyl-3,4-dinitroindole (30%), and the

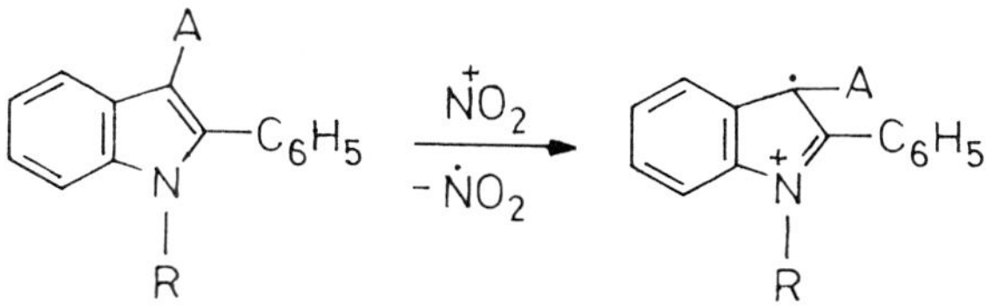

Nitration of Condensed Ring Systems

(1-37)

dinitroindole (**1-127h**) (30%); from **1-133b** the same products were obtained: **1-127b** (26%), 1-methyl-2-phenyl-3,4-dinitroindole (24%), and the dinitroindole (**1-127h**) (30%), Eq. (1-38). From the indole (**1-133b**) benzaldehyde was also isolated. An e.s.r. signal from the nitro radical was obtained from both **1-133a,b**.[95,96]

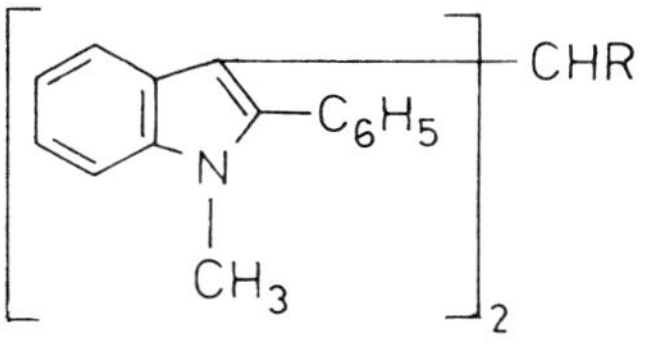

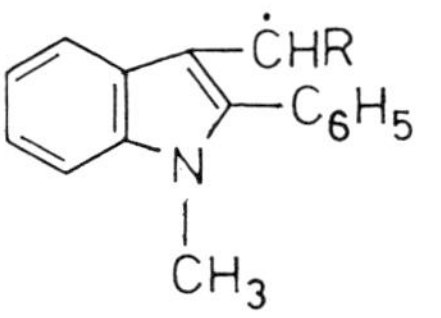

1-133

a, R = H
b, R = C$_6$H$_5$

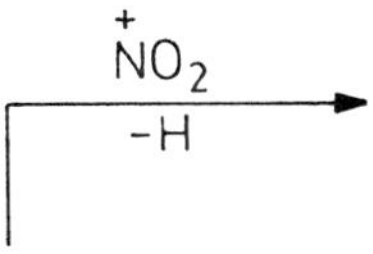

(1-38)

A straightforward attack by tetranitromethane gave an intermediate adduct with indole. Loss of nitrous acid produced an example of a nitro enamine (**1-134**) (10%), Eq. (1-39a).[97]

In a recent total synthesis of the coenzyme methoxatin, a deoxymethoxatin triester in a mixture of fuming nitric and sulfuric acids at 0° for 10 min gave its 3,5-dinitro derivative (**1-135a**) (R = CO$_2$CH$_3$) (94%).[95,98]

In 1983 2-phenylindole was nitrated by 2-cyano-2-propyl nitrate in a phase transfer reaction catalyzed by potassium hydroxide in acetonitrile containing dibenzo-18-crown-6 ether to the 3-nitro derivative (**1-127a**) (60%).[99]

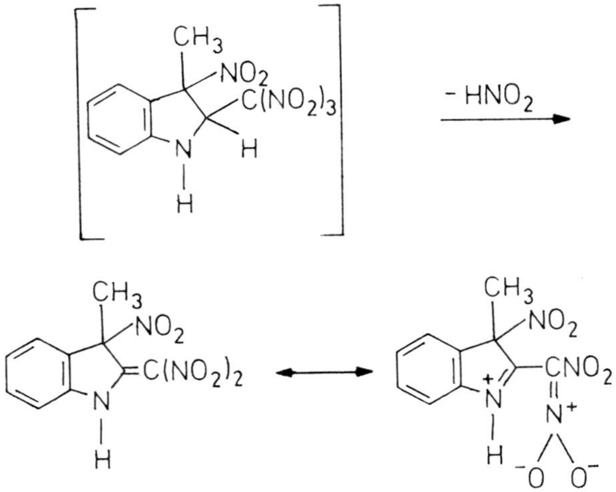

1-134

(1-39a)

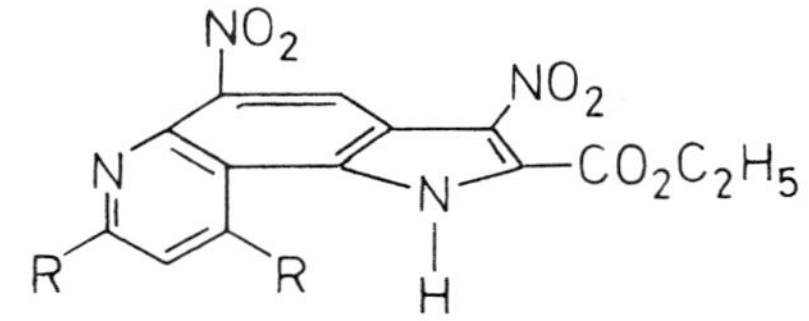

1-135a 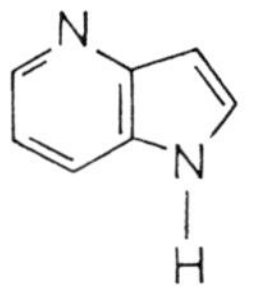 **1-136**

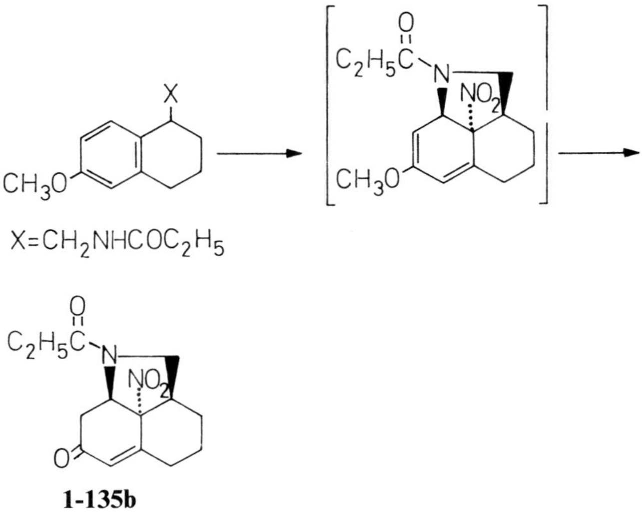

X=CH$_2$NHCOC$_2$H$_5$

1-135b

(1-39b)

An example of a nitropyrrolidine was discovered in the conversion of N-[(6-methoxy-1,2,3,4-tetrahydronaphthalen-1-yl)methyl]propanamide to the tricyclic nitro compound (**1-135b**) (13%), Eq. (1-39b), on treatment with nitric acid in trifluoroacetic acid. The reaction also gave two products of nitration *ortho* to the 6-methoxy substituent. The structure assignment to compound **1-135b** was confirmed by spectroscopy and an X-ray analysis. An intermediate, Eq. (1-39b), following an *ipso*-nitration and ring closure, was suggested; however, nitration in propionic anhydride did not give the tricyclic pyrrolidine (**1-135b**), and treatment of β-*p*-methoxyphenethylpropanamide with nitric acid in trifluoroacetic acid failed to give the analogous *ipso*-nitration product, Eq. (1-39c).[100]

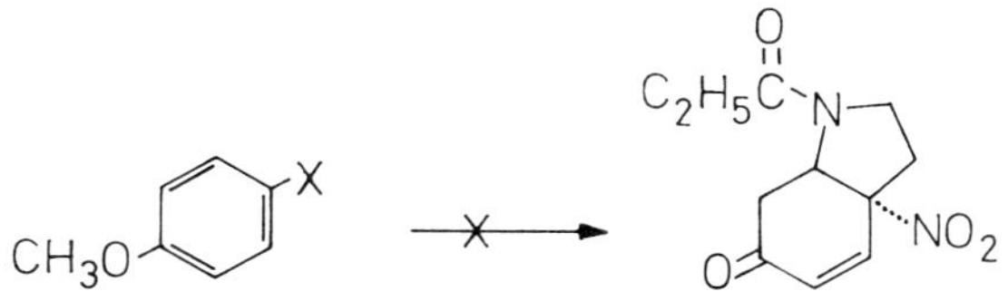

$$X = CH_2NHCOC_2H_5 \tag{1-39c}$$

Azaindoles

Nitration of 4-azaindole (**1-136**) in nitric acid (d 1.52) at $-15°-0°$ for 1 h gave the 3-nitro derivative (84%).[101] Similar nitration converted 7-azaindole (**1-137**) (X = Y = Z = H) to its 3-nitro derivative (**1-138**) (X = Y = H) (83%), Eq. (1-40).[102] Conversions of other 7-azaindoles into their 3-nitro derivatives are described in Table 1-5.[38,103,104]

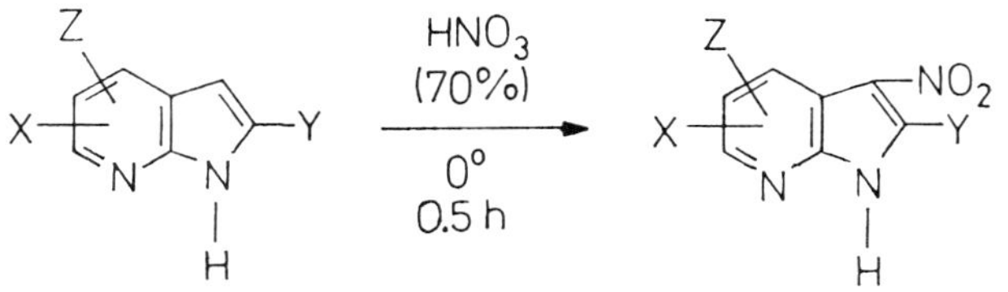

$$\textbf{1-137} \xrightarrow[\substack{0° \\ 0.5\,h}]{HNO_3\ (70\%)} \textbf{1-138} \tag{1-40}$$

Table 1-5. Nitration of Derivatives of 7-Azaindole (**1-137**) into Derivatives of 3-Nitro-7-azaindole (**1-138**)[a]

1-137 and 1-138	X	H	6-Cl	6-OCH$_3$	H	H	H	H
	Z	H	4-CH$_3$	4-CH$_3$	H	H	H	4-CH$_3$
	Y	H	H	H	CH$_3$	–C$_5$H$_4$N	C$_6$H$_5$	H
Product yield(%)		83	67	60	74	71	73	56

[a] Refs. 36, 90, 91.

Nitration with fuming nitric acid converted 2-phenyl-7-azaindole to 3-nitro-2-(p-nitrophenyl)-7-azaindole (**1-139**) (69%), but with concentrated nitric acid mononitration gave 3-nitro-2-phenyl-7-azaindole (**1-140**) (73%). In the first example of an intermolecular electrophilic substitution at the 2-position of a 7-azaindole, nitration of 4-methyl-3-phenyl-7-azaindole with fuming nitric acid produced 4-methyl-2-nitro-3-(p-nitrophenyl)-7-azaindole (**1-141**) (66%). A mixture of the dinitro derivative (**1-141**) and 4-methyl-2-nitro-3-phenyl-7-azaindole was obtained by nitration with concentrated nitric acid but could not be isolated.[38]

A mixture of nitric and sulfuric acids at 0° for 2 h converted 7-azaindole-7-oxide(1H-pyrrolo[2,3-b]pyridine-7-oxide) to the 3-nitro derivative (**1-142**) (48%); nitration with nitronium tetrafluoroborate increased the yield to 74%. None of the expected 4-nitro derivative was obtained.[104]

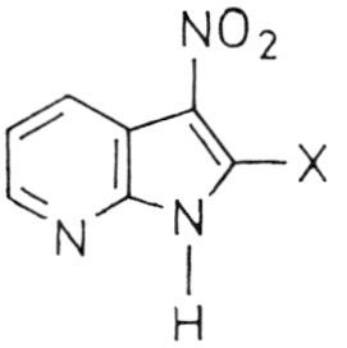

1-139, X = p-O$_2$NC$_6$H$_4$
1-140, X = C$_6$H$_5$

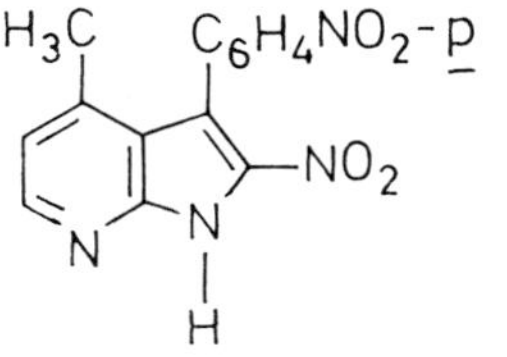

1-141

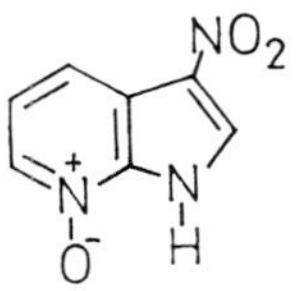

1-142

After 1 h at 0°, nitric acid (d 1.52) converted 4-aza- and 5-azaindoles to the corresponding 3-nitro derivatives (**1-143a**) (99%) and (**1-143b**) (99%).[105]

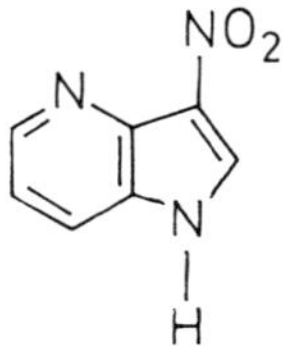

1-143a

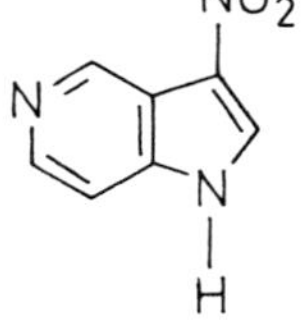

1-143b

Pyrrolo[2,3-d]pyrimidinyl-4-one and concentrated nitric acid at 0°–20° for 30 min gave the 5-nitro derivative (**1-144a**) (52%).[106,107] The related nucleoside antibiotic, tubercidin (7-deazaadenosine), as its triacetyl derivative gave a mixture of its 7- and 8-nitro derivatives (**1-144b**).[107]

A mixture of concentrated nitric and sulfuric acids nitrated 1-benzyl-6-methoxy-7-cyano-5-azaindole at −10° for 30 min to a mixture of the 3-nitro- and the 1-p-nitrobenzyl-3-nitro-derivatives (**1-145a,b**) (>80%, ratio of 2.5:1), characterized by M$^+$ for each compound in the mass spectrum of the mixture and by NMR analysis. When the nitrating mixture stood for 1 h at −15° there was extensive degradation of the pyrrole ring.[108]

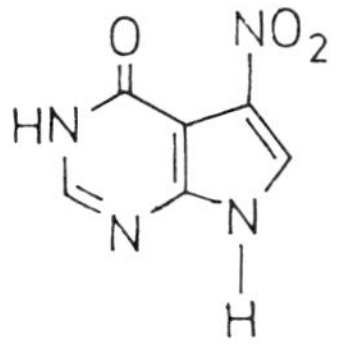

1-144a

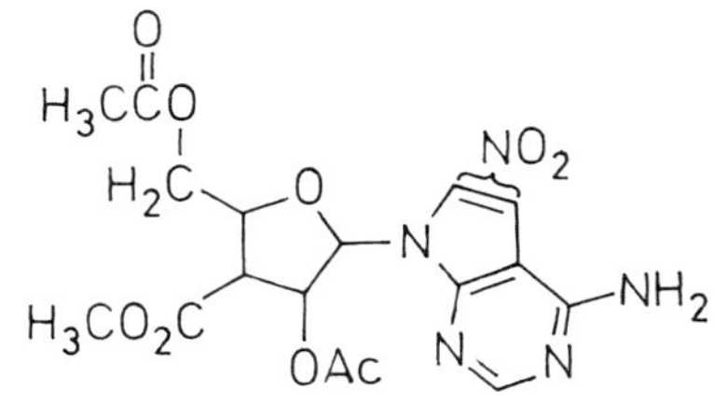

1-144b

1-145a X = H
1-145b X = NO$_2$

Indolizines

In a mixture of nitric and sulfuric acids 2-methylindolizine (**1-146a**) was nitrated at the 1-position but nitration occurred at the 3-position when the solvent was acetic anhydride. Nitration of 2-phenylindolizine (**1-146b**) in sulfuric acid with a 2 M excess of nitric acid gave 2-p-nitrophenylindolizine, and the 1-nitro and 3-nitro-2-p-nitrophenylindolizine. In contrast, 1-methyl-2-phenylindolizine (**1-146c**) underwent mononitration in the phenyl substituent and resisted dinitration.[109-115]

A more extensive study of the nitration of 2-phenylindolizines was recently reported. Warming a solution of 2-phenylindolizine (**1-146b**) in a 14 M excess of nitric acid (d = 1.4) gave the 1,3-dinitro derivative (14%), whereas the addition of a molar excess of nitric acid (d 1.4) over 5 min to the base (**1-146b**) in concentrated sulfuric acid at 0° gave 2-p-nitrophenylindolizine (50%), and further nitration with a 0.5 M excess of nitric acid gave 1-nitro-2-p-nitrophenylindolizine (3%). The nitration of a series of 2-m- and p- substituted phenylindolizines (**1-147a**) is summarized in Table 1-6, Eq. (1-41).[110-112]

Nitrolysis of the indolizines (**1-148a,b**) gave the 3-nitro derivatives (**1-148c,d**).[113-115]

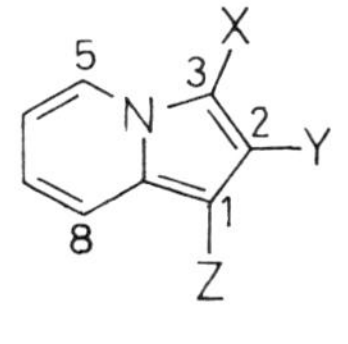

1-146a X = Z = H, Y = CH$_3$
1-146b X = Z = H, Y = C$_6$H$_5$
1-146c X = CH$_3$, Y = C$_6$H$_5$, Z = H

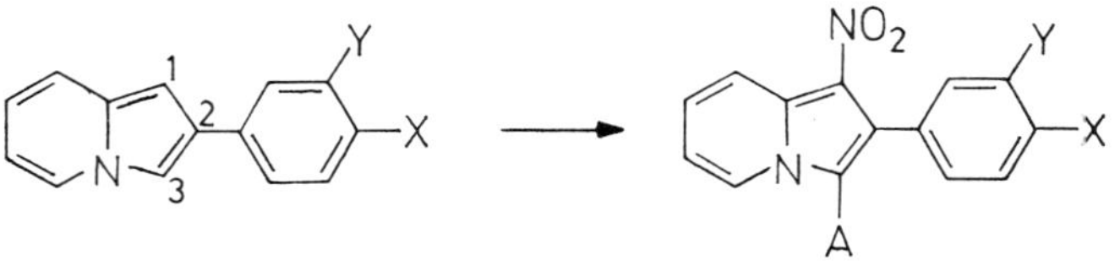

1-147a → 1-147b   (1-41)

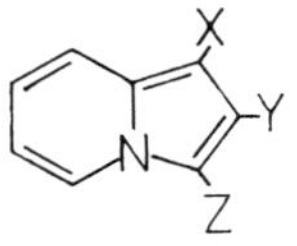

1-148a X = CHY, Y = Z = CO_2CH_3 (with OCH₃ on CH)
1-148b* X = S–,   Y = Z = α-pyridyl
1-148c X = NO_2, Y = Z = CO_2CH_3
1-148d X = NO_2, Y = Z = α-pyridyl
*Disulfide of the type ArssAr.

Table 1-6. Nitration of 2-Arylindolizines (1-147a)[a]

| 1-147a X | Br | Br | H | Cl | H | CH_3O | H | CH_3 | CH_3 | H |
Y	H	H	Br	H	Cl	H	CH_3O	H	H	CH_3
1-147b A	H	NO_2	H	NO_2	H	NO_2	NO_2	H	NO_2	H
Yield(%)	3	15	14	3	18	2	17	5	12	5

[a]Nitric acid, d 1.4, sulfuric acid, 0°, 40 min.

A study of the electrophilic substitution reactions of 2,7-dimethylpyrrolo-[1,2-a]quinoline (1-149a) revealed position 1 to be the most reactive in the free base. Nitric acid (d = 1.4) and a few drops of sulfuric acid in acetic acid transformed the base (1-149a) into its 1-nitro derivative (1-149b) (10%); in acetic anhydride the yield rose to 33%. In a mixture of nitric (d 1.4) and concentrated sulfuric acid, nitration was attributed to the protonated base to account for the formation of the 6-nitro derivative (1-149c) (70%), and in the presence of a molar excess of nitric acid, the 3,6-dinitro derivative (1-149d) (25%).[98] The latter two (1-149c,d) were also prepared (6% and 39%) from 1-acetonyl-6-methyl-5-nitroquinolinium bromide (1-149e) and nitromethane, Eq. (1-42).[116]

A ring transformation afforded ethyl 1-nitro-2-methylthiopyrrolo[2,1-a]isoquinoline-3-carboxylate (1-150) by a reaction between a thiazoloisoquinolinium bisulfate and nitromethane, Eq. (1-43).[117]

A 1,5-dipolar cyclization apparently produced a nitro derivative (1-151) of a 1,9-dihydroindolizine as an intermediate, Eq. (1-44a).[118] The reaction has been extended by replacing the pyridine moiety with other heterocyclic groups.[119]

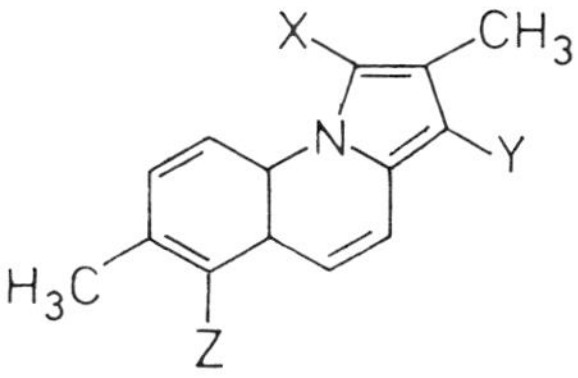

1-149a X = Y = Z = H
1-149b X = NO_2, Y = Z = H
1-149c X = Y = H, Z = NO_2
1-149d X = H, Y = Z = NO_2

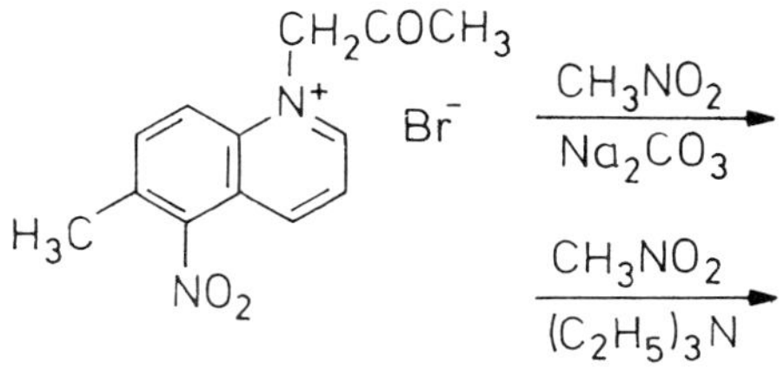

1-149c

1-149d

1-149e (1-42)

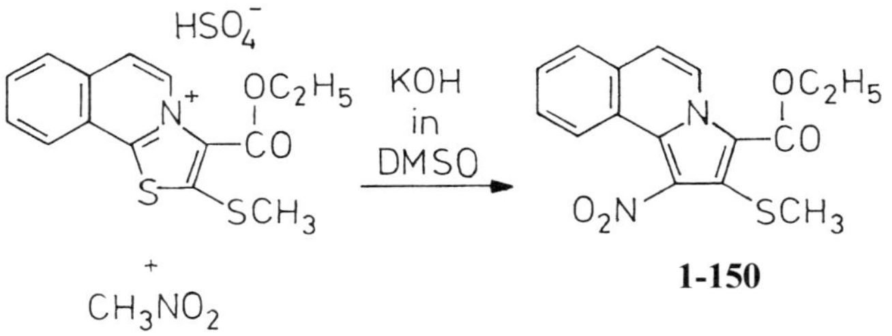

1-150 (1-43)

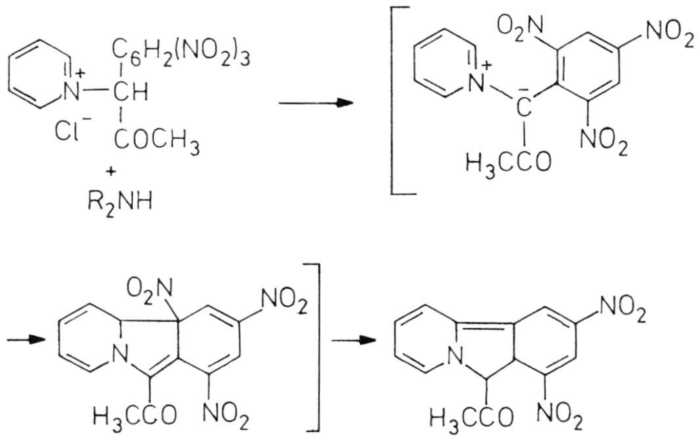

1-151 (1-44a)

Neither potassium nitrate in trifluoroacetic acid nor fuming nitric acid alone or in acetic anhydride nitrated pyrrolo[1,2-*a*]quinoxaline (**1-152**); however, the reaction with potassium nitrate in concentrated sulfuric acid at 0° gave a mixture (86%) of mononitro derivatives separated into the 1-nitro (25% yield) and 3-nitropyrrolo[1,2-a]quinoxaline (48% yield). Nitration of the 3-sulfonic acid derivative of **1-152** gave 1-nitropyrrolo-[1,2-a]quinoxaline-3-sulfonic acid which was also obtained from a freshly prepared solution of the pyrrole (**1-152**) in concentrated sulfuric acid to which powdered potassium nitrate was added.[120]

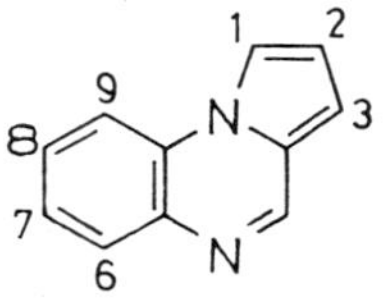

1-152

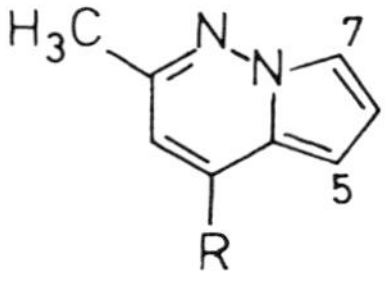

1-153 R = CH$_3$
1-154 R = C$_6$H$_5$

Nitric acid (6 M) and 6,9-dimethoxypyrrolo[1,2-a]quinoxaline at 25° for 1 h gave the 6,9-dimethoxy-3-nitro derivative (80%) of compound **1-152**.[121]

1-Nitro, 3-nitro, and 1,3-dinitropyrrolo[1,2-a]pyrazines have been reported.[122]

Nitric acid (d = 1.5) added to a cold solution of 2,4-dimethylpyrrolo[1,2-b]-pyridazine (**1-153**) in concentrated sulfuric acid converted the base into its 5,7-dinitro derivative (31%) and converted 2-methyl-4-phenylpyrrolo[1,2-b]-pyridazine (**1-154**) into its 5,7-dinitro-4-p-nitrophenyl derivative (12%). The 5- and 7-positions were shown to be the most receptive to electrophilic attack by the calculated total π- and frontier electron densities for the unsubstituted heterocycle.[123]

Potassium nitrate added to a solution of pyrrolo[2,1-c]-[1,2,4]benzotriazine (**1-155**) in concentrated sulfuric acid at 25° converted the base into its 3-nitro derivative (42%).[124] Apparently the 1-iodo derivative of **1-155** on treatment with concentrated nitric acid gave the 3-iodo-1-nitro derivative via conversion to the free base, nitration, and iodination. Attempts to nitrate and sulfonate 4-oxo-4H-pyrrolo[2,1-c][1,4]benzoxazine (**1-156**) were unsuccessful; however, bromination readily gave the 1-bromo and 1,2-dibromo derivatives.

One equivalent of fuming nitric acid and 1,2-dihydro-1-oxopyrrolo[1,2-d]-1,2,4-triazine (**1-157**) in sulfuric acid at −25° for 1 h gave the 6-nitro (55%) and 8-nitro (28%) derivatives; with an excess of fuming nitric acid in sulfuric acid at 0° the 6,8-dinitro derivative (71%) was obtained.[125]

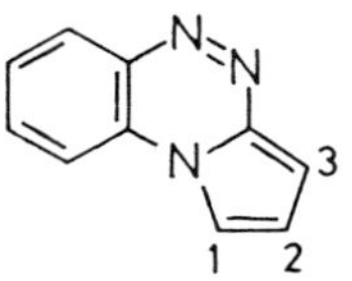

1-155 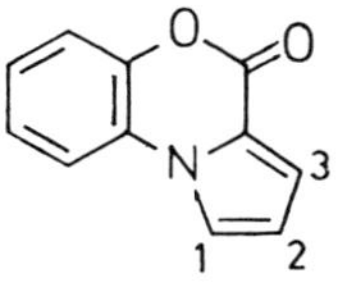   **1-156** 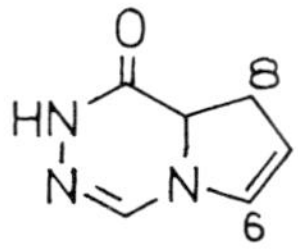   **1-157** 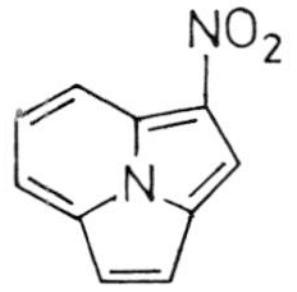   **1-158**

In the pyrrolo[1,2-b]pyridazine ring system (see **1-154**) the reactivity of the 5- and 7-positions resemble that of the 1- and 3-positions in **1-152** and the 6- and 8-positions in **1-157** toward electrophilic substitution. Electron densities and localization energies were determined by HMO calculations with the Coulomb and resonance integrals for hetero-atoms recommended by Streitwieser for pyrrolo[1,2-a]quinoxaline (**1-152**), pyrrolo[2,1-c][1,2,4]-benzotriazine (**1-155**), and 4-oxo-4H-pyrrolo[2,1-c][1,4]benzoxazine (**1-156**). The results based on electron densities indicated the following orders of reactivity in electrophilic attack at positions in **1-152**: 2 > 3 > 9; in **1-155**: 3 > 2 > 9; and in **1-156**: 2 > 6 > 9. Different sequences were based on localization energies—in **1-152**: 1 > 3 > 6; in **1-155**: 1 > 3 > 6; and in **1-156**: 1 > 2 > 3.[124]

1-Nitropyrrolo[2,1,5-cd]indolizine (**1-158**) (83%) was obtained from the unsubstituted base and cupric nitrate in acetic anhydride.[126]

Pyrrolo[2,1-b]thiazoles

Tetranitromethane in a mixture of ethanol and pyridine nitrated 6-methylpyrrolo[2,1-b]thiazole at 25° for 10 min to give an ether believed to be either compound **1-159a** or the isomer **1-159b**. Attempted nitration by cupric nitrate in acetic anhydride led instead to degradation.[127]

Thiazolo[2,3,4-c,d]pyrrolizines

A periselective endo [3 + 2] cycloaddition of trans-p-methoxy-β-nitrostyrene to 3-phenylthiazolo[4,3-a]isoindole gave the trans,trans-3,4-benzo-5,6-dihydro-5-p-methoxyphenyl-6-nitro-6a-phenylthiazolo[2,3,4,-cd]pyrrolizine (60%), Eq. (1-44b).[128] This provided one of the few examples of a nitropyrrolidine.

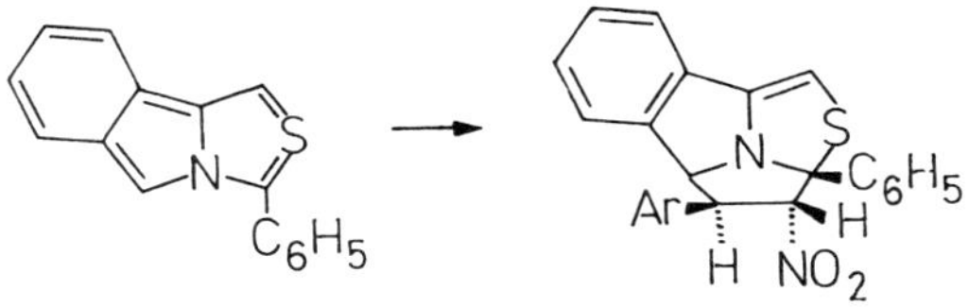

(1-44b)

5H-Pyrrolo[2,1-a]isoindoles

A compound thought to be the 1-nitro derivative (**1-160**) was obtained from a reaction between 2,3-dihydro-2,5-dioxo-3,3-dimethyl-5H-pyrrolo[2,1a]isoindole and concentrated nitric acid at 20°.[129]

CHEMICAL PROPERTIES

pK_a Values for Nitropyrrolecarboxylic Acids

Electron withdrawal by the nitro substituent was revealed in the dissociation constants of carboxylic acids. Acidity increased from pK_a = 4.50 for pyrrole-2-carboxylic acid to pK_a = 3.37 for its 4-nitro derivative and to pK_a = 3.22 for its 5-nitro derivative.[130] A similar enhancement from pK_a = 7.83 for 2,4-dimethylpyrrole-3-carboxylic acid to pK_a = 6.58 for its 5-nitro derivative was noted.[131] An indication from spectral data that 2-nitropyrroles were protonated on the nitro group was not unexpected.[132] A mathematical model was developed to predict pK_a values of heterocyclic carboxylic acids. The predicted values for 4- and 5-nitropyrrole-2-carboxylic acids were pK_a = 3.76 and 3.66, respectively in fairly close agreement with observed values cited.[133]

Reduction

Reduction of nitropyrroles provides an attractive route to aminopyrroles needed for the construction of fused ring heterocyclic compounds. Although catalytic hydrogenation can be recommended for its efficiency, mildness, and simplicity, a variety of chemical reduction methods are also useful. Aminopyrroles are generally sensitive to oxidation and other degradative reactions so they are best isolated as derivatives in which the amine function is protected. α-Aminopyrroles tend to be less stable than their β-isomers.

Hydrogenation over palladium has been generally successful and, for example, converted 1-methyl-4-nitropyrrole-2-carboxylic acid (**1-29**) (from its ethyl ester (**1-24**) by saponification) to 1-methyl-4-amino-pyrrole-2-carboxylic acid (**1-161a**), the starting material for an attractive preparation of distamycin A.[134] The acid (**1-29**) in 1.0 M aqueous sodium carbonate was hydrogenated (60 psi, 25°) over palladium (5%) on carbon until hydrogen absorption ceased (3–4 h). After the catalyst was removed the filtrate was treated with a solution of *tert*-butoxycarbonyl fluoride in ether to give the *tert*-butoxycarboxamide (**1-161b**) (76%). Alternatively the filtrate was treated with formic anhydride (prepared at −78° to −40°) to form the formamide (**1-161c**) (76%). Other examples of the hydrogenation of nitropyrroles to aminopyrroles isolated as the thiourea derivatives, N-acetyl derivatives, and the free amines are recorded.[135,136]

1-159a X = A
1-159b X = B

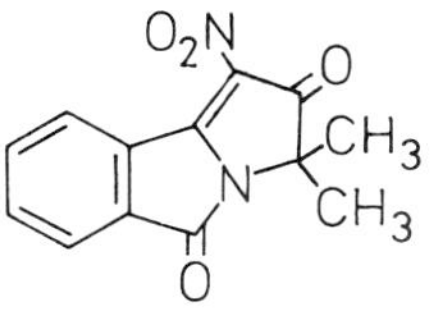

Chemical Properties

[Structure 1-160]

[Structures 1-161a X = H, 1-161b X = (CH₃)₃COCO, 1-161c X = HCO]

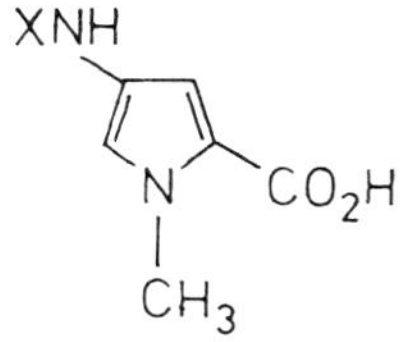

1-161a X = H
1-161b X = (CH$_3$)$_3$COCO
1-161c X = HCO

Although 3-aminopyrrole could not be isolated from the reduction of a 3-nitropyrrole, 2-acetyl-3-nitropyrrole (**1-162**) in methanol was hydrogenated over 5% Pd/C at 25° to 2-acetyl-3-aminopyrrole (**1-163**) (96%). Sodium borohydride reduced the ketone (**1-162**) to a secondary alcohol (**1-164**) (99%) as expected.[83] In related reactions diborane in tetrahydrofuran reduced each of 3-, 4-, and 5-nitro-1-methyl-2-formylpyrroles (**1-165–1-167**) to the expected nitropyrrolemethanols (69%–77%)[15,26] and gave a quantitative yield in the reduction of 2-formyl-4-nitropyrrole to 2-hydroxymethyl-4-nitropyrrole.[137]

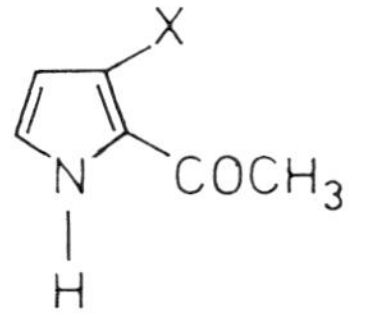

[Structures]

1-162, X = NO$_2$
1-163, X = NH$_2$

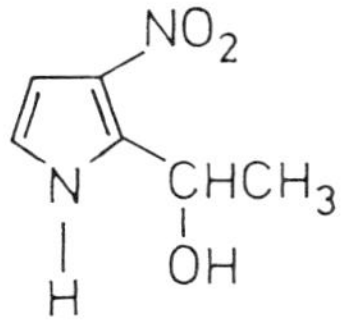

1-164

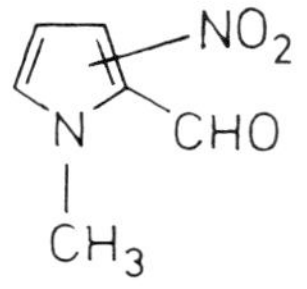

1-165, 3-NO$_2$
1-166, 4-NO$_2$
1-167, 5-NO$_2$

An interesting modification called for the reduction of methyl 1-methyl-4-nitropyrrole-2-carboxylate and isolation of the amine as the amide derivative (**1-168**) (75%) of 1-methyl-4-nitropyrrole-2-carboxylic acid. The amide was an intermediate in a preparation of distamycin A.[29,138]

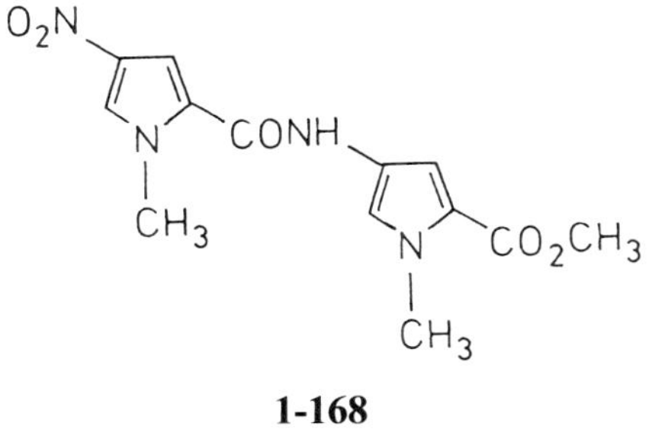

[Structure 1-168]

1-168

Nitrodipyrrole intermediates (**1-169**, **1-170**) (X = CH$_2$, CH$_2$CH$_2$, CH(CH$_3$)CH$_2$) were similarly reduced to corresponding amines (10%–28%).[138,139]

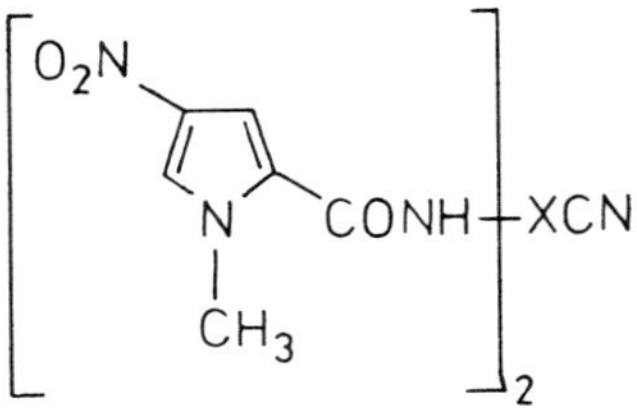

1-169

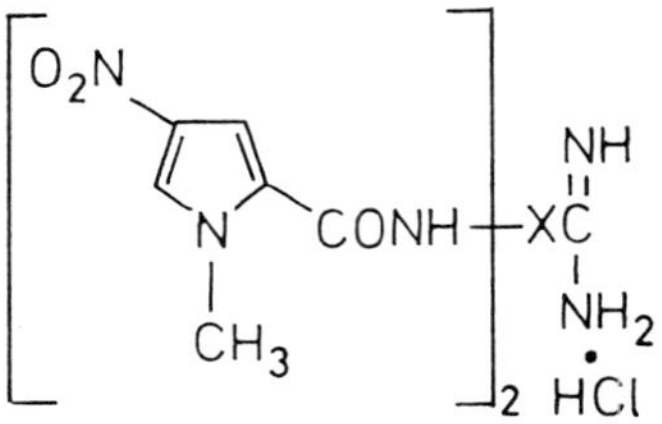

1-170

In addition to palladium (Tables 1-8 and 1-9) and platinum (last entry in Table 1-9), Raney nickel has been an effective catalyst. It promoted hydrogenation and hydrogenolysis of 1-hydroxy-2-methyl-3-acetyl-4-nitro-5-(5-methyl-3-isoxazolyl) pyrrole **(1-34)** to 2,5-dimethyl-3-acetyl-7-amino-1H-pyrrolo-[3,2-b]pyridine **(1-172)** (40%). The intermediacy of an aminopyrrole **(1-171)**, Eq. (1-45), was assumed.[32]

Iron powder in acetic acid also brought about the conversion **1-34 → 1-172** (10%)[32] and in similar reactions transformed the nitropyrrole derivatives **(1-31, 1-32,** and **1-33)** to pyrrolopyridones **(1-173, 1-174)** and a pyrrolothiazine **(1-175)**.[30,31] Zinc and ammonium chloride effected the change **1-31 → 1-173** (80%) whereas Raney nickel hydrogenation gave poor yields (10%) for **1-31 → 1-173** and **1-32 → 1-174**.[30]

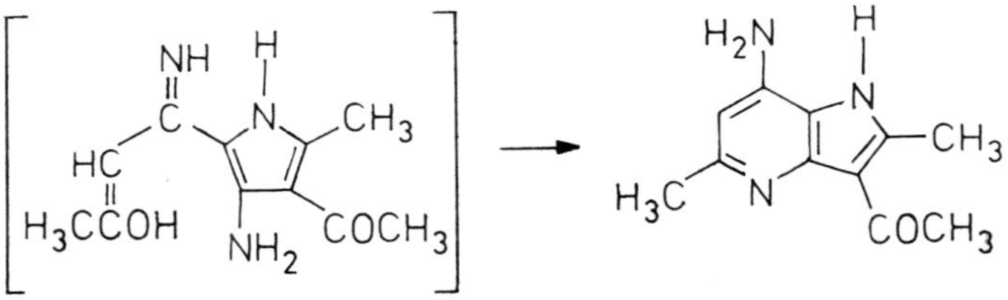

1-171   **1-172**   (1-45)

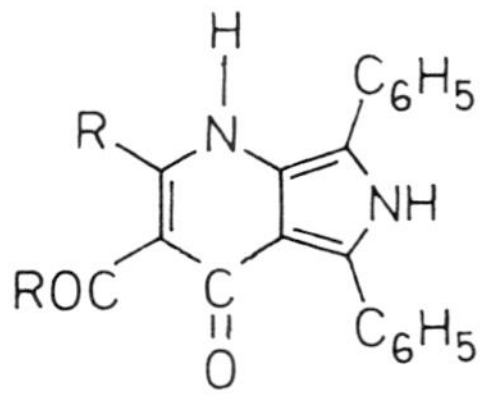

1-173 R = CH$_3$
1-174 R = C$_6$H$_5$

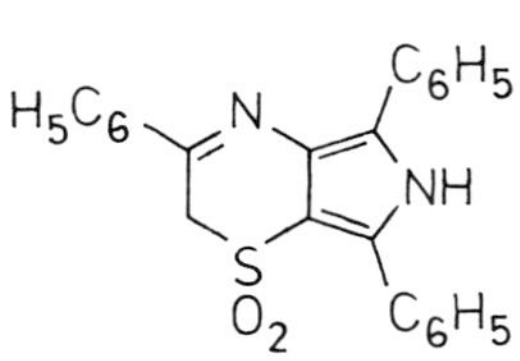

1-175

A preferential Zinin reduction with sodium disulfide hydrate nearly quantitatively converted the 3,5-dinitroindole **1-135a** to a 3-nitro-5-aminoindole **(1-176)** (R = CO$_2$CH$_3$, R' = CO$_2$H).[98] That the "less conjugated 5-nitro group" was preferentially reduced can be attributed to an interaction with the neighboring basic nitrogen function to lower its resonance stabilization, see **1-177**, and enhance its reactivity toward hydrogenation.

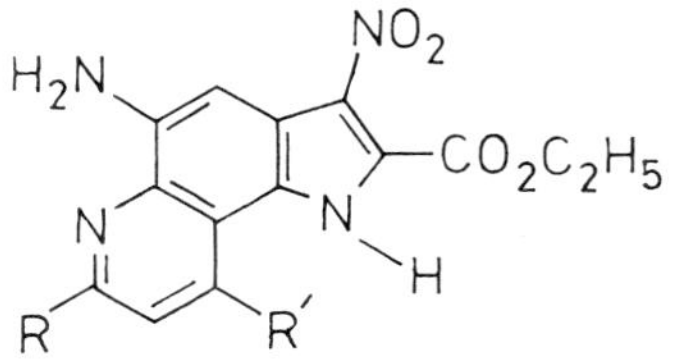

1-176

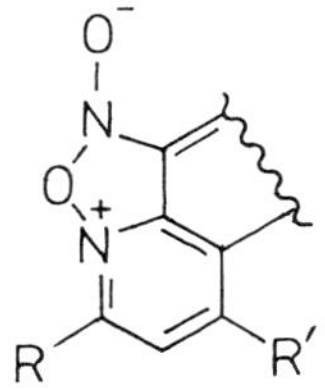

1-177

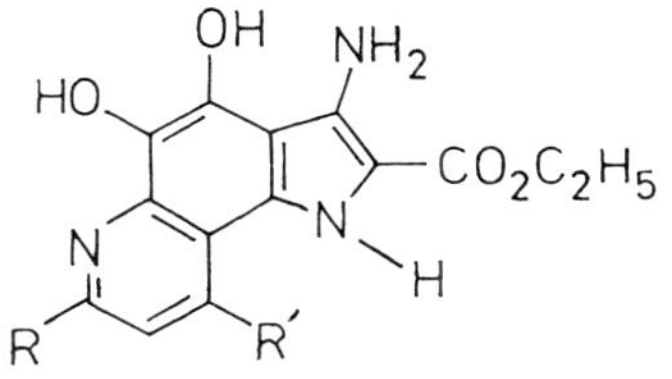

1-178 R = CO_2CH_3

After oxidation of the 5-aminoindole (**1-176**), R = R′ = CO_2CH_3, by manganese dioxide in sulfuric acid to an *o*-quinone, hydrogenation (10% Pd/C, 15 psi, 4 h) gave the 3-aminocatechol (**1-178**) quantitatively.[98]

Stannous chloride dihydrate in concentrated hydrochloric acid converted 1-nitropyrrolo[1,2-*a*]quinoxaline (see **1-152**) to the primary amine (41%) and a similar conversion of the 3-nitro isomer was reported.[120] Titanium trichloride reduction of aromatic and aliphatic nitro compounds has been developed into a quantitative determination of the nitro group, Eq. (1-46).[140] The nitro content (58.58%) in 2,4-dinitropyrrole was found to be 58.10%.

$$6TiCl_3 + RNO_2 + 6HCl \rightarrow 6TiCl_4 + RNH_2 + 2H_2O \qquad (1\text{-}46)$$

Aluminum powder in potassium hydroxide reduced 3-nitro-1,2,4,5-tetraphenylpyrrole to the amine.[53]

One-nitro-2-methyl-4-aminopyrrole (**1-180**), Eq. (1-47), was isolated "from a large scale reaction mixture of 1,4-dinitro-2-methylpyrrole (**1-179**) and ascorbic acid."[141] Although there were no further details for the reaction, the product was adequately characterized.

A least negative half-wave potential (-0.10 V *vs.* S.C.E.) for the dinitropyrrole (**1-179**) was assigned to the reduction of the C-NO_2 function since the nitramine (**1-180**) showed no reduction wave.[141] Other $E_{1/2}$ values for nitropyrroles were reported: 2-nitropyrrole, -0.07; 3-nitropyrrole, -0.18; 1-methyl-2-nitropyrrole, -0.08; 1-methyl-3-nitropyrrole, -0.25.[45]

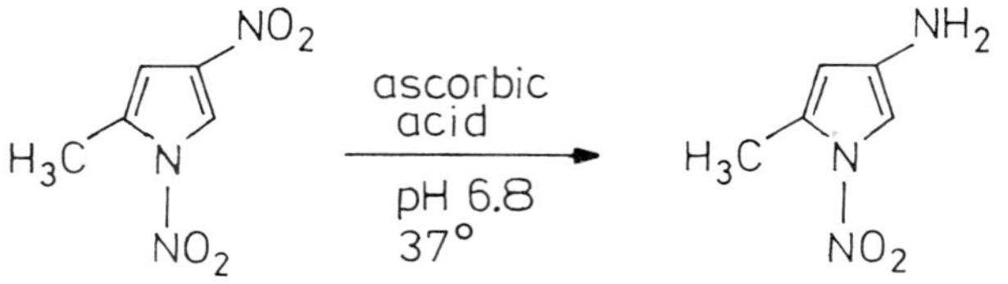

Nitropyrroles

$$\text{1-179} \xrightarrow[\text{pH 6.8, 37°}]{\text{ascorbic acid}} \text{1-180} \quad (1\text{-}47)$$

Copper in acetic acid replaced the iodo substituent in ethyl 5-methyl-4-nitro-3-iodopyrrole-5-carboxylate (**1-49d**) with hydrogen.[35]

Zinc in a mixture of acetic acid and acetic anhydride converted 1-benzyl-3-nitro-2,5-diphenylpyrrole (**1-181**) and 2,4,5-triphenyl-3-nitropyrrole (**1-182**) to the expected amines.[142,143]

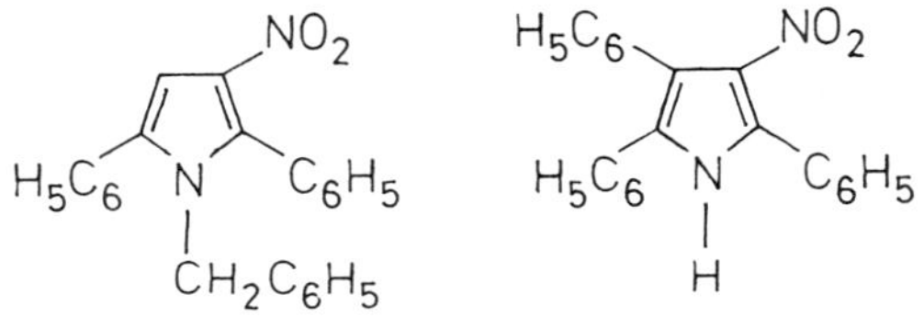

1-181   **1-182**

Thirteen simple nitroheterocyclic compounds were selected for an investigation of a correlation between cellular radiosensitization with electron affinities and redox potentials obtained in pulse radiolysis and ESR studies. As each nitroheterocycle accepted the hydrated electron the electronic spectrum of the nitroheterocycle was replaced by the spectrum of the radical anion (Table 1-7).

Table 1-7. Electronic Absorption for Nitroheterocycle Radical Anions

Compound	λ_{max}^{a} (nm)	ε_{max} ($M^{-1}\text{cm}^{-1} \times 10^{-3}$)
5-Nitro-2-furaldehyde	400	3.6
2-Bromo-5-nitrothiazole	365	2.5
1-Nitropyrazole	380	0.1
2-Nitroimidazole	410	1.1
4-Nitroisothiazole	510	2.0
2-Nitrothiophene	330	2.2
3-Nitrothiophene	415	1.3
3-Nitropyrazole	340	2.1
2-Methyl-5-nitroimidazole	390	1.7
4-Nitroimidazole	380	0.1
2-Nitropyrrole (**1**)	420	2.5
3-Nitropyrrole (**2**)	415	1.6

[a] Absorption < 300 nm excluded.

Chemical Properties

The rate of electron transfer from these radical anions to nitroheterocycles was dependent on the redox potential difference between the donor and the acceptor. Similar results described electron transfers from nitroheterocycle radical anions to either *anti*-5-nitrofuraldoxime or *p*-benzoquinone (Tables 1-8, 1-9).

Nitroheterocycle radical anions gave an inverse correlation between the spin density on the nitro group and their half-wave redox potential (Table 1-10). This provided an alternative method for screening potential cellular radiosensitizers. A further correlation was noted in the comparison of polarographic data for the nitroheterocycles and the direct measure of their electron affinity using one electron redox-potentials.[144]

Table 1-8. Reduction of Nitroheterocycles to Radical Anions.[a]

Compounds[b]	k_1^c	k_2^d	k_3^e
anti-5-Nitrofuraldoxime	38	5.5	3.5
2-Bromo-5-nitrothiazole	20	4.9	3.0
1-Nitropyrazole	27		
2-Nitroimidazole	37	4.0	3.5
4-Nitroisothiazole	33		
2-Nitrothiophene	30	3.5	3.0
4-Amino-5-nitrothiazole		2.0	2.0
3-Nitrothiophene	33	2.5	2.0
3-Nitropyrazole	35		
2-Methyl-5-nitroimidazole	30	4.0	2.5
4-Nitroimidazole	31	4.0	3.5
2-Nitropyrrole	28	3.0	2.0
3-Nitropyrrole	33	3.0	2.0

[a] Rate constants, M^{-1} sec^{-1} × 10^9.
[b] Listed in the order of decreasing electron affinity.
[c] For attack by the hydrated electron.
[d] For electron transfer from the thymine radical anion.
[e] For electron transfer from the 2-propanol radical anion.

Table 1-9. Oxidation of the Radical Anions of Nitroheterocycles[a]

Compound[b]	k_1^c	k_2^d
2-Bromo-5-nitrothiazole	0.2	
2-Nitroimidazole	0.24	4.0
2-Nitrothiophene	0.32	3.8
3-Nitrothiophene	0.30	3.8
2-Methyl-5-nitroimidazole	0.27	3.6
4-Nitroimidazole	0.30	2.9
2-Nitropyrrole	0.35	
3-Nitropyrrole	0.40	3.4

[a] Rate constants, M^{-1} sec^{-1} × 10^9.
[b] Listed in the order of decreasing electron affinity.
[c] For electron transfer from the heterocycle radical anion to *anti*-5-nitrofuraldoxime.
[d] For electron transfer from the heterocycle radical anion to *p*-benzoquinone.

Table 1-10. Nitroheterocyclic Compounds Correlation

Compound	$E_{1/2}$, V vs. SCE[a]	a^N, gauss[b]
5-Nitrofurfuraldehyde	−0.19	9.73
anti-5-Nitrofurfuraldoxime	−0.25	11.43
5-Nitro-2-furoic acid	−0.35	12.20
5-Nitrofurfural diacetate	−0.32	12.79
2-Nitrothiophene	−0.46	13.10
2-Amino-5-nitrothiazole	−0.50	13.49
2-Nitroimidazole	−0.40	13.90
4-Nitroisothiazole	−0.45	14.79
3-Nitrothiophene	−0.52	15.01
3-Nitropyrazole	−0.53	15.16
2-Nitropyrrole	−0.62	15.22
2-Methyl-5-nitroimidazole	−0.56	15.87
4-Nitroimidazole	−0.60	15.97
3-Nitropyrrole	−0.79	16.96

[a] Polarographic half-wave potential determined at ph 7 in 0.05 M tris buffer with a PAR model 174 polarographic analyzer.
[b] Hyperfine splitting constants.

Deoxygenation

Triethyl phosphate deoxygenated ethyl 1-methyl-3-nitro-5-phenyl-4-(2-pyridyl)-pyrrole-2-carboxylate (**1-183a**) to ethyl 8-methyl-9-phenylpyrrolo[3′,4′:2,3]-pyrazolo[1,5-*a*]pyridine-7-carboxylate (**1-84**) (40%), a 10π-electron triazapentalene. Deoxygenation of the related 1-methyl-3-nitro-4-(2-pyridyl)2,5-

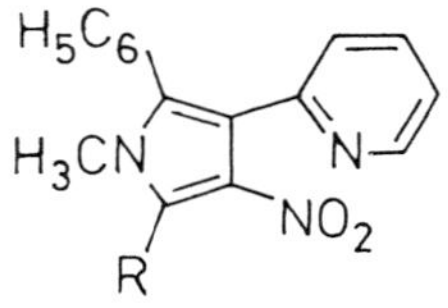

1-183a R = CO$_2$C$_2$H$_5$
1-185 R = C$_6$H$_5$

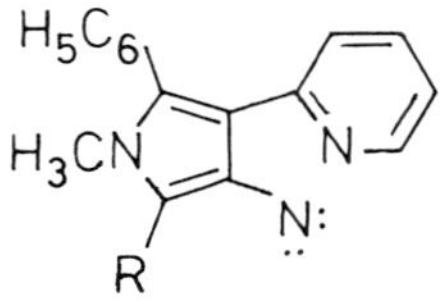

1-183b

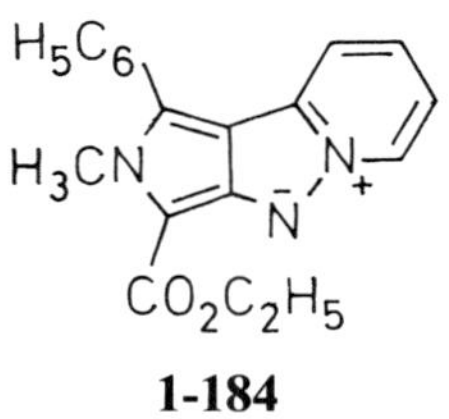

1-184

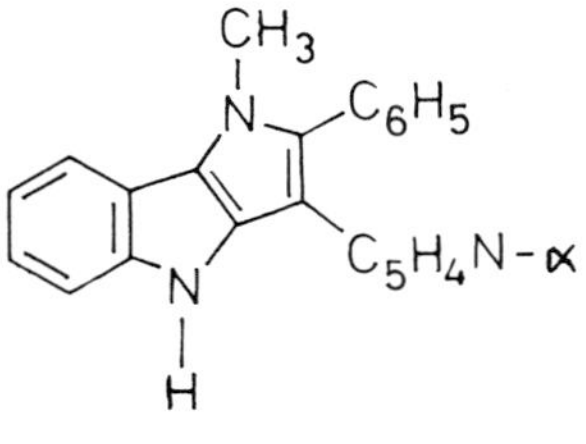

1-186

diphenylpyrrole (**1-185**) followed a different course to give the indole (**1-186**) in low yield. An intermediate nitrene (**1-183b**) was assumed for each reaction.[54]

Products from oxidation at nitrogen of an amino- or nitreno-pyrrole have not been identified; however, other aryl amines and nitrenes have been oxidized into nitroarenes.[145]

Oxidation

Pyrrole oxidation by nitric acid associated with nitration has occasionally led to discrete products. Dilute nitric acid oxidized 2-acetyl-4- and 5-nitropyrroles (**1-187**) to diacylfuroxans (**1-189**) and 4-acetyl-2-nitropyrrole (**1-188**) to 2-nitropyrrole-4-glyoxalic acid (**1-190**).[18] Since their discovery[18,146] these reactions have not been explained; nevertheless they appear related to the conversion of acetophenone in nitric acid to dibenzoylfuroxan (**1-192**). Just as the latter was seen to be a straightforward dimerization of benzoylnitrile oxide (**1-191**), Eq. (1-48),[147] the furoxan (**1-189**) is the corresponding dimer of the intermediate nitrile oxide (**1-194**), Eqs. (1-49, 1-50). The generation of benzoylnitrile oxide by nitration of acetophenone followed by dehydration of α-nitroacetophenone was proposed, Eq. (1-48).[148] Formation of the glyoxalic acid (**1-190**) is the product expected from the nitroacetylpyrole intermediate (**1-193**) by acid hydrolysis, Eq. (1-49); dehydration of **1-193** can afford the nitrile oxide (**1-194**).

A patent reported an unspecified oxidation of 2-methylthio-3-nitropyrrole (**1-110**) to the sulfoxide (85%).[77]

In an example of the haloform reaction alkaline hypochlorite converted 2-acetyl-4-nitro-1-methylpyrrole (**1-195**) to 4-nitro-1-methylpyrrole-2-carboxylic acid (38%). The latter decarboxylated in quinoline with suspended copper powder at 175°–190° for 10 min to 3-nitro-1-methylpyrrole (**1-5a**) (80%).[8]

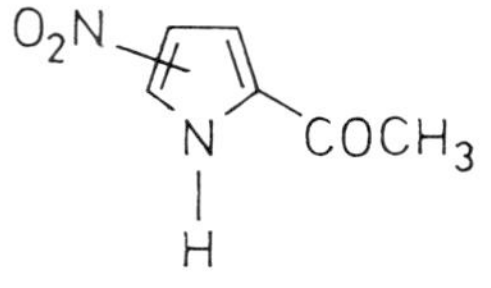

1-187

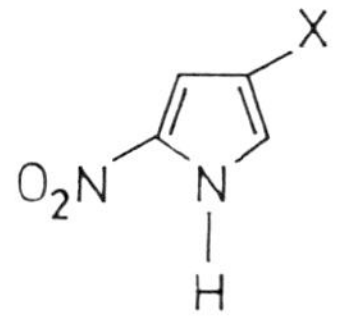

1-188 X = COCH$_3$
1-190 X = COCO$_2$H

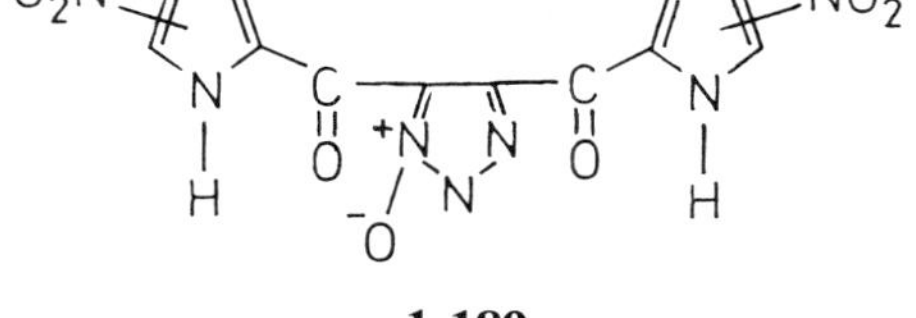

1-189

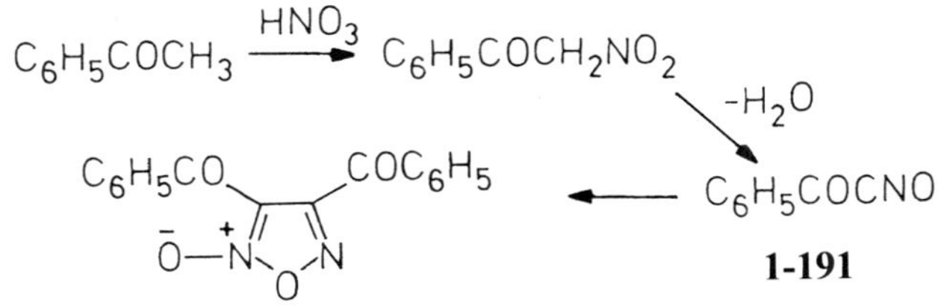

$$C_6H_5COCH_3 \xrightarrow{HNO_3} C_6H_5COCH_2NO_2$$

$$\underset{\textbf{1-192}}{\underset{\bar{O}-\overset{+}{N}\diagdown_O\diagup N}{C_6H_5CO\diagdown\diagup COC_6H_5}} \xleftarrow{-H_2O} \underset{\textbf{1-191}}{C_6H_5COCNO}$$

(1-48)

$$\textbf{1-187} \longrightarrow \underset{\textbf{1-193}}{ArCOCH_2NO_2} \xrightarrow{H_3O^+} \textbf{1-190}$$

Ar = 4(5)-$O_2NC_4H_3N$

(1-49)

$$\textbf{1-193} \xrightarrow{-H_2O} \underset{\textbf{1-194}}{ArCOCNO} \longrightarrow \textbf{1-189}$$

(1-50)

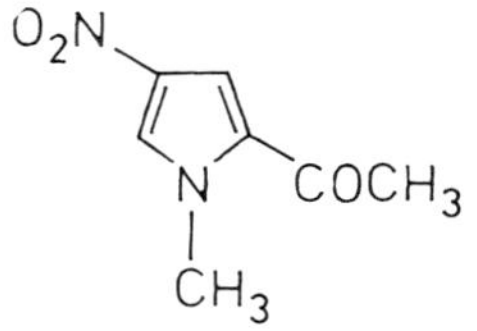

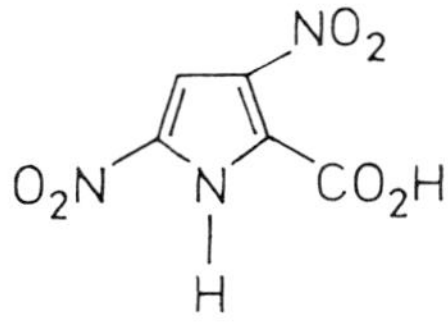

1-195: 4-nitro-1-methyl-2-acetylpyrrole (O$_2$N, N-CH$_3$, COCH$_3$)

1-196: 3,5-dinitropyrrole-2-carboxylic acid (O$_2$N, NO$_2$, NH, CO$_2$H)

Decarboxylation

Decarboxylation of 3,5-dinitropyrrole-2-carboxylic acid (**1-196**) (mp = 161°, dec) (from ethyl 5-nitropyrrole-2-carboxylate (**1-43b**) by nitration followed by hydrolysis of the ester in sulfuric acid) in quinoline at 160°–170° gave 2,4-dinitropyrrole (**1-66**) (30%). Although saponification of the methyl ester of the acid (**1-196**) brought about decomposition (the odor of ammonia was detected)[33] without the formation of the acid (**1-196**), ethyl 2-methyl-5-nitropyrrole-3-carboxylate (**1-197**) gave the expected acid quantitatively upon saponification with potassium hydroxide followed by acidification. The acid decarboxylated in quinoline with suspended copper chromite at 170°–190° for 3 h under nitrogen to 2-methyl-5-nitropyrrole (**1-7**) (71%). Similar treatment converted ethyl 5-methyl-4-nitropyrrole-2-carboxylate to 2-methyl-3-nitropyrrole (**1-8**) (69% overall).[14] Decarboxylation of 4-nitropyrrole-2-carboxylic acid afforded a recommended preparation of 3-nitropyrrole (**1-2**).[9]

Ethyl 4-nitro-5-formylpyrrole-2-carboxylate (**1-43c**) was similarly converted to 2-formyl-3-nitropyrrole; the ester (**1-43d**) to 2-formyl-4-nitropyrrole; the ester

1-197

1-198

(**1-44b**) to 3-formyl-2-nitropyrrole; and the ester (**1-44d**) to 3-formyl-4-nitropyrrole.[28]

Nucleophilic Reactions

Information on the reactivity of pyrrole substrates toward nucleophiles began to be organized in 1974 with an investigation of the interaction between dinitropyrroles and the methoxide anion. An unexpected mobility of hydrogen at the 5-position in 1-methyl-2,4-dinitropyrrole (**1-84**) permitted a base-catalyzed deuterium exchange with perdeuteromethanol in the presence of methoxide anion. Since the exchange rate on an unactivated five-membered aromatic heterocycle is low, a necessary activation was attributed to the electron withdrawing nitro groups.[149] A similar deuterium exchange with H-5 in 1,4-dinitro-2-methylpyrrole (**1-111**) was more remarkable since base catalysis was not required. That the pyrrole (**1-111**) gave a Meisenheimer adduct with one equivalent of methoxide anion in methanol[78] was consistent with the suggestion that nucleophilic attack at a pyrrole α-position was facilitated by a strong electron-withdrawing substituent at the 1-position. Accordingly the formation of a Meisenheimer adduct (**1-198**) from the methoxide anion and 1-(*p*-nitrophenyl)-2,4-dinitropyrrole apparently obscured proton exchange.[149] These reaction differences attributed to the electronegativity of the 1-substituent need more thorough examination.

Nucleophilic aromatic substitution in 2,5-dinitro-1-methyl-pyrrole (**1-82**) by a 3 *M* excess of piperidine in dimethyl-sulfoxide at 25° was complete in 3 days while substitution by an equimolar amount of methoxide ion in methanol at 40° was complete in 2 min. The products were 1-methyl-2-nitro-5-piperidinopyrrole (**1-199a**) and the 5-methoxypyrrole (**1-199b**).[150] An intramolecular base-catalyzed displacement of the 5-nitro group in 2-acetyl-1-(2-hydroxyethyl)-5-nitropyrrole (**1-200**) gave 5-acetyl-2,3-dihydropyrrolo[2,1-*b*]oxazole (**1-201**).[151] These displacements were assisted not only by blocking groups at the 1-position, but also by activation derived from electron-withdrawing groups in conjugated positions, and by the high capability of the nitro group to depart. For comparison, piperidino denitration of 2,5-dinitrofuran, 2,5-dinitrothiophene, and 1,4-dinitrobenzene occurred more rapidly than it did for 1-methyl-2,4-dinitropyrrole (**1-82**) by the factors 1.24×10^6, 4.4×10^3 and 9.6.[152] Denitration by the *p*-tolylthio anion showed the same order of reactivity toward the three heterocycles but occurred more rapidly for 2,5-dinitro-1-methylpyrrole than for 1,4-dinitrobenzene.[153] This inversion of order in reactivity was associated with both the polarizability of the leaving group and the increase in polarizability (through conjugation) of the bond between the reaction center in the heterocycle and the nitro substituent.[153]

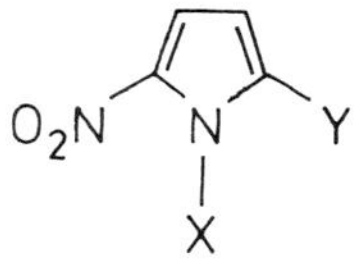

1-199a X = CH$_3$, Y = NC$_5$H$_{10}$
1-199b X = CH$_3$, Y = OCH$_3$
1-200 X = CH$_2$CH$_2$OH
Y = COCH$_3$

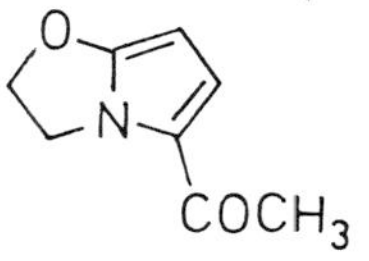

1-201

A structural correlation between 2,3-dinitronaphthalene (**1-202**), 3,4-dinitrothiophene (**1-203**), and a 3,4-dinitropyrrole (**1-87a**) suggested an investigation on the interaction between the latter and nucleophiles. It was known that *cine*-substitution can compete with or replace nucleophilic substitution in reactions of **1-202** and **1-203**. Sodium methoxide in methanol reacted with 1-methyl-3,4-dinitropyrrole (**1-87a**) to produce *trans*-4,5-dimethoxy-1-methyl-3-nitro-2-pyrroline (**1-204**). A regiospecific acid-promoted elimination of methanol from the pyrroline led to the formation of 2-methoxy-1-methyl-4-nitropyrrole (**1-205**), the formal product of *cine*-substitution of **1-87a**.[154] A base-promoted regiospecific elimination led to the formation of the 3-methoxy isomer (**1-206**).[155]

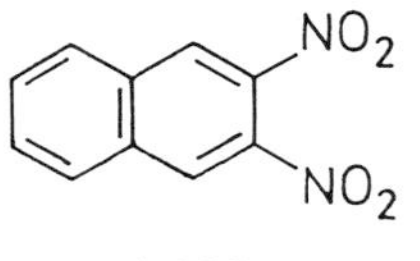

1-202 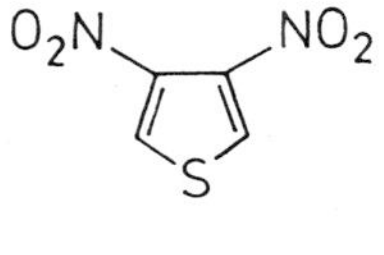   **1-203**

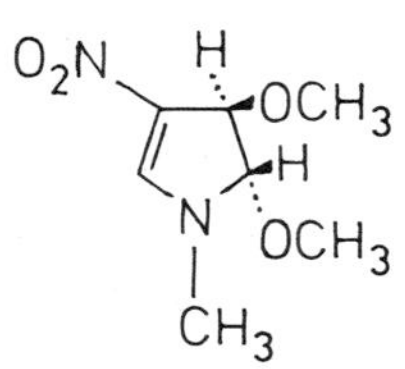

1-204 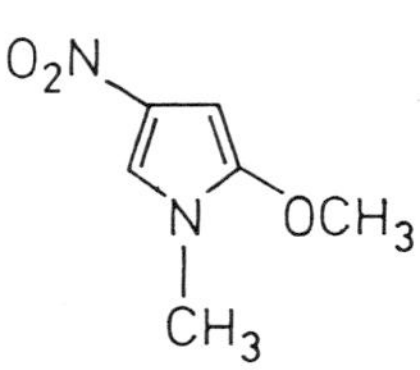   **1-205** 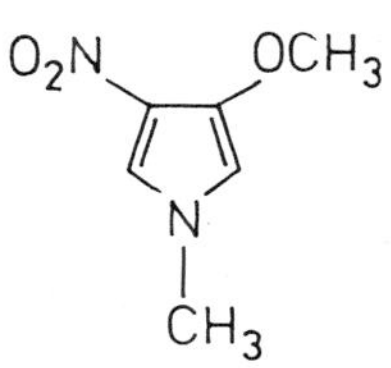   **1-206**

The *cine*-substitution product, 1-methyl-2-piperidino-4-nitropyrrole (**1-207**) (90%), and 2,3-dinitro-1,4-dipiperidinobuta-1,3-diene (**1-208**) (3%) were formed in the reaction between piperidine (1 M excess) and 1-methyl-3,4-dinitropyrrole (**1-87a**) in refluxing acetonitrile for 30 h. A similar reaction between 1-*t*-butyl-

3,4-dinitropyrrole (**1-87c**) and dimethylamine (25 equivalents) in a mixture of methanol and acetonitrile at 80° (sealed tube) gave 1-*t*-butyl-2-dimethylamino-4-nitropyrrole (**1-209**).[156,157]

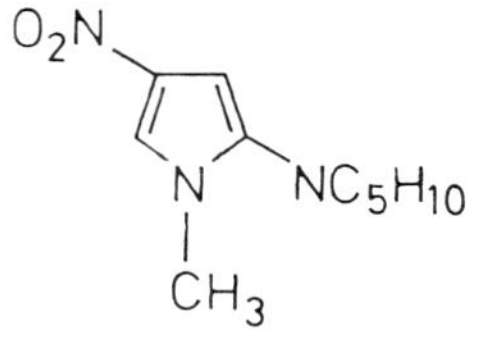

1-207

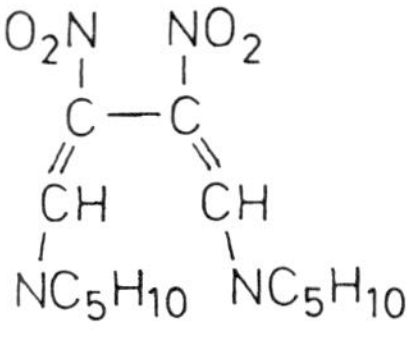

1-208

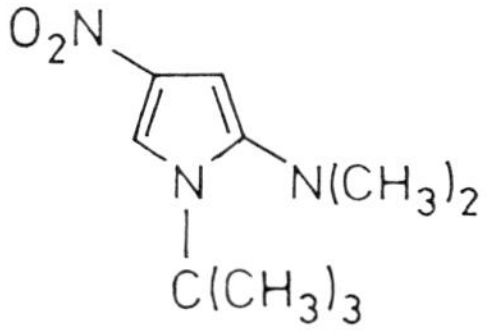

1-209

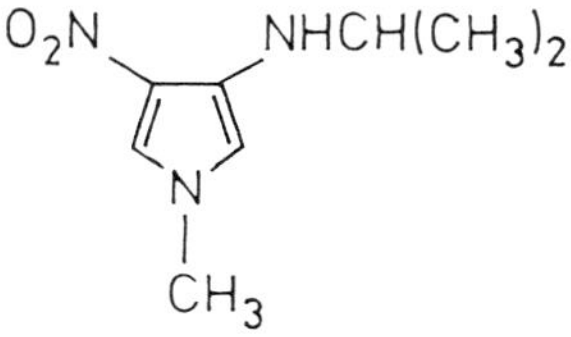

1-210

In contrast isopropylamine gave 1-methyl-3-isopropylamino-4-nitropyrrole (**1-210**) by formal direct displacement and two products of ring-opening and ring closure: 1-isopropyl-3-methyl-amino-4-nitropyrrole (**1-211a**) and 1-isopropyl-3-isopropylamino-4-nitropyrrole (**1-211b**). *t*-Butylamine gave products corresponding to **1-211a** and **b** in 15% and 23% yields, respectively.[156,157]

A short report in 1984 announced the conversion of 1-alkyl-2,4-dinitropyrroles by treatment with piperidine in dimethylsulfoxide or acetonitrile at 25° for 2 h to *trans*-4,5-dipiperidino-3-nitro-2-pyrrolines (**1-212**).[158]

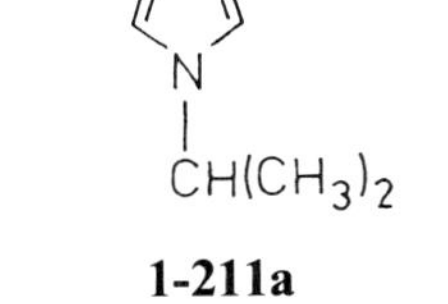

1-211a

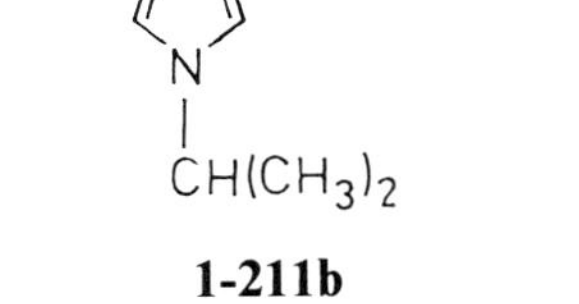

1-211b

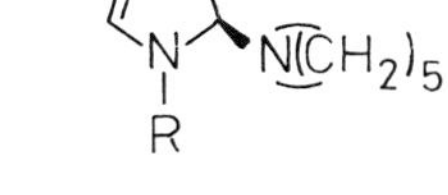

1-212

Efficient conversions of 3-methoxy-4-nitro-5-phenyl-3-pyrrolin-2-one (**1-213**) (from the enol (**1-105**) and diazomethane) to 3-amino derivatives (**1-214**), Eq. (1-51), in reactions with ammonia, primary amines (cyclohexyl, benzyl, and phenyl), and pyrrolidine were accounted for by the intermediacy of compound **1-215**.[74] Perhaps an alternative route to the amines (**1-214**) via an intermediate hydroxypyrrole (**1-216**) and pyrroline (**1-217**) (compare **1-204**) should now be investigated.

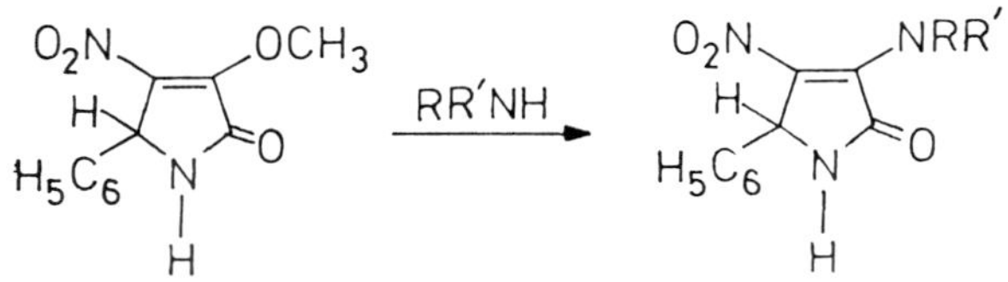

1-213 **1-214**

$RR' = (CH_2)_4$
$R = H$ $R' = cyclo\text{-}C_6H_{11}$,
$C_6H_5CH_2$, C_6H_5 (1-51)

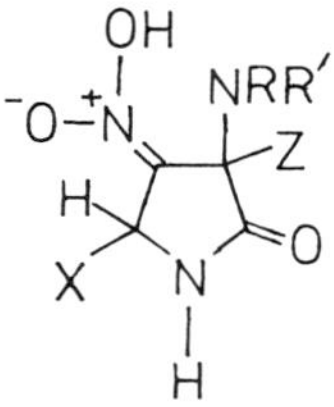

1-215

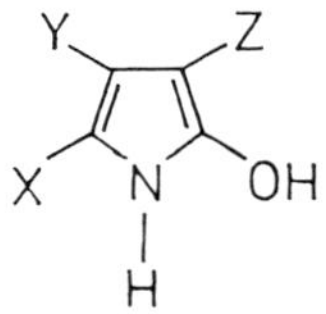

1-216

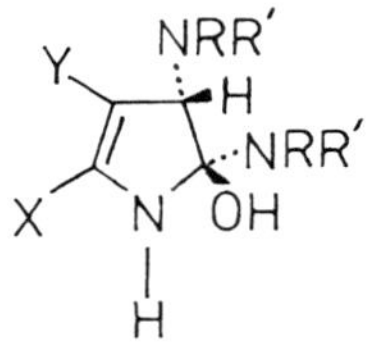

1-217

$X = C_6H_5$, $Y = NO_2$, $Z = OCH_3$

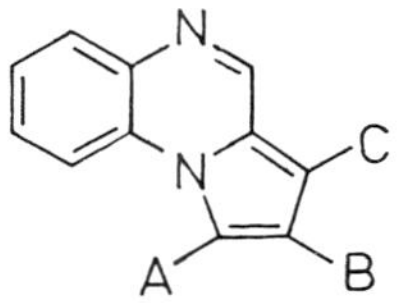

1-152 $A = B = C = H$
1-218 $A = NO_2$, $B = C = H$
1-219 $A = NO_2$, $B = H$, $C = SO_3H$
1-220 $A = Cl$, $B = C = H$
1-221 $A = H$, $B = C = Br$
1-222 $A = B = C = Br$
1-223 $A = Br$, $B = H$, $C = SO_3H$
1-224 $A = B = H$, $C = Br$
1-225 $A = Br$, $B = C = H$

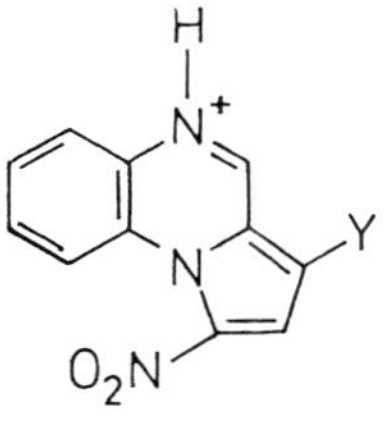

1-226a $Y = H$
1-226b $Y = SO_3^-$

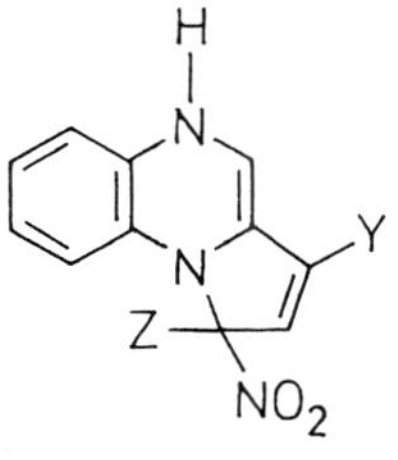

1-227a $Y = H$
$Z = Cl/Br$
1-227b $Y = SO_3^-$
$Z = Cl/Br$

Replacement of the nitro group in 1-nitropyrolo[1,2-*a*]quinoxaline (**1-218**) and its 3-sulfonic acid (**1-219**) was brought about by hydrochloric or hydrobromic acid and by lithium chloride or bromide. Concentrated hydrochloric acid (110°, 6 h) or lithium chloride in dimethylformamide (DMF, 150°, 1 h) produced the 1-chloro derivative (**1-220**) (90%, 88%) from the nitro-sulfonic acid (**1-219**). Hydrochloric acid (110°, 6 h) gave the 1-chloro derivative (**1-220**) (50% based on recovery of the nitropyrrole 32%) from the nitropyrrole (**1-218**) but lithium chloride in DMF failed to react with **1-218**. It also failed to react with the sodium salt of the sulfonic acid (**1-219**).[120,159]

Concentrated hydrobromic acid (126°, 6 h) converted the nitropyrrolesulfonic acid (**1-219**) to a mixture (70%) of 2,3-dibromo- (**1-221**) (44%) and 1,2,3-tribromopyrrolo[1,2-*a*]quinoxaline (**1-222**) (56%) whereas the 1-nitropyrrolo[1,2-*a*]quinoxaline (**1-218**) gave a mixture (about 40%) of **1-221** (5%) and **1-222** (95%). Lithium bromide (DMF, 150°, 10 min) converted **1-219** to 1-bromopyrrolo[1,2-*a*]quinoxaline-3-sulfonic acid (**1-223**) (34%) whereas the product was 3-bromopyrrolo[1,2-*a*]quinoxaline (**1-224**) (36%) after 1 h. Lithium bromide also failed to react with the nitro compound (**1-218**).[120,159]

Observations[120,159] from these investigations supported the occurrence of nucleophilic displacement of the nitrite ion by chloride and bromide ions:

1. The 1-nitro compound (**1-218**) was slowly converted to the 1-chloro compound (**1-220**) in hot hydrochloric acid but failed to react with either lithium chloride or bromide in hot DMF. An acid-catalyzed activation of the 1-nitro group in **1-218** was attributed to protonation at N-5. Acid-catalyzed displacement reactions of halogenoheterocycles were analogous.
2. A zwitterionic form (**1-226b**) of the nitrosulfonic acid (**1-219**) accounted for the reactions with lithium chloride and bromide in DMF.
3. Both lithium chloride and bromide failed to react with the sodium salt of **1-219** in DMF.
4. In these examples the replacement of a nitro group as a nitronium ion (NO_2^+) by an electrophilic attack did not occur. The nitro group in compound (**1-218**) was less reactive than it was in the nitrosulfonic acid (**1-219**) but the opposite order of reactivity was expected in an electrophilic attack at the C-NO_2 position. On treatment with sulfuric acid (50%) the nitrosulfonic acid (**1-219**) gave **1-218** (removal of the sulfonic acid group) and failed to convert to either the unsubstituted base (**1-152**) or its 3-sulfonic acid derivative.
5. Reactions between compounds (**1-218**) and (**1-219**) with hydrobromic acid and with lithium bromide paralleled those with hydrochloric acid and lithium chloride. Rearrangement of 1-bromopyrrolo[1,2-*a*]quinoxaline (**1-225**) to its 3-bromo isomer (**1-224**) in hydrobromic acid or by lithium bromide in DMF was independently established. Di- and tribrominated products (**1-221**) and (**1-222**) were produced by further bromination of the monobromo compounds (**1-224**) and (**1-225**).
6. Oxidation of the bromide anion by nitrous acid afforded bromine needed for

bromination. The polybrominated products were also obtained from the monobromide (**1-225**) and sodium nitrite in concentrated hydrobromic acid.

These observations and the results shown above supported the nucleophilic displacement reactions, Eq. (1-52).[120,159]

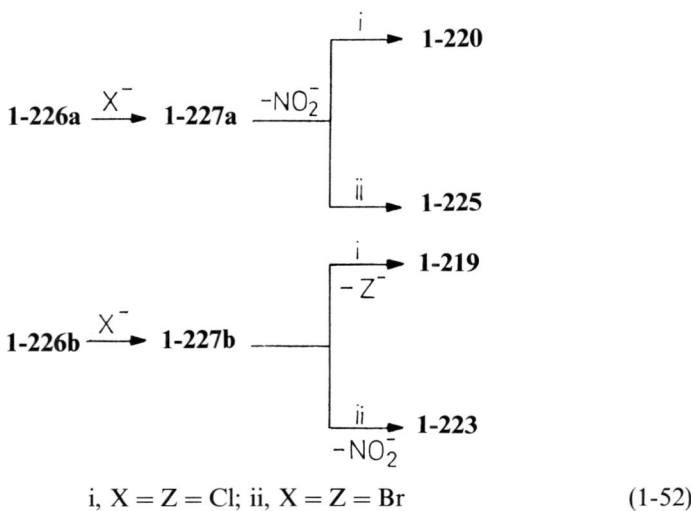

i, X = Z = Cl; ii, X = Z = Br   (1-52)

Mobility of NH and CH Bonds

Acidity assigned to the NH function increased from pK_a 17.51 for the unsubstituted pyrrole[160] to pK_a 10.60 for 2-nitropyrrole (**1-1**).[161] A larger increase brought about by an α-nitro relative to a β-nitro substituent was shown in the acidities of 2-formylpyrrole.[162] A similar effect was seen in 2,4-dinitro- (**1-66**) ($pK_a = 6.3$) and 2,5-dinitropyrrole (**1-67**) ($pK_a = 3.6$).[161] These results were compatible with a CNDO/2 calculated increase in the positive charge on H-1 by either a 2- or 3-nitro substituent.[163]

The yellow, water soluble sodium salts of the nitropyrroles (**1-1**), (**1-66**), and (**1-67**) were obtained from sodium methoxide in dioxane (**1-1** and **1-66**) or methanol (**1-67**) in yields of 83%, quantitative, and 98% and decomposed at 240°–260°, 235°, and 230°. Each pyrrole (**1-1**), (**1-66**), and (**1-67**) gave ammonium salts (62%, 75%, 88%). From its salt the pyrrole (**1-1**) was regenerated on melting (40°), but the other two salts decomposed on melting (155°, odor of ammonia; 215°–245°, sublimation). A hydrazine salt of 2-nitropyrrole (**1-1**) was not isolated but pyrroles (**1-66**) and (**1-67**) quantitatively gave monohydrazine salts decomposing at 185° and about 220°. Each nitropyrrole was quantitatively recovered on acidification of its sodium, ammonium, and hydrazinium salts.[161]

Cyanoethylation of 2-nitropyrrole (**1-1**) (catalyzed by sodium methoxide) and 2,4-dinitropyrrole (**1-66**) (catalyzed by sodium bicarbonate or by sodium acetate) gave the expected 1-cyanoethyl derivative (50% and 64%) but a similar derivative from 2,5-dinitropyrrole (**1-67**) could not be obtained. N-Methylation with dimethyl sulfate and sodium hydroxide was correspondingly easier with the pyrroles (**1-1**), (**1-2**), and (**1-66**)[48,49] while the N-methyl derivative of **1-67** was obtained in 37% yield. On recrystallization from dilute nitric acid the latter was only partially recovered (42%).[161] Eight derivatives (**1-228**) were obtained from the sodium salt of 2,4-dinitro pyrrole (**1-66**) and the appropriate alkylating agent.[164] A competitive methylation at an oxygen atom of the nitro substituent to form a nitronate ester (**1-101**) rather than the isomeric 1-methylpyrrole appears exceptional.[67]

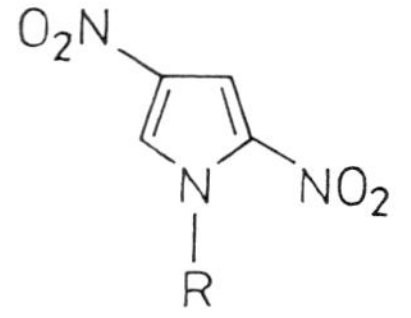

1-228 R

a $n\text{-}C_3H_7$
b $n\text{-}C_4H_9$
c $sec\text{-}C_4H_9$
d $iso\text{-}C_5H_{11}$
e CH_2COCH_3
f CH_2CONH_2
g $CH_2CO_2C_2H_5$
h CH_2CHCH_2OH
    |
    OH

1-228a–h

Nonalkaline alkylations of a nitropyrrole are also known. Ethylene oxide and 2-cyano-5-nitropyrrole (**1-65**) in an inert solvent gave the imine (**1-229**) of a pyrrolooxazine, Eq. (1-53).[165] Cyclization also afforded the imine (**1-229**) when the potassium salt of 5-nitro-2-(2-chloroethoxycarbonyl)pyrrole was stirred in dimethylsulfoxide.[166]

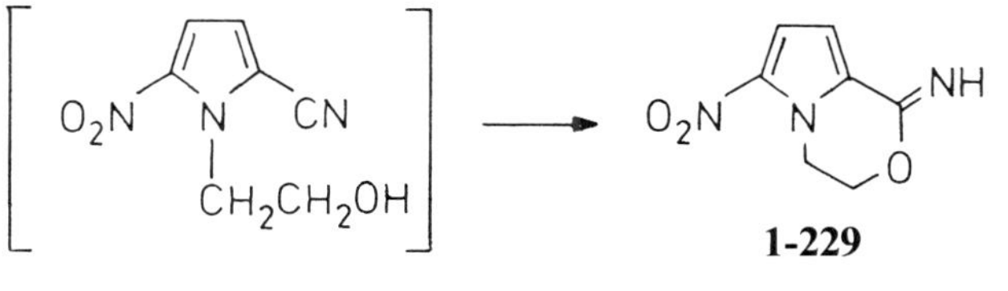

1-229   (1-53)

Aminoalkylation (catalyzed by sodium hydride) of 2-dichloroacetyl-5-nitropyrrole in dioxane gave the derivatives (**1-230a–d**) (49%–60%).[24]

Diazoalkanes in ether converted the corresponding pyrroles into the 1-alkyl derivatives (**1-231**), (**1-232**), and (**1-233**) in yields above 60% except for **1-231i** (49%) and **1-233** (30%).[24] These and other reactions tend to negate an older generalization that substitution on nitrogen in a pyrrole required alkaline conditions.[167]

A unique methylation of 1,4-dinitro-2-methylpyrrole (**1-111**) by diazomethane in ether gave 1,3-dinitro-2,5-dimethylpyrrole (20%) and revealed a remarkable proton mobility at the 2-position. This reactivity was also responsible for the

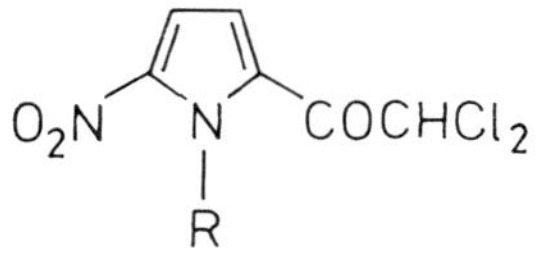

1-230   R

a CH$_2$CH$_2$N(CH$_2$CH$_2$)$_2$O
b (CH$_2$)$_3$N(CH$_2$CH$_2$)$_2$O
c (CH$_2$)$_2$N(C$_2$H$_5$)$_2$
d (CH$_2$)$_2$N⟨morpholin-3-one⟩

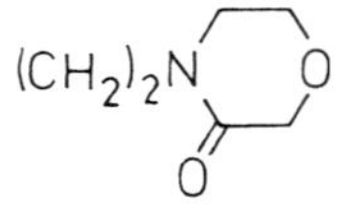

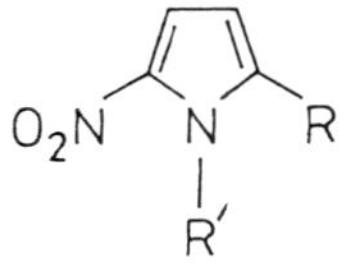

1-230a–d

1-231

	R	R'
a	COCH$_2$Cl	CH$_3$
b	COCHCl$_2$	CH$_3$
c	COCCl$_3$	CH$_3$
d	COCON(CH$_2$)$_2$O	CH$_3$
e	COCH$_2$Cl	C$_2$H$_5$
f	COCHCl$_2$	C$_2$H$_5$
g	COCCl$_3$	C$_2$H$_5$
h	COCON(CH$_2$)$_2$O	C$_2$H$_5$
i	CO(CH$_2$)$_3$Cl	C$_2$H$_5$

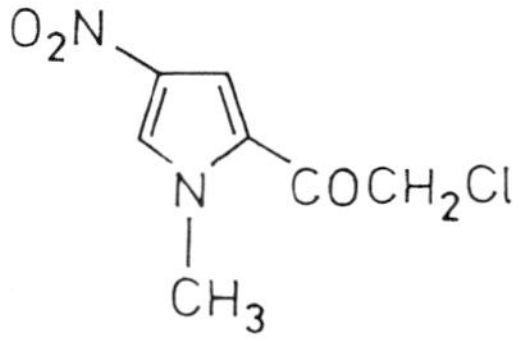

1-231a–i

1-232 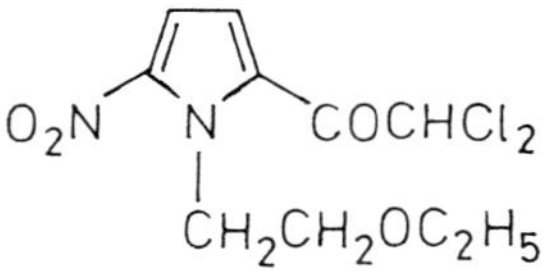   **1-233**

non-base-catalyzed hydrogen-deuterium exchange in deuterium oxide.[78] A base-catalyzed hydrogen exchange at the open α-position in 1-methyl-2,4-dinitropyrrole (**1-84**) was recently reported.[149]

Triethylamine in chloroform promoted dehydrochlorination of 4-nitropyrrole-2-carbonyl chloride (**1-234**) with dimerization of the hypothetical ketene intermediate (**1-235**) to a dinitropyrrocoll (**1-237**) (85%). Consideration was given to the intermediacy of the pyrrole conjugate base (**1-236**) but without recognition of its competitive reaction as a nitronate. Conversion of pyrrole (**1-234**) to the formal dipolar adduct (**1-238**) from the ketene (**1-235**) and benzal-

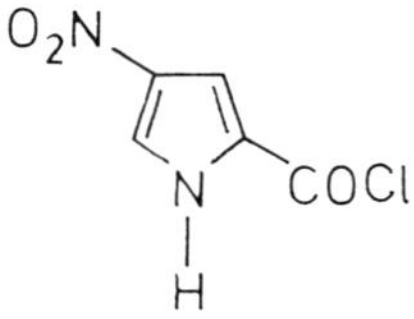

1-234 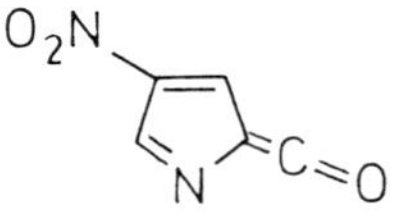   **1-235**

or *p*-methoxybenzalaniline was also reported.[168] The unsubstituted ketene related to intermediate (**1-235**) can be invoked for an older dehydration of pyrrole-2-carboxylic acid by acetic anhydride to pyrrocoll.[169]

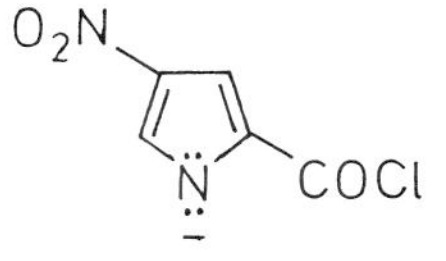

1-236

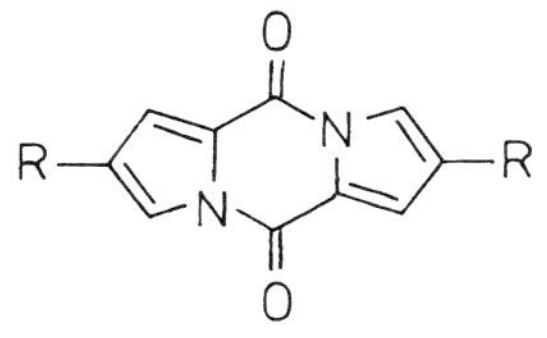

1-237 R = NO$_2$

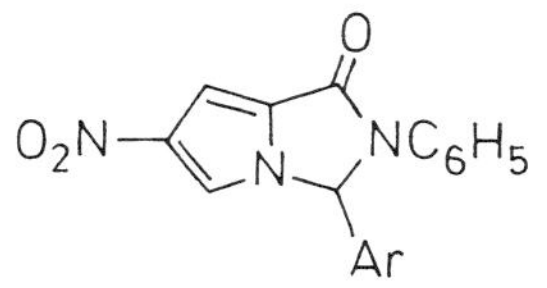

1-238
Ar = C$_6$H$_5$, *p*-CH$_3$OC$_6$H$_4$

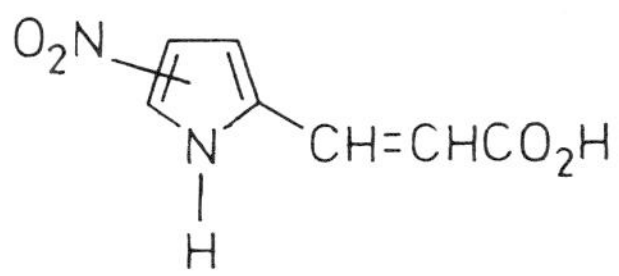

1-239a 4-NO$_2$
1-239b 5-NO$_2$

Straightforward condensations of 4- **1-98a** and 5-nitro-2-formylpyrrole (**1-98b**) with malonic acid in pyridine gave the corresponding acrylic acids (**1-239**).[170] Related condensations afforded α-aryl-(1-methyl-4-nitropyrrol-2-yl)acrylic acids and α-aryl-(1-methyl-5-nitropyrrol-2-yl)acrylic acids.[171] An acid catalyzed condensation of a dipyrrylmethane and two moles of 2-formyl-3,5-dimethyl-4-nitropyrrole gave the biladiene (**1-240**).[55]

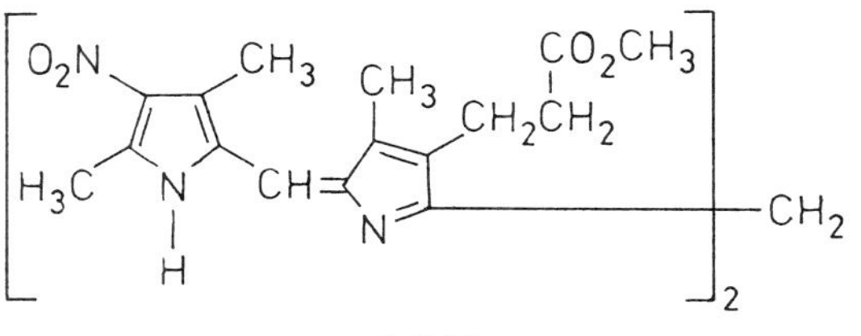

1-240

Nitropyrrolenines (**1-241**, **1-242**) were prepared from 4-nitro (**1-98a**) or 5-nitro-2-formylpyrrole (**1-98b**) with an appropriate secondary amine at 25°.[172] The aldehydes (**1-98a,b**) and large numbers of their 1-substituted derivatives

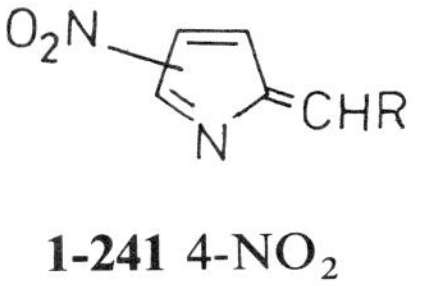

1-241 4-NO$_2$
1-242 5-NO$_2$

R = (CH$_3$)$_2$N,
(CH$_2$)$_5$N,
O(CH$_2$CH$_2$)$_2$N

have been converted to oximes and a variety of hydrazone derivatives by standard operations.[20,27,34] Reduction of the nitro group was not competitive.

Electrophilic Reactions

Nitrations are described in earlier sections. Inability to couple with diazotized p-nitroaniline placed ethyl pyrrole-1-carboxylate, the 2- and the 4-methyl derivatives of diethyl pyrrole-2,4-dicarboxylate, and ethyl 4-nitropyrrole-2-carboxylate (**1-44c**) amongst the pyrroles least reactive toward an electrophilic attack.[173] This observation was upheld in the resistance to nitration shown by the 4-nitro derivatives of 2-acetyl-, 1-methyl-2-carboethoxy- (**1-24**), and 1-methyl-2-nitropyrrole (**1-84**). As expected, the 3- and 5- positions became deactivated toward electrophilic attack when positions 2- and 4- were occupied by electron withdrawing groups. An effective deactivation of the pyrrole ring to electrophilic attack by an electron-withdrawing substituent at the 1-position was shown in the resistance of ethyl pyrrole-1-carboxylate to couple with diazotized p-nitro-aniline.[173] A complementary enhanced electrophilicity for the pyrrole ring with an electron-withdrawing group at the 1-position was seen in the formation of a Meisenheimer adduct (**1-198**).[149]

Boron trifluoride etherate catalyzed the acetylation of 2-nitro-1-methylpyrrole (**1-4a**) and the 3-nitro isomer (**1-5a**) to single products 4-acetyl-2-nitro- (**1-243**) (74%) and 2-acetyl-4-nitro-1-methylpyrrole (**1-244**) (50%).[8] These were reported as the first acylations of a nitropyrrole.

1-243 X = NO$_2$, Y = COCH$_3$
1-244 X = COCH$_3$, Y = NO$_2$

Formylation of 3-nitropyrrole (**1-2**) by dimethylformamide and phosphorus oxychloride in a Villsmeier-Haack reaction gave a 1:1 complex between **1-2** and 4-nitro-2-formylpyrrole (**1-98a**) which was extracted nearly quantitatively from an ether solution of the reaction mixture by aqueous sodium carbonate. Although an attempt to formylate 2-nitropyrrole (**1-1**) by a similar reaction was unsuccessful, a Friedel-Crafts reaction between 2-nitropyrrole, benzoyl chloride, and aluminum chloride in nitromethane gave 4-benzoyl-2-nitropyrrole (41%); 3-nitropyrrole gave 5-benzoyl-3-nitropyrrole (28%).[115] An extension of the reaction to 3-nitro-1,2-dimethylpyrrole (**1-10**) gave 5-acetyl-3-nitro-1,2-dimethylpyrrole (26%). No other acylpyrrole was detected. On the other hand, 1,2-dimethyl-4-nitropyrrole (**1-11**) under the same conditions gave a complex and variable mixture of labile compounds from which 3,5-diacetyl-1,2-dimethyl-4-nitropyrrole could be isolated in trace amounts after extensive chromatography.[15]

In marked contrast, an attempted Friedel-Crafts alkylation of 3-nitro-1,2-dimethylpyrrole (**1-10**) failed. When the pyrrole (**1-10**) in dry refluxing dichloromethane was treated with methyl iodide (excess) in the presence of aluminum chloride the desired 3-nitro-1,2,5-trimethylpyrrole was not detected and some starting material was recovered. In another attempt the pyrrole (**1-10**) in dry 1,2-dichloroethane was treated with dimethyl sulfate with boron trifluoride etherate as catalyst but the desired methylation was not detected after heating at reflux for 24 h; prolonged heating destroyed the starting material and a complex tar was obtained.[15]

Equimolar portions of bromine and 2-nitropyrrole (**1-1**) with a trace of iodine in carbon tetrachloride at 0°–5° for 1 h gave 4-bromo-2-nitropyrrole (90%), identified solely on the basis of the magnitude of H3–H5 coupling.[137]

Under similar conditions equimolar portions of bromine and 1-methyl-2-acetyl-4-nitropyrrole (**1-245a**) in chloroform for 18 h at 25° gave the 2-bromoacetyl derivative (**1-245b**) (90%); a similar yield of the 5-nitro isomer (**1-245d**) was obtained from 1-methyl-2-acetyl-5-nitropyrrole (**1-245c**).[27]

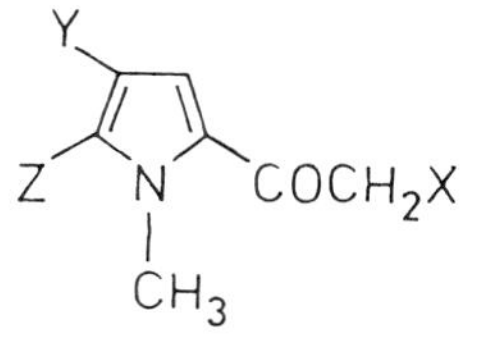

1-245
a $Y = NO_2, Z = X = H$
b $Y = NO_2, Z = H, X = Br$
c $Z = NO_2, Y = X = H$
d $Z = NO_2, Y = H, X = Br$

Although alkaline hypochlorite converted 1-methyl-4-nitro-2-acetylpyrrole (**1-195**) into the corresponding carboxylic acid, an alkaline solution of iodine and potassium iodide converted 2-acetylpyrrole to its 3,4,5-triiodo derivative (**1-246**) with apparently no competition from the iodoform reaction.[43] Ethyl 5-methylpyrrole-2-carboxylate and iodine with potassium iodide in potassium carbonate solution at about $-50°$ gave the 3-iodo derivative (**1-247**).[39]

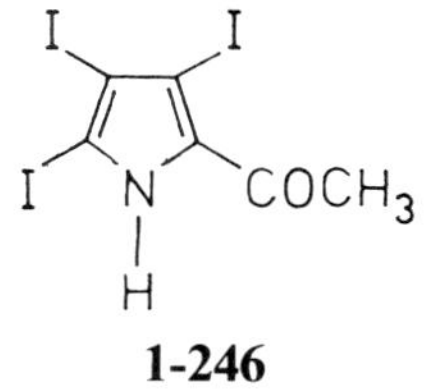

1-246 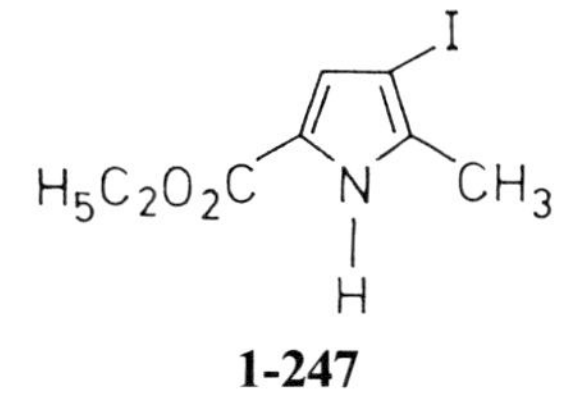   **1-247**

Although other pyrroles condensed with formaldehyde in the presence of air and hydrobromic acid, 3-nitro-4-phenylpyrrole failed to do so.[174]

Free Radical Reactions

Equimolar portions of N-bromosuccinimide and 3-nitro-1,2-dimethylpyrrole (**1-10**) in refluxing benzene containing a small amount of dibenzoyl peroxide gave the 5-bromo derivative (63%) and a trace amount of 5-bromo-3-nitro-2-bromomethyl-1-methylpyrrole (**1-248**) after 15 min. Two equivalents of N-bromosuccinimide brought about the formation of the 4,5-dibromo derivative (31%) and the isomer (**1-248**) (37%) isolated after hydrolysis as 5-bromo-3-nitro-2-hydroxymethyl-1-methylpyrrole (**1-249**). Trace amounts of 4,5-dibromo-3-nitro-2-hydroxymethyl-1-methylpyrrole were also detected. For spectroscopic comparison with **1-249**, its isomer 5-nitro-3-bromo-2-hydroxymethyl-1-methylpyrrole (**1-250**) was prepared by the bromination of 5-nitro-2-hydroxymethyl-1-methylpyrrole in dichloromethane containing a slight molar excess of pyridine and a 2 M excess of bromine at 25° for 10 min; the yield was 59%.[26]

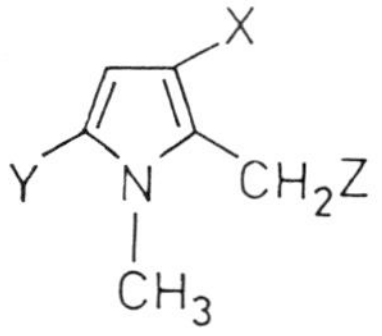

1-10 $X = NO_2$, $Y = Z = H$
1-248 $X = NO_2$, $Y = Z = Br$
1-249 $X = NO_2$, $Y = Br$, $Z = OH$
1-250 $X = Br$, $Y = NO_2$, $Z = OH$

Methyl radicals and 2-nitropyrrole gave the 5-methyl (30%) and 3,5-dimethyl derivatives (12%).[175]

Unclassified Reactions

Pyrrolomycin A (**1-252**), a new antifungal antibiotic, was prepared from 3-nitropyrrole (**1-2**) and sulfuryl chloride in acetic acid. In the initial step the reaction gave 2-chloro-4-nitropyrrole (**1-251**), followed by a mixture of 2,3-dichloro- (**1-252**) and 2,5-dichloro-4-nitropyrrole (**1-253**); and finally 2,3,5-trichloro-4-nitropyrrole (**1-254**). The 3,5-dichloro isomer (**1-255**) was indirectly obtained by decarboxylation of 3,5-dichloro-4-nitropyrrole-2-carboxylic acid (**1-256**), in turn a product from methyl 4-nitropyrrole-2-carboxylate and sulfuryl chloride followed by saponification.[176–182]

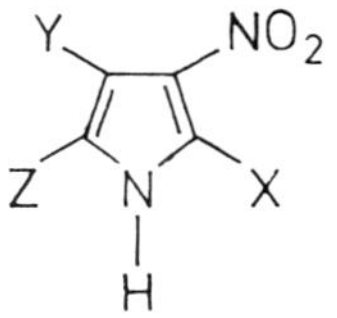

	X	Y	Z
1-251	H	H	Cl
1-252	H	Cl	Cl
1-253	Cl	H	Cl
1-254	Cl	Cl	Cl
1-255	Cl	Cl	H
1-256	Cl	Cl	CO_2H

Thermal and Photochemistry

Kinetic information on the thermal degradation of 2-nitro- (**1-1**), 2,5-dinitro- (**1-67**), 1-methyl-2-nitro- (**1-4a**), 1-methyl-3-nitro- (**1-5a**), and 1-methyl-2,4-

dinitropyrrole (**1-257**) and the sodium (**1-258a**) and potassium salts (**1-258b**) of 2-nitropyrrole is presented in Table 1-11.[183,184]

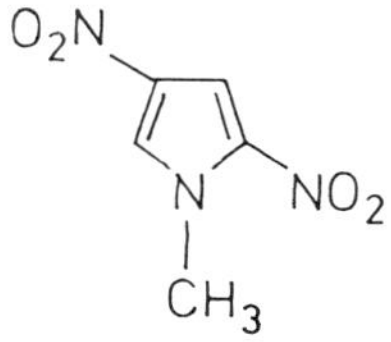

1-258a M = Na
1-258b M = K

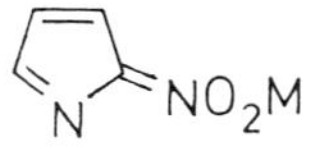

1-257

Table 1-11. Thermal Degradation of Nitropyrroles

Pyrrole derivative	T,°C	p_o,torr	E_A, kcal/mol	log A	$10^9 k$, sec^{-1} at 200°C
1-1[a]	260–290	399–410	41.2	11.03	10.00
1-67[a]	260–290	135–150	35.1	9.89	40.00
1-4[a]	310–340	227–297	57.2	15.53	0.013
1-5[a]	310–340	341–489	57.5	15.58	0.010
1-257[a]	320–350	145–235	57.0	15.12	0.006
1-258a[b]	210–250		41.9	12.7	228.00
1-258b[b]	230–260		41.8	11.8	34.00

[a]Gas phase.
[b]Solid phase.

An assignment of a molecular mechanism to the thermal degradations of 2-nitro- (**1-1**), $E_A = 41.2$ kcal/mole, and 2,5-dinitropyrrole (**1-67**), $E_A = 35.1$ kcal/mole, was supported by an investigation of the thermal degradation of 2-nitro- (**1-1**) and 3-nitropyrrole (**1-2**) at low pressures with analysis of thermolysis products by mass spectrometry. Cleavage of the hydroxyl group was characteristic of pyrrole (**1-1**), Eq. (1-54), while cleavage of the nitro group was characteristic of pyrrole (**1-2**), Eq. (1-55). A mass spectrum of 2-nitropyrrole did not show a mass peak at 95 (M-17).

It was consistent then to assign a radical mechanism to the thermal degradation of the nitro derivatives (**1-4a**), (**1-5a**), and (**1-257**) of 1-methylpyrrole, each of which required the higher E_A of 57 kcal/mole.

Although irradiation (Hanovia photochemical reactor with a 100 W medium pressure mercury arc) converted 2-nitropyrrole (**1-1**) in acetone to 3-hy-

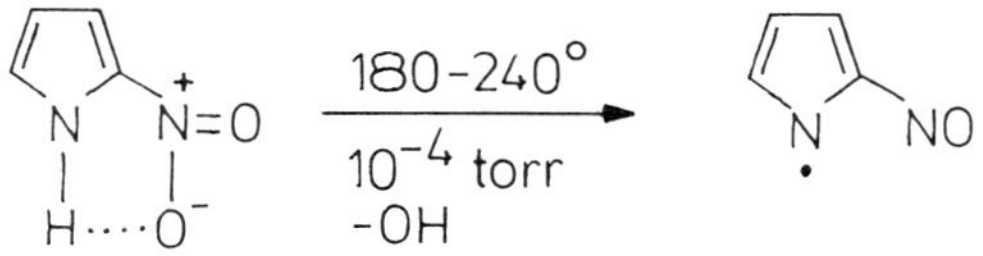

1-1

(1-54)

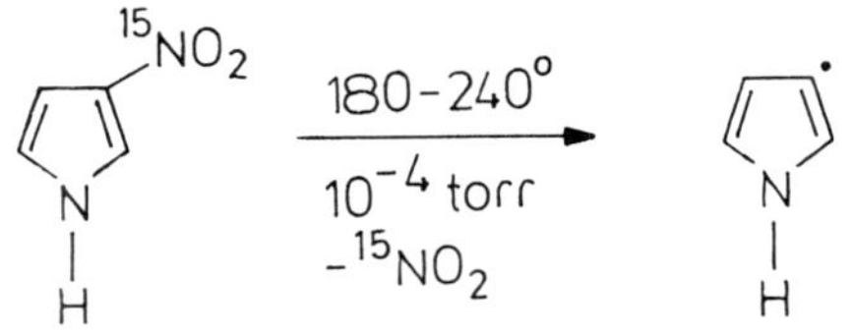

droxyiminopyrrol-2(3H)-one (**1-259**),[185] the nitropyrrole (**1-1**) was photostable to hydrogen abstraction[186] and to nucleophilic photosubstitution.[187] The product (**1-259**) was accounted for by proposing an initial rearrangement to a nitrite, cleavage of the O—NO bond, and radical recombination at the 3-position, Eq. (1-56).

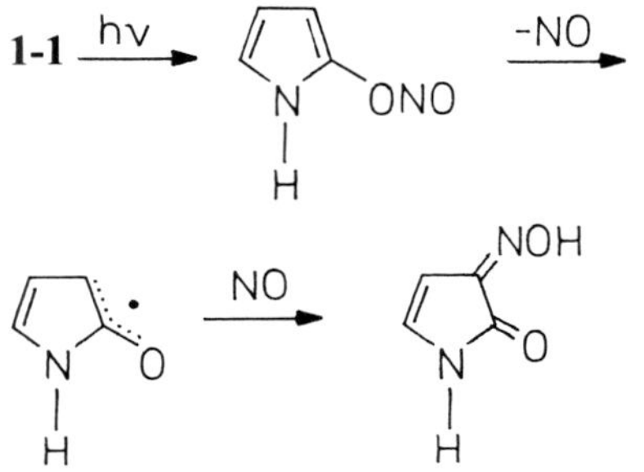

(1-56)

PHYSICAL PROPERTIES

Ultraviolet Spectroscopy

Derivatives of 2-nitro and 3-nitropyrroles are typical aromatic compounds insofar as their ultraviolet (uv) absorption spectra show maxima in two ranges, 225–280 and 300–365 nm, with a tendency for shorter wavelengths in the lower range and longer wavelengths in the higher range to be associated with 2-nitropyrrole systems.[106,188,189] A pale yellow color of a nitropyrrole became a deeper yellow or red in alkaline solution as a nitronate anion, eg, **1-260** or **1-261**, developed. This reversible change was followed for 2-nitropyrrole (**1-1**)

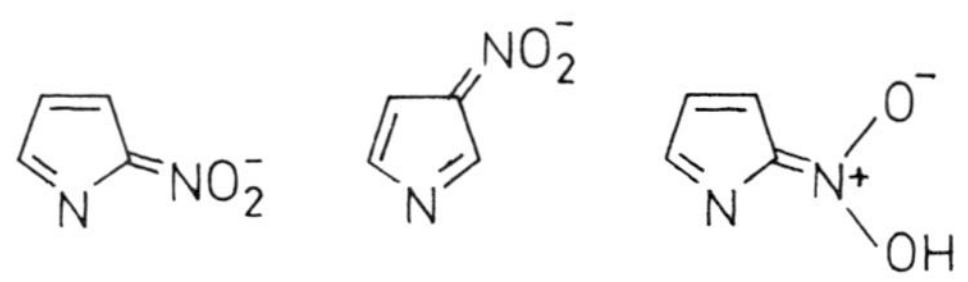

1-260   **1-261**   **1-262**

in water as the pH change from 2 to 13 brought about a red hyperchromic shift from 370 (log ε 3.60) to 390 (log ε 4.20) nm. Larger shifts of 75 and 52 nm accompanied similar changes in the pH of water solutions of 2,4-dinitro (**1-66**) and 2,5-dinitropyrrole (**1-67**).[190-193]

Immediately after acidification of the nitronate (**1-260**) the nitronic acid (**1-262**) was expected to be present.[158-160] Since the nitro form (**1-1**) was detected 8 min after acidification, the nitronic acid was either bypassed on acidification or rapidly changed to the nitro form. An irreversibility in the spectral changes in 1,4-dinitro-2-methylpyrrole (**1-111**) brought about by alkali was interpreted as evidence for alkaline degradation of the pyrrole. A similar alkaline solution brought about no change in the spectra for 1,3-dinitro-2,5-dimethylpyrrole.[78]

A green chloroform solution of the red biladiene (**1-240**) gave absorption, λ_{max} (log ε), at 739 (4.85), 677 (4.58), 620 sh (4.08), 508 sh (3.99), 414 (4.85), and 395 sh (4.62) nm.[55]

Corresponding to the hyperchromic red shifts associated with an increase in pH for 2-nitropyrrole and its 4- and 5-nitro derivatives, hypochromic red shifts were found for six nitropyrrolo[2,3-*d*]pyrimidines as their solutions changed in pH from 2 to 11.[106]

Two 1-nitroindolizines (**1-148**),[110] (Table 1-6), the benzoindolizine (**1-150**),[116] the nitrocyclazine (**1-158**),[126] the nitropyrocoll (**1-237**),[168] and the bicyclic 3-nitropyrrole (**1-238**)[168] showed absorption typical of related nitropyrrole systems [λ_{max} (solvent) in nm (log ε)]:

1-148, X = A = H, Y = CH$_3$ (CH$_3$OH): 210(4.07), 240(4.39), 294(4.06), and 350(4.07).

1-148, A = NO$_2$ X = Y = H (CH$_3$OH): 220(4.27), 252(4.55), 300(3.85), and 355(4.25).

1-150, (C$_2$H$_5$OH): 220(4.22), 257(4.45), 279(4.55), 340(3.89), and 390(3.58).

1-158, (solvent not given): 227(4.45), 261(4.12), 338(3.82), 415(4.14), and 425(4.18).

1-237, (CHCl$_3$): 266(4.54) and 288(4.45).

1-238, (CHCl$_3$): 256(4.20) and 303(4.04).

For ten nitroindolizines (**1-147b**) (Table 1-6) absorption below but not above 300 nm was reported.[110]

An absorption maximum near 360 nm, characteristic of a nitroenamine,[193] was also present in the spectra for the 3-nitro-2-pyrroline (**1-204**), λ_{max}(CH$_3$OH) 366 nm (log ε 4.5),[154] and for the 5-phenyl-4-nitro-3-cyclohexylamino-3-pyrrolin-2-one (**1-214**) (RR′N = cyclo-C$_6$H$_{11}$NH), (λ_{max}(C$_2$H$_5$OH) 372 nm (log ε 4.2)); but the 4-nitro-3-pyrrolidino-3-pyrrolin-2-one (**1-214**) (RR′N = (CH$_2$)$_4$N) and 4-nitro-3-anilino-3-pyrrolin-2-one (**1-214**) (RR′N = C$_6$H$_5$NH) absorbed at 402 nm (log ε 4.2) and 385 nm (log ε 4.0).[74] Although absorption was reported for 4-nitro-3-methoxy-3-pyrrolin-2-one (**1-213**) (a nitroenol ether) at λ_{max}(C$_2$H$_5$OH) 257 nm (log ε 3.8), an expected absorption near 360 nm was not; in contrast its nitroenol precursor (**1-105**) absorbed at λ_{max}(C$_2$H$_5$OH) 358 (log ε 4.0)[74] in good correspondence with the above values for nitroenamines.

A structure proof for 2-propionyl-4-nitropyrrole (from a nitration of 2-propionylpyrrole) λ_{max} 247 (log ε 4.16) and 300 nm (log ε 3.89), was partially based on the close comparison of its electronic spectra with that of 2-acetyl-4-nitropyrrole, λ_{max} 245 (log ε 4.15) and 299 nm (log ε 3.81). A similar comparison was observed for 5-nitro-2-propionylpyrrole and 5-nitro-2-acetylpyrrole: λ_{max} 240 (log ε 4.03) and 328 nm (log ε 4.15) for one and λ_{max} 239 (log ε 4.00) and 328 nm (log ε 4.11) for the other.[194]

Nuclear Magnetic Resonance Spectroscopy

Downfield increments in pyrrole CH proton nuclear magnetic resonance (NMR) signals have been attributed to both α- and β-substituents.[15,17,26,28,135,155,171,195] Increments arising from the 2-nitro substituent were 1.06, 0.24, and 0.45 for H-3, H-4, and H-5, respectively; from a 3-nitro-substituent 1.04, 0.60, and 0.15 for H-2, H-4, and H-5, respectively. The largest interactions came from 2-NO$_2$-3-H (1.06) and 3-NO$_2$-2-H (1.04); the smallest from 2-NO$_2$-4-H (0.24) and 3-NO$_2$-5-H (0.15). An interaction was only slightly larger for 3-NO$_2$-4-H (0.60) than for 2-NO$_2$-5-H (0.45).

The lowest signal, δ 8.98, for a nitropyrrole ring proton was reported for 2-methyl-1,4-dinitropyrrole (**1-111**);[78] δ 8.53 was found for 1-isopropyl-2,3,4-trinitropyrrole (**1-88b**).[11] A summation of three increments in the latter example gave 1.64, in qualitative agreement with 1.80, the observed value. Apparently the 1- and 3-nitro substituents have contributed nearly equally to the increment of 2.15 for the chemical shift for H-2 in compound (**1-111**).

Pyrrole amine substituents gave expected upfield increments to the chemical shifts of ring protons. Increments for H-5 in five 1-alkyl-2-(3)-amino-4-nitropyrrole ranged from 0.37 to 0.75 downfield from 1.04, the value expected in the absence of the amino substituent by ignoring a contribution from the alkyl substituent.[156] A similar effect was shown in the shift at δ 5.49 for H-4 in 1-methyl-2-nitro-5-piperidinopyrrole.[152] The largest upfield increment resulted in δ 5.37, a shift for pyrrole H-3 in E-α-phenyl-β-[2-(1-methyl-5-nitro)-pyrrolyl]acrylic acid.[171]

The pyrroline (**1-204**) olefinic proton was detected by its signal at δ 7.87 (s, methanol).[154]

^{13}C NMR spectroscopy for over 50 nitropyrroles has been published.[15,17,26,135,195]

NO-*cis*-**1-263**   NO-*trans*-**1-263**   NO-*cis*-**1-264**   NO-*trans*-**1-264**

Preferred conformation for 2- and 3-formylpyrrole (**1-263** and **1-264**) and certain ring-substituted derivatives were established in a study of the stereospecificity of the long-range couplings ^4J and ^5J (Table 1-12). A population ratio ($100^nJ_{obs}/^nJ_{max}$) was determined on the assumption that $^nJ_{max}$ corresponded to

Table 1-12. A. 2-Formylpyrrole (**1-263**)

		Coupling constants		
Substituent	Solvent	^5J	^4J	% trans
—	$(CD_3)_2CO$	CHO-H5, 1.05		5
3-I	$(CD_3)_2CO$	CHO-H5, 1.11		0
3-$CO_2C_2H_5$	$(CD_3)_2CO$	CHO-H5, 0.95		15
3-NO_2	$(CD_3)_2CO$	CHO-H5, 0.88		20
4-I	$(CD_3)_2CO$	CHO-H5, 1.02		8
4-NO_2	$(CD_3)_2CO$	CHO-H5, 1.03		7
5-$CO_2C_2H_5$	$(CD_3)_2SO$	CHO-H4, 0.40–0.45	CHO-H3, ~0.12	45
	$(CD_3)_2CO$	CHO-H4, ≤0.20–0.25		25
5-NO_2	$(CD_3)_2SO$	CHO-H4, 0.45	CHO-H3, ~0.13	50
	$(CH_3)_2NCHO$	CHO-H4, 0.32		35–40
	$(CO_3)_2CO$	CHO-H4, 0.24		25–30
	$(CH_2)_4O$	CHO-H4, ≤0.20		20–25

B. 3-Formylpyrrole (**1-264**)

		Coupling constants		
Substituent	Solvent	^5J	^4J	% trans
—	$(CD_3)_2SO$	0.79	CHO-H4, 0.42	95
	$(CD_3)_2CO$	0.77	CHO-H4, 0.40	95
2-I	$(CD_3)_2CO$	0.80	CHO-H4, 0.40	95
2-NO_2	$(CD_3)_2CO$	0.76	CHO-H4, 0.42	95
2-$CO_2C_2H_5$	$(CD_3)_2CO$	0.80	CHO-H4, 0.46	100
5-I	$(CD_3)_2SO$		CHO-H4, 0.42	90
	$(CD_3)_2CO$		CHO-H4, 0.37	80
5-NO_2	$(CD_3)_2SO$		CHO-H4, 0.26	55
	$(CD_3)_2CO$		CHO-H4, ~0.23	50
5-$CO_2C_2H_5$	$(CD_3)_2SO$		CHO-H4, 0.32	70
	$(CD_3)_2CO$		CHO-H4, ~0.22	50
4-NO_2	$(CD_3)_2SO$		CHO-H2, 0.40	0
	$(CD_3)_2CO$		CHO-H2, 0.40	0
	$(CH_2)_4O$		CHO-H2, 0.40	0
4-I	$(CD_3)_2SO$	0.52		65
	$(CH_2)_4O$	0.36	CHO-H2, 0.23	45
	$HCON(CH_3)_2$	0.48		60
	$(CD_3)_2CO$	~0.30	CHO-H2 < CHO-H5	40

100% of the stereoisomer required by stereospecificity rules. The rules predicted a ^5J coupling in NO-cis- and trans- (**1-263**) and in NO-trans- (**1-264**); and a ^4J coupling in NO-trans- (**1-263**) and in NO-cis- and trans- (**1-264**). A decoupling of a ^4J in the NO-cis- (**1-263**) and ^5J in the NO-cis- (**1-264**) was attributed to rapid N–H exchange.[196]

Substituent control of the formyl conformation was erratic. An NO-cis-conformation was preferred for the aldehyde (**1-263**) and its iodo, nitro, and carboethoxy derivatives. It was not suppressed at all by a 3-iodo-substituent, slightly suppressed by 3-carboethoxy, 3-nitro, 4-iodo, and 4-nitro substituents, and more effectively suppressed by the 5-carboethoxy and 5-nitro substituents. An NO-trans-conformation is preferred for the aldehyde (**1-264**) and its 2-iodo, 2-carboethoxy, 2-nitro, 5-iodo, 5-carboethoxy, and 5-nitro derivatives. The NO-cis- and trans-conformations approach equal populations in the 4-iodo derivative in the four solvents investigated and in the 5-nitro and 5-carboethoxy derivatives in acetone. A complete suppression of the NO-trans-conformation for **1-264** with a 4-nitro substituent was noted.[197]

In acetone-d_6, pyrrole showed an ^{14}N or ^{15}N chemical shift of 120.7 ppm downfield from the $[NH_4]^+$ ion[168] or 229 ± 2 ppm upfield from nitromethane.[199] For 2-nitropyrrole and 1-methyl-2-nitropyrrole similar shifts (± 5) of 227 and 228 were obtained along with a shift of 22.5 ± 1 ppm upfield from nitromethane for ^{14}N and ^{15}N of the nitro group in each molecule and shifts of 85 (± 10) and 69 (± 10) ppm upfield from nitromethane for ^{17}O in each molecule.[199,200] A greater accuracy was claimed in reporting pyrrole ring nitrogen chemical shifts (ppm downfield from $[NH_4]^+$) for 2-nitropyrrole (121.25), 3-nitropyrrole (125.54), methyl 4-nitropyrrole-2-carboxylate (128.3), and methyl 5-nitropyrrole-2-carboxylate (120.5).[201] The latter was the only one of these four nitropyrroles to show an upfield shift from 120.7, the value for pyrrole; a related upfield shift of 7 ppm was determined for 2,5-dinitropyrrole.[199,200]

Signs of the J_{HH} and J_{NH} coupling constants were determined for nitropyrroles and their salts containing ^{15}N in their nitro groups (Table 1-13).[202]

Table 1-13. J_{HH} and J_{15NH} Constants (± 0.05 Hz)^a

Pyrrole derivative	J_{23}	J_{24}	J_{25}	J_{34}	J_{35}	J_{45}
2-Nitro	−0.30	−0.80	−0.85	+4.10	+1.70	+2.70
salt	+0.35	−0.60	−0.50	+3.85	+1.30	+1.40
3-Nitro	∼0.2	+1.70	+2.10	∼0	0.80	+3.25
salt	+0.45	+1.40	+1.40	∼0	−0.70	+2.85
2,4-Dinitro	−0.40		−0.95	−0.45	+1.95	−0.30
salt	∼0		∼0.4	∼0	1.55	∼0

^{a}Ref. 172.

Mass Spectroscopy

Vicinal C-methyl-C-nitropyrroles resemble o-aromatic compounds in displaying peaks in their mass spectra (70 eV) for the $[M-OH]^+$ fragments. For 1,2-dimethyl-3-nitropyrrole and five of its derivatives: 4-nitro, 5-nitro, 5-methyl, 5-acetyl, and 5-carboethoxy, the ratios $[M-OH]^+/[M]^+$ lie between 0.44 and 0.62. When C-methyl and C-nitro are not vicinal the ratio dropped to a barely significant value of about 0.03; however, 1,2-dimethyl-3,5-diacetyl-4-nitropyrrole (1-265) showed a ratio of 0.66 and ethyl 1,2,4-trimethyl-3-nitro-5-carboxylate (1-266), 1.17.[15]

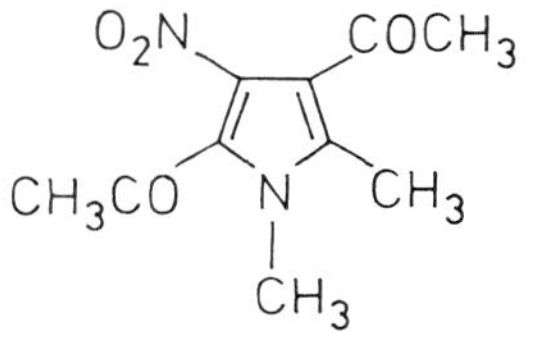

O_2N $COCH_3$ O_2N CH_3
CH_3CO N CH_3 H_3C N $CO_2C_2H_5$
          CH_3 CH_3
1-265 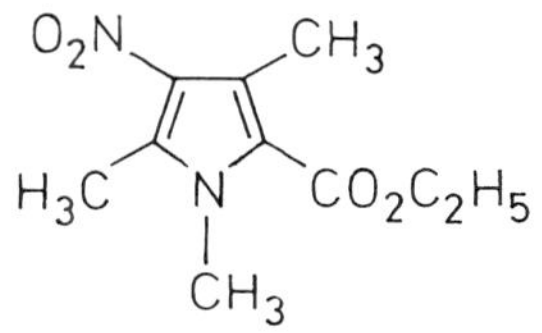   1-266

An absence of a peak at 168 for the fragment $[M-OH]^+$ in the spectra[15] for 2,5-dimethyl-1,3-dinitropyrrole (1-111) was cited for its nonconformity (a peak at 169 with a ratio of 0.06 for $[M-O]^+/[M]^+$ was noted).[78]

Infrared Spectroscopy

Nitropyrroles are typically aromatic with their infrared (ir) absorption bands at 1550–1520 (v_{as}) and at 1360–1320 cm^{-1} (v_s) characteristic of the nitro group.[34,51,54,55,78,83,85,116,123] Low values of 1490 cm^{-1} for 1-methyl-2,5-diphenyl-3-α-pyridyl-4-nitropyrrole,[54] and for 1-methyl-2,5-diphenyl-3-nitropyrrole[85] were reported. Each showed the expected absorption at 1350–1320 cm^{-1}.

Absorption between 1494 and 1472 cm^{-1} was reported for six 4-nitropyrroline-2-ones in which nitroenol and nitroenamine structures are present (see 1-105).[74]

Correlation Methods

Investigations on dipole moments also revealed conformation preferences for aldehydes (1-263 and 1-264).[197] An NO-*cis*-conformation for the aldehyde (1-263) (90%), its 3-iodo (100%), 4-iodo (90%), 5-iodo (70%), 3-nitro (100%), 4-nitro (95%), and 5-nitro (70%) derivatives and an NO-*trans*-conformation for the aldehyde (1-264) (100%), its 4-iodo (47%), 2-nitro (100%), and 4-nitro (10%) derivatives were found. There is qualitative agreement between these results and those obtained from the analysis of long-range coupling constants.

A determination of conformational preferences by *ab initio* molecular orbital theory with the STO-3G basis set found NO-*cis* to be lower in energy than NO-*trans* for both aldehydes (**1-263** (7.3 kJ mol^{-1} difference) and **1-264** (1.3 kJ mol^{-1} difference)).[151] The study also determined internal rotational barriers of 32.3 and 24.2 kJ mol^{-1} for 2- and 3-nitropyrrole (compare 24.0 kJ mol^{-1} for nitrobenzene). A stabilization of the planar form of 2-nitropyrrole by electrostatic interaction between pyrrole NH and NO_2 groups was suggested to account for the differences.[203]

Calculations by the extended Hückel method and the α,π method of Dewar also demonstrated the stability of NO-*cis* (**1-263**) and of the α-formyl derivatives of other five-membered ring heterocycles. A calculated value for the barrier to rotation of the formyl group was directly affected by electron donation from a 5-substituent (CH_3, Cl, Br, NH_2, NO_2). An insignificant effect was noted when the substituent was at the 4-position.[204]

A quantitative relationship between the dissociation constants of pyrrole-2- (**1-267**) and 3- (**1-268**) carboxylic acids and the electronic effect of ring substituents followed an application of the Hammett equation. This led to an investigation of the correlation of pK_a values with the energies of the highest filled molecular orbital (E) and π-electron densities at the hydroxyl oxygen atom (q_o) calculated for nine pyrrolecarboxylic acids. A direct relationship between the calculated q_o values and the experimental pK_a values was found (Table 1-14) and there was an inverse relationship between E and pK_a (Table 1-14).[205]

An order of intensity of transmission of electronic effects was shown to be: –CH = CH– < –S– < –Se– < –O– < –NH– by comparing reaction constants ρ in an application of the Hammett equation to five-membered aromatic heterocycles (aromatic carboxylic acid, ρ): benzoic, 1; 2-thiophene, 1.10; 2-selenophene, 1.23; 2-furoic, 1.40; 2-pyrrole, 1.65. This order had previously been observed in the polarographic reductions of nitro groups in the series. The ρ value of 1.65 for

1-269 der.	$\sigma_{i\Sigma}$	pK$_a$ calc	exp
—	0	6.35	6.35
1-C$_6$H$_5$	0.10	5.89	5.68
3-Br	0.44	4.33	5.41
3-NO$_2$	0.60	3.59	3.58

Table 1-14. Pyrrole 2- and 3-Carboxylic Acids, **1-268** and **1-269**[a]

	1-268			1-269		
R	pK_a	$m^{b,c}$	q_o^d	pK_a	$m^{b,c}$	q_o^d
CH_3	7.30	0.774	1.8972			
H	7.01	0.774	1.8972	7.83	0.645	1.8963
$CO_2C_2H_5$	6.40	0.791	1.8968	7.30	0.791	1.8958
$COCH_3$	6.17	0.792	1.8967	6.96	0.794	1.8957
NO_2	5.61	0.796	1.8960	6.58	0.803	1.8951

[a] Ref. 175.
[b] $m = (E - \alpha^0)/\beta^0$ where α^0 is the Coulomb integral of the carbon atom and β^0 is the resonance integral of the C—C bond in benzene.
[c] Calculations were carried out with parameters proposed by Streitwieser with allowance for the hyperconjugation of the methyl radical. For calculations based on other assumptions see the original paper.[175]
[d] π-Electron densities at the hydroxyl oxygen atom.

pyrrole-2-carboxylic acids was determined by plotting log K/K_o values against σ_m, σ_p constants (substituent in pyrrole-2-carboxylic acid, pK_a): 5-NO_2, 3.22; 4-NO_2, 3.37; 4-Br, 4.06; 4-Cl, 4.07; 5-Br, 4.17; 5-Cl, 4.32; H, 4.50; 4-CH_3, 4.60; 5-CH_3, 4.88.[206] A linear plot was also obtained for the correlation of pK_a values of 4- and 5-substituted pyrrole-2-carboxylic acids and their ionization energies determined by ultraviolet photoelectron spectroscopy (correlation coefficients 0.97 and 0.98).[207] Correlation methods were also applied to the pK_a values of derivatives of 5-azaindole. Experimental values were determined spectrophotometrically and/or potentiometrically, whereas calculated values were provided by the equation $-pK_a = -6.35 + 4.60\ \sigma i_\Sigma$ in which σi_Σ is the summation of induction constants of the substituents.[208]

A correlation between molecular ionization potentials (determined by the electron impact technique) and reactivities toward electrophiles for five-membered aromatic heterocycles was rationalized by the similarity between the molecular ion (**1-270**) obtained in the first ionization process and the intermediate (**1-271**) in an electrophilic substitution reaction. Heterocycle sensitivity to substituent effects on the ionization followed the order: furan > pyrrole > thiophene > benzene (the reverse of ground-state aromaticity). Comparisons between the ionization and electrophilic substitution reactions were facilitated by ρ values for thiophenes (-16.5), furans (-20.2), and pyrroles (-18.2) derived from plots of ionization potentials for monosubstituted derivatives of each heterocycle vs. σ^+ constants. These values are considerably more negative than those for the most selective electrophilic substitutions. Ionization potentials (eV) were determined for pyrrole (8.40) and its derivatives: 2-methyl, 8.01; 2-ethyl,

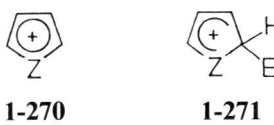

1-270   **1-271**

7.97; 2-t-butyl, 7.95; 2-acetyl, 8.72; 2-carbomethoxy, 8.65; 2-formyl, 8.93; 2-trifluoroacetyl, 9.18; and 2-nitro, 9.30.[209]

In a study of the dielectric relaxation and dipole moments of 17 pyrroles, the dipole moment for 2-nitropyrrole (**1-1**) in benzene was determined to be 4.33 and in dioxane 4.57.[210] Dipole moments of 4.67 and 6.15 were observed for the 2- and the 3-nitro derivatives of 1-methylpyrrole in benzene.[211]

Energies of formation of 2-substituted and 3-substituted pyrroles were determined to be nearly the same in an experimental and theoretical investigation of monosubstituted pyrroles (substituents: H, CH_3, CHO, CH = NOH, CO_2H, CN, NO_2).[212]

BIOLOGICAL PROPERTIES

Numerous derivatives of nitroheterocycles have chemotherapeutic effectiveness.[110]

A mutagen produced in mixtures of sorbic acid and sodium nitrite was shown to be 2-methyl-1,4-dinitropyrrole (**1-111**). Its mutagenicity was destroyed by ascorbic acid, which reduced only the C-nitro function to give 1-nitro-2-methyl-4-aminopyrrole.[78,141,213,214] Compound **1-111** appears to be the only known example of an N-nitropyrrole and the resistance of the nitramine function to reduction by ascorbic acid the only known reduction of a $C-NO_2$ function in the presence of a nitramino group.

Cellular radiosensitization by various nitroheterocyclic compounds has been correlated with their electron affinities and with electron-spin densities on the nitro groups (Table 1-10). Simple nitroheterocyclic compounds reacted rapidly with potential target radicals and radical anions, and those with higher electron affinities were marginally more reactive.[144,215,216]

Histamine H_2-antagonist activity in various derivatives of 3(4)-nitropyrrole has been studied.[30,77,217] The structurally related 4-nitro-1-cyclohexyl-3-ethoxy-2-oxo-3-pyrroline (**1-272**) was investigated for skin sensitization and was found ineffective as an adjuvant for the production of delayed hypersensitivity to tubercilin purified protein derivatives or to ovalbumin.[218]

The 3-nitro function is also present in pyrrolomycin (**1-252**), a new antifungal antibiotic.[176-182]

1-272

1-273a R = CH_3
1-273b R = $CH_2CH_2CH(CH_3)_2$

$$O_2N-\underset{\underset{CH_3}{|}}{N}-CH=NNHCOC_6H_4OH\text{-}\underline{p}$$

1-274

1-275 (quinazoline-HNCH$_2$CONHN=pyrrole-NO$_2$ structure)

Generally, greater biological activity was found for derivatives of 2(5)-nitropyrroles than for their 3(4)-nitro isomers. Thus antibacterial, antifungal, and antiprotozoal activity for 1-unsubstituted or 1-alkyl-2-acyl-5-nitropyrroles was reduced when the nitro group was shifted to the 4-position or replaced by the methylthio group. The compound 1-ethyl-2-dichloroacetyl-5-nitropyrrole showed equal or higher activity on Gram-positive bacteria than obtained for the reference drugs Sterosan and Vioform and was recommended for testing on intestinal bacterial diseases.[24]

Numerous antiprotozoal agents are derivatives of 1-β-hydroxyethyl-2-nitropyrrole.[165,166,219] Antitrichromonadal activity was associated with 2-nitro-5-substituted pyrroles[220,221] and with 1-substituted 2,4-dinitropyrroles.[164] 5-Nitropyrrole-2-sulfonamides were active antifungal, antibacterial, and phytotoxic compounds.[222] Two hydrazone derivatives (**1-273a,b**) of 5-nitro-1-alkyl-2-formylpyrrole were inhibitory at the 1000 or 100 μg/ml level against *Staphylococcus albus, Escherichia coli, Serralia marcescens, Klebsiella aerobacter, Saccharomyces cereviseae, Penicillium notatum,* and *Sporobolomyces salmoncolor,* but 33 structurally related 4- and 5-nitro-2-formylpyrrole derivatives had no activity and were ineffective in an antimalarial screening test.[20] Slight *in vivo* activity at 200 mg/kg p.o. against *Escherischia histolytica* infections in golden hamsters was reported for the p-hydroxybenzoylhydrazone (**1-274**) of 5-nitro-2-formylpyrrole.[27] A group of 13 derivatives of 1- and 3-nitro-2-phenylindolizines (Table 1-6) were evaluated *in vitro* for antibacterial activity on Gram-negative *Escherischia coli* and Gram-positive *Staphylococcus aureus* bacteria with negative results.[110] Nitro pyrrolo[2,3-b]pyridines (**1-143**) and pyrimidines (**1-144**) were studied not only for their biological properties but also for their structural relationship to nucleoside antibiotics tubercidin, toyocamycin, and sangivamycin and their application in the synthesis of related nucleoside derivatives.[36,106,108]

An intriguing structure (**1-275**) related to nitroazacyclopentadienone was reported without a description of its preparation as an effective antibacterial agent.[223]

Nitropyrrole intermediates have been utilized in the syntheses of methoxatin[84] (a coenzyme), distamycin A[29,80,132,139] (an antiviral agent), analogues of the related congocidine,[138,224] and certain pyrrolo[3,2-b]pyridines as possible metabolite antiagonists.[32]

Nitropolyzonamine (1-276) was found in the defense secretion of the milliped *Polyzonium rosalbum*. A partial synthesis was reported.[225]

1-276

REFERENCES

1. Schofield, K., "Hetero-Aromatic Nitrogen Compounds: Pyrroles and Pyridines," Plenum Press, New York, 1967.
2. (a) Jones, R. A., Bean, G. P., "The Chemistry of Pyrroles," Academic Press, New York, 1977. (b) R. A. Jones, *Adv. in Heterocyclic Chem.*, 1970, *11*, 383.
3. Gossauer, A., "Die Chemie der Pyrrole," Springer-Verlag, West Berlin, 1974.
4. Fischer, H., Orth, H., "Die Chemie des Pyrrols," Bd. I, Akademische Verlagsgesellschaft, Leipzig, 1934; Bd. II, 1 Hälfte, 1937; Bd. II, 2 Hälfte, 1940.
5. Livingstone, R., "Compounds Containing a Five-membered Ring with one Hetero Atom from Group V: Nitrogen," in "Rodd's Chemistry of Carbon Compounds," second ed., edited by S. Coffey, Vol. IV, Part A, Elsevier Scientific Publishing Co., Amsterdam, 1973, pp. 329–397.
6. Corwin, A. H., "The Chemistry of Pyrrole and its Derivatives," in R. C. Elderfield, "Heterocyclic Compounds," Vol. 1, John Wiley and Sons, Inc., New York, 1950, pp. 277–343.
7. Cooksey, A. R., Morgan, K. J., Morrey, D. P., *Tetrahedron*, 1970, *26*, 5101.
8. Anderson, H. J., *Can. J. Chem.* 1957, *35*, 21.
9. Morgan, K. J., Morrey, D. P., *Tetrahedron*, 1966, *22*, 57.
10. Dhout, J., Wibaut, J. P., *Rec. Trav. Chim.*, 1943, *62*, 177.
11. Doddi, G., Mencareli, P., Razzine, A., Stegel, F., *J. Org. Chem.*, 1979, *44*, 2321.
12. Sunder, S., Blanton, C. D., Jr., *J. Chem. Eng. Data*, 1970, *15*, 592.
13. Muchowski, J. M., Solas, D. R., *Tetrahedron Lett.*, 1983, *24*, 3455.
14. Sonnet, P. E., *J. Heterocycl. Chem.*, 1970, *7*, 399.
15. Grehn, L., *Chem. Scr.*, 1979, *13*, 67.
16. Astakhova, L. N., Skvortov, I. M., *Tezisy Vses. Soveshch. Khim. Nitrosoedinenii, 5th*, 1974, 36. *Chem. Abstr.*, 1977, *87*, 39212s.
17. Grehn, L., *Chem. Scr.*, 1980, *16*, 85.
18. Morgan, K. J., Morrey, D. P., *Tetrahedron*, 1971, *27*, 245.
19. Xu, R. X., Anderson, H. J., Gogan, N. J., Loader, C. E., McDonald, R., *Tetrahedron Lett.*, 1981, *22*, 4899.
20. Colwell, W. T., Lange, J. H., Henry, D. W., *J. Med. Chem.*, 1968, *11*, 282.
21. Fournari, P., Tirouflet, J., *Bull. Soc. Chim. Fr.*, 1963, 484.
22. Rinkes, I. J., *Rec. Trav. Chim.*, 1938, *57*, 423.
23. Sonnet, P. E., *J. Org. Chem.*, 1972, *37*, 925.
24. Vecchietti, V., Torre, A. D., Lauria, F., Castellين, S., Monti, G., Trane, F., De Carneri, I., *Eur. J. Med. Chem.—Chim. Ther.*, 1974, *9*, 76.
25. Bélanger, P., *Tetrahedron Lett.*, 1979, 2505.
26. Grehn, L., *Chem. Scr.*, 1980, *16*, 72.
27. Arya, V. P., Honkan, V., *Indian J. Chem., Sect. B*, 1976, *14B*, 752.
28. Fournari, P., Farnier, M., Fournier, C., *Bull. Soc. Chim. Fr.*, 1972, 283.
29. Bialer, M., Yagen, B., Mechoulam, R., *Tetrahedron*, 1978, *34*, 2389.
30. Dattolo, G., Aiello, E., Plescia, S., Cirrincione, G., Daidone, G., *J. Heterocyclic Chem.*, 1977, *14*, 1021.

References

31. Aiello, E., Dattolo, G., Plescia, S., *J. Heterocyclic Chem.*, **1976**, *13*, 645.
32. Aiello, E., Dattolo, G., Cirrincione, G., Plescia, S., Daidone, G., *J. Heterocyclic Chem.*, **1978**, *15*, 537.
33. Rinkes, I. J., *Rec. Trav. Chim.*, **1937**, *56*, 1142.
34. Fernandez-Bolaños, J., Mota, J. F., Certventula, A., Calero, M. J. M., *An. Quim.*, **1979**, *75*, 778.
35. Sonnet, P. E., *Chem. Ind. (London)*, **1970**, 156.
36. Rinkes, I. J., *Rec. Trav. Chim.*, **1942**, *60*, 937.
37. Cordell, G. A., *J. Org. Chem.*, **1975**, *40*, 3161.
38. Herbert, R., Wibberley, D. G., *J. Chem. Soc., C*, **1969**, 1505.
39. Ref. 2, pp. 126–127.
40. Rinkes, I. J., *Rec. Trav. Chim.*, **1941**, *60*, 303.
41. Treibs, A., Kolm, H. G., *Ann.*, **1958**, *614*, 176.
42. Ref. 1, p. 80.
43. Rinkes, I. J., *Rec. Trav. Chim.*, **1941**, *60*, 650.
44. Effi, Y., de Sevricourt, M. C., Rault, S., Robba, M., *Heterocycles*, **1981**, *16*, 1519.
45. Tirouflet, J., Fournari, P., *Bull. Soc. Chim. Fr.*, **1963**, 1651.
46. Anderson, H. J., *Can. J. Chem.*, **1959**, *37*, 2053.
47. Safonova, E. N., Belikov, V. M., Novikov, S. S., *Izv. Akad. Nauk S.S.S.R., Otel. Khim. Nauk*, **1959**, 1307. *Chem. Abstr.*, **1960**, *54*, 1486f.
48. Fournari, P., *Bull. Soc. Chim. Fr.*, **1963**, 488.
49. Rinkes, I. J., *Rec. Trav. Chim.* **1934**, *53*, 1167.
50. Sharnin, G. P., Falyakhov, I. F., Butovetskii, D. N., *Khim. Geterotsikl. Soedin.*, **1975**, 655. *Chem. Abstr.* **1975**, *83*, 78062e.
51. Duffy, T. D., Wibberley, G., *J. Chem. Soc., Perkin Trans. I*, **1974**, 1921.
52. Spiro, V., Fabra, I., *Ann. Chim. (Rome)*, **1956**, *46*, 263; Giambrono, S., Fabra, I., *ibid.*, **1960**, *50*, 237.
53. Mineo, A., *Corriere Farm.*, **1966**, *21*, 318.
54. Potts, K. T., Datta, S. K., Marshall, J. L., *J. Org. Chem.*, **1979**, *44*, 622.
55. Clezy, P. S., Liepa, A. J., Webb, N. W., *Aust. J. Chem.*, **1972**, *25*, 2687.
56. Sprio, V., Fabra, I., *Ann. Chim. (Rome)*, **1956**, *46*, 263.
57. Fischer, H., Zerweck, W., *Ber.*, **1922**, *55*, 1949.
58. Aiello, E., Dattolo, G., Cirrincione, G., *J. Chem. Soc., Perkin Trans. I*, **1981**, 1.
59. Moon, M. W., *J. Org. Chem.*, **1977**, *42*, 2219.
60. Kuhn, R., Tainer, H., *J. Liebigs Ann. Chem.*, **1952**, *578*, 227.
61. Cichra, D. A., Holden, J. R., Dickinson, C., "Estimation of 'Normal' Densities of Explosive Compounds from Empirical Atomic Volumes," Naval Service Weapons Center, Silver Springs, Maryland 20910, TR 79-273, 1980, pp. 15-16.
62. Treibs, A., Bader, H., *Chem. Ber.*, **1958**, *91*, 2615.
63. Fischer, H., Zerweck, W., *Ber.*, **1922**, *55B*, 1949.
64. Dedonatis, S., Giardi, M. T., Sleiter, G., *J. Org. Chem.*, **1982**, *47*, 1750.
65. King, M. M., Brown, R. H., *Tetrahedron Lett.*, **1975**, 3995.
66. Kira, M. A., Bruckner, A., Ruff, F., Borsy, Jozef, *Acta Chim. (Budapest)*, **1968**, *56*, 189. *Chem. Abstr.* **1968**, *69*, 86743d.
67. Ajello, T., *Gazz. Chim. Ital.*, **1939**, *69*, 315.
68. Ref. 1, p. 108.
69. Tedder, J. M., Webster, B., *J. Chem. Soc.*, **1960**, 3270.
70. Sprio, V., Fabra, I., *Ann. Chim. (Rome)*, **1959**, *49*, 2053.
71. Sprio, V., Fabra, J., *Ric. Sci., Rend. Sez. B*, **1964**, *4*, 581; *Chem. Abstr.*, **1965**, *62*, 510b.
72. Feuer, H., Blecker, L. R., Jans, R. W., Jr., Frost, J. W., *J. Heterocyclic Chem.*, **1979**, *16*, 481.
73. Feuer, H., Panda, S. S., Hou, L., Bevinakatti, H. S., *Synthesis*, **1983**, 187.
74. Southwick, P. L., Fitzgerald, J. A., Madhav, R., Welsh, D. A., *J. Org. Chem.*, **1969**, *34*, 3279.
75. Novikov, S. S., Belikov, V. M., *Izv. Akad. Nauk S.S.S.R., Otdel. Khim. Nauk*, **1959**, 1098; Eng 1059.
76. Novikov, S. S., Safonova, E. N., Belikov, V. M., *Izv. Akad. Nauk S.S.S.R., Otdel. Khim. Nauk*, **1960**, 1053; Eng 984.
77. Roantree, M. L., Young, R. C., *Eur. Pat. Appl.* 5, 985; *Chem. Abstr.*, **1980**, *93*, 46411h.
78. Kito, Y., Namiki, M., *Tetrahedron*, **1978**, *34*, 505.
79. Osawa, T., Kito, Y., Namiki, M., Tsuji, K., *Tetrahedron Lett.*, **1979**, *45*, 4399.
80. Turchin, K. F., Grokhovskii, S. L., Zhuse, A. L., Gottikh, B. P., *Bioorg. Khim.*, **1978**, *4*, 1065; *Chem. Abstr.*, **1978**, *89*, 179107c.

81. Hale, W. J., Hoyt, W. V., *J. Am. Chem. Soc.*, **1915**, *37*, 2538.
82. Buevich, V. A., Rudchenko, V. V., Perekalin, V. V., *Khim. Geterotsikl, Soedin.*, **1976**, 1429; *Chem. Abstr.*, **1977**, *86*, 72357v.
83. Stepanova, S. V., L'vova, S. D., Belikov, A. B., Gunar, V. I., *Zh. Org. Khim.*, **1977**, *13*, 889; Eng. p. 812.
84. Stepanova, S. V., L'vova, S. D., El'yanov, B. S., Gunar, V. I., *Khim.-Farm. Zh.*, **1977**, *11*, 92; *Chem. Abstr.*, **1978**, *87*, 117803b.
85. Verbruggen, R., Viehe, H. G., *Chimia*, **1975**, *29*, 350.
86. Abramovitch, R. A., Cue, B. W., Jr., *J. Am. Chem. Soc.*, **1976**, *98*, 1478.
87. Rajappa, S., Advani, B. G., Kartha, G., Hartloff, H., *J. Chem. Soc., Perkin Trans. 1*, **1983**, 1953.
88. Hurst, J., Melton, T., Wibberly, D. G., *J. Chem. Soc.*, **1965**, 2948.
89. Kiel, W., Kröhnke, F., *Chem. Ber.*, **1972**, *105*, 3709.
90. Angelico, F., Verladi, G., *Gazz. Chim. Ital.*, **1904**, *34* II, 57.
91. Noland, W. E., Smith, L. R., Johnson, D. C., *J. Org. Chem.*, **1962**, *28*, 2262.
92. Noland, W. E., Smith, L. R., Rush, K. R., *J. Org. Chem.*, **1965**, *30*, 3457.
93. Noland, W. E., Rush, K. R., Smith, L. R., *J. Org. Chem.*, **1966**, *31*, 65.
94. Perkin, W. H., Jr., Plant, S. G. P., *J. Chem. Soc.*, **1923**, *123*, 676.
95. Collonna, M., Greci, L., Poloni, M., *J. Chem. Soc., Perkin Trans. II*, **1981**, 628.
96. Colonna, M., Poloni, M., *Atti Accad. Sci. Ist. Bologna, Cl. Fis., Rend.*, **1980**, 7, 147; *Chem. Abstr.*, **1982**, *97*, 215175z.
97. Spande, T. F., Fontana, A., Witkop, B., *J. Am. Chem. Soc.*, **1969**, *91*, 6199.
98. Hendrickson, J. B., DeVries, J. G., *J. Org. Chem.*, **1982**, *47*, 1148.
99. Gonzalez, A., Galvez, C., *Synthesis*, **1983**, 212.
100. Beeley, N. R. A., Cremer, G., Dorlhéne, A., Mompon, B., Pascard, C., Dau, E. T. H., *J. Chem. Soc., Chem. Commun.*, **1983**, 1046.
101. Yakhontov, L. N., Azimov, V. A., *Khim. Geterotsikl. Soedin.*, **1970**, 32; *Chem. Abstr.*, **1970**, *72*, 121396d.
102. Robinson, M. M., Robinson, B. L., Butler, F. P., *J. Am. Chem. Soc.*, **1959**, *81*, 743.
103. Yakhontov, L. N., Uritskaya, M. Ya., Rubtsov, M. V., *Zh. Org. Khim.*, **1965**, *1*, 2032; Eng. p. 2072.
104. Schneller, S. W., Luo, J.-K., *J. Org. Chem.*, **1980**, *45*, 4045.
105. Yakhontov, L. N., Azimov, V. A., Lapan, E. I., *Tetrahedron Lett.*, **1969**, 1909.
106. Gerster, J. F., Hinshaw, B. C., Robins, R. K., Townsend, L. B., *J. Heterocycl. Chem.*, **1969**, *6*, 207.
107. Watanabe, S., Ueda, T., *Nucleic Acids Symp. Ser.*, **1980**, *8*, s21; *Chem. Abstr.*, **1981**, *94*, 175411c.
108. Bychikhina, N. N.,Azimov, V. A., Yakhontov, L. N., *Khim. Geterotsikl. Soedin.*, **1982**, 356; Eng. p. 268.
109. Greci, L., Ridd, J. H., *J. Chem. Soc., Perkin Trans. II*, **1979**, 312.
110. Borrows, E. T., Holland, D. O., Kenyon, J., *J. Chem. Soc.*, **1962**, 2627.
111. Hickman, J. A., Wibberley, D. G., *J. Chem. Soc., Perkin Trans. 1*, **1972**, 2954.
112. Lins, C. L. K., Block, J. H., Doerge, R. F., *J. Pharm. Sci.*, **1982**, *71*, 556.
113. Borrows, E. T., Holland, D. O., *J. Chem. Soc.*, **1947**, 672.
114. Diels, O., Meyer, R., *Ann.*, **1934**, *513*, 129.
115. Emmert, B., Groll, M., *Chem. Ber.*, **1953**, *86*, 205.
116. Kuo, H.-S., Yoshina, S., Tung, Y.-C., *J. Heterocyclic Chem.*, **1979**, *16*, 393.
117. Mizuyama, K., Matsuo, Y., Tominaga, Y., Matsuda, Y., Kobayashi, G., *Chem. Pharm. Bull.*, **1976**, *24*, 1299.
118. Augstein, W., Kröhnke, F., *Justus Liebig's Ann. Chem.*, **1966**, *697*, 158.
119. Reuschling, D. B., Kröhnke, F., *Chem. Ber.*, **1971**, *104*, 2103, 2110.
120. Cheeseman, G. W. H., Tuck, B., *J. Chem. Soc. (C)*, **1967**, 1164.
121. Al-Sammerrai, D. A.-J., Ralph, J. T., West, D. E., *J. Heterocyclic Chem.*, **1980**, *17*, 1705.
122. Dunham, D. E., Ph.D. Thesis, University of Ohio, Athens, Ohio, 1967.
123. Zupan, M., Stanovnik, B., Tisler, M., *J. Heterocyclic Chem.*, **1971**, *8*, 1.
124. Cheeseman, G. W. H., Rafiq, M., Roy, P. D., Turner, C. J., Boyd, G. V., *J. Chem. Soc. (C)*, **1971**, 2018.
125. Lancelot, J.-C., Maume, D., Robba, M., *J. Heterocyclic Chem.*, **1980**, *17*, 631.
126. Boekelheide, V., Small, T., *J. Am. Chem. Soc.*, **1961**, *83*, 462.
127. McKenzie, S., Molloy, B. B., Reid, D. H., *J. Chem. Soc. (C)*, **1966**, 1908.
128. Kanemasa, S., Ikeda, S., Shimoharada, H., Kajigaeshi, S., *Chem. Lett.*, **1982**, 1533.
129. Gabriel, S., *Ber.*, **1911**, *44*, 70.
130. Fringuelli, F., Marino, G., Savelli, G., *Tetrahedron*, **1969**, *25*, 5815.

131. Melent'eva, T. A., Kazanskaya, L. V., Berezovskii, V. M., *Dokl. Akad. Nauk S.S.S.R.*, **1967**, *175*, 354; *Chem. Abstr.*, **1967**, *67*, 120484r.
132. Cookson, G. H., *J. Chem. Soc.*, **1953**, 2789.
133. Dash, S. C., Mishra, B. K., Behera, G. B., *Indian J. Chem.*, **1982**, *21A*, 195.
134. Grehn, L., Ragnarsson, U., *J. Org. Chem.*, **1981**, *46*, 3492.
135. Grehn, L., *Chem. Scr.*, **1979**, *13*, 78.
136. Britten, A. Z., Griffiths, G. W. G., *Chem. Ind. (London)*, **1973**, 278.
137. Sonnet, P. E., *J. Heterocyclic Chem.*, **1970**, *7*, 1101.
138. Bailer, M., Yagen, B., Mechoulam, R., Becker, Y., *J. Med. Chem.*, **1980**, *23*, 1144.
139. Grokhovskii, S. L., Zhuze, A. L., Gottikh, B. P., *Bioorg. Khim.*, **1975**, *1*, 1616; *Chem. Abstr.*, **1976**, *84*, 150434t.
140. Klimova, V. A., Dubinskii, R. A., *Izv. Akad. Nauk S.S.S.R.*, *Ser. Khim.*, **1974**, 640; Eng. p. 604.
141. Osawa, T., Ishibashi, H., Namiki, M., Kada, T., *Biochem. Biophys. Res. Commun.*, **1980**, *95*, 835.
142. Spiro, V., Aiello, T., Fabra, I., *Ann. Chim. Rome*, **1966**, *56*, 866.
143. Aiello, T., Sigillo, G., *Gazz. Chim. Ital.*, **1938**, *68*, 681.
144. Greenstock, C. L., Ruddock, G. W., Neta, P., *Radiat. Res.*, **1976**, *66*, 472.
145. Boyer, J. H., *Chem. Rev.*, **1980**, *80*, 495.
146. Rinkes, I. J., *Rec. Trav. Chim.*, **1933**, *52*, 538.
147. Boyer, J. H., Chang, M. S., *J. Am. Chem. Soc.*, **1960**, *82*, 2220.
148. Boyer, N. E., Czerniak, G. M., Gutowsky, H. S., Snyder, H. R., *J. Am. Chem. Soc.*, **1955**, *77*, 4238.
149. De Santis, F., Stegel, F., *Tetrahedron Lett.*, **1974**, 1079.
150. Doddi, G., Mencarelli, P., Stegel, F., *J. Chem. Soc., Chem. Commun.*, **1975**, 273.
151. Vecchietti, V., Dradi, E., Lauria, F., *J. Chem. Soc. (C)*, **1971**, 2554.
152. Doddi, G., Illuminati, G., Mencarelli, P., Stegel, F., *J. Org. Chem.*, **1976**, *41*, 2824.
153. Mencarelli, P., Stegel, F., *J. Org. Chem.*, **1977**, *42*, 3550.
154. Mencarelli, P., Stegel, F., *J. Chem. Soc., Chem. Commun.*, **1978**, 564.
155. Bonaccina, L., Mencarelli, P., Stegel, F., *J. Org. Chem.*, **1979**, *44*, 4420.
156. Mencarelli, P., Stegel, F., *J. Chem. Soc., Chem. Commun.*, **1980**, 123.
157. Devincenzis, G., Mencarelli, P., Stegel, F., *J. Org. Chem.*, **1983**, *48*, 162.
158. Mencarelli, P., Stegel, F., *J. Chem. Res., Synop.*, **1984**, 18.
159. Cheeseman, G. W. H., Roy, P. D., *J. Chem. Soc. (C)*, **1969**, 956.
160. Yagil, G., *Tetrahedron*, **1967**, *23*, 2855.
161. Safonova, E. N., Belikov, V. M., Novikov, S. S., *Izv. Akad. Nauk S.S.S.R., Otdel Khim. Nauk*, **1959**, 1130; Eng. p. 1094.
162. Fournari, P., Person, M., Watelle-Marian. G., Delepine, M., *Compt. Rend.*, **1961**, *253*, 1059.
163. Ref. 2, pp. 26–27.
164. Ger. 1,252,204, *Chem. Abstr.*, **1968**, *68*, 104972p.
165. Farge, D., Messer, M. N., *S. African* 68 00,809, *Chem. Abstr.*, **1969**, *70*, 87825q.
166. Farge, D., Messer, M. N., *Fr.* 1,592,066, *Chem. Abstr.*, **1972**, *77*, 19654c.
167. Ref. 6, p. 337.
168. Boatman, R. J., Whitlock, H. W., *J. Org. Chem.*, **1976**, *41*, 3050.
169. Ciamician, G., Silber, P., *Ber.*, **1884**, *17*, 103.
170. Rahman, M. O. A., Osman, A. I., Kira, M. A., *Egypt. J. Chem.*, **1975**, *18*, 175; *Chem. Abstr.*, 1977, *86*, 155430q.
171. Bottino, F. A., Mineri, G., Sciotto, D., *Tetrahedron*, **1978**, *34*, 1557.
172. Kirra, M. A., Shoeb, H. A., Korkor, M. I., *Egypt. J. Chem.*, **1972**, *15*, 609; *Chem. Abstr.*, **1974**, *80*, 120682x.
173. Treibs, A., Fritz, G., *Justus Liebigs Ann. Chem.*, **1958**, *611*, 162.
174. Le Goff, E., Cheng, D. O., *Porphyrin Chem. Adv. [Pap. Porphyrin Symp.]*, **1977** (Pub. 1979), 153, ed. F. R. Longo, Ann Arbor Sci., Ann Arbor, Mich.; *Chem. Abstr.*, **1979**, 91, 56973r.
175. Rudqvist, U., Torssell, K., *Acta Chem. Scand.*, **1971**, *25*, 2183.
176. Koyama, M., Kodama, Y., Tsuruoka, T., Ezaki, N., Niwa, T., Inouye, S., *J. Antibiot., Tokyo*, **1981**, *34*, 1569.
177. Jpn. Kokai Tokkyo JP 82 67,557; *Chem. Abstr.*, **1982**, *97*, 182203b.
178. Koyama, M., Ezaki, N., Tsuruoka, T., Inouye, S., *J. Antibiot. Tokyo*, **1983**, *36*, 1483.
179. Ezaki, N., Shomura, T., Koyama, M., Niwa, T., Kojimo, M., Inouye, S., Ito, T., Niida, T., *J. Antibiot.*, **1981**, *34*, 1363.
180. Ezaki, N., Shomura, T., Niwa, T., Kojima, M., Inouye, S., Ito, T., *Fr. Demande FR* 2,472,611; *Chem. Abstr.*, **1982**, *96*, 4942p.

181. Koyama, M., Kodama, Y., Tsuruoka, T., Ezaki, N., Nirva, T., Inouye, S., *J. Antibiot.*, **981**, *34*, 1569.
182. Umezawa, K., Mimura, S., Matsushima, T., Muramatsu, S., Sawa, T., Takeuchi, T., *Biochem. Biophys. Res. Commun.*, **1982**, *105*, 82.
183. Sharin, G. P., Khabirov, R. A., Nurgatin, V. V., Khmel'nitskii, L. I., Lebedev, O. V., Prikhod'ko, A. S., *Izv. Akad. Nauk S.S.S.R., Ser. Khim.*, **1977**, 2711; Eng. p. 2506.
184. Sharnin, G. P., Nurgatin, V. V., Khabirov, R. A., Kolosov, V. D., Khmel'nitskii, L. I., Lebedev, O. V., Prikhod'ko, A. S., *Izv. Adad. Nauk S.S.S.R., Ser. Khim.*, **1977**, 447; Eng. p. 404.
185. Hunt, R., Reid, S. T., *J. Chem. Soc., Perkin Trans. I*, **1972**, 2527.
186. Sleight, R. B., Sutcliffe, L. H., *Trans. Faraday Soc.*, **1971**, *67*, 2195.
187. Groen, M. B., Havinga, E., *Mol. Photochem.*, **1974**, *6*, 9.
188. Yoshida, Z., Kobayashi, T., Yamada, H., *Bull. Chem. Soc. Jap.*, **1972**, *45*, 313.
189. Anderson, H. J., Griffiths, S. J., *Can. J. Chem.*, **1967**, *45*, 2227.
190. Novikov, S. S., Belikov, V. M., *Izv. Akad. Nauk S.S.S.R., Otdel. Khim.*, **1959**, 1098; Eng. p. 1059.
191. Ref. 1, p. 56.
192. Novikov, S. S., Belikov, V. M., Egorov, Yu. P., Safonova, E. N., Semenov, L. V., *Izv. Akad. Nauk S.S.S.R., Otdel. Khim. Nauk*, **1959**, 1438; Eng. p. 1386.
193. Marchetti, L., Passalacqua, V., *Ann. Chim. (Rome)*, **1967**, *57*, 1266.
194. Gardner, T. S., Wenis, E., Lee, J., *J. Org. Chem.*, **1959**, *24*, 570.
195. Grehn, L., *Chem. Scr.*, **1980**, *16*, 77.
196. Farnier, M., Drakenberg, T., *J. Chem. Soc., Perkin Trans. II*, **1975**, 333.
197. Bertin, D. M., Farnier, M., Liegeois, C., *Bull. Soc. Chim. Fr.*, **1974**, 2677.
198. Saito, H., Nukada, K., *J. Am. Chem. Soc.*, **1971**, *93*, 1072.
199. Lippmaa, E., Mägi, M., Novikov, S. S., Khmelnitski, L. I., Prihodko, A. S., Lebedev, O. V., Epishina, L. V., *Org. Magn. Reson.*, **1972**, *4*, 153.
200. Lippmaa, E., Mägi, M., Novikov, S. S., Khmelnitski, L. I., Prihodko, A. S., Lebedev, O. V., Epishina, L. V., *Org. Magn. Reson.*, **1972**, *4*, 197.
201. King, M. M., Yeh, H. J. C., Dudek, G. O., *Org. Magn. Reson.*, **1976**, *8*, 208.
202. Negrebetskii, V. V., Kessenikh, A. V., Novikov, S. S., Khmel'nitskii, L. I., Prikhod'ko A. S., Lebedev, O. V., *Izv. Akad. Nauk S.S.S.R., Ser. Khim.*, **1971**, 2613; Eng. p. 2489.
203. Kao, J., Hinde, A. L., Radom, L., *Nouv. J. De Chim.*, **1979**, *3*, 473.
204. Minyaev, R. M., Minkin, V. I., Sheinker, V. N., *Zh. Org. Khim.*, **1975**, *11*, 1950; Eng. p. 1968.
205. Melent'eva, T. A., Landau, M. A., Kazanskaya, L. V., Berezovskii, V. M., *Zh. Org. Khim.*, **1972**, *8*, 191; Eng. p. 194.
206. Fringuelli, F., Marino, G., Savelli, G., *Tetrahedron*, **1969**, *25*, 5815.
207. Cauletti, C., Giancaspro, C., Monaci, A., Piancastelli, M. N., *J. Chem. Soc., Perkin Trans. II*, **1981**, 656.
208. Yakhontov, L. N., Portnov, M. A., Azimov, V. A., Lapan, E. I., *Zh. Org. Khim.*, **1969**, *5*, 956; Eng. p. 942.
209. Linda, P., Marino, G., Pignataro, S., *J. Chem. Soc. (B)*, **1971**, 1585.
210. Cumper, C. W. N., Wood, J. W. M., *J. Chem. Soc. (B)*, **1971**, 1811.
211. Lumbroso, H., Carpanelli, C., *Bull. Soc. Chim. Fr.*, **1964**, 3198.
212. Marey, T., Arriau, J., *Compte Rend Acad. Sci., Ser. C*, **1971**, *272*, 850.
213. Namiki, M., Osawa, T., Ishibashi, H., Namiki, K., Tsuji, K., *J. Agric. Food Chem.*, **1981**, *29*, 407; *Chem. Abstr.*, **1981**, *94*, 138115p.
214. Namiki, M., Udaka, S., Osawa, T., Tsuji, K., Kada, T., *Mutat. Res.*, **1980**, *73*, 21; *Chem. Abstr.*, **1981**, *94*, 14088n.
215. Raleigh, J. A., Chapman, J. D., Reuvers, A. P., Biaglow, J. E., Durand, R. E., Rauth, A. M., *Br. J. Cancer, Suppl.*, **1978**, *37*, 6.
216. Chin, J. B., Sheinin, D. M. K., Rauth, A. M., *Mutat. Res.*, **1978**, *58*, 1.
217. Roantree, M. L., Young, R. C., *Eur. Pat. Appl.* 28,117: *Chem. Abstr.*, **1981**, *95*, 115529h.
218. Reif, A. E., Southwick, P. L., Judd, K. P., *J. Surg. Oncol.*, **1980**, *13*, 135; *Chem. Abstr.*, **1980**, *93*, 142839z.
219. Lauria, F., Vechietti, V., De Carneri, I., *Farmaco, Ed. Sci.*, **1967**, *22*, 479; *Chem. Abstr.*, **1968**, *68*, 29508m.
220. Albrecht, R., Schroeder, E., *Arch. Pharm.*, **1975**, *308*, 588; *Chem. Abstr.*, **1975**, *83*, 193044n.
221. Yamabe, S., Shimizu, M., Yamamoto, T., *Ger. Offen.* 1,965,267; *Chem. Abstr.*, **1970**, *73*, 66412z.
222. Carter, G. A., Dawson, G. W., Garraway, J. L., *Pestic. Sci.*, **1975**, *6*, 43; *Chem. Abstr.*, **1975**, *82*, 165736a.

223. Mazur, I. A., Sinyak, R. S., Katkevich, R. I., Steblyuk, P. N., *Farm. Zh. (Kiev)*, **1980**, 34; *Chem. Abstr.*, **1981**, *94*, 47258w.
224. Bialer, M., Yagen, B., Mechoulam, R., Becker, Y., *J. Pharm. Sci.*, **1980**, *69*, 1334; *Chem. Abstr.*, **1981**, *94*, 192034z.
225. Meinwald, J., Smolanoff, J., McPhail, A. T., Miller, R. W., Eisner, T., Hicks, K., *Tetrahedron Lett.*, **1975**, 2367.

BIBLIOGRAPHY

Aiello, E., Dattolo, G., Cirrincione, G., Almerico, A. M., Polycondensed nitrogen heterocycles. 13. Pyrrolo (3,2-B) indole by intramolecular nucleophilic substitution reaction in the pyrrole series. *J. Heterocycl. Chem.*, **1984**, *21*, 721.
Annuli, A., Mencarelli, P., Stegel, F., Nucleophilic Aromatic Substitution in the Pyrrole Ring: Leaving Group Effect, *J. Org. Chem.*, **1984**, *49*, 4065.
Bazzano, F., Mencarelli, P., Stegel, F., Cyano and Nitro Group Effect on the Rate of Methoxydenitration Reaction in the Pyrrole and Benzene Ring. *J. Org. Chem.*, **1984**, *49*, 2375.
Catalano, M. M., Crossley, M. J., King, L. G., Efficient Synthesis of 2-Oxy-5,10,15,20-Tetraphenylporphyrins from a Nitroporphyrin by a Novel Multi-Step Cine-Substitution Sequence, *J. Chem. Soc., Chem. Commun.*, **1984**, 1537.
Colonna, M., Greci, L., Poloni, M., Electrophilic Ipso Substitutions. PART 3. Reactions of 3-Substituted Indoles, 4-Substituted N,N-Dimethylanilines, and 1 and 3-Substituted Indolizines with Nitrous Acid. *J. Chem. Soc., Perkin Trans. 2*, **1984**, 165.
Cooper, D. G., Sach, G. S., Pyridine Derivatives. Eur. Pat. Appl. EP 89,153; *Chem. Abstr.*, **1984**, *100*, 68176a.
Dattolo, G., Cirrincione, G., Almerico, A. M., Presti, G., Aiello, E., Reactivity of 3-Diazopyrroles. PART 2. *Heterocycles*, **1983**, *20*, 829.
Girard, Y., Atkinson, J. G., Bélanger, P. C., Furentis, J. J., Rokach, J., Rooney, C. S., Remy, D. C., Hunt, C. A., Synthesis, Chemistry and Photochemical Substitutions of 6,11-Dihydro-5H-pyrrolo [2,1-b][3]Benzazepin-11-ones. *J. Org. Chem.*, **1983**, *48*, 3220.
Gorb, L. G., Morozova, I. M., Belen'kii, L. I., Abronin, I. A., Relations between Activity and Selectivity in Electrophilic Substitution Reactions of Five-Membered Heteroaromatic Compounds. 3. Electronic Effects of Substituents. *Izv. Akad. Nauk S.S.S.R., Ser. Khim.*, **1983**, 828; *Chem. Abstr.*, **1983**, *99*, 4894m.
Grehn, L., Ragnarsson, U., Eriksson, B., Oeberg, B., Synthesis and Antiviral Activity of Distamycin a Analogs: Substitutions on the Different Pyrrole Nitrogens and in the Amidine Function. *J. Med. Chem.*, **1983**, *26*, 1042.
Gorelik, M. V., Alimova, R. A., Synthesis and Reaction of Pyrrolophenanthrone and Thiophenanthrone with 1,10-Anthraquinoid Structure, *Zh. Org. Khim.*, **1984**, *20*, 1553; *Current Abstr. of Chem.*, **1984**, 356095.
Grehn, L., Ragnarsson, U., A Convenient Method for the Preparation of 1-(Tert-butyloxycarbonyl) Pyrroles. *Angew. Chem.*, **1984**, *96*, 291.

Hartman, P. E., Review: Putative Mutagens and Carcinogens in Foods. II: Sorbate and Sorbate Interactions. *Environ. Mutagen*, **1983**, *5*, 217.

Hurtel, P., Decroix, B., Morel, J., Terrier, F., Triflluoromethyylsulfonyl Meisenheimer Complexes. PART IV. Five-Membered Ring Heterocyclic Triflones and Related σ-Complexes. Kinetic Evidence for an Anomalous Effect of the Trifluoromethylsulfonyl Group. *J. Chem. Res., Synop.*, **1983**, 58.

Koyama, M., Ezaki, N., Tsuruoka, T., Inouye, S., Structure Studies on Pyrrolomycins C, D, and E. *J. Antibiot.*, **1983**, *36*, 1483.

Koyama, M., Kai, E., Tsuruoka, T., Miyauchi, K., Matsumoto, K., Akita, E., Inouye, S., Niida, T., Heterocyclic Compounds and Antibacterial and Antifungal Compositions Containing Them as Active Ingredients. *Eur. Pat. Appl.* EP 80,051; *Chem. Abstr.*, **1983**, *99*, 194964e.

Makosza, M., Slomka, E., Reactions of Organic Anions. 118. Vicarious Nucleophilic Substitution of Hydrogen in Nitro Derivatives of 5-Membered Aromatic Heterocycles, *Bull. Pol. Acad. Sci., Chem.*, **1984**, *32*, 69 (Eng); *Chem. Abstr.*, **1984**, *101*, 230283c.

Mencarrreli, P., Stegel, F., Novel Reaction Course for 1-Alkyl-2,4-dinitropyrroles: Formation of Pyrrolines upon Reaction with Piperidine. *J. Chem. Res.*, **1984**, 518.

Mencarrreli, P., Stegel, F., Novel Reaction Course for 1-Alkyl-2,4-dinitropyrroles: Formation of Pyrrolines upon Reaction with Piperidine. *J. Chem. Res., Synop.*, **1984**, 18.

Muchowski, J. M., Solas, D. R., β-Substituted Pyrroles via Electrophilic Substitution of N-Triisopropylsilylpyrrole. *Tetrahedron Lett.*, **1983**, *24*, 3455.

Pyrrole Derivatives. *Jpn. Kokai Tokkyo Koho JP* 58 62,159 (83 62,159): *Chem. Abstr.*, **1984**, *100*, 6326r.

Chapter 2

Nitroimidazoles

INTRODUCTION

Over the decade 1920–1930 Pyman and Fargher and their collaborators examined nitro derivatives in their extensive research on imidazoles.[1a-s] Elsewhere these compounds received only moderate attention following the discoveries of 8-nitrocaffeine (**2-1**) (1867),[2] 1-methyl- and 1-phenyl-2-thiomethyl-5-nitroimidazoles (**2-2**) and (**2-3**) (1889),[3] 4(5)-nitroimidazole (**2-4**) (1892),[4,5] and 4(5)-methyl-5(4)-nitroimidazole (1909)[6] until the discovery (1953) of the natural occurrence of 2-nitroimidazole (**2-5**), otherwise known as the antibiotic azomycin.[7] Demand for information on the biological activity of nitroimidazoles has grown rapidly (more than 1500 publications over the 6.5 year period 1977–June 1983).

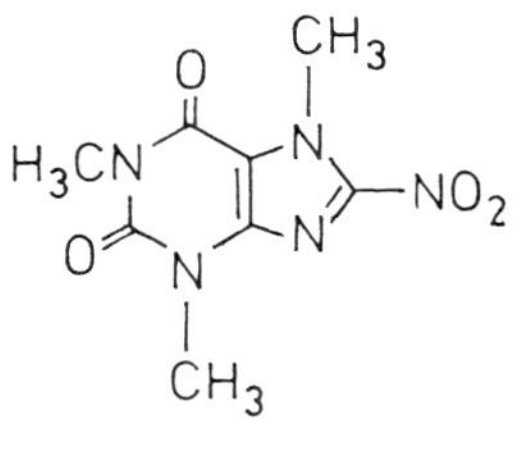

2-1

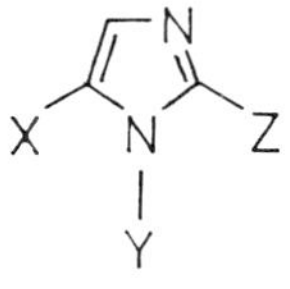

2-2, X = NO$_2$, Y = CH$_3$, Z = SCH$_3$
2-3, X = NO$_2$, Y = C$_6$H$_5$, Z = SCH$_3$
2-4, X = NO$_2$, Y = Z = H
2-5, X = Y = H, Z = NO$_2$

Nitroimidazoles were reviewed in two books: "Imidazole and Its Derivatives" (1953)[8] and "The Azoles" (1976).[9] Also two chapters present selected preparations and chemical properties of nitroimidazoles of biological interest in a book (1982) which also reviewed their pharmacology and clinical application.[10,11]

A Polish review on the chemistry and pharmacology of nitroimidazoles also appeared in 1982.[12] Other reviews on the biological properties of nitroimidazoles are cited in the section on Biological Properties. Nitroimidazole derivatives of polycyclic systems, nitroimidazolines, and nitroimidazolidines are occasionally mentioned in other reviews of the heterocycles.[8-14]

In this review the literature has been covered through the listings in Chemical Abstracts, **1983**, *98*; there is a bibliographic addendum for volumes *99* and *100*. The large volume of literature on biochemical properties was considered to be outside of the scope of this chapter and is given a superficial presentation. Chemical reactions in nitroimidazole substituents were omitted when the reaction site was remote from the nitro group.

PREPARATIONS OF SIMPLE NITROIMIDAZOLES

Nitration and Nitrolysis

Imidazole and Alkylimidazoles

Nitric acid in sulfuric acid (mixed acid) converted imidazole to 4(5)-nitroimidazole (**2-4**)[1a,4,5,15-22] Two reports described dependence on the acid medium. Nitric acid (d 1.42) in concentrated sulfuric acid gave the mononitration product (**2-4**) (73%) and a similar treatment with nitric acid (d 1.5) in concentrated sulfuric acid gave both the mononitration product (**2-4**) (21%) and 4,5-dinitroimidazole (31%).[16] The second report described a decrease in yield from 46% to 19% for mononitration as the concentration of sulfuric acid increased from 83.7% to 98.8% and then a dramatic increase in yield to 90% for the reaction in oleum (1% sulfur trioxide). Second-order kinetics supported nitration of an imidazolium cation (**2-6**) in a reaction complicated by side reactions. A calculated partial rate factor of 3.0×10^{-9} for the C-4(5)-position during nitration in 98% sulfuric acid indicated a low reactivity; however, the correlation between product formation and the acidity of the medium made further interpretation unreliable.[17]

An uncertainty of the intermediacy of 1-nitroimidazole (**2-7**) or its conjugate acid (**2-8**) remains unresolved. Although the unknown nitramine (**2-7**) is presumably unstable,[23-25] 1,4-dinitroimidazole (**2-9**) has been prepared.[16]

2-6   **2-7**   **2-8**

Treatment with concentrated sulfuric acid or an amine brought about denitration to the mononitroimidazole (**2-4**) (66%) at 25° and rearrangement to the 4,5-dinitro isomer (**2-10**) at 120°;[26,27] however the latter reaction was not shown to be intramolecular. In chlorobenzene at 120° the dinitroimidazole (**2-9**) rearranged to 2,4(5)-dinitroimidazole (**2-11**).[26] Similar reactions for 2-methyl-1,4-dinitroimidazole in sulfuric acid were reported.[16]

Imidazole was also nitrated to its 4(5)-nitro derivative (**2-4**) (58%) by potassium nitrate in concentrated sulfuric acid.[18] The absence of mononitration of imidazole to 2-nitroimidazole (**2-5**) has been repeatedly noted.[1a,4,5,15-22]

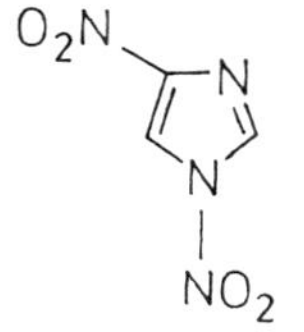

2-9 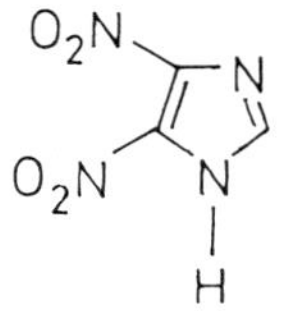   **2-10** 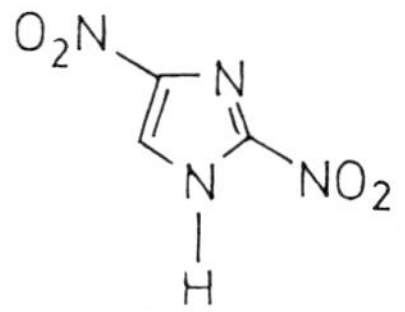   **2-11**

In a different approach mixed acid nitrated pentamine(imidazolato)cobalt(III) perchlorate (**2-12**) at 0°–10° to a hydrated complex chloride (**2-13**) (97%) of 4-nitroimidazole. The structure (**2-13**) was determined by X-ray analysis.[28] If the nitroheterocycle (**2-4**) can be easily liberated from its complex this appears to be a promising new preparative method.

$$[Co(NH_3)_5(ImH)](ClO_4)_3$$
2-12

$$[Co(NH_3)_5(4-O_2NIm)]Cl_2 \cdot H_2O$$
2-13

Mixed acid nitrated 1-methylimidazole to a mixture (80%) of the 4- and 5-nitro derivatives (**2-14, 2-15**) (5.5:1);[1h,29] 2-methylimidazole to its 4(5)-nitro derivative **2-16** (84%);[1a,16,30-34] and 4(5)-methylimidazole to its 5(4)-nitro derivative (**2-17**) (90%).[1a,35,36] Warm fuming nitric acid also gave product **2-17** but in lower yield.[6] Mixed acid afforded homologous 2-alkyl-4(5)-nitroimidazoles (**2-18**)[34,37,38] with yields up to 75% from reactions (6–12 min, 210°–245°) in the presence of urea.[37]

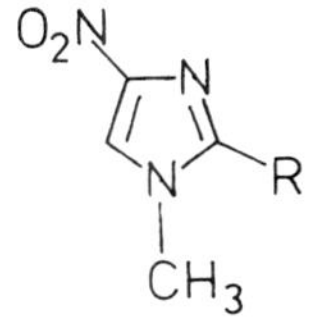

2-14 R = H   **2-15** R = H
2-19 R = CH$_3$ **2-20** R = CH$_3$

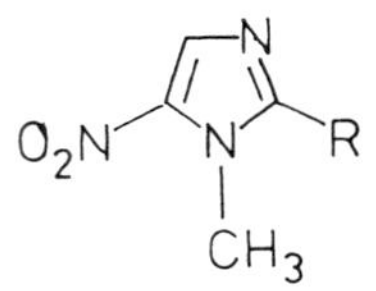

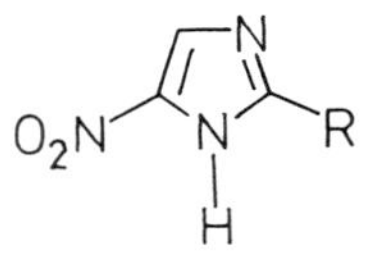

2-16 R = CH₃
2-18 R = C$_n$H$_{2n+1}$, n = 2,3,4

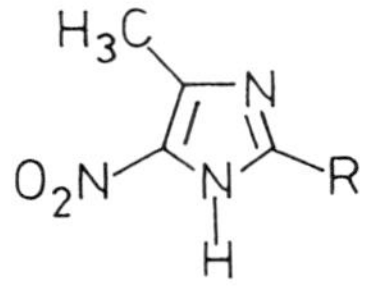

2-17 R = H
2-21 R = CH₃
2-22 R = C₂H₅

Mixed acid converted 1,2-dimethylimidazole to its 4- (20%) and 5-nitro- (10%) derivatives (**2-19, 2-20**);[1k] 1,4- and 1,5-dimethylimidazoles to the 5-nitro- (100%) and 4-nitro- (83%) derivatives;[1d] and 2,4(5)-dimethylimidazole to its 5(4)-nitro derivative (**2-21**).[37] A similar nitration afforded 2-ethyl-4(5)-methyl-5(4)-nitro imidazole (**2-22**).[34,37] Mixed acid failed to nitrate 4,5-dimethylimidazole and instead gave an inefficient oxidation to 5(4)-methyl imidazole-4(5)-carboxylic acid.[1a] The nitrate salt of N-acetylhistamine in concentrated sulfuric acid heated at 90° for 2 h followed by hydrolysis gave 5(4)-nitrohistamine (**2-23a**) nearly quantitatively. In a similar reaction nitrohistidine (**2-23b**) was obtained.[39,40]

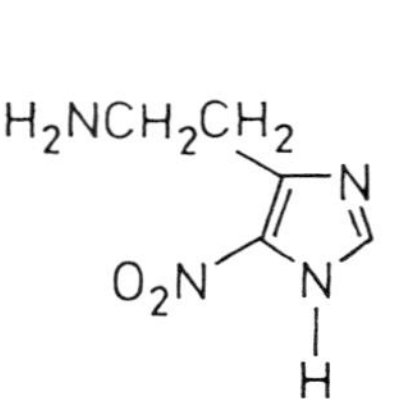

2-23a

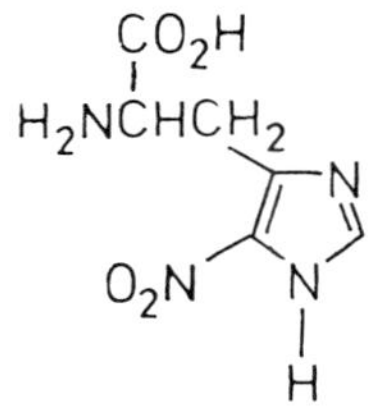

2-23b

Sodium[41,42] or potassium nitrate[18] in sulfuric acid converted 2-methylimidazole to its 4(5)-nitro derivative (**2-16**) (49%) and mixed acid converted a zinc hydroxide complex of the imidazole to the derivative (**2-16**) contaminated with zinc oxide.[43-47]

Mixed acid nitrated 1-ethyl-2,4-dimethyl- and 1,2-dimethyl-4-isopropylimidazoles to 5-nitro derivatives[48,49] and an unspecified nitration converted 1,2,5-trialkylimidazole to its 4-nitro derivative.[50] Nitration of 1,4,5-trialkylimidazoles has not been reported.

α-Hydroxyalkyl and Acylimidazoles

Although nitration of 4(5)-acetoxymethylimidazole followed by hydrolysis gave 5(4)-nitro-4(5)-hydroxymethylimidazole (**2-24a**),[51] when unprotected the carbinol was oxidized to a mixture of an aldehyde and a carboxylic acid.[1d] Potassium nitrate in concentrated sulfuric acid converted the unprotected 2-hydroxymethylimidazole to its 4(5)-nitro derivative (**2-24b**),[18] and 1-alkyl-2-

hydroxymethylimidazoles were nitrated and oxidized to 1-alkyl-5-nitroimidazole-2-carboxylic acids (**2-25**).[52,53] A large excess of warm concentrated nitric acid oxidized 5-hydroxymethyl-1,2-dimethylimidazole to the aldehyde, 5-formyl-1,2-dimethylimidazole.[54] By lowering the temperature to $-10°$ a mixture of concentrated nitric and sulfuric acids nitrated 4(5)-hydroxymethylimidazole to the 5(4)-nitro derivative (67%) without oxidation of the methylol side chain.[55]

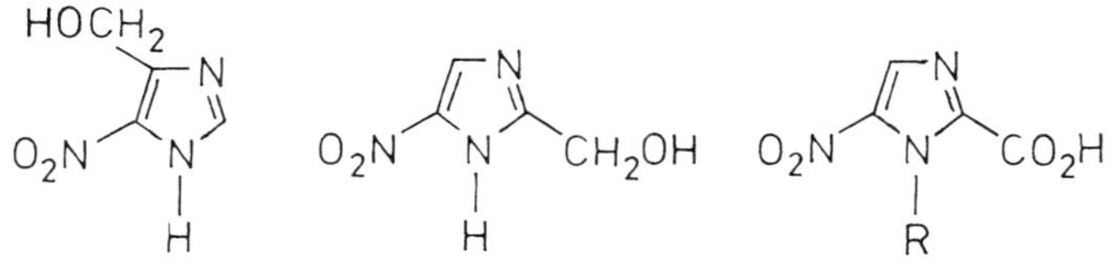

2-24a   **2-24b**   **2-25**

Ring nitration of imidazole aldehydes, ketones, carboxylic acids, and derivatives is unknown.

Arylimidazoles

Mixed acid converted the three isomeric phenylimidazoles predominantly to 1- (58%),[1n,56-58] 2- (**2-26**) (50%),[1i,59,60] and 4(5)-(4-nitrophenyl)imidazole (69%);[1b,61] and by further nitration 4(5)-nitro-2-(4-nitrophenyl)imidazole (**2-27**) was obtained from the mononitration product (**2-26**).[37] Dinitration by fuming nitric acid afforded 2-(4-nitrophenyl)-4(5)-nitro-5(4)-methylimidazole (**2-28**).[61] A kinetic study showed the conjugate acid of 1-phenylimidazole to be involved in the nitration by mixed acid.[56] A recent patent claim[31] that a mononitration gave 4(5)-nitro-2-phenylimidazole (**2-29**) instead of the isomer (**2-26**) was apparently an error.

In similar reactions a series of 5(4)-nitro-4(5)-nitroarylimidazoles (**2-30**) was prepared.[62]

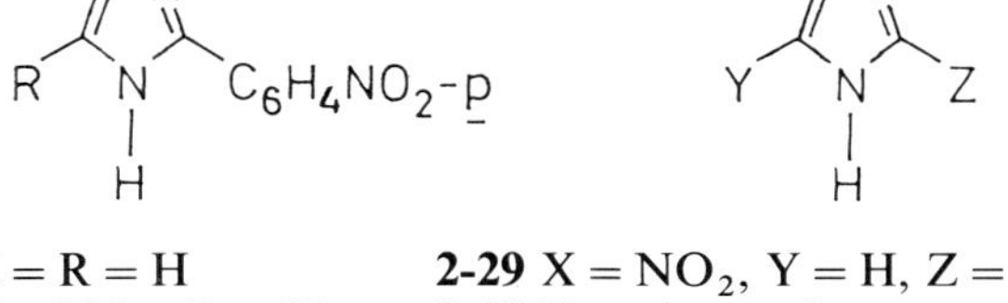

2-26 X = R = H
2-27 X = NO$_2$, R = H
2-28 X = NO$_2$, R = CH$_3$

2-29 X = NO$_2$, Y = H, Z = C$_6$H$_5$
2-30 X = nitroaryl, Y = NO$_2$, Z = H

Grant and Pyman discovered that nitration of 4(5)-(4-chlorophenyl)imidazole gave primarily 4(5)-(4-chlorophenyl)-5(4)-nitroimidazole.[1b] More recently nitration in oleum at $-10°$ to $25°$ produced 4(5)-nitro-2-(4-fluorophenyl)- (**2-31**)

(80%) and 4(5)-nitro-2-(3-nitro-4-fluorophenyl)imidazole (**2-32**) (90%).[63] At 25° nitrosation–oxidation gave 2-(4-fluorophenyl)-5-hydroxyiminoimidazolin-4-one (**2-33**) (44%) when 2-(4-fluorophenyl)imidazole was treated with mixed acid; but a similar treatment of the mononitro compound (**2-31**) gave the dinitro compound (**2-32**).[63] A nitration by nitric acid in acetic acid of 2-(3,4,5-trichlorophenyl)imidazole to 4(5)-nitro-2-(3,4,5-trichlorophenyl)imidazole was disclosed in a patent.[64]

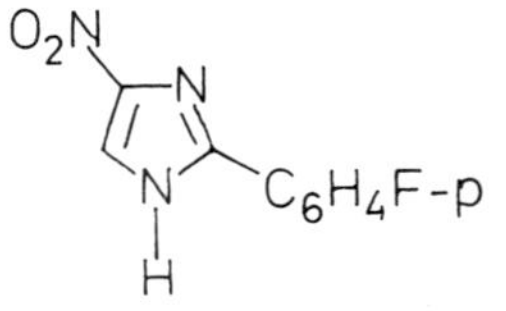

2-31

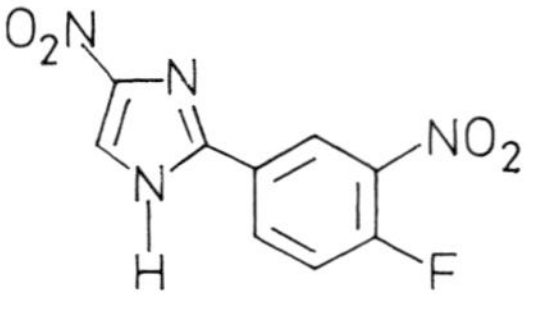

2-32

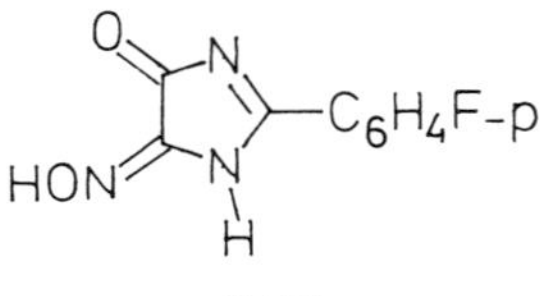

2-33

Each ring in a 4(5)-(4-alkoxyphenyl)imidazole was nitrated by 3 N nitric acid to give a 4(5)-(3-nitro-4-alkoxyphenyl)imidazole (**2-34**) (17–18%) and a 4(5)-(4-alkoxyphenyl)-5(4)-nitroimidazole (**2-35**) (55–60%); similar treatment with nitric acid (d 1.4) produced 4(5)-(3-nitro-4-alkoxyphenyl)-5(4)-nitroimidazoles (**2-36**) (83%) and (**2-37**) (95%); whereas nitric acid (d 1.46) gave 4(5)-(3,5-dinitro-4-alkoxyphenyl)-5(4)-nitroimidazole (**2-38**) (72%).[65]

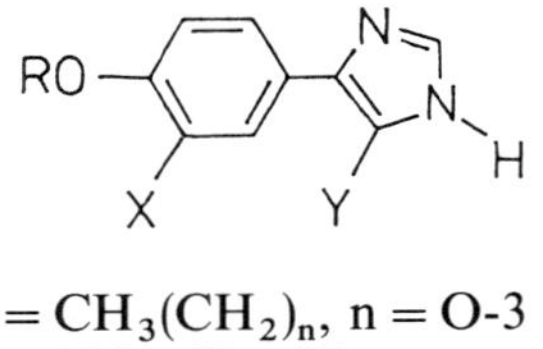

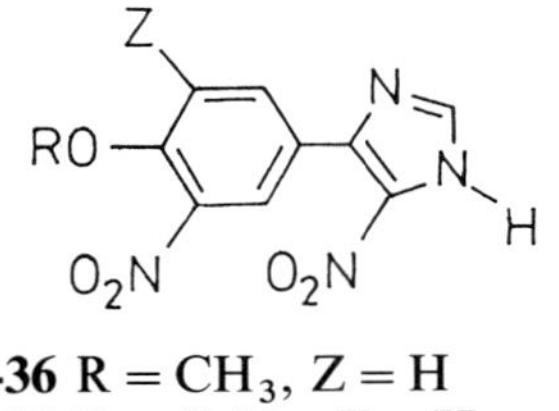

2-34 R = $CH_3(CH_2)_n$, n = 0–3
X = NO_2, Y = H
2-35 R = $CH_3(CH_2)_n$, n = 0–3
X = H, Y = NO_2

2-36 R = CH_3, Z = H
2-37 R = C_2H_5, Z = H
2-38 R = $CH_3CH_2CH_2$, Z = NO_2

Mixed acid nitrated only the phenyl rings in converting 4- and 5-phenyl-1-methylimidazole to the corresponding *p*-nitrophenyl derivatives (56%, 64%) along with the formation of some of the *ortho* isomer.[1h] Further nitration of 2-(4-nitrophenyl)-4(5)-methylimidazole gave the dinitro compound (**2-39**).[1b,37] Although nitration also occurred only in the phenyl ring to produce

2-(4-nitrophenyl)-4(5)-imidazolecarboxylic acid (**2-40**) (52%) orientation in substitution changed from the *op*-pattern to give predominantly *m*-substitution in the formation of 2-(3-nitrophenyl)-4,5-imidazoledicarboxylic acid (**2-41**) (52%) by nitration in mixed acids.[1i,m]

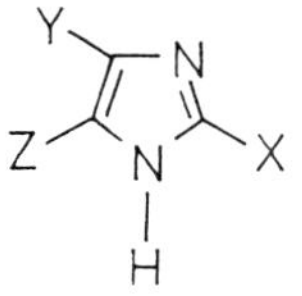

2-39 X = $C_6H_4NO_2$-*p*, Y = NO_2, Z = CH_3
2-40 X = $C_6H_4NO_2$-*p*, Y = CO_2H, Z = H
2-41 X = $C_6H_4NO_2$-*p*, Y = Z = CO_2H

Nitroimidazoles with imidazolyl and 1,3,4-thiadiazolyl substituents are known.[66,67]

Imidazole Halides

Mixed acid efficiently nitrated 4(5)-bromoimidazole,[1c] its 2-methyl and 2-ethyl derivatives,[68] and 2-iodoimidazole[69] to the appropriate 5(4)-nitro compounds, e.g., **2-42** (87%) and **2-43** (40%). Mixed acid nitrolyzed 4,5-diiodoimidazole (a structure assignment[70,71] in 1979 corrected an earlier assignment of 2,4(5)-diiodoimidazole) to 4(5)-iodo-5(4)-nitroimidazole (**2-44**), (85%).[70-72] Derivatives thought to be related to 2,4(5)-diiodoimidazole prior to 1979 need to be reexamined for correlations with 4,5-diiodoimidazole.[71-74]

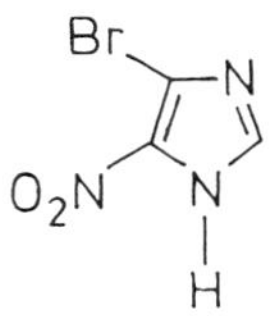

2-42 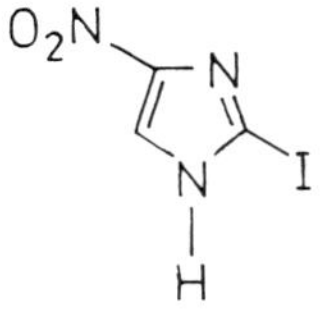   **2-43** 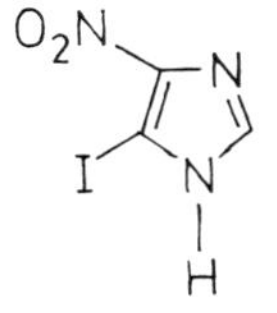   **2-44**

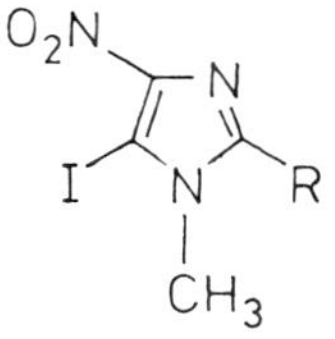

2-45 R = H   **2-46** R = H   **2-47**
2-48 R = CH_3 **2-49** R = CH_3

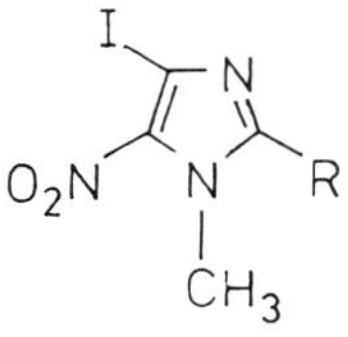

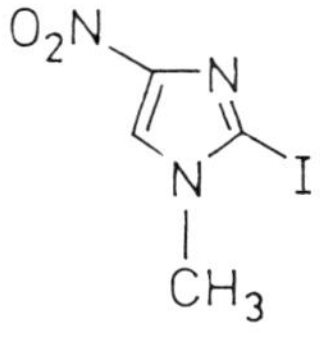

By N-methylation the imidazole (**2-44**) was converted to the expected pair of derivatives (**2-45**) and (**2-46**) partially confirmed by an independent mixed acid nitration of 1-methyl-2-iodoimidazole to its 4-nitro derivative (**2-47**) (30%).[71] In related reactions mixed acids converted 2-methyl-4,5-diiodoimidazole to 2-methyl-4(5)-nitro-5(4)-iodoimidazole (83%) and 1,2-dimethyl-4,5-diiodoimidazole to the pair of nitroimidazoles (50%) (**2-48** and **2-49**) (2:1).[72] The data is insufficient to verify a suggestion of preferential nitrolysis of a C-4 iodo group.

Mixed acid converted 1-methyl-2-bromoimidazole to a mixture (50% yield) of the 4-nitro- and 5-nitro- derivatives (roughly 2:1). In a similar way 1-methyl-5-bromoimidazole was nitrated to its 4-nitro derivative (80%).[75] Mixed acid nearly quantitatively converted 1-methyl-4- and 5-chloroimidazole to the 5- and 4-nitro derivatives (**2-50** and **2-51**).[76-81] Earlier investigations found that similar nitrations afforded 2-methyl-4(5)-bromo-5(4)-nitroimidazole (**2-52**) (82%)[1e] and 2-bromo-4(5)-methyl-5(4)-nitroimidazole (**2-53**) (70%).[1f] Five 1,2-dialkyl-4-chloro and eight 1,2-dialkyl-5-chloroimidazoles were nitrated to the 5- and 4-nitro derivatives (**2-54** and **2-55**) (74%–98%).[80] Similar nitrations afforded 1-ethyl-2-n-propyl-4-nitro-5-chloro and 1-n-butyl-2-methyl-4-nitro-5-chloroimidazoles.[82]

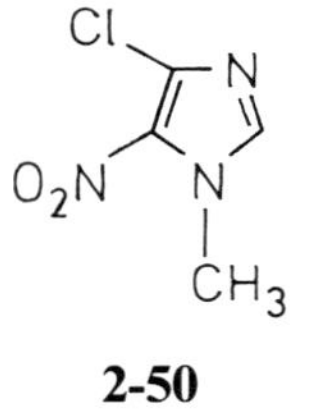

2-50

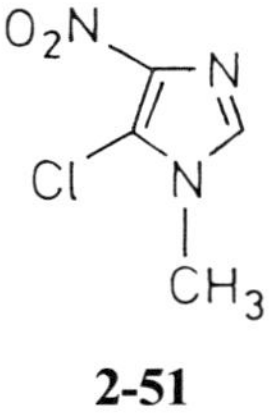

2-51

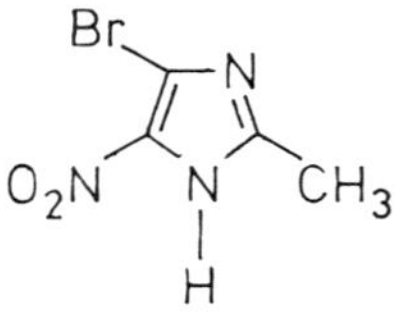

2-52

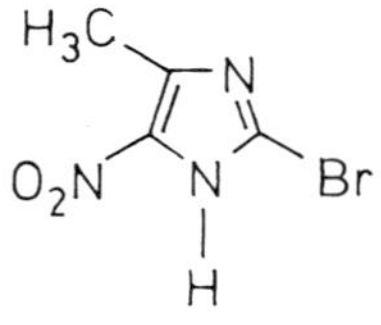

2-53

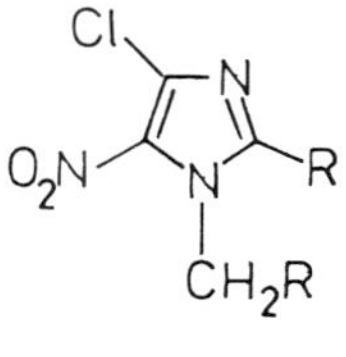

2-54
R = H, CH_3, C_2H_5,
n-C_3H_7, i-C_4H_9

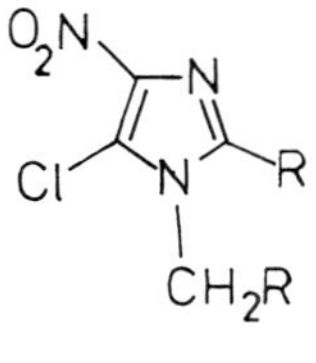

2-55
R = H, CH_3, C_2H_5,
n-C_3H_7, i-C_3H_7,
n-C_4H_9, i-C_4H_9,
n-C_5H_{11}

Unexpected reactions were encountered in the conversion of 1,2,4,5-tetraiodoimidazole and boiling nitric acid (d 1.32) to 1,2,4-triiodo-5-nitroimidazole (**2-56**) also obtained from 2,4,5-triiodoimidazole and nitric acid (d 1.32).[69] Aqueous potassium iodide reduced the N-iodo compound (**2-56**) to 2,4(5)-diiodo-5(4)-nitroimidazole (**2-57**) also obtained (55%) from 2,4,5-triiodoimidazole and mixed acid.

Attempts to nitrate 4,5-dibromoimidazole were unsuccessful.[1c]

A facile room temperature replacement of a 4(5)-iodo substituent occurred in the formation of 2-(1-methyl-1-hydroxyethyl)-4(5)-nitro-5(4)-iodoimidazole (**2-58**) by a nitrolysis of the appropriate diiodo derivative in mixed acid.[83,84]

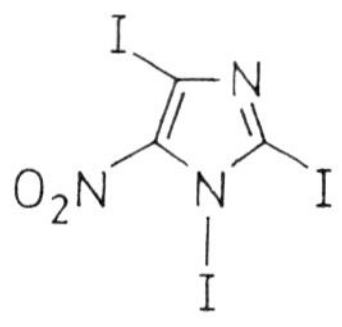

2-56 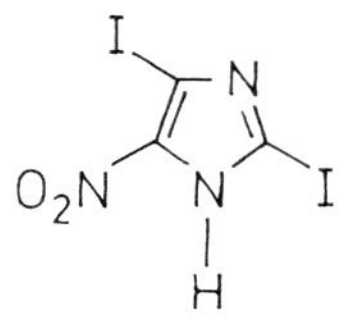   **2-57** 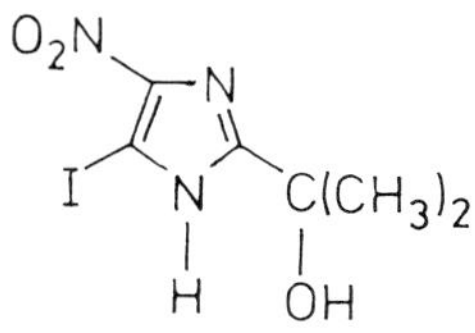   **2-58**

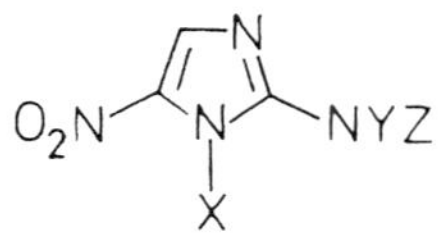

2-59a–t

	X	NYZ
a	CH_3	pyrrolidino
b	CH_3	piperidino
c	CH_3	1,2,5,6-tetrahydro-1-pyridyl
d	CH_3	hexahydroazepino
e	CH_3	octahydroazocino
f	CH_3	3-azabicyclo[3,2,2]nonyl
g	CH_3	morpholino
h	CH_3	2,6-dimethylmorpholino
i	CH_3	4-acetylpiperazino
j	CH_3	4-carbethoxypiperazino
k	CH_3	4-diethylcarbamoylpiperazino
l	CH_3	4-benzylpiperazino
m	CH_3	4-(4-chlorophenyl)piperazino
n	CH_3	4-dimethylsulfamoylpiperazino
o	CH_3	4-thiamorpholino(S,S-dioxide)
p	C_2H_5	piperidino
q	C_2H_5	pyrrolidino
r	cyclo C_6H_{11}	pyrrolidino
s	$C_6H_5CH_2$	pyrrolidino
t	CH_3	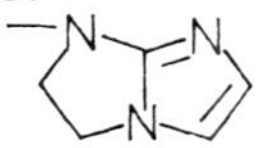 (fused bicyclic imidazoline structure)

Imidazole Amines, Ethers, and Sulfides

The 5-nitro derivatives (**2-59a–t**) (4%–32%) were obtained from twenty 1-alkyl-2-dialkylaminoimidazoles by nitration in glacial acetic acid at 0°–15°.[85]

Concentrated nitric acid in trifluoroacetic acid converted ethyl 4[6-(1-methyl-2-imidazolylthio)hexyloxy]benzoate to the 4(5)-nitroimidazole derivative (**2-60**) (8%) and the nitrophenyl derivative (**2-61**) (10%).[86]

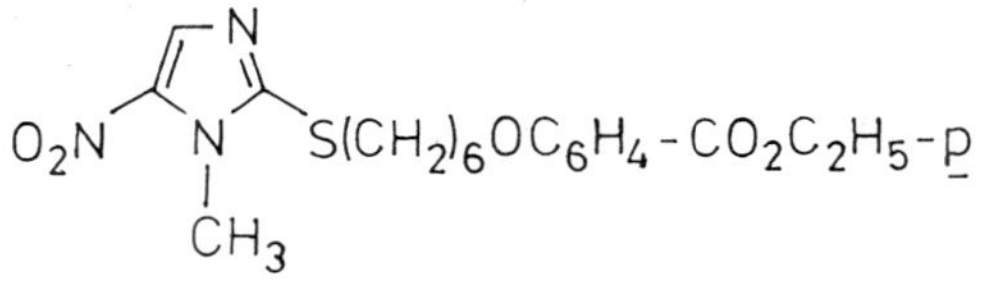

$$O_2N \underset{\underset{CH_3}{|}}{\overset{\displaystyle\frown_N}{N}} S(CH_2)_6 OC_6H_4 - CO_2C_2H_5\text{-}\underline{p}$$

2-60

$$\begin{array}{c} CO_2C_2H_5 \\ \bigcirc \!\!\!\!\!\!\!\!\!\!\!\text{-}NO_2 \\ O(CH_2)_6 SC_3H_2N_3O_2 \end{array}$$

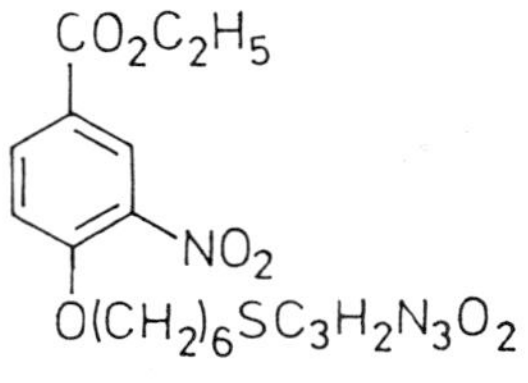

2-61

To avoid violent reaction with mixed acid Marckwald's procedure[3] with dilute nitric acid was followed to bring about nitration and oxidation of 1-alkylimidazolyl-2-sulfides to 4- and 5-nitro-1-alkylimidazolyl-2-sulfides, sulfoxides, and/or sulfones. The reaction was detailed for the conversion of 2-hexadecylthio-1-methylimidazole to its 5-nitro (**2-62**) (30%) and 4-nitro derivative (**2-63**) (1%), and to 2-hexadecylsulfinyl-1-methyl-5-nitroimidazole (**2-64**) (5%).[87]

Concentrated nitric acid at 90° nitrated 1-methyl-2-methylthioimidazole to the 5-nitro derivative.[88]

A mixture of sodium nitrite and nitric acid at 30° removed sulfur in 1-alkyl-2-sulfhydrylimidazole-5-carboxylic acid and its ethyl ester.[89]

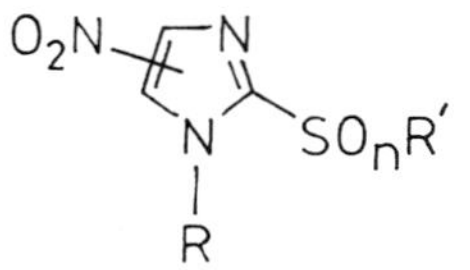

$$O_2N \underset{\underset{R}{|}}{\overset{\displaystyle\frown_N}{N}} SO_nR'$$

2-62 NO_2 at C-5, $R = CH_3$, $n = O$, $R' = (CH_2)_{15}CH_3$
2-63 NO_2 at C-4, $R = CH_3$, $n = O$, $R' = (CH_2)_{15}CH_3$
2-64 NO_2 at C-5, $R = CH_3$, $n = 1$, $R' = (CH_2)_{15}CH_3$

The nitration of alkoxyimidazoles has not been explored.

Nitronium Tetrafluoroborate

A reaction of 1-phenylimidazole in chloroform with nitronium tetrafluoroborate for 0.5 h at 25° gave 1-phenyl-5-nitroimidazole. Similar preparations afforded 5-nitroimidazoles with these 1-substituents: *p*-nitrophenyl-, 2-hydroxyethyl, 2-oxopropyl, 2-acetoxypropyl, 2-acetoxyethyl, 2-benzoyloxyethyl, and ethoxyethyl.[90]

Hot fuming nitric acid failed to nitrate 2-trideuteriomethylimidazole and instead brought about a major replacement of deuterium with hydrogen. The 4(5)-nitro derivative (**2-16**) (R = CD$_3$) (12%) was obtained by nitrating with nitronium tetrafluoroborate in sulfolane.[91]

Dinitrogen Tetroxide

Because other attempts to nitrate an imidazole at C-2 have been unsuccessful a considerable interest in the nitration of the oxime (**2-65**) of 4(5)-methyl-5(4)-formylimidazole by dinitrogen tetroxide in acetonitrile at 2°–3° followed by heating at 70° to give the 2-nitro derivative (**2-66**) can be anticipated.[92,93]

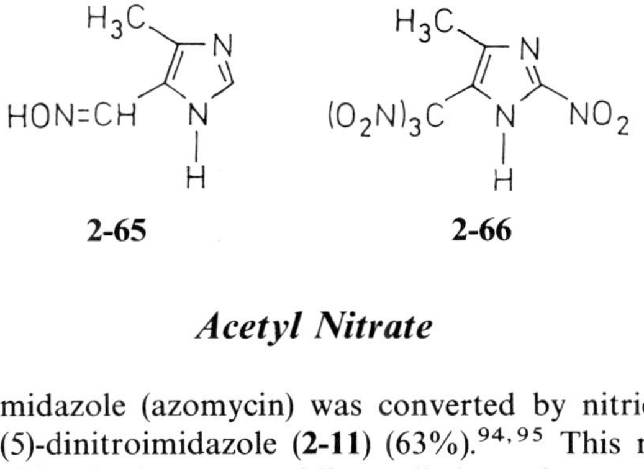

Acetyl Nitrate

In 1963, 2-nitroimidazole (azomycin) was converted by nitric acid in acetic anhydride to 2,4(5)-dinitroimidazole (**2-11**) (63%).[94,95] This new preparative nitration of the imidazole ring was rapidly applied to other examples. A variety of 2-aryl-4(5)-nitroimidazoles were similarly obtained and patented as parasiticides.[96] Since 1-methylimidazoles generally nitrated at C-4, the conversion of 2-amino-5-(1-methyl-2-imidazolyl)-1,3,4-thiadiazole to the 5-nitroimidazole derivative (**2-67**) (27%) in a reaction with nitric acid in acetic anhydride was exceptional. A similar reaction on the demethyl homolog gave a 4(5)-nitro derivative (**2-68**) (30%).[67]

In a mixture of nitric and acetic acids in acetic anhydride 2,2'-bisimidazole gave the 4(5)-nitro derivative (**2-69**) (67%) from one equivalent of nitric acid and the 4(5),4'(5')-dinitro derivative (**2-70**) (39%) from an excess of nitric acid. The mononitration was also brought about in a reaction with one equivalent of nitric acid in sulfuric acid.[97] These results confirmed corrections by Russian investigators[8] of earlier nitrations of bisimidazoles by Lehmstedt.[98]

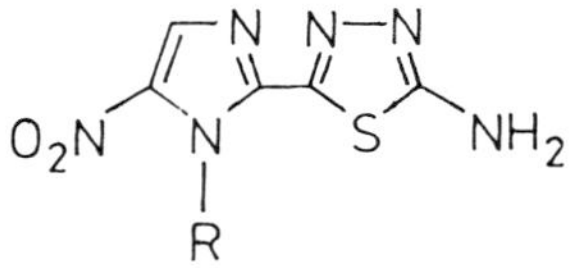

2-67 R = CH$_3$
2-68 R = H

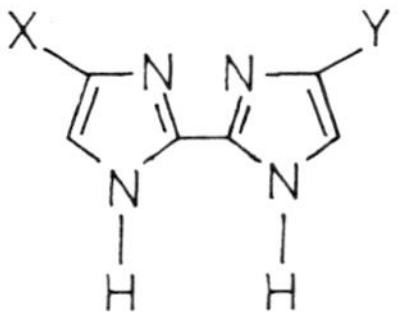

2-69 X = NO$_2$, Y = H
2-70 X = Y = NO$_2$

An unexpected formation of 1-methyl-4′(5′)-nitro-2,2′-bisimidazole (**2-71**) (49%) resulted from the treatment of 1-methylbisimidazole with the mixture of one equivalent of nitric acid in acetic acid and acetic anhydride. The expected nitration in the activated ring was not detected. Nitration with an excess of nitric acid gave 1-methyl-5,4′(5′)-dinitro-2,2′-bisimidazole (**2-72**) (40%). Similarly mononitration of 1,1′-dimethyl-2,2′-bisimidazole gave the 5-nitro derivative (**2-73**) (83%) and nitration of compound (**2-73**) afforded a mixture of the 4′-(**2-74**) (25%) and the 5′-nitro derivative (**2-75**) (12.5%).[97]

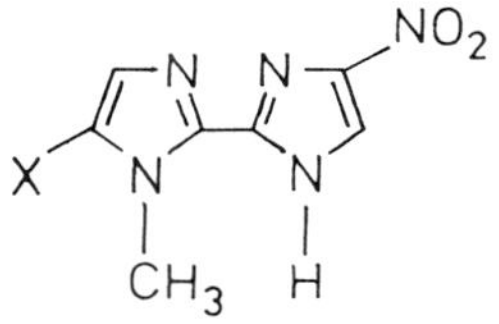

2-71 X = H
2-72 X = NO$_2$

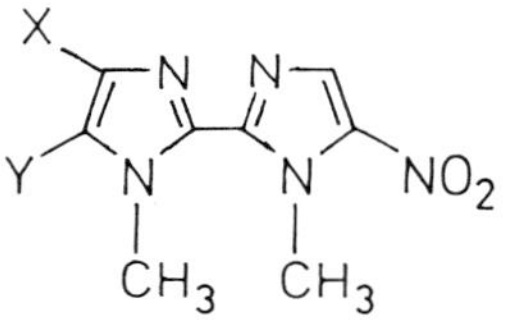

2-73 X = Y = H
2-74 X = NO$_2$, Y = H
2-75 X = H, Y = NO$_2$

Structure proofs were offered. Treatment with sulfuric acid cyclized 1-methyl-2-N-β-dimethoxyethylamidino-5-nitroimidazole (**2-76**) to 1-methyl-5-nitro-2,2′-bisimidazole (**2-77**) (89%). Nitration of compound (**2-77**) gave 1-methyl-5,4′(5′)-dinitro-2,2′-bisimidazole (**2-72**). Diazomethane converted compound (**2-72**) to a mixture of **2-74** and **2-75** which were differentiated by nmr.[97]

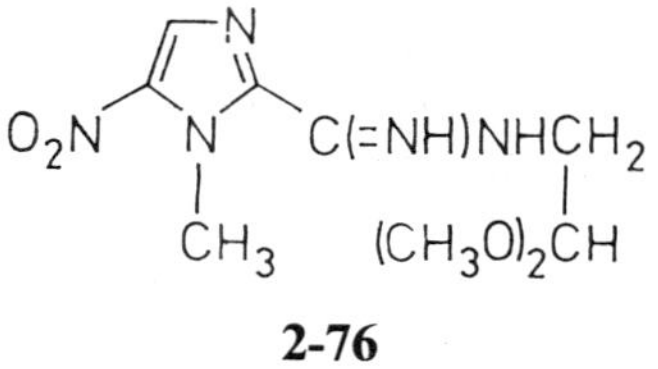

2-76

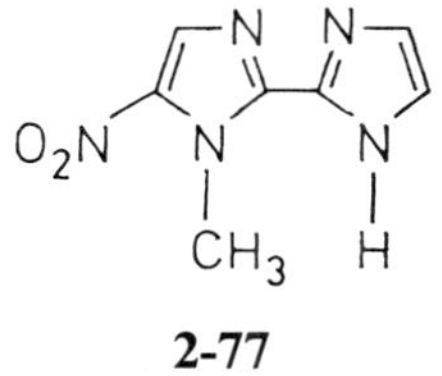

2-77

The repeated successes[16,67,96,97] in nitrating 2-arylimidazoles were not shared in attempted nitration of 1-phenylimidazole by nitric acid in acetic anhydride where the only product detected was 1-phenylimidazole nitrate.[56]

Nitrolysis of a Diazonium Group

Nearly simultaneous preparations of 2-nitroimidazole (azomycin) (**2-5**) by the replacement of a diazonium group by a nitro group were discovered in Italy,[99] Russia,[100] and the United States.[101,102] The presence of copper was not required but apparently improved the yield.

The method was rapidly extended to the preparation of 2-nitroimidazole derivatives: 1-methyl-,[103,104] 4(5)-methyl-,[104,105] 4,5-dimethyl-,[104,105] 1,5-dimethyl-,[105] 5-(1-hydroxy-1-methylethyl)-1-methyl- (**2-78**) and its methyl ether derivative **2-79**,[107,109] 5-(2-hydroxyl-1-methylethyl)-1-methyl- (**2-80**) and selected ester derivatives,[109] 5-isopropyl-1-methyl-,[108] 5-(2-chloroethyl)-1-methyl- (**2-81**),[108] 5-carboethoxy-1-methyl- (**2-82**),[110] and others with various alkyl and functionalized alkyl substituents.[111] The method was also successful in the preparation of 2-nitrobenzimidazole from the amine.[112]

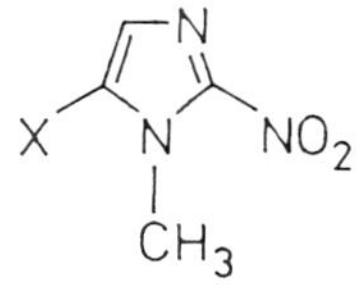

2-78 X = $C(CH_3)_2OH$
2-79 X = $C(CH_3)_2OCH_3$
2-80 X = $CH(CH_3)CH_2OH$
2-81 X = CH_2CH_2Cl
2-82 X = $CO_2C_2H_5$

A minor competitive conversion of 2-amino-4(5)-methyl-5(4)-nitroimidazole to 2-azido-4(5)-methyl-5(4)-nitroimidazole accompanied displacement of the diazonium group on treatment with sodium nitrite. It became an unexpected and unaccounted for major product in one report.[113] A suggested explanation calls for the dissociation of a triazene intermediate, Eq. (2-1).

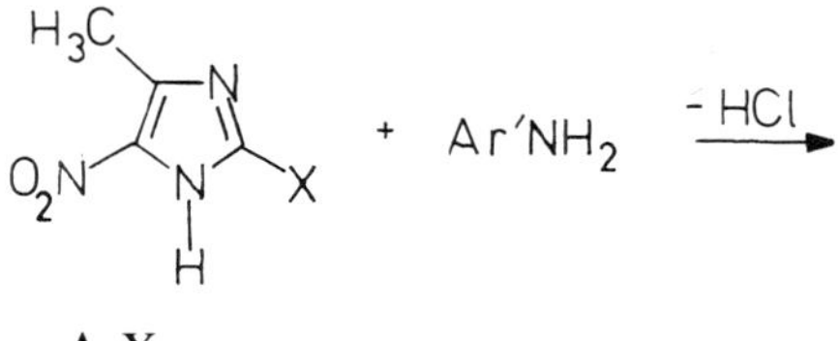

ArX
X = N_2Cl
ArN = NNHAr' → ArN_3 + Ar'H   (2-1)

Nitration by Nitrites

Amyl nitrite nitrated 4,5-diphenylimidazole and nitrolyzed 2-bromo-4,5-diphenylimidazole in unexpected preparations of 2-nitro-4,5-diphenylimidazole, Eq. (2-2); similar reactions afforded 8-nitrocaffeine (**2-1**).[114] These unexplained conversions may be related to incompletely understood redox reactions between benzene compounds and "lower level" nitrogen oxide derivatives which gave nitric oxide and a nitrophenyl compound without the intermediacy of a nitrosophenyl compound, Eq. (2-3).[115,116]

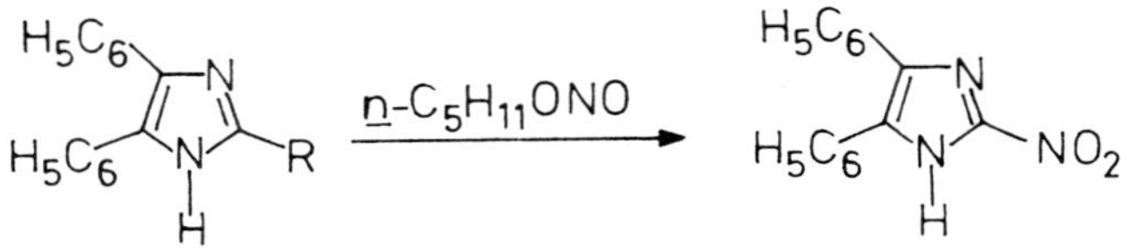

$$R = H, Br \quad (2-2)$$

$$ArH + 3HNO_2 \rightarrow ArNO_2 + 2NO + 2H_2O \quad (2-3)$$

Nitration of 2-lithioimidazoles

Preparations of 1-methyl and 1-phenyl-2-nitroiomidazoles were discovered in reactions between 2-lithioimidazoles and dinitrogen tetroxide at $-78°$ to $-110°$, Eq. (2-4).[117,118] In similar reactions an unsubstituted imidazole nitrogen atom was protected by temporary conversion to a tetrahydro-2-pyranyl or a trityl derivative.[113,119] The less acidic *n*-propyl nitrate was recommended to avoid an untimely removal of the trityl group, Eq. (2-5).[113]

In a reaction related to Eq. (2-2) amyl nitrate converted 2,4(5)-diphenylimidazole to the 5(4)-nitroso- and 5(4)-nitro derivatives (**2-83**, **2-84**) (45%),

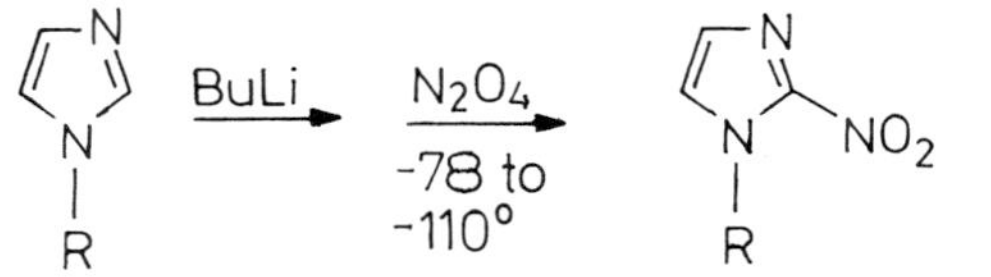

$$R = CH_3, C_6H_5 \quad (2-4)$$

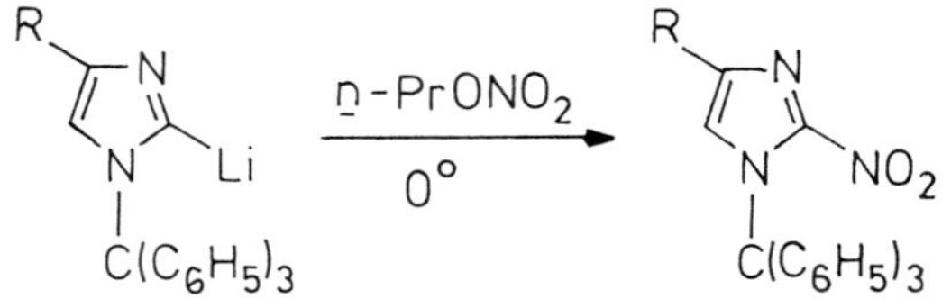

$$R = H, CH_3, CH_2OLi \quad (2-5)$$

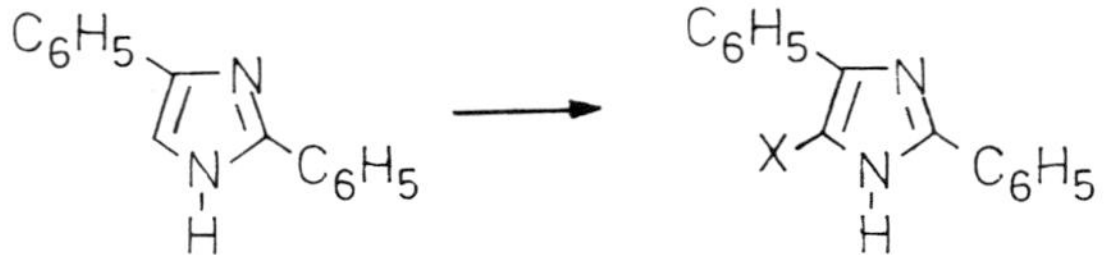

2-83 X = NO
2-84 X = NO$_2$ \quad (2-6)

Eq. (2-6).[120] Similar treatment converted the nitroso compound (2-83) to the nitro compound (2-84).

Polynitrations

In 1963 2,4(5)-dinitroimidazole (2-11), the first polynitroimidazole, was obtained from 2-nitroimidazole (2-5) and fuming nitric acid in acetic anhydride at 100°.[94] Soon thereafter mixed acid polynitration of imidazole and 2-methylimidazole was discovered.[16] The procedure called for heating the heterocycle in nitric acid (d 1.42) alone or preferably with concentrated sulfuric acid at reflux temperature for 1 h, cooling, the addition of a mixture (1:1) of nitric acid and sulfuric acids, and heating again at reflux temperature for 1 h. Imidazole was converted to its 4(5)-nitro derivative (2-4) (73%) and its 4,5-dinitro derivative (2-10); 2-methylimidazole to its 4(5)-nitro derivative (2-16) (60%) and its 4,5-dinitro derivative. Mixed acid from nitric acid (d 1.5) was recommended for the conversion of a 4(5)-nitro- to a 4,5-dinitroimidazole and 2,4(5)-dinitro- (2-11) to 2,4,5-trinitroimidazole (2-85) (26%).[16]

2-85 X = Y = NO$_2$
2-86 X = NO$_2$, Y = I
2-87 X = C$_6$H$_4$F-*p*, Y = NO$_2$

Hot nitric acid (d 1.32) converted 2,4(5)diiodo- (probably 4,5-diiodo, see Imidazole Halides), 2,4,5-triiodo, 1,2,4,5-tetraiodo-, and 2,4(5)-diiodo-5(4)-nitroimidazoles to 2,4(5)-dinitro-5(4)-iodoimidazole (2-86), and nitric acid (d 1.5) nitrolyzed the iodoimidazole (2-86) to the trinitro compound (2-85) (67%). Nitric acid (50%) converted 1,2,4,5-tetraiodoimidazole first to 1,2,4-triiodo-5-nitroimidazole (2-56) and then to the dinitroiodoimidazole (2-86). The results supported nitrolysis of a 4(5)-iodo substituent to be the easiest, followed by nitrolysis of a 2-iodo substituent, and nitrolysis of the second 4(5)-iodo substituent to be the most difficult.[69] A confirmation of the nitrolysis of 2,4,5-triiodoimidazole to 2,4,5 trinitroimidazole (2-85) (95%) by treatment with nitric acid (100%) at 100° for 1 h (product destroyed after 2 h) was reported.[121,122]

In contrast with the nitration of 2-(4-fluorophenyl)-4(5)-nitroimidazole (2-31) to 2-(3-nitro-4-fluorophenyl)-4(5)-nitroimidazole (2-32) (90%) by nitric acid (90%) in oleum (20%) at 25°, nitric acid (90%) in acetic acid at 95° afforded 4,5-dinitro-2-(4-fluorophenyl)imidazole (2-87) (80%).[63]

NITROIMIDAZOLES BY RING FORMATION

A condensation reaction between 1-piperidino-2-nitroethylene and N-chloro-N'-methylisobutyramidine in a mixture of carbon tetrachloride and pyridine at reflux temperature for 6 h gave 1-methyl-2-isopropyl-5-nitroimidazole (**2-88a**).[123] Similar reactions provided a variety in the 1- and 2-substituents.

In refluxing glacial acetic acid for 6 h 1-(5-nitro-2-thiazolyl)-2-imidazolidinone (**2-88b**) and ammonium acetate gave 1-(5-nitro-2-imidazolyl)-2-imidazolidinone (**2-88c**) (50%). A similar reaction with hydrazine in ethanol gave 1-(1-amino-5-nitro-2-imidazolyl)-2-imidazolidinone (**2-88d**).[124]

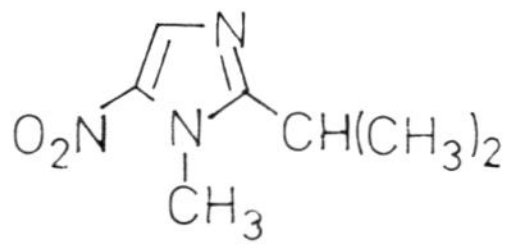

2-88a

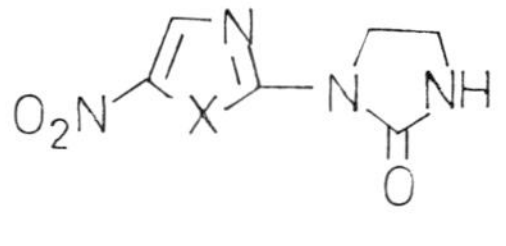

2-88b X = S
2-88c X = NH
2-88d X = NNH$_2$

OXIDATION OF 2-AMINOIMIDAZOLES

A generalization (1970) that an amino group cannot be oxidized to a nitro group by peroxytrifluoroacetic acid when it is attached to a heterocyclic ring carbon atom flanked by two heteroatoms was based on oxidation failures with melamine, 2-aminopyrimidine and 5,5'-diamino-3,3'-bi-1,2,4-oxadiazolyl.[125] Apparently 2-aminoimidazole was not oxidized by peroxytrifluoroacetic acid,[85] but Caro's acid (H$_2$SO$_5$) oxidized 2-aminoimidazoledicarboxylic acid to the 2-nitro compound, Eq. (2-7).[10,126] Also autooxidation of 2-aminobenzimidazoles gave 2-nitrobenzimidazoles and azo compounds, Eq. (2-8).[127]

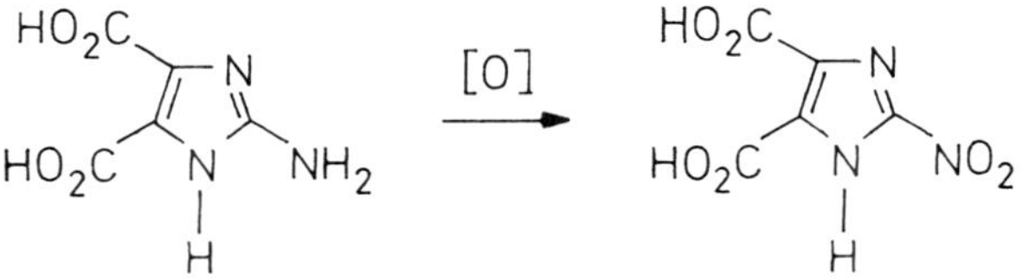

(2-7)

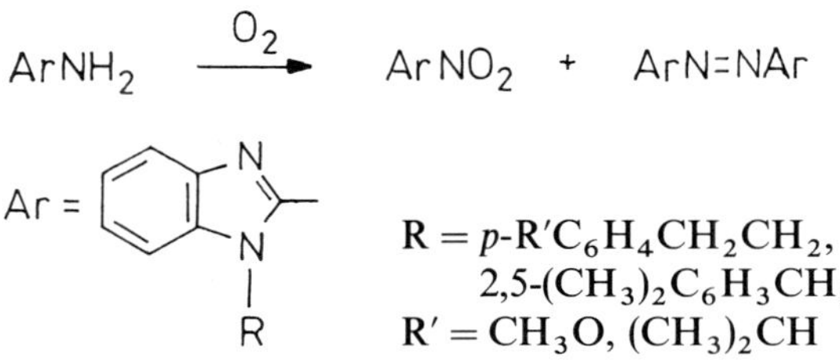

$$ArNH_2 \xrightarrow{O_2} ArNO_2 + ArN=NAr$$

Ar =

R = p-R'C$_6$H$_4$CH$_2$CH$_2$,
2,5-(CH$_3$)$_2$C$_6$H$_3$CH$_2$
R' = CH$_3$O, (CH$_3$)$_2$CH

(2-8)

BIOCHEMICAL CONVERSION OF 2-AMINOIMIDAZOLES

Conversion of 2-aminoimidazole to 2-nitroimidazole by a *Streptomyces* strain known to produce azomycin was established in the period 1966–1970.[128-131] Similar conversions to various alkyl derivatives (**2-89**) of 2-nitroimidazole are shown with yields.

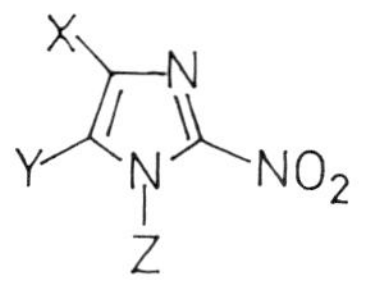

2-89

X	Y	Z	yield%
H	H	H	32
CH_3	H	H	36
C_2H_5	H	H	38
$n-C_3H_7$	H	H	30
$i-C_3H_7$	H	H	6
$n-C_4H_9$	H	H	7
CH_3	CH_3	H	8
H	H	CH_3	0

CONDENSED RING SYSTEMS

Nitration and Nitrolysis

Imidazo[2,1-*b*]thiazoles

At room temperature and below mixed acids converted 6-chloro- and 6-arylimidazo[2,1-*b*]thiazoles to the 5-nitro derivatives (**2-90**) (87%) and (**2-91a–d**) (62%–96%).[132-134] When the *p*-position was blocked mononitration occurred at the 5-position; when both positions were blocked nitration was directed to the 2-position in the thiazole ring.

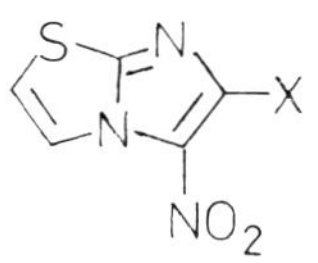

2-90 X = Cl
2-91a X = p-BrC_6H_4
2-91b X = p-ClC_6H_4
2-91c X = p-$CH_3C_6H_4$
2-91d X = p-$NO_2C_6H_4$

Nitrosation by sodium nitrite in acetic acid converted 6-phenylimidazo-[2,1-*b*]thiazole to the 5-nitroso derivative, or to the 6-*p*-nitrosophenyl derivative when the 5-position was blocked; when both the 5- and *p*-positions were blocked nitration occurred in the thiazole ring to give the 2-nitro derivative.[135]

Similar treatment mono- and dinitrated 2-phenylimidazo[2,1-*b*]-benzothiazole to give 2-*p*-nitrophenyl-3-nitro[2,1-*b*]benzothiazole (**2-92**).[134]

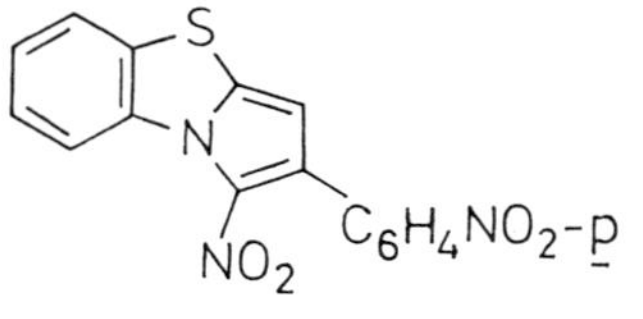

2-92

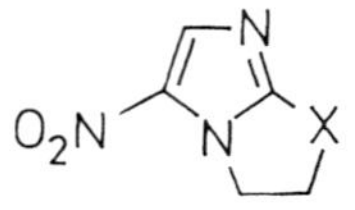

2-93a X = S
2-93b X = NCH$_3$

2,3-Dihydroimidazo[2,1-*b*]thiazole was nitrated by concentrated nitric acid in a nitrogen atmosphere at 80° for 40 min to the 5-nitro derivative (**2-93a**) (20%). The product was oxidized to a sulfone by monoperpthalic acid.[39]

Imidazo[1,2-*a*]imidazoles

Mononitration (66%) of 1,6-dimethyl-1*H*-imidazo[1,2-*a*]imidazole in concentrated sulfuric acid at −20° was brought about by treatment with ethyl nitrate. Two equivalents of ethyl nitrate afforded 1,6-dimethyl-2(3),5-dinitro-1*H*-imidazo[1,2-*a*]imidazole (20%), Eq. (2-9a).[136]

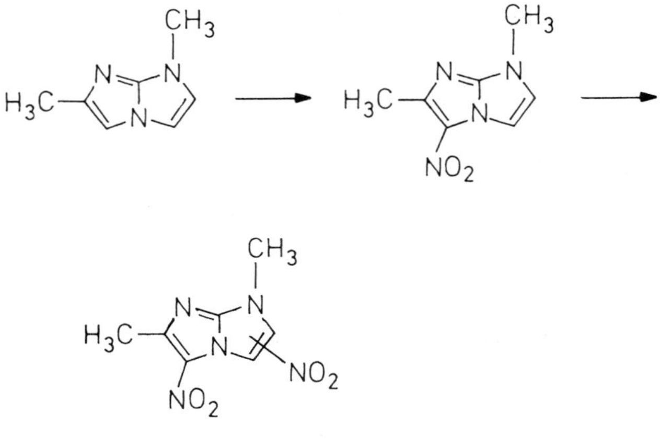

(2-9a)

The isolated product from a nitration of 5,6-dihydro-7-methyl-7*H*-imidazo[1,2-*a*]imidazole by nitric acid in acetic acid at 0° for 1 h was assumed to be the 3-nitro derivative (**2-93b**) (12%).[39]

Imidazo[1,2-*a*]pyridines

Mixed acid at 25° for 20–30 min converted imidazo[1,2-*a*]pyridine and its 2-chloro- derivative to 3-nitro derivatives (**2-94**) (84%) and (**2-95**) (65%).[137] Similarly nitric acid in trifluoroacetic acid at 25° converted 2-phenylimidazo[1,2-*a*]pyridine to the 3-nitro derivative (30%).[138] The 3-nitro derivative of methylimidazo[1,2-*a*]pyridine-5-carboxylate was also produced in a nitration with mixed acids at 25°. Further treatment with nitric acid brought about ring degradation to give the pyridinium dinitro methylide, Eq. (2-9b) and the nitroimidazole ring was restored on treatment with hot dimethyl formamide (DMF).[139] Frontier MO and CNDO/2 calculations agreed with these exclusive electrophilic substitutions at the 3-position[140-142] and correlated with the X-ray structure determinations of the 5-ethoxy-, 8-methyl-2-carboethoxy-, and 6-methyl-3-nitro-2-carboethoxy derivatives (**2-96**–**2-98**) of the heterocycle.[143] On nitration, compound **2-96** gave the 8-nitro- (**2-99**) (1%) and the 3,6- (**2-100**) (45%) and the 3,8-dinitro- (**2-101**) (40%) derivatives; compound **2-97** gave the 3-nitro (**2-102**) (70%) and the 5-nitro- (**2-103**) (30%) derivatives; and compound **2-98** (80%) was obtained from ethyl 6-methylimidazo[1,2,*a*]pyridine-2-carboxylate (**2-104**) (a lesser amount (20%) of the 3,5-dinitro derivative (**2-105**) was detected). Mononitration at the 3-position gave the 2-carboethoxy-, 6-chloro-, 6-methyl, and 8-methyl derivatives (**2-106**–**2-109**); however, ethyl 6-chloroimidazo[1,2,*a*]pyridine-2-carboxylate gave both 3- and 5-nitro derivatives (**2-110**) (98%) and (**2-111**) (2%).[144,145] A variety of ether, sulfide, amino, and nitrile

2-94, 3-NO_2
2-95, 2-Cl-3-NO_2
2-96, 5-OC_2H_5
2-97, 2-$CO_2C_2H_5$-8-CH_3
2-98, 2-$CO_2C_2H_5$-3-NO_2-6-CH_3
2-99, 5-OC_2H_5-8-NO_2
2-100, 3-NO_2-5-OC_2H_5-6-NO_2
2-101, 3-NO_2-5-OC_2H_5-8-NO_2
2-102, 2-$CO_2C_2H_5$-3-NO_2-8-CH_3
2-103, 2-$CO_2C_2H_5$-5-NO_2-8-CH_3
2-104, 2-$CO_2C_2H_5$-6-CH_3
2-105, 2-$CO_2C_2H_5$-3-NO_2-5-NO_2-6-CH_3
2-106, 2-$CO_2C_2H_5$-3-NO_2
2-107, 3-NO_2-6-Cl
2-108, 3-NO_2-6-CH_3
2-109, 3-NO_2-8-CH_3
2-110, 2-$CO_2C_2H_5$-3-NO_2-6-Cl
2-111, 2-$CO_2C_2H_5$-5-NO_2-6-Cl

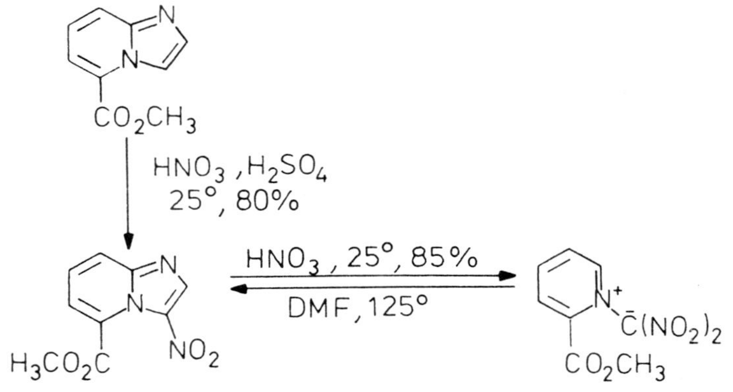

derivatives of the heterocycle (**2-94**) were similarly obtained by nitration.[146–148]

The nitration of 2-(2-furyl)imidazo[1,2,*a*]pyridine (**2-112**) and related compounds has been investigated in Russia.[149–153] Product formation depended on the nitrating agent, the order of mixing the reactants, the substituents already present, and steric factors. Just as electrophilic attack led to substitution at the 3-position in the free base and bromination led to substitution in the 5-furyl position of the conjugate acid and substitution in both positions occurred when an excess of reagent was present, mixed acid [from 1 mole of nitric acid (70%)] afforded 2-(5-nitro-2-furyl)imidazo[1,2,*a*]pyridine (**2-113**) (15%) and, with an excess of nitric acid, 3-nitro-2-(5-nitro-2-furyl)imidazo[1,2,*a*]-pyridine (**2-114**) (71%).[149–152] Nitric acid in phosphoric acid also produced the mononitro derivative (**2-113**).[149]

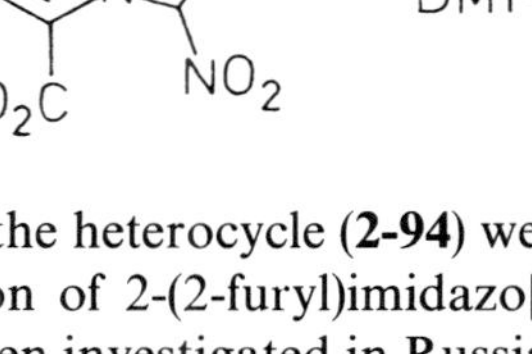

2-112 X = Y = H
2-113 X = H, Y = NO$_2$
2-114 X = Y = NO$_2$
2-115 X = Y = SO$_3$H

When the heterocycle (**2-112**) was added to the mixed acid (reverse order) the dinitro derivative (**2-114**) was produced even in the presence of one equivalent of nitric acid. Similar circumstances led to the dinitration of 2-phenyl-imidazo[1,2,*a*]pyridine. Mixed acid converted the disulfonic acid (**2-115**) to the dinitro compound (**2-114**) (30%) and oxidized the furan ring.[149]

Fuming nitric acid in chloroform or methanol converted the heterocycle (**2-112**) to its anhydrous mononitrate salt, and to compounds **2-113** and **2-114**.[151] Although other furylazoles were nitrated in the furan ring by nitric acid in acetic anhydride the heterocycle (**2-112**) was converted to intractable material presumably brought about by oxidation.[149]

Instead of further nitration mixed acid oxidized the dinitro compound (**2-114**), to a mixture of the unstable *cis*-isomer, isolated as the lactol (**2-116a**), and the stable *trans* modification (**2-116b**) of 3-nitro-2-(3-carboxyacryloyl)imidazo-

[1,2-*a*]pyridine. Nitrolysis of the 3,5'-dibromo derivative of compound **2-112**, as well as the corresponding diiodide gave the dinitro compound (**2-114**).[152]

A complicated reaction between mixed acid and 2-(5-methyl-2-furyl)imidazo[1,2,*a*]pyridine (**2-117**) was believed to give nitro polymers.[152,153]

Mixed acid nitration of 2[α-chloro-β(5-nitro-2-furyl)-vinyl]imidazo[1,2,*a*]pyridine gave the dinitro derivative (**2-118**) (24%) and the chloronitro derivative (**2-119**) (10%).[153]

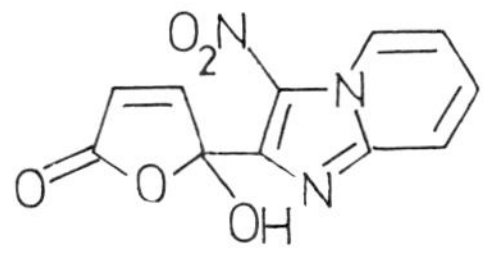

2-116a

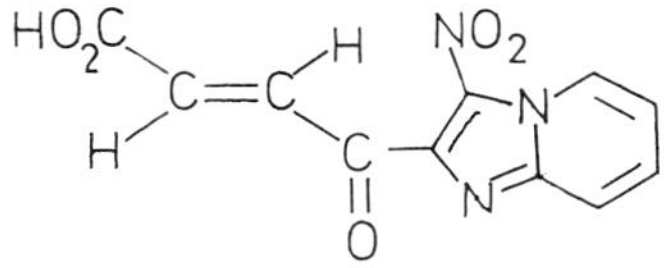

2-116b

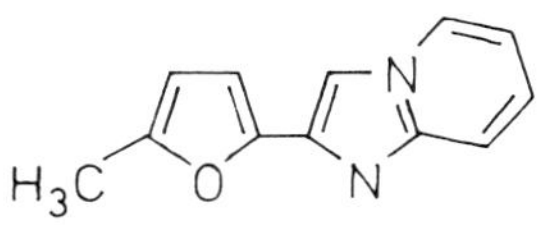

2-117

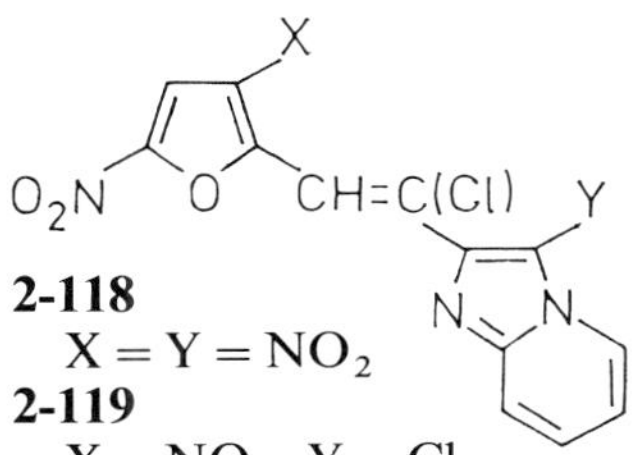

2-118
X = Y = NO₂

2-119
X = NO₂, Y = Cl

Mixed acid at 25° for 30 min converted the 5,5'-dehydrodimer of imidazo-[1,2-a]pyridine to the expected 3,3'-dinitro derivative (**2-120**) (96%).[142]

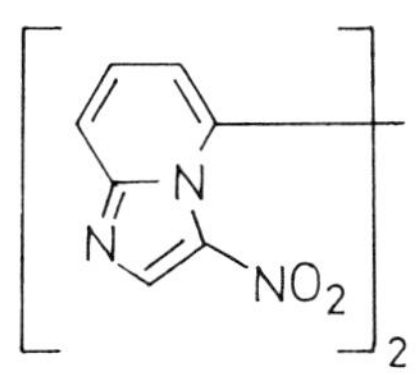

2-120

Imidazo[1,5-*a*]pyridines

Nitration of the bisulfate salts of the parent heterocycle and its 1-phenyl derivative by mixed acid in acetic acid gave 1-nitroimidazo[1,5-*a*]pyridine (**2-121**) (21%) and 1-phenyl-3-nitroimidazo[1,5-*a*]pyridine (**2-124**) (24%). Similar nitration of the appropriate free bases afforded 1-nitro-3-methyl- and 3-phenyl derivatives (**2-122**) (34%) and (**2-123**) (33%) and the 1-methyl-3-nitro derivatives (**2-125**) (20%).[154]

Mixed acid in acetic acid converted the 1-phenyl and the 1-phenyl-3-nitro

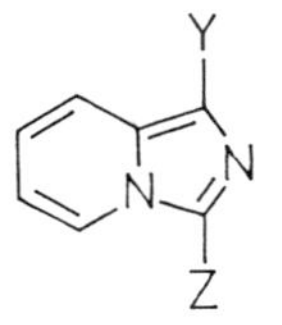

2-121, Y = NO$_2$, Z = H
2-122, Y = NO$_2$, Z = CH$_3$
2-123, Y = NO$_2$, Z = C$_6$H$_5$
2-124, Y = C$_6$H$_5$, Z = NO$_2$
2-125, Y = CH$_3$, Z = NO$_2$
2-126, Y = C$_6$H$_4$NO$_2$-p, Z = NO$_2$
2-127, Y = Z = NO$_2$

derivatives of the heterocycle to the dinitro compound (**2-126**) (22% and 70%) upon warming for a few minutes; similar treatment afforded the dinitro heterocycle (**2-127**) (28%) from the 1-nitro derivative (**2-121**).[154]

Formation of the nitroheterocycles (**2-122**, **2-123**) was accompanied with an oxidative ring-opening and cyclization to the oxadiazoles (**2-128**). An explanation (Eq. 2-10) required the generation of a nitrosating agent by reduction of nitric acid.[154]

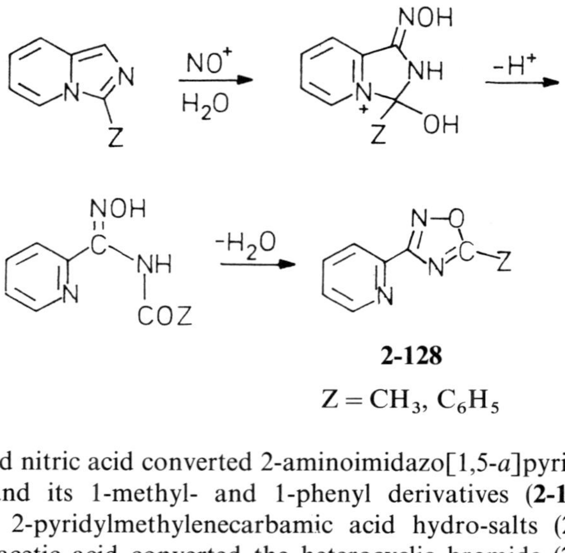

2-128

Z = CH$_3$, C$_6$H$_5$ (2-10)

Concentrated nitric acid converted 2-aminoimidazo[1,5-*a*]pyridinium bromide (**2-129a**) and its 1-methyl- and 1-phenyl derivatives (**2-129b,c**) to the corresponding 2-pyridylmethylenecarbamic acid hydro-salts (**2-130a–c**) but nitric acid in acetic acid converted the heterocyclic bromide (**2-129a**) to the 1-nitro-3-bromo derivative (**2-131**) (49%) and converted the 3-methyl- and the 3-phenyl derivatives (**2-129d,e**) to the 3-methyl and 3-phenyl derivatives (**2-132**)

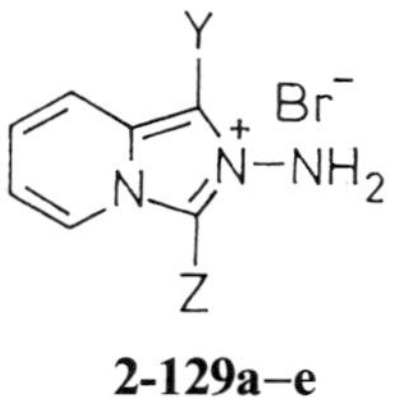

a, Y = Z = H
b, Y = CH$_3$, Z = H
c, Y = C$_6$H$_5$, Z = H
d, Y = H, Z = CH$_3$
e, Y = H, Z = C$_6$H$_5$
f, Y = Br, Z = CH$_3$

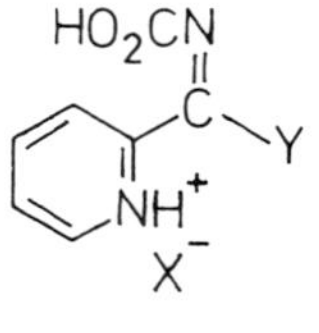

2-129a–e   **2-130a–c**

(38%) and (**2-133**) (41%) of 1-nitroimidazo[1,5-*a*]pyridine. An insufficiency of nitric acid in acetic acid converted the methyl derivative (**2-129d**) to the 1-bromo derivative (**2-129f**). The results were compatible with a sequence of (1) nitric acid oxidation of the bromide anion to bromine, (2) bromination at C-1, (3) oxidative deamination, and (4) nitrolysis.[155]

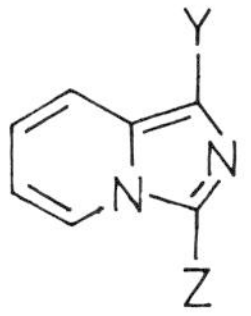

2-131, Y = NO$_2$, Z = Br
2-132, Y = NO$_2$, Z = CH$_3$
2-133, Y = NO$_2$, Z = C$_6$H$_5$
2-134, Y = Z = Br
2-135, Y = Br, Z = CH$_3$

The sequence was supported by the unreactivity of 3-methylimidazo-[1,5-*a*]pyridinium *p*-toluenesulfonate in acetic acid to concentrated nitric acid in the absence of added bromide anion and by the independent nitrolyses of compounds **2-134** and **2-135** to the 1-nitro compounds (**2-131** and **2-132**).[155]

Imidazo[1,2-*b*]pyridazines

A mixture of benzoyl chloride and silver nitrate in chloroform at 25° for 40 days converted 2-methyl-6-phenyl-imidazo[1,2-*b*]pyridazine-1-oxide to the 3-nitro derivative (**2-136**) (11%). Mixed acid nitration at 25° for two days afforded the derivatives **2-137a** (20%) and **2-137b** (13%).[156] A similar preparation gave compound **2-138**.[157]

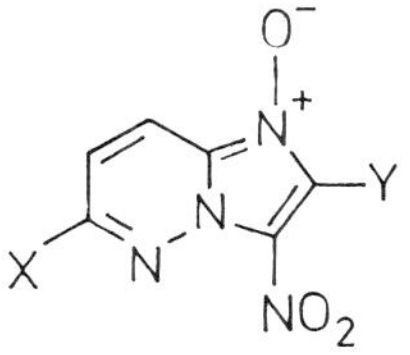

2-136, X = C$_6$H$_5$, Y = CH$_3$
2-137a, X = CH$_3$, Y = *p*-O$_2$NC$_6$H$_4$
2-137b, X = *m*-O$_2$NC$_6$H$_4$, Y = CH$_3$
2-138, X = Cl, Y = H

Imidazo[1,2-*a*]pyrimidines

A product from nitrating 5,6,7,8,-tetrahydro-8-methylimidazo[1,2-*a*]pyrimidine by nitric acid in acetic acid at 0° for 1 h was assumed to be the 3-nitro derivative (**2-139**) (8%).[39]

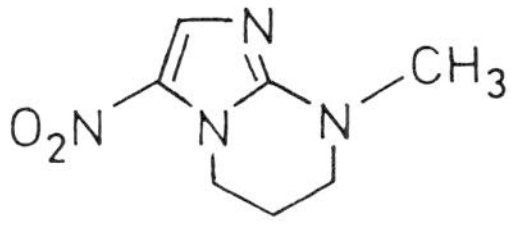

2-139

Imidazo[4,5-*d*]pyrimidines (purines)

Caffeine,[2,158] theobromine,[159,160] and theophylline[161-163] gave the 8-nitropurines (**2-1, 2-140,** and **2-141a**) on treatment with nitric acid.[164] Similar nitrations of 7-benzyl- and 7-cyanobenzyltheophylline were reported.[163] Although xanthine did not react, both 9-methyl and 8-sulfhydryl-9-methylxanthine were converted to 8-nitro-9-methylxanthine (**2-141b**) on heating in nitric acid (50%). Other 9-alkyl-8-thiopurines were similarly converted to 8-nitropurines.[165,166]

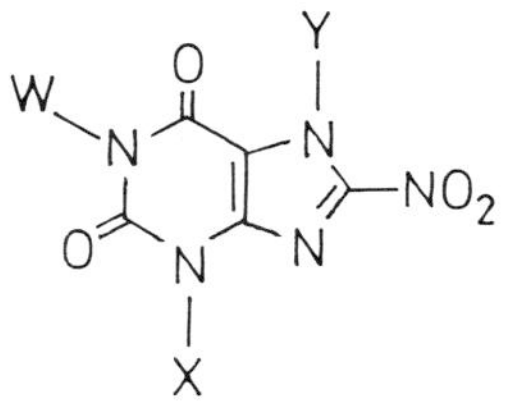

2-140, W = H, X = Y = CH_3
2-141a, W = X = CH_3, Y = H
2-141b, W = X = H, Y = CH_3

Imidazo[1,2-*a*]benzimidazoles

In refluxing dimethylformamide for 1 h sodium nitrite converted 2-phenyl-3-bromo-9-methylimidazo[1,2-*a*]benzimidazole to the corresponding 3-nitroheterocycle (**2-142**) (80%).[167] Bromine lability was sufficiently lower when 2-methyl replaced the 2-phenyl substituent to cause the reaction with sodium nitrite to fail; however, the desired nitration of 2,9-dimethylimidazo[1,2-*a*]benzimidazole to the 3-nitro derivative (**2-143**) (70%) was brought about by potassium nitrate in concentrated sulfuric acid at room temperature for 1 h. A higher yield (88%) was realized when the mononitrate salt of the 2,9-dimethyl base was stored in concentrated sulfuric acid at −5° to −10° for 1 h.[168]

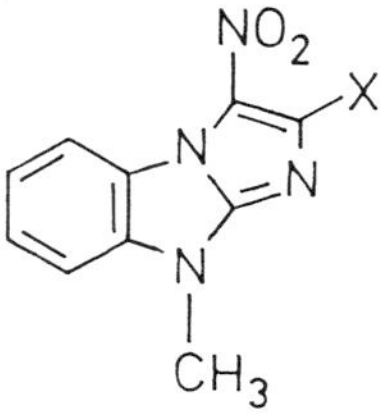

2-142, X = C_6H_5
2-143, X = CH_3
2-144a, X = p-BrC_6H_4
2-144b, X = p-$O_2NC_6H_4$

In appropriate reactions 2-(*p*-bromophenyl)-3-nitro-9-methylimidazo[1,2-*a*]-benzimidazole (**2-144a**) was obtained from the parent base by potassium nitrate in sulfuric acid or from the 3-bromo derivative and potassium nitrite in yields of 87% and 65%. By a similar exchange of a labile bromine atom potassium nitrite afforded 2-*p*-nitrophenyl-3-nitro-9-methylimidazo[1,2-*a*]-benzimidazole (**2-144b**).[168]

Dinitration of 2-phenyl-9-methylimidazo[1,2-*a*]benzimidazole by potassium nitrate (one equivalent) in sulfuric acid gave a product (47%) of unknown structure. The yield became nearly quantitative with two equivalents of potassium nitrate. Since the same unknown product was obtained from 2-*p*-nitrophenyl-9-methylimidazo[1,2-*a*]benzimidazole and potassium nitrate in sulfuric acid the location of a *p*-nitro substituent in the unknown product was established. Apparently the second nitro group was located in the benzimidazole ring since the product was different from the dinitro compound (**2-144b**).[168]

Imidazo[1,2-*f*]xanthines

Nitric acid (d 1.5) in acetic acid or sulfuric acid or potassium nitrate in sulfuric acid converted seven 6,8-dimethylimidazo[1,2-*f*]xanthines to the 3-nitro derivatives (**2-145a–g**) in yields from 70% to 90%. Potassium nitrate in sulfuric acid also converted the 3-nitro derivative (**2-145d**) to the dinitro derivative (**2-145f**) (90%) and converted mononitrated (**2-145e**) to the dinitro derivative (**2-145g**) (93%).[169]

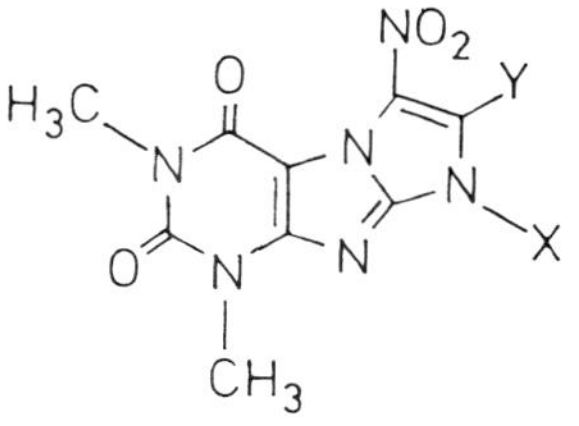

a, $X = Y = H$
b, $X = n\text{-}C_4H_9$, $Y = CH_3$
c, $X = \text{cyclo-}C_6H_{11}$, $Y = CH_3$
d, $X = CH_3$, $Y = C_6H_5$
e, $X = n\text{-}C_4H_9$, $Y = C_6H_5$
f, $X = CH_3$, $Y = p\text{-}O_2NC_6H_5$
g, $X = n\text{-}C_4H_9$, $Y = p\text{-}O_2NC_6H_5$

2-145

Ring Formation

Imidazo[5,1-*b*]thiazoles

Nitro derivatives of this lesser known ring system were obtained when 4-[4(5)-nitroimidazolyl-5(4)thioacetyl]pyridine hydrobromide was heated at reflux temperature in trifluoroacetic anhydride for six days to produce a mixture of 7-nitro-2-trifluoroacetyl-3-(4-pyridyl)imidazo[5,1-*b*]thiazole (**2-146a**), 3-trifluoromethyl-7-nitroimidazo[5,1-*b*]thiazole (**2-146b**), and 7-nitro-3-(4-pyridyl)imidazo[5,1-*b*]thiazole (**2-146c**).[170]

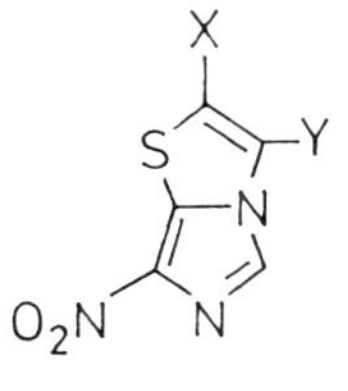

2-146a, $X = COCF_3$, $Y = \gamma\text{-}C_5H_4N$
2-146b, $X = H$, $Y = CF_3$
2-146c, $X = H$, $Y = \gamma\text{-}C_5H_4N$

Imidazo[1,2-*a*]pyrimidines

Both an imidazo[1,2-*a*]pyrimidine derivative (**2-147a**) and a 1,3-diazepine derivative (**2-148**) (minor amount) were produced by a reaction between a dialkylacetylenedicarboxylate and 2-aminoimidazole;[171] but 7-carbomethoxy-1-methyl-2-nitro-4,5-dihydroimidazo[1,2-a]pyrimidin-5-one **(2-147b)** (50%) without an azepine co-product was obtained from 1-methyl-2-amino-5-nitroimidazole and dimethylacetylenedicarboxylate.[172]

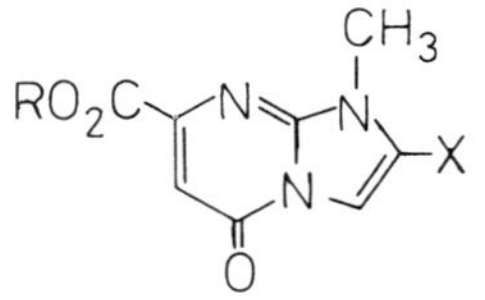

2-147a R = CH$_3$, C$_2$H$_5$
X = H
2-147b R = CH$_3$, C$_2$H$_5$
X = NO$_2$

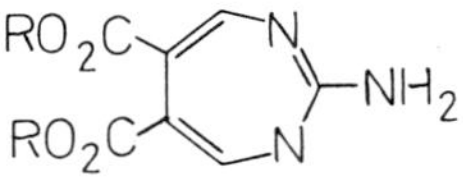

2-148
R = CH$_3$, C$_2$H$_5$

Cyclization of diethyl N-(1-methyl-5-nitroimidazol-2-yl)-aminomethyl-enemalonate in acetic anhydride heated at reflux temperature for 16 h gave the 6-carboethoxy derivative **(2-149)**, a position isomer of **2-147b** (R = C$_2$H$_5$), Eq. (2-11).[172]

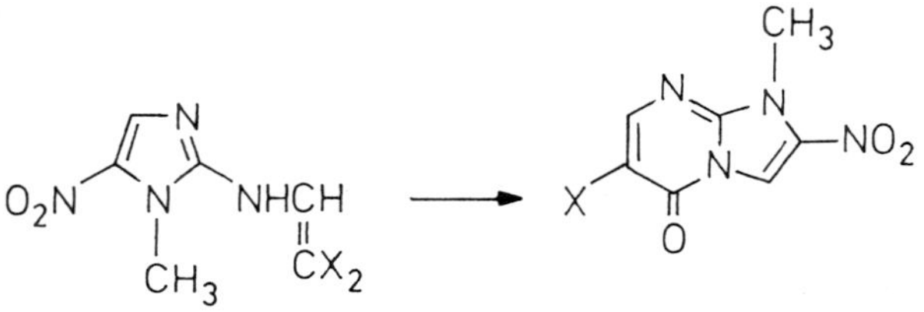

2-149

X = CO$_2$C$_2$H$_5$ (2-11)

2,3-Dihydroimidazo[2,1-b]oxazoles

In the preparation of 1-(2-hydroxyethyl)-2,4-dinitroimidazole **(2-150a)** (50%) from ethylene oxide and 2,4(5)-dinitroimidazole in ethanol heated at the reflux temperature for 48 h a by-product of 5-nitro-2,3-dihydroimidazo[2,1-b)oxazole **(2-151a)** (18%) was obtained. The structure of the latter was confirmed by X-ray, ir, nmr, and mass spectroscopy. Propylene oxide afforded the expected homologs **(2-150b)** (53%) and **(2-151b)** (23%); epichlorohydrin gave the chlorides **(2-150c)** (15%), **(2-151c)** (25%), and the 5-nitro isomer **(2-152y)** (37%); 1,2-epoxy-3-methoxypropane gave the methyl ethers **(2-150d)** (14%), **(2-151d)** (26%), and **(2-152z)** (35%). At 25° 2,4(5)-dinitroimidazole and epichlorohydrin or 1,2-epoxy-3-methoxypropane without a solvent gradually became homogenous after 96 and 48 h and gave higher yields of **2-150c** (58%) and **2-150d** (51%).[95, 173, 174]

It was discovered that ethanol solutions of the alcohol **(2-150b,c,d)** and the

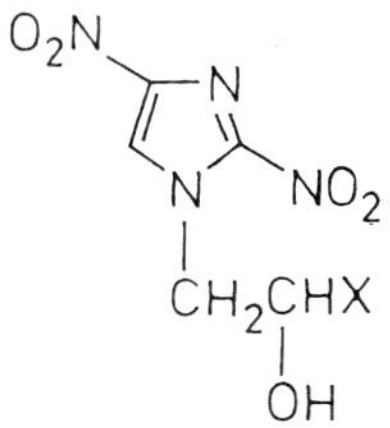

2-150a, X = H
2-150b, X = CH$_3$
2-150c, X = Cl
2-150d, X = OCH$_3$

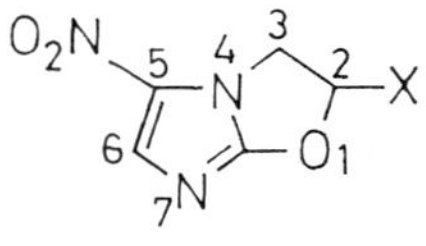

2-151a, X = H
2-151b, X = CH$_3$
2-151c, X = Cl
2-151d, X = OCH$_3$

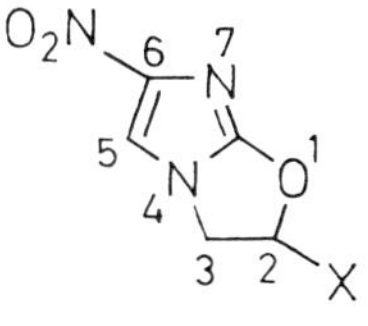

2-152x, X = CH$_3$
2-152y, X = Cl
2-152z, X = OCH$_3$

appropriate epoxide precursor heated at reflux brought about cyclization to the 6-nitro compound (**2-152x,y,z**). Apparently an intramolecular nucleophilic displacement of the 2-nitro group provided a reasonable explanation for conversions of 2,4(5)-dinitroimidazole to the **2-152** compounds via the **2-150** alcohols.[130,131] Alkylation of the conjugate base of 2,4(5)-dinitroimidazole was involved to explain the formation of the **2-151** products and, when appropriate, the co-products **2-152y,z**, Eq. (2-12).[173,174]

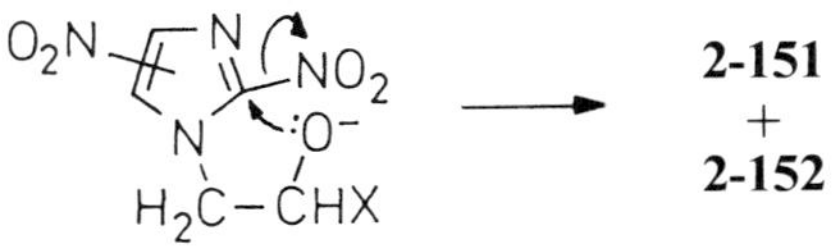

$$\text{2-151} + \text{2-152} \tag{2-12}$$

There is an alternative explanation based on a dialkylated intermediate in which the two N-alkyl moieties become equivalent, Eq. (2-13).

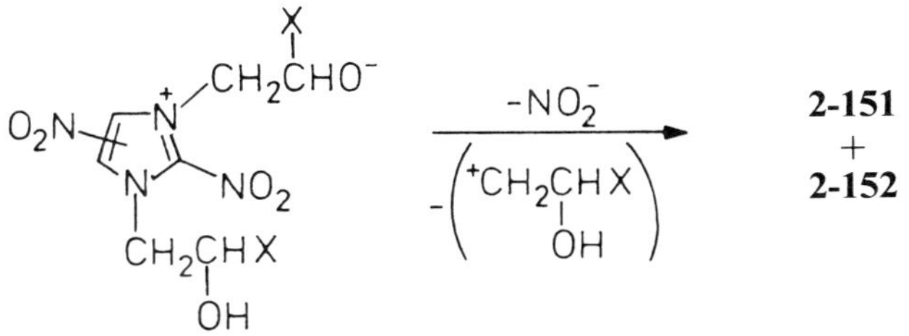

$$\text{2-151} + \text{2-152} \tag{2-13}$$

The two explanations anticipate different mechanisms for the alkylation of 4(5)-nitroimidazole: S_E2cB for alkaline solutions with an intermediate imidazole anion and a slower S_E2' path in a neutral or an acidic environment. A dialkylquaternary imidazolium salt was produced in the latter but not the former reaction. This consideration tends to favor Eq. (2-13) over Eq. (2-12) as the expected pathway. Another example of a thermolysis of an imidazole quaternary salt is discussed elsewhere.

Di- and tetrahydroimidazo[1,5-c]-1,2,4-triazines

Hydrazine hydrate and 1-β-bromoethyl-2-methyl-4-nitro-5-bromoimidazole in methanol at 160°–180° in a sealed tube for 5 h gave 1,2,3,4-tetrahydro-6-methyl-8-nitroimidazo[1,5-c]-1,2,4-triazine (60%), Eq. (2-14). The related compound, 1,2,3,4-tetrahydro-3-oxo-6-methyl-8-nitroimidazo[1,5-c]-1,2,4-triazine was obtained in 78% yield from a condensation between hydrazine hydrate and 1-carbethoxymethyl-2-methyl-4-nitro-5-bromoimidazole in dimethylformamide containing pyridine at reflux temperature for 2 h, (Eq. (2-15).[175]

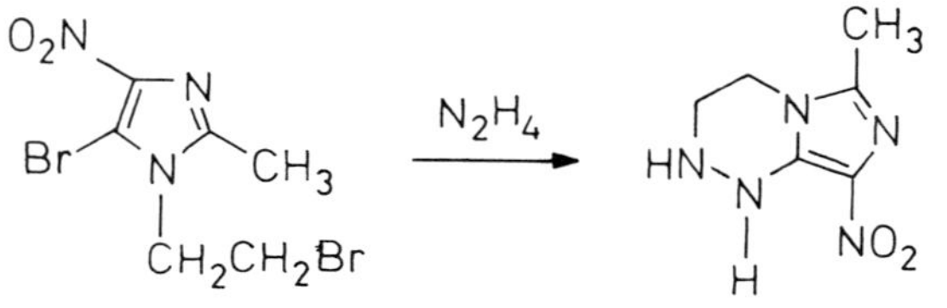

(2-14)

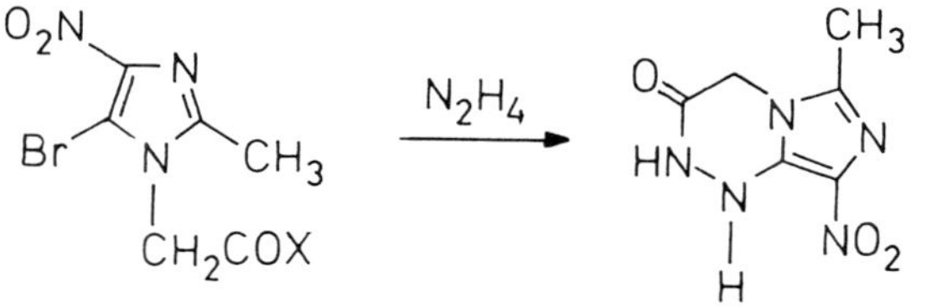

(2-15)

X = OC₂H₅

A reaction similar to Eq. (2-15) for ketones (X = C₆H₅) gave 1,4-dihydroimidazo[1,5-c]-1,2,4-triazines.[176]

Imidazo[2,1-d]-1,4-oxazepines

See REACTIONS OF NITROIMIDAZOLES for examples of this ring system.

Imidazo[2,1-a]isoquinolines

When heated at 200° for 20 min 2-(β-o-chlorophenylethyl)-4(5)-nitroimidazole gave 3-nitro-5,6-dihydroimidazo[2,1-a]isoquinoline, Eq. (2-16).[96]

$$X = CH_2CH_2C_6H_4Cl\text{-}o \qquad (2\text{-}16)$$

9H-Imidazo[1,2-a]indoles

In a reaction similar to Eq. (2-16) 2-o-chlorobenzyl-4(5)-nitroimidazole at 152° for 3 min gave 3-nitro-9H-imidazo[1,2-a]indole, Eq. (2-17) (erroneously reported as 3-nitroimidazo[2,1-a]isoindole). An uncertainty about the reported 3,7(or 8)dinitroimidazo[2,1-a]isoindole cannot be as easily resolved; probably it should be described as 3,x-dinitro-9H-imidazo[1,2-a]indole, Eq. (2-17).[96]

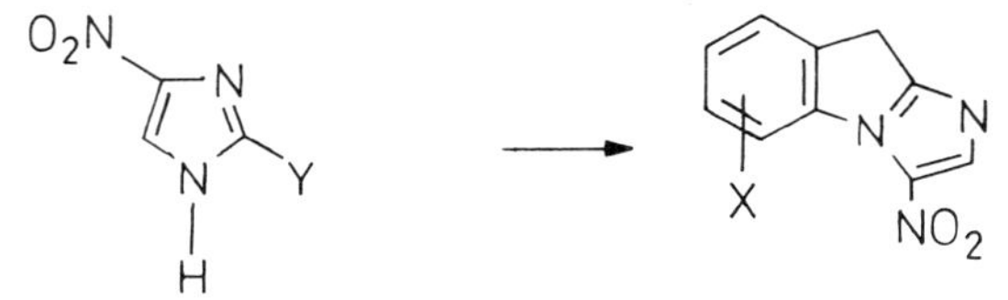

$$Y = CH_2C_6H_3(o\text{-}Cl)X$$
$$X = H, NO_2 \qquad (2\text{-}17)$$

REACTIONS OF NITROIMIDAZOLES

Halogenation

Electrophilic substitution at carbon in the 2- and 5(4)-positions has been established for reactions between nitroimidazoles and bromine. The calculated amount of bromine added dropwise to 4(5)-nitroimidazole in dilute sodium hydroxide at 50° gave 2,5(4)-dibromo-4(5)-nitroimidazole (75%).[177] In an aqueous solution bromine converted 4(5)-methyl-5(4)-nitroimidazole to 2-bromo-4(5)-methyl-5(4)-nitroimidazole (**2-153**).[1f, 6]

A curious difference in reactivity between 4-nitro-1,5-dimethyl- and 5-nitro-1,4-dimethylimidazoles was discovered when bromine readily combined with the former to produce 2-bromo-4-nitro-1,5-dimethylimidazole (**2-154**) and the latter was unreactive.[1f] Similar conversions gave 2-alkyl-4(5)-nitro-5(4)-bromo and 1,2-dialkyl-4-nitro-5-bromoimidazoles (**2-155a,b** and **2-156a–d**).[178-189] On occasion bromination in sodium hydroxide was recommended for the preparation of the monoalkyl derivatives (**2-155**).[182, 183] The yields were good: **2-155a**, 59%; **2-155b**, 95%; **2-156a**, 72%; **2-156b**, 92%; **2-156c**, 82%; **2-156d**, 68%.[180, 182]

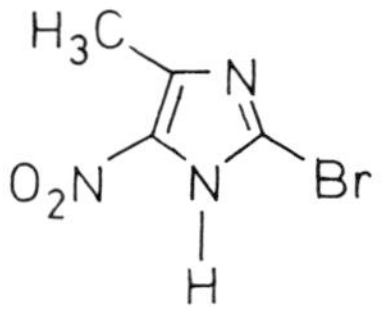

2-153

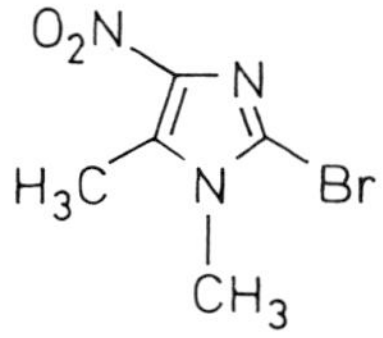

2-154

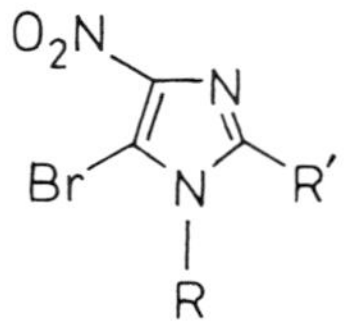

2-155a, R = H, R = CH$_3$
2-155b, R = H, R' = C$_2$H$_5$
2-156a, R = CH$_2$CH(OH)CH$_2$OH, R' = CH$_3$
2-156b, R = CH$_2$CH$_2$OH, R' = CH$_3$
2-156c, R = R' = CH$_3$
2-156d, R = C$_6$H$_5$CH$_2$, R' = CH$_3$

A lower reactivity at the C-4 position toward electrophilic substitution was discovered in the inability of 1-alkyl-2-methyl-5-nitroimidazoles to react with bromine under comparable conditions for the formation of compounds **2-156b–d**.[183] The formation of 2,4(5)-dibromo-5(4)-nitroimidazole (**2-157**) from 5(4)-nitro-4(5)-imidazolecarboxylic acid and bromine in aqueous alkali at 0°–30° for 20 min may be similar to the Hunsdiecker reaction.[190]

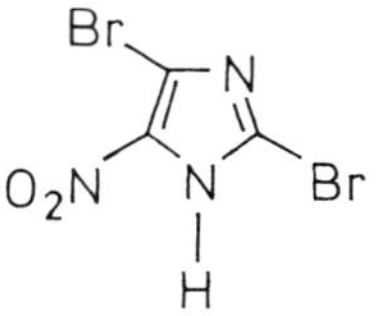

2-157

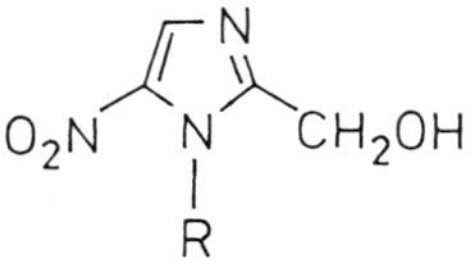

2-158
a R = CH$_3$
b R = CH$_2$CH$_2$
         |
         OH

Bromine in dimethylformamide at 60°–65° for 2 h converted 1-benzyl-4-nitroimidazoles to their 5-bromo derivatives (24%–69%).[180,189]

C-Alkylation

An incomplete account[1s] of a reaction between 1-methyl-5-nitroimidazole and formaldehyde was confirmed in patent reports that the heterocycle and paraformaldehyde in dimethylsulfoxide in a sealed tube at 110° for 24 h gave 1-methyl-2-hydroxymethyl-5-nitroimidazole (**2-158a**) (70%).[191–195] Similar reactions with acetaldehyde and benzaldehyde were claimed. On heating at 140°

for 6 h a mixture of aqueous formaldehyde (40%) and the heterocycle gave the hydroxymethyl derivative (**2-158a**) but in a lower yield of 30%. This latter method applied to 1-(2-hydroxyethyl)-5-nitroimidazole produced the desired 2-hydroxymethyl derivative (**2-158b**) (20%); a quantitative yield was obtained when the reaction was carried out with paraformaldehyde in dimethylsulfoxide at 130° for 48 h.

Mono- and disubstituted Mannich bases (**2-159a,b**) were obtained from 2-nitroimidazole.[196] In contrast attempts to hydroxymethylate 2-nitro and 4(5)-nitroimidazoles were unsuccessful.[195]

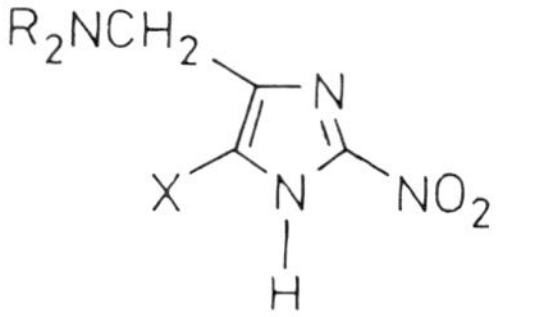

2-159a, X = H
2-159b, X = CH_2NR_2

C-Acylation

Aroyl chlorides in the presence of triethylamine converted 1-methyl-5-nitroimidazole to its 2-aroyl derivative (**2-160a**) (28% to 86%); in addition to the expected product (**2-160a**) p-nitrobenzoyl chloride gave a minor product (**2-160b**). Methyloxalyl chloride also gave two products (**2-160c,d**).[197]

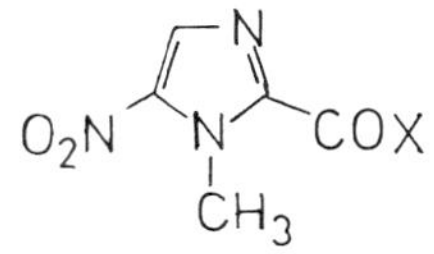

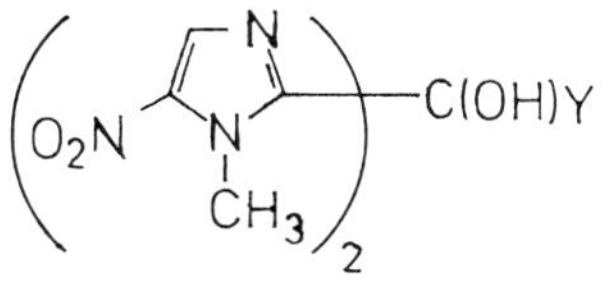

2-160a X = C_6H_5,
α-C_4H_3S,
p-ClC_6H_4,
p-$O_2NC_6H_4$
2-160c X = CO_2CH_3

2-160b Y = p-$O_2NC_6H_4$
2-160d Y = CO_2CH_3

N-Alkylation, Amination, and Arylation

Alkylations

Selective alkylation of 4(5)-nitroimidazoles has shown a dependency on the alkalinity of the medium. Alkyl halides and sulfates gave predominantly 1-alkyl-4-nitroimidazoles when a base was present, and 1-alkyl-5-nitroimidazoles when the medium was neutral or acidic.[10] There have been numerous reports in support of this generality.[1h-j, 19, 20, 32, 33, 66, 198-204] Alkylation of a nitroimidazole to form a nitronate ester has not been reported.

N-Methylation of 4(5)-nitroimidazole (**2-4**) by either a heterogeneous or a homogeneous reaction depended on the pH of the solution. Under preparative conditions the reaction with methyl sulfate (equimolar with the nitroimidazole) in aqueous sodium hydroxide (2.5 M) consumed 83% of the heterocycle (**2-4**) to give 1-methyl-4-nitroimidazole (**2-14**) (53%) and 1-methyl-5-nitroimidazole (**2-15**) (47%).[1j] On the other hand, a preparative reaction in boiling water for 30 min consumed 88% of the nitroimidazole (**2-4**) to give the methylated derivatives (**2-14** and **2-15**) in a ratio of 1:350.[1h]

A kinetic study revealed two paths for the reaction in dilute homogenous solutions. In dilute aqueous sodium hydroxide (0.09 M) containing ethanol (10%) at 25° a homogeneous S_EcB reaction between methyl sulfate (large excess) and 4(5)-nitroimidazole (**2-4**) gave the products **2-14** and **2-15** in a ratio of 89:11,[205,206] in accordance with methylation at the more nucleophilic nitrogen atom (further from the nitro groups) in the nitroimidazole anion.

A slow S_E2' reaction with methyl sulfate (0.1 M) and the nitroimidazole (**2-4**) (*ca*. 0.006 M) in anhydrous formic acid containing sodium formate (1.0 M) at 50° consumed about 10% of the heterocycle (**2-4**) to produce 1-methyl-5-nitroimidazole (**2-15**) and its 3-methyl quaternary derivative (**2-161a**).[205,206] A transition between the two mechanisms was expected to occur somewhere in the range of pH 6–11.[206] The formation of 1-alkyl-5-nitroimidazoles was also favored in strongly acidic media.

2-161a **2-161b**

Orientation in N-methylation (S_E2') of a 4(5)-nitroimidazole was considered to be dependent on protomeric tautomerization in the substrate, Eq. (2-18). It was shown that the relative concentrations of neutral **2-4a** and **2-4b** gave a ratio of about 400:1;[154–156] should similar bonding stabilization persist in the transition states, Eqs. (2-19, 20), then cation **2-4c** would predominate over **2-4d** and favor the formation of product **2-15** after proton removal by base. A transition-state energy difference greater than 2 kcal/mol (25°) could afford the regiospecific alkylation. Although methylation and related reactions were found to be in qualitative agreement with this correlation of ratios of tautomers and of alkylated products, quantitative aspects were neither expected nor observed.[207]

Methylation of 4(5)-nitroimidazole (1 of 15 protomeric ambident systems studied) by methyl fluorosulfonate (5–10 molar excess) in methylene chloride at 25° for 1–2 hours gave the methylated product **2-15** (71%) after treatment with aqueous sodium hydroxide (1 N). The intermediate quaternary salt (**2-161b**) precipitated on formation and was characterized by nmr.[207]

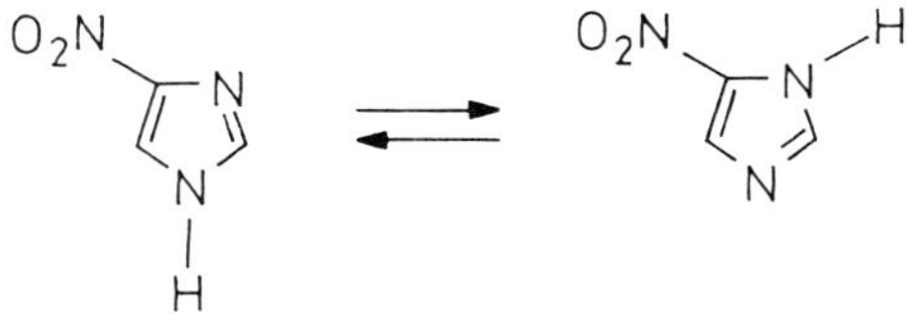

$$\begin{array}{c}\text{(Structure: 4-nitroimidazole with N-H)} \rightleftarrows \text{(Structure: 4-nitroimidazole tautomer)}\end{array} \qquad (2\text{-}18)$$

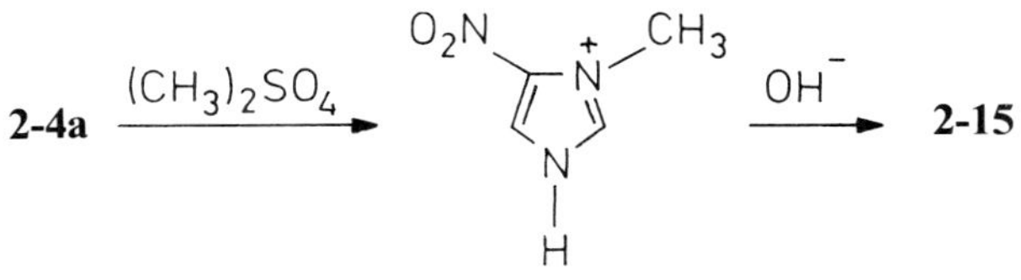

$$\mathbf{2\text{-}4a} \xrightarrow{(CH_3)_2SO_4} \mathbf{2\text{-}4c} \xrightarrow{OH^-} \mathbf{2\text{-}15} \qquad (2\text{-}19)$$

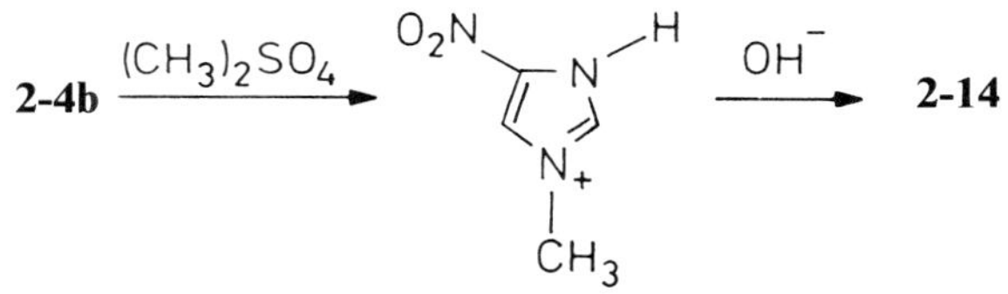

$$\mathbf{2\text{-}4b} \xrightarrow{(CH_3)_2SO_4} \mathbf{2\text{-}4d} \xrightarrow{OH^-} \mathbf{2\text{-}14} \qquad (2\text{-}20)$$

Dimethyl sulfate in the presence of sodium hydroxide converted 4(5)-iodo-5(4)-nitroimidazole to a mixture (72%) of 1-methyl-4-nitro-5-iodoimidazole (**2-162**) and 1-methyl-4-iodo-5-nitroimidazole (**2-163**) in the ratio of 13:1; as expected, **2-163** (50%) and only a trace of **2-162** were produced when the reaction, in the absence of base, was carried out in dioxan heated at reflux temperature for one hour.[200]

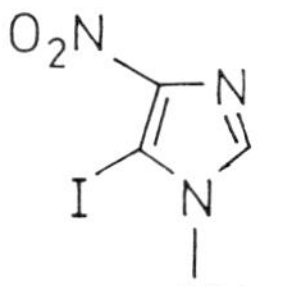

2-162 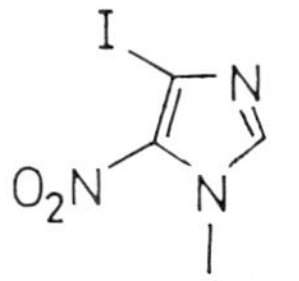   **2-163**

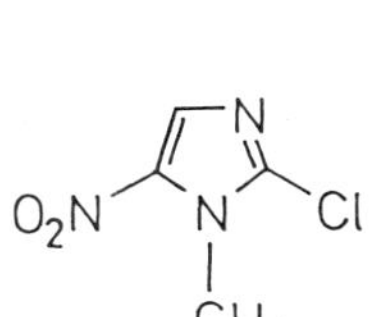

2-164 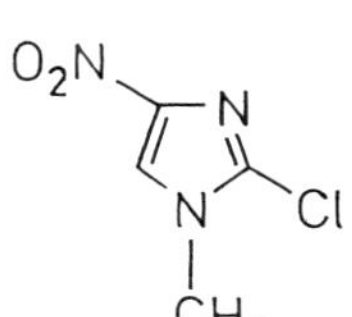   **2-165**

The presence of benzyltrimethylammonium chloride as a phase transfer catalyst promoted a reaction between 2-chloro-4(5)-nitroimidazole and dimethylsulfate in toluene at 125° for 8 h to give a mixture (40%) of 2-chloro-1-methyl-5-nitroimidazole (**2-164**) and 2-chloro-1-methyl-4-nitroimidazole (**2-165**) in a ratio of 5:1. A similar mixture (60%) was provided by methylation with diazomethane to form **2-164** and **2-165** in the ratio 3:2.[70]

Methylation of 2,4(5)-dinitroimidazole by methyl iodide in the presence of benzyltrimethylammonium chloride gave 1-methyl-2,4-dinitroimidazole (**2-166**) (30%) and minor amounts of 2-chloro- and 2-iodo-4-nitroimidazoles. The formation of the minor products was attributed to reactions of chloride and iodide anions with either of the two 2-nitroimidazole derivatives.[70]

Diazomethane methylated the dinitroimidazole (**2-11**) to a mixture (50%) of 1-methyl-2,5-dinitroimidazole (**2-167**) and its isomer (**2-166**) in nearly equal amounts.[70]

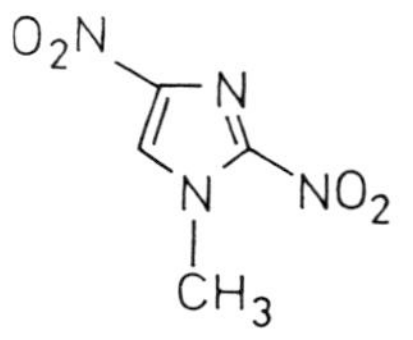

2-166

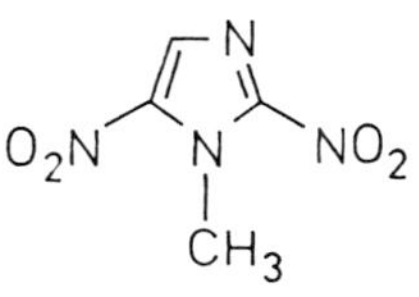

2-167

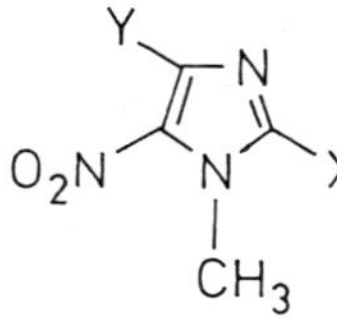

2-168 X = H, CH_3, $(CH_3)_2CH$, n-C_8H_{17}
        Y = H
2-169 X = H
        Y = CH_3

Other N-methylations have been brought about by treating a nitroimidazole with sulfuryl chloride in methanol,[208] methyl pyrophosphate at 140°–160°,[209] and methyl sulfate in either formic or acetic acid. By the latter method a group of 1-methyl-2-alkyl-5-nitroimidazoles (**2-168** and **2-169**) was obtained.[210]

Attempts to methylate at the amide nitrogen atom in the imidazole (**2-170**) led to the discovery of an unusual reaction. Although the desired product (**2-171**) was obtained in low yield the predominant reaction required cycloaddition of diazomethane to the C-4–C-5 double bond to give imidazo[4,5-c]pyrazoles (**2-173**, **2-174**) and other products (**2-172**, **2-175**–**2-177**). A rationalization for the product formations was offered, Eq. (2-21a,b).[211]

An additional product (**2-178**) was occasionally detected. Its formation was

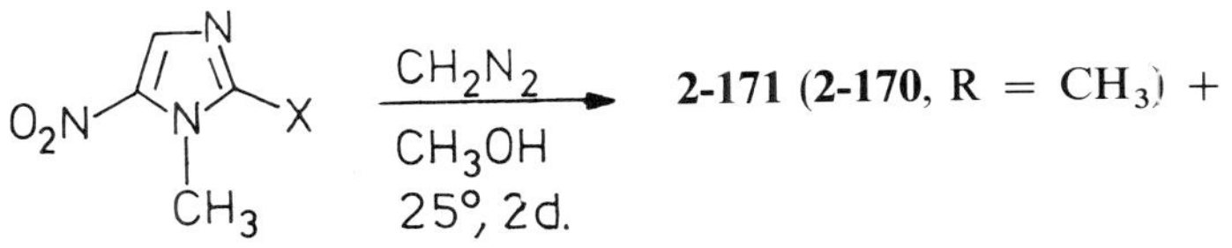

$$\underset{\underset{2\text{-}170}{}}{\underset{\overset{|}{CH_3}}{O_2N-\text{imidazole}-X}} \xrightarrow[\underset{25°,\ 2d.}{CH_3OH}]{CH_2N_2} \mathbf{2\text{-}171}\ (\mathbf{2\text{-}170},\ R = CH_3) +$$

X = N(R)COCHCl$_2$
R = H

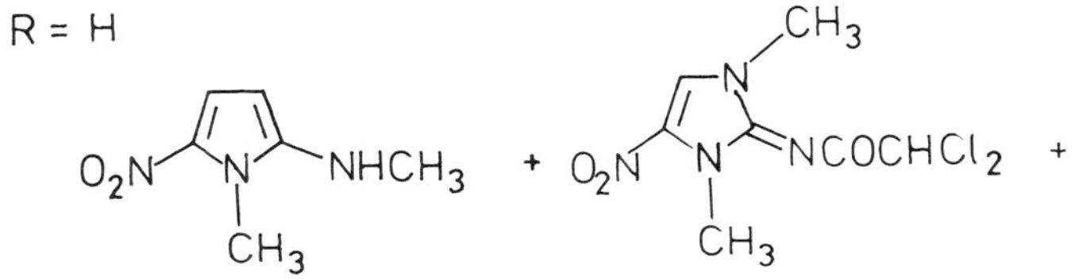

R = H

2-172 $\quad O_2N\text{-pyrrole-}NHCH_3$ (N-CH$_3$) $\quad + \quad$ **2-175** $\quad O_2N\text{-imidazole-}NCOCHCl_2$ (N-CH$_3$, N-CH$_3$) $\quad +$

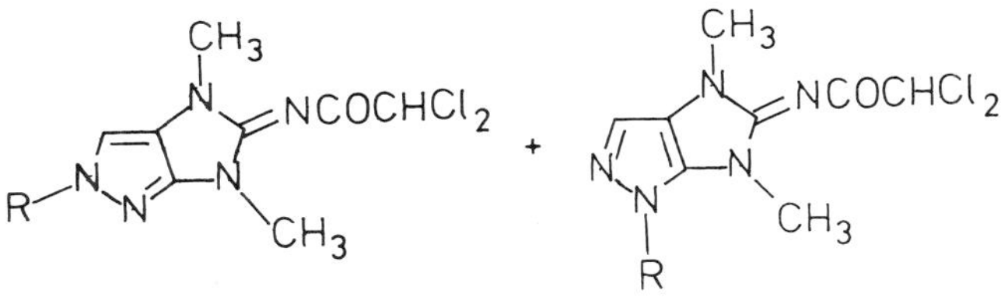

2-173, R = CH$_3$
2-176, R = H

2-174, R = CH$_3$
2-177, R = H

(2-21a)

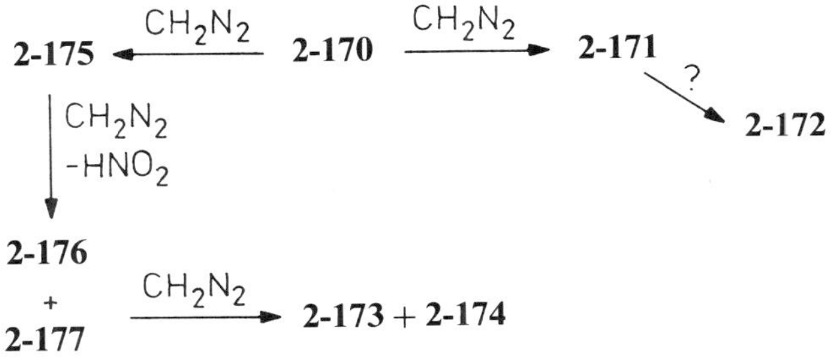

$$\mathbf{2\text{-}175} \xleftarrow{CH_2N_2} \mathbf{2\text{-}170} \xrightarrow{CH_2N_2} \mathbf{2\text{-}171}$$

$$\Big\downarrow \begin{array}{c} CH_2N_2 \\ -HNO_2 \end{array} \qquad\qquad\qquad \searrow \mathbf{2\text{-}172}$$

$$\mathbf{2\text{-}176}$$
$$+ \xrightarrow{CH_2N_2} \mathbf{2\text{-}173} + \mathbf{2\text{-}174} \qquad\qquad (2\text{-}21b)$$
$$\mathbf{2\text{-}177}$$

explained by a sequence initiated by a cycloaddition between the intermediate (**2-175**) and diazomethane, Eq. (2-22).[211]

Ethylation with ethyl orthoformate in the presence of trifluoroacetic acid converted the dinitroimidazole (**2-11**) to a mixture (90%) of 1-ethyl-2,4-dinitroimidazole (**2-179**) and 1-ethyl-2,5-dinitroimidazole (**2-180**) in a ratio of 8:1.[70]

Ethylation with diazoethane converted the dinitroimidazole (**2-11**) to the expected N-ethyl derivatives (**2-179**) (24%) and (**2-180**) (18%) and trace amounts

2-178 (2-22)

2-179

2-180

of 1,5-diethyl-2,4-dinitroimidazole (**2-182**). Formation of the latter product was attributed to the intermediacy of a dipolar adduct (**2-181**) followed by a loss of nitrogen.[70] This formation of compound **2-182** may be related to the methylation by diazomethane of 1,3-dinitro-2-methylpyrrole to give its 5-methyl derivative (Chapter 1).

2-181

2-182

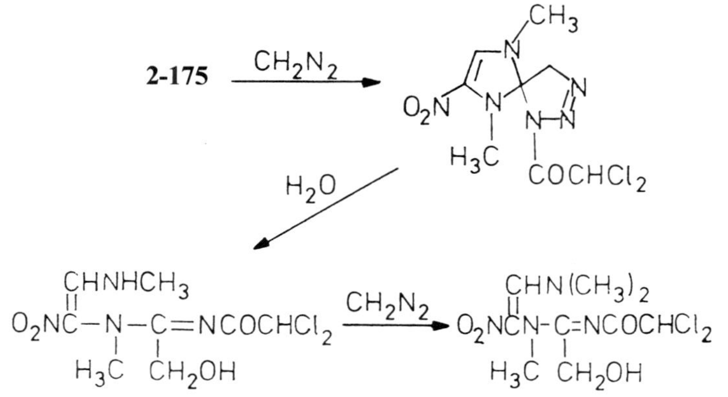

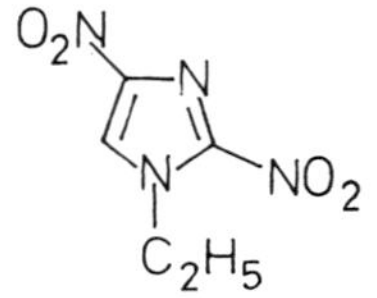

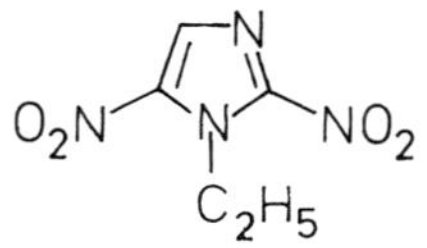

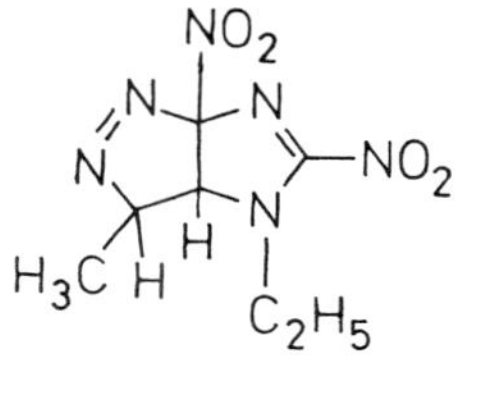

N-alkyl substituents, C_nH_{2n+1} with $n > 1$, are less frequently encountered. A series of 1-alkyl-2-nitroimidazoles (**2-183**) was obtained by treating 2-nitroimidazole or its sodium salt with appropriate alkylating agents.[212,213]

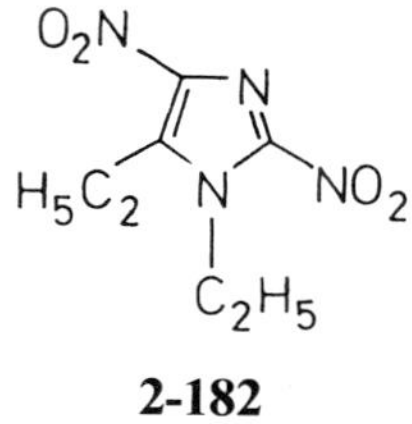

Aminations

O-Diphenylphosphinylhydroxylamine (from H_2NOH and $(C_6H_5)_2POCl$) aminated the sodium-salt of 2-nitroimidazole to give after workup 1-amino-2-

nitroimidazole (40%). A similar treatment afforded 1-amino-2-methyl-4-nitroimidazole (30%).[214]

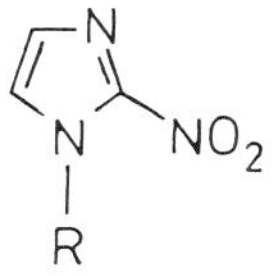

2-183

Arylations

An interesting preparation of a pyrazolylimidazole was discovered in the reaction between 4(5)-nitroimidazole and 1-phenyl-3-methyl-4-nitro-5-chloropyrazole in dimethyl sulfoxide containing an excess of sodium hydride at 25° for 24 h to give 1-(1-phenyl-3-methyl-4-nitro-5-pyrazolyl)-4(or 5)-nitroimidazole (**2-184**) (53%).[215] A related reaction between 1-methyl-2-methylsulfonyl-5-nitroimidazole and imidazole gave 1-methyl-2-(1-imidazolyl)-5-nitroimidazole (**2-185**).[216] Displacement of an imidazole sulfone group is discussed later.

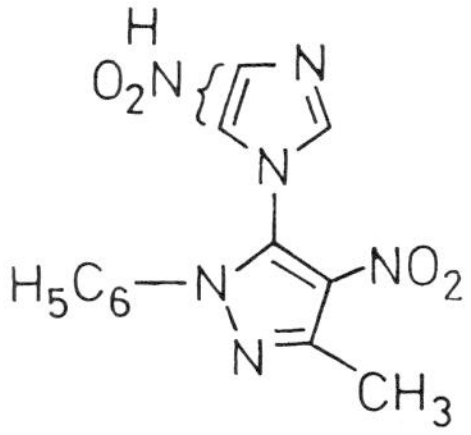

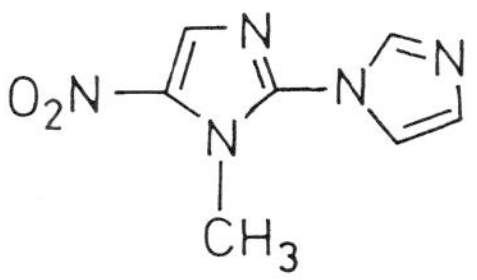

2-184   **2-185**

Arylalkylations

The formation of 1-benzyl-4-nitroimidazole-5-sulfonamide (**2-186**) (43%) and 1-benzyl-5-nitroimidazole-4-sulfonamide (**2-187**) (41%) was brought about by treating 4(5)-nitroimidazole-5(4)-sulfonamide with an equimolar amount of benzyl bromide and finely ground sodium bicarbonate in dimethylformamide at 20° for 12 h.[217] Other benzylations at N-1 in 4(5)-nitro- or 2-nitroimidazoles were carried out on the heterocycle or its sodium salt. Yields (%) are shown with the product formulas (**2-188–2-192**).[189,218-220]

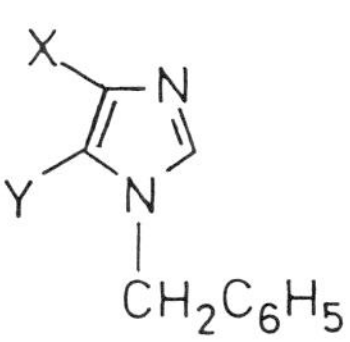

2-186, X = NO$_2$, Y = SO$_2$NH$_2$
2-187, X = SO$_2$NH$_2$, Y = NO$_2$

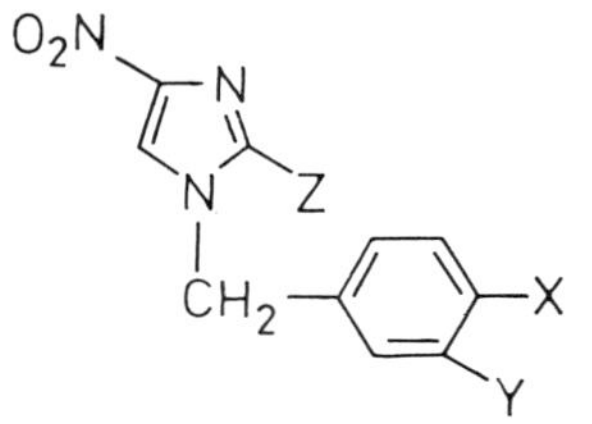

2-188
X = H, F, Cl, OCH$_3$
Y = H, Br, Cl
Z = H, CH$_3$
60%–73%

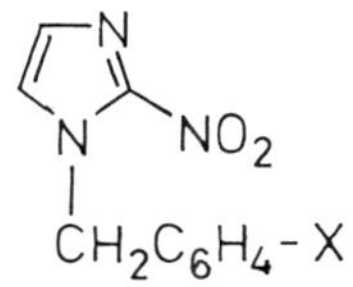

2-189 X = p-NO$_2$, p-Cl, p-F
2-190a X = m-NO$_2$
2-190b X = o-NO$_2$

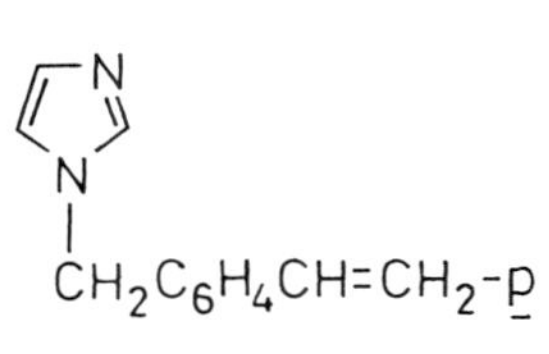

2-191 88%

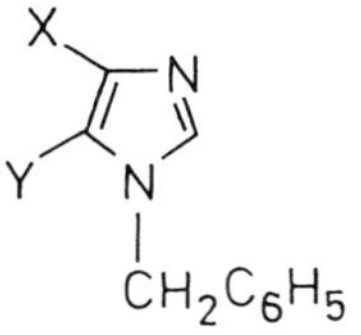

2-192
X = C$_6$H$_5$CH = CH$_2$, Y = NO$_2$, 75%
X = NO$_2$, Y = C$_6$H$_5$CH = CH$_2$, 25%

A reaction occurred readily in an alcohol or dimethylformamide between 2-methyl-4(5)-nitro-5(4)-bromoimidazole and 1,2-dibromoethane in the presence of a base to give 1-β-bromoethyl-2-methyl-4-nitro-5-bromoimidazole (**2-193**) (41%).[132,221] Two series of homologs (**2-194, 2-195**) were similarly prepared from ω,ω'-dihaloalkanes and 2-methyl-4(5)-nitroimidazole in ethanol containing sodium methoxide.[222]

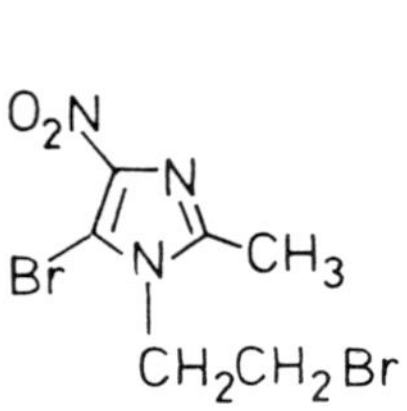

2-193

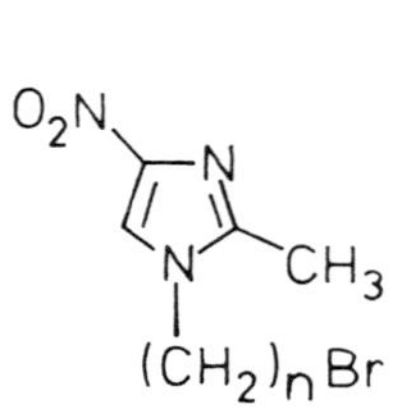

2-194a n = 2
2-194b n = 3 (32%)
2-194c n = 4 (24%)

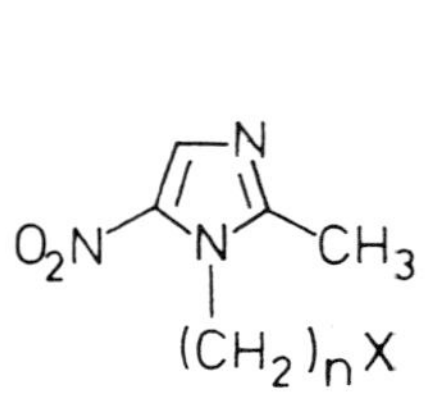

2-195a n = 2, X = Cl
2-195b n = 3, X = Br
2-195c n = 4, X = Br

In the presence of 2-methyl-4(5)-nitroimidazole the halides (**2-194, 2-195**) react further to give bisimidazoles (**2-196, 2-197**). Bisimidazoles (**2-198** and **2-199**) were also obtained from dichloromethylether and dichloroacetone.[223]

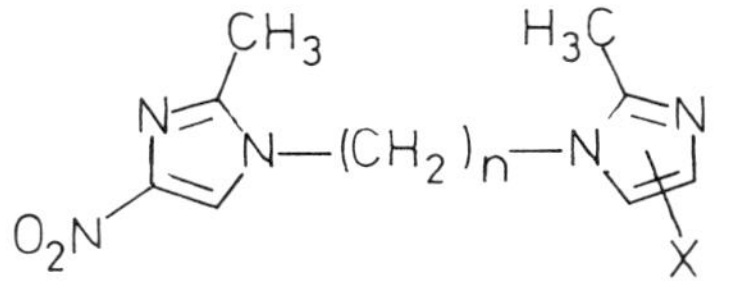

2-196 X = 4-NO$_2$
n = 2 (38%)
n = 3 (29%)
n = 4 (27%)

2-197 X = 5-NO$_2$
n = 2 (22%)
n = 3 (19%)

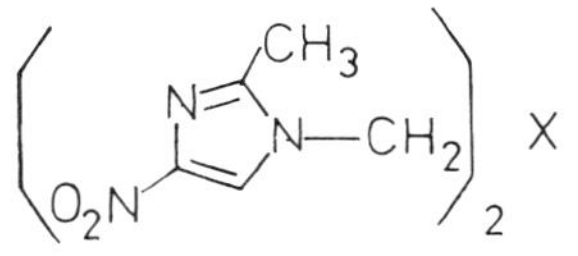

2-198 X = O
2-199 X = CO

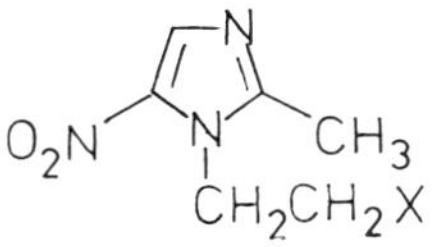

2-200 X = Br, I, OH

In acetic acid at 110° for 48 h 2-methyl-4(5)-nitroimidazole and the appropriate bifunctional halide gave the 5-nitro derivatives (**2-200**).[224]

Alkylations with Functionalized Alkyl Groups

N-Alkyl substituents which contain functional groups are well known and have been obtained in straightforward alkylations with β-hydroxyethyl chloride, bromide, or tosylate,[225-229] β-alkoxyethyltosylate,[230] β-thioalkylethyl chloride,[231] β-acetoimidoethyl phosphate,[232] and other alkylating agents. Related alkylations by epoxides have afforded a variety of N-alkylimidazoles.[233-239] Cyanoethylation has been reported.[226,239]

Epichlorohydrin and 2-methyl-4(5)-nitroimidazole in the presence of a base afforded 1-(3-chloro-2-hydroxypropyl)-2-methyl-4-nitroimidazole (**2-201**) (60%) but in the presence of acid the 5-nitro isomer (**2-202**) (42%) was obtained.[240] A similar reaction afforded the 2-isopropyl-5-nitro derivative (**2-203**) (50%). Epichlorohydrin and 2-methyl-4(5)-iodo-5(4)-nitroimidazole in ethanol heated at the reflux temperature gave a separable mixture of the iodinated derivatives (**2-204**) (32%) and (**2-205**) (16%). Conversion of 2-nitroimidazole to 1-(3-chloro-

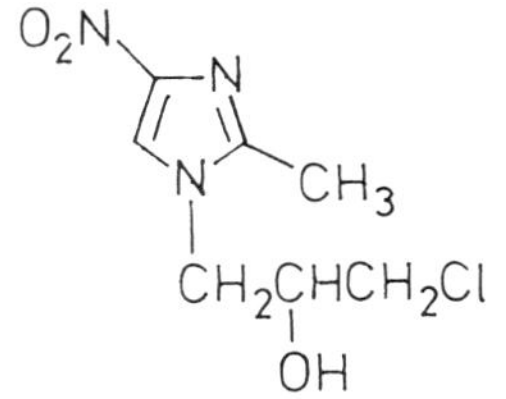

2-201

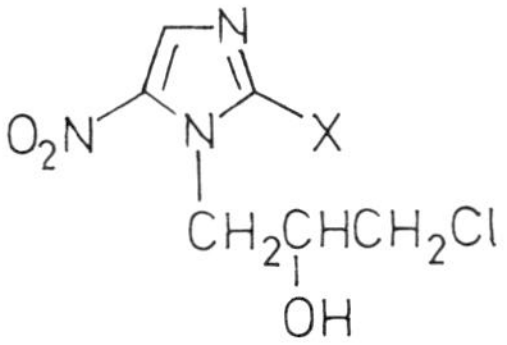

2-202 X = CH$_3$
2-203 X = (CH$_3$)$_2$CH

2-hydroxypropyl)-2-nitroimidazole (**2-206**) (65%) was also brought about by a reaction with epichlorohydrin in ethanol.

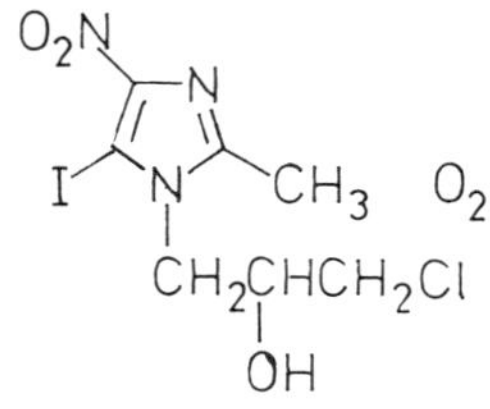

2-204

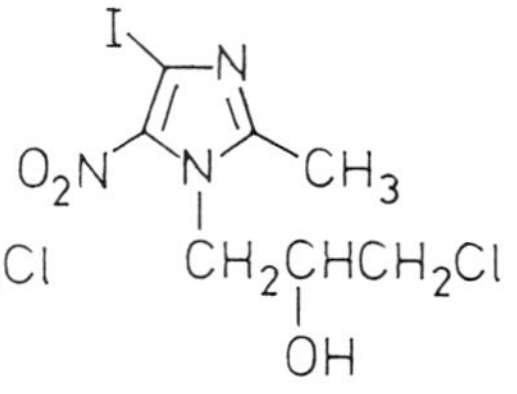

2-205

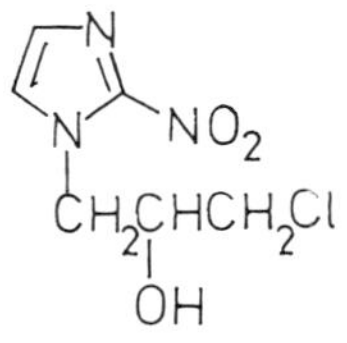

2-206

In one account N-benzoylaziridine converted 2-methyl-4(5)-nitroimidazole to 1-β-benzoylamidoethyl-2-methyl-5-nitroimidazole (**2-207**).[241,242]

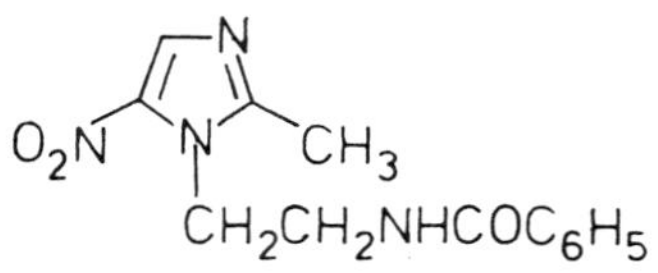

2-207

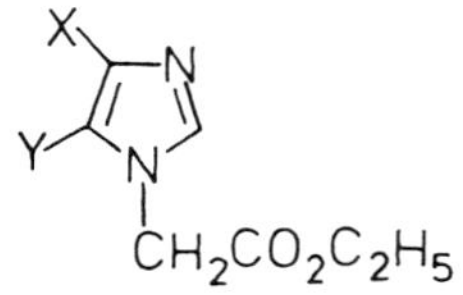

2-208
X = C_6H_4Z, NO_2
Y = H, NO_2, C_6H_4Z
Z = p-Cl, Br, F, NO_2, OR

Alkylation of 4(5)-nitro-5(4)arylimidazoles with ethyl chloroacetate gave the expected derivatives of ethyl 1-imidazolylacetates (**2-208**) (21–93%).[243] Similar products (**2-209**) (70%–88%) were obtained with phenacyl bromide and derivatives.[244] Methyl chloroacetate converted 2-nitroimidazole to methyl 2-nitro-1-imidazoleacetate (**2-210**).[245] Similarly ethylene bromohydrin afforded the N-β-hydroxyethyl derivative (**2-111**).[246] When a solution of 2-methyl-4(5)-nitroimidazole in chloromethyl ether was heated at reflux temperature for 26 h 1-methoxymethyl-2-methyl-4-nitroimidazole was obtained.[221]

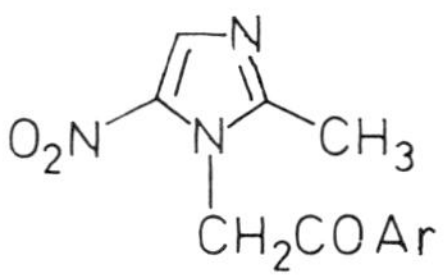

2-209
Ar = C_6H_4X
X = H, p-Br, Cl, NO_2

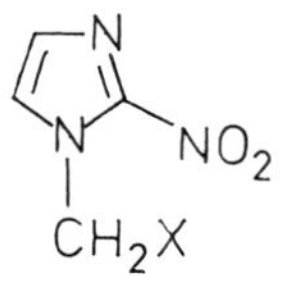

2-210 X = CO_2CH_3
2-211 X = CH_2OH

Aminoalkylation of 4(5)-nitroimidazole was achieved by heating its sodium salt with 1-methyl-4-(3-chloropropyl)piperazine in refluxing toluene for 48 h to give 1-methyl-4-[3-(5-nitroimidazole)propyl]piperazine (**2-212a**) (63%). Other aminoalkyl derivatives (**2-212b**) were also prepared.[247]

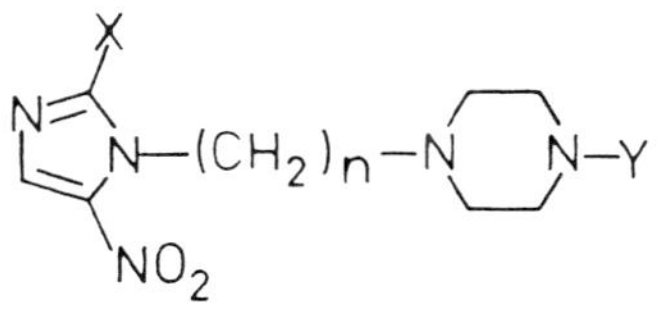

2-212a,b

2-212a n = 3, X = H, Y = CH_3
2-212b n = 2, 3
X = CH_3, C_2H_5, C_6H_5
Y = CH_3, C_2H_5, CO_2CH_3, $CO_2C_2H_5$

In a practical application methoxymethylation of 2-p-fluorophenyl-4-nitroimidazole in toluene at 25° in the presence of an excess of triethylamine gave a mixture of the 4-nitro and 5-nitroderivatives (3:1) of 1-methoxymethyl-2-p-fluorophenylimidazole. By refluxing the reaction mixture for 1 h the thermodynamically more stable 4-nitro isomer (**2-213a**) (97%) was obtained. An acetoxyethylimidazolium tetrafluoroborate (**2-213b**) was then prepared by treatment with 2-methyl-1,3-dioxolanium tetrafluoroborate and thermolyzed in pyridine at the reflux temperature to give 1-acetoxyethyl-2-p-fluorophenyl-5-nitroimidazole (**2-213c**). Hydrolysis of the ester gave flunidazole (overall yield 90%), an antiprotozoan agent.[198]

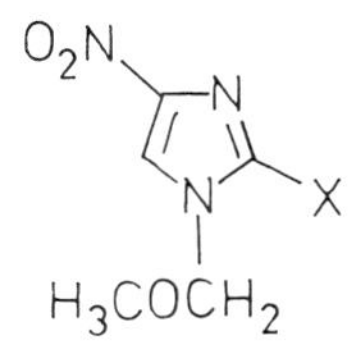

2-213a X = C_6H_4F-p 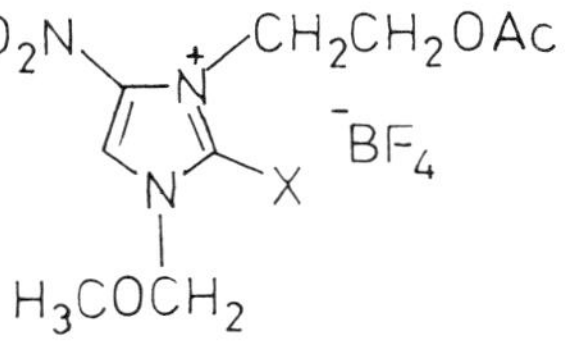   **2-213b** X = C_6H_4F-p

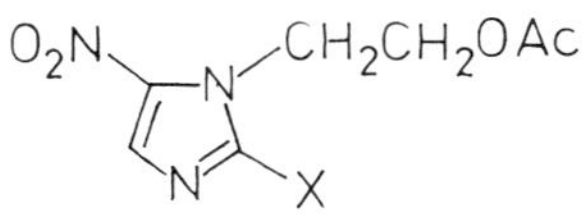

2-213c X = C_6H_4F-p

Complementary results come from a study of the N-alkylations of 4(5)-nitroimidazole (**2-4**), its 2,5(4)-dimethyl, 2-methyl, and 5(4)-methyl derivatives (**2-21, 2-16,** and **2-17**). The bases and N-(β-chloroethyl)acetamide in acetic acid for 24 h gave the corresponding pairs of alkylation products (**2-214a, 2-215a–2-214d, 2-215d**).[248]

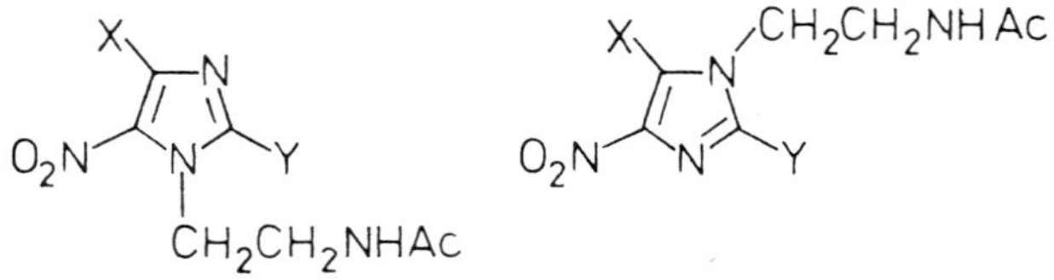

2-214   **2-215**

2-214 / 2-215	a	b	c	d
X	H	CH_3	CH_3	H
Y	H	CH_3	H	CH_3

Preparative chromatography showed greater accumulation of product **2-214c** or **d** throughout the 24-h period; however, after a maximum effect during the first 3–5 h, the rate of accumulation remained constant or decreased slightly. The accumulation of product **2-215c** or **d** steadily increased during the monitoring period. Product isomerization [**2-214c(d)** ⇌ **2-215c(d)**] did not occur in the absence of an alkylating agent and product **2-215c(d)** was formed from isomer **2-214c(d)** (or its conjugate acid) in reaction with a second mole of the alkylating agent at a higher rate than that observed for its formation by the alkylation of an imidazole (**2-16** or **2-17**). The results were consistent with the intermediacy of a quaternary dialkylimidazolium salt (**2-214e**) and its dealkylation.[248]

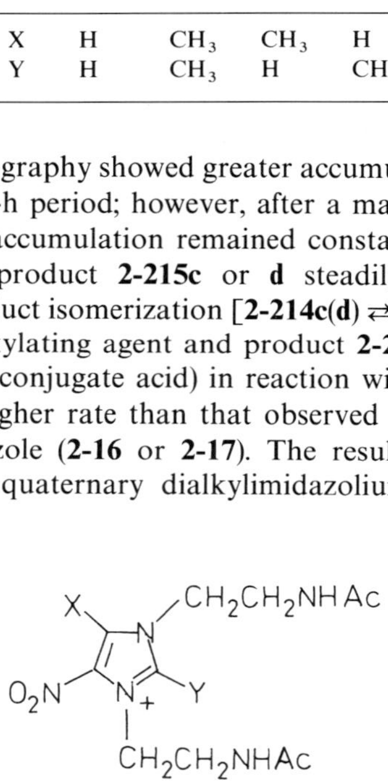

2-214e

In a similar reaction the hydrochloride of a **2-214** product thermolyzed to its **2-215** counterpart but only a trace of thermolysis of a **2-215** product hydrochloride to the corresponding **2-214** base was detected.[248]

A nitroimidazolyl steriod (**2-216**) (65%) was obtained from 24-bromo-5-β-cholane and the sodium salt of 4(5)-nitroimidazole in dimethylformamide heated at reflux temperature for 36 h. The product was either a 1-alkyl-5-nitro- or a 1-alkyl-4-nitroimidazole but not an equimolar mixture of the two as was claimed. Similar reactions provided corresponding imidazole derivatives of the 3-oxosteroid and its dioxolane acetal.[249]

Propargylation occurred when a mixture of 2-methyl-4(5)-nitroimidazole and

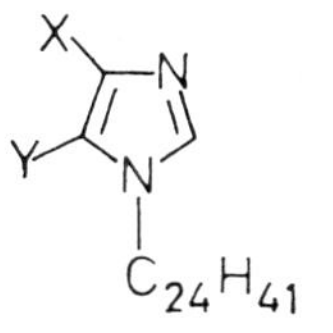

2-216
X = NO$_2$, Y = H
X = H, Y = NO$_2$

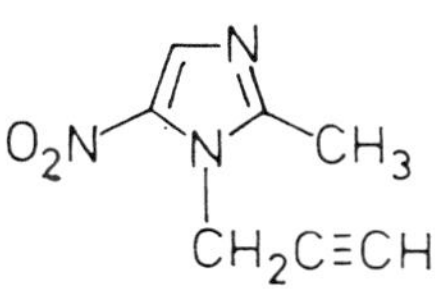

2-217

propargyl *p*-toluenesulfonate (excess) was heated with stirring to a final temperature of 140° within 3–4 h to give 1-propargyl-2-methyl-5-nitroimidazole (**2-217**). Other 1-alkynyl derivatives prepared included butynyl and a hydroxybutynyl.[250]

Additional examples of N-substitution of 4(5)-nitroimidazoles with alkyl or vinyl substituents at C-2 and/or C-5(4) positions have been described.[251–254] A preliminary examination of the effects of steric hindrance by a C-2 substituent was made.[200] Investigations on alkylating agents have also included: tosylates;[19] β-chlorosulfides;[255] haloacetals;[256] onium salts;[199] haloalkyl ethers, amines, and esters;[199,201,252] epoxides,[251] and lactones.[32]

Glycosylations

Glycosylation of 2- and 4(5)-nitroimidazoles is known. A less efficient acid catalyzed fusion of 2-nitroimidazole (**2-5**) with tetra-O-acetyl-β-D-ribofuranose followed by ester hydrolysis to give 1-β-D-ribofuranosyl-2-nitroimidazole (**2-218**) (17.5%)[257] was improved upon by a condensation between **2-5** and 2,3,5-tri-O-benzoyl-D-ribofuranosyl bromide in acetonitrile containing mercuric cyanide at 60° followed by ester hydrolysis to give the imidazole (**2-218**) in 54% yield.[258,259] An unexpected formation of the α-anomer was discovered when a condensation between **2-5** or its 1-trimethylsilyl derivative and 1-O-acetyl-2,3,5-tri-O-benzoyl-β-D-ribofuranose in the presence of stannic chloride and mercuric cyanide gave 1-(2,3,5-tri-O-benzoyl-α-D-ribofuranosyl)-2-nitroimidazole (**2-219**) (61%) after ester hydrolysis. The anomaly may be restricted to the furanose system since 1-(2,3,4,6-tetra-O-acetyl-β-D-glucopyranosyl)-2-nitroimidazole (61%) was obtained from a condensation between the nitro-

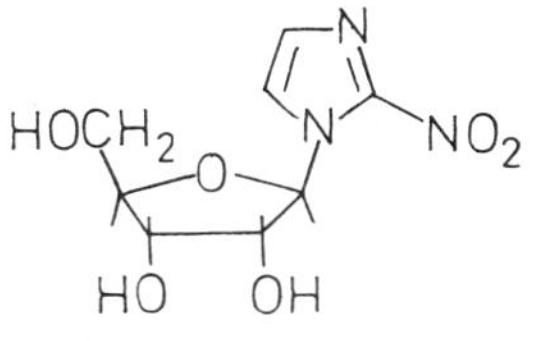

2-218

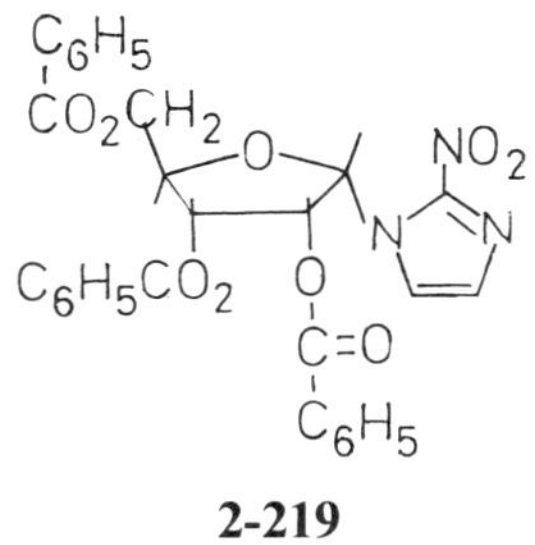

2-219

imidazole (**2-5**) and 1,2,3,4,6-penta-O-acetyl-β-D-glucopyranose. It remains unexplained but a suggested rearrangement of an intermediate nitronate ester may be involved.[258]

Ribosylation of 4-nitroimidazoles gave 1-(D-ribofuranosyl)-4-nitroimidazoles in a 1:9 ratio of α- and β-anomers by a fusion with a 1-O-acetylribose in the presence of iodine.[260] Low yields of only the β-anomers were obtained from condensations between acylglycosyl halides and the mercury(II) acetate complex of 4(5)-imidazole in a nonpolar solvent.[261] A similar reaction was described for acylglycosyl halides and the silver salts of 4(5)-nitro and 2-methyl-4(5)-nitroimidazoles[262] and of methyl 5(4)-nitroimidazole-4(5)-carboxylate.[263,264]

N-Acylation

Carbamoylation with N,N-dimethylcarbamoyl chloride converted 2-methyl-4(5)-nitroimidazole to its 1-N,N-dimethylcarbamoyl derivative (**2-220**).[265] Related derivatives were also reported.

2-220

Isocyanates reacted with imidazoles, e.g., 4(5)-nitro and 4,5-dinitro derivatives, to give a variety of N-carbamoyl compounds.[266]

Condensations

Methylene Groups

The base-catalyzed exchange of C-methyl protons in the six isomers of C,N-dimethylnitroimidazoles was investigated in 0.01 N NaOD in a mixture (1:1) of D_2O and CD_3OD at 60°. Half-life determinations showed 1,5-dimethyl-4-nitroimidazole (**2-221a**) ($t_{1/2}$ 2.8 s) to have the greatest kinetic acidity, comparable to the exchange for nitromethane ($t_{1/2}$ 0.1 s) and 5×10^6 greater than that for o-nitrotoluene ($t_{1/2}$ 15.5 days). For the other five isomers $t_{1/2}$ values were: **2-221b**, 11.8 s; **2-221c**, 30.4 s; **2-221d**, 116 min; **2-221e**, 33 h; **2-221f**, 38 h.[267]

Since the isomer **2-221a** showed a closer comparison with nitromethane than with o-nitrotoluene it was suggested that this isomer be considered a simple vinylog of nitromethane for the formation of its conjugate carbanion; however, the comparable $t_{1/2}$ for proton exchange in an α,β-unsaturated nitroalkane was not given. By attributing a partial retardation in the development of the carbanion from 1,4-dimethyl-5-nitroimidazole (**2-221c**) to a second order ALP (adjacent lone pair) effect[243] the slower exchange (10 times slower than that for isomer **2-221a**) was partially explained.[267]

Dienoid coupling of nitro and C-methyl groups presumably contributed to stabilization of carbanions from the isomers (**2-221b,d**). A higher exchange rate for **2-221b** was attributed to an inductive effect from imidazole ring nitrogen atoms where both contribute in **2-221b**, one in **2-221d**. This order of reactivity in vinyl carbanion formation at C-2 and C-5 was also observed in nitro- and fluoroimidazoles.[268] The lower reactivities in isomers **2-221e,f** was correlated with the absence of vinylogous coupling of nitro and C-methyl groups.[267]

A correlation[269] between the rate of isotopic exchange of hydrogen in C-methyl groups of 39 five-membered heterocycles and the electronic effects of ring heteroatoms and substituents may need a reexamination since the validity of the kinetic data for the 4- and 5-nitro derivatives of 1,2-dimethylimidazole has been questioned.[267] A quantum chemical study of the CH acidities of methyl derivatives of five-membered aromatic heterocycles correlated rate constants with deprotonation energies and charges on hydrogen atoms in methyl groups under attack.[270] How a reexamination of the kinetic data would affect these conclusions is uncertain.

Activation provided by a nitro group has been noted in aldol condensations of C-methylimidazoles (**2-221a**)[35,36] (**2-221b**),[271,272] (**2-221c**),[35,36] and (**2-221d**).[110,273] Reaction failures were reported for each of the isomers (**2-221b,f**).[1k] Condensation occurred selectively at the 5-methyl group in 1,2,5-trimethyl-4-nitroimidazole and at the 4-methyl group in 1,2,4-trimethyl-5-nitroimidazole.[274]

2-221	X	Y	Z
a	NO_2	CH_3	H
b	H	NO_2	CH_3
c	CH_3	NO_2	H
d	H	CH_3	NO_2
e	CH_3	H	NO_2
f	NO_2	H	CH_3

2-221

Aroyl chlorides and 1,2-dialkyl- or 1,2-dimethyl-5-nitroimidazole in the presence of a base gave 2-aroylmethylimidazole derivative (**2-222a,b**).[272,275-278] In a similar condensation diethyl oxalate gave ethyl 1-methyl-5-nitroimidazol-2-ylpyruvate **2-222c**.[279] Treatment of the latter with chlorine in sulfuric acid followed by hydrolysis with sulfuric acid gave 1-methyl-2-formyl-5-nitroimidazole.

Methyl 4-nitroimidazolylacetate and *n*-amyl nitrite in methanol containing sodium methoxide gave, after acidification, the Z-oxime (**2-223**) (70%), amyl 4-nitro-1-imidazoylacetate, and 4(5)-nitroimidazole.[280]

Conversion of a trinitromethyl group to the potassium salt of the dinitro-

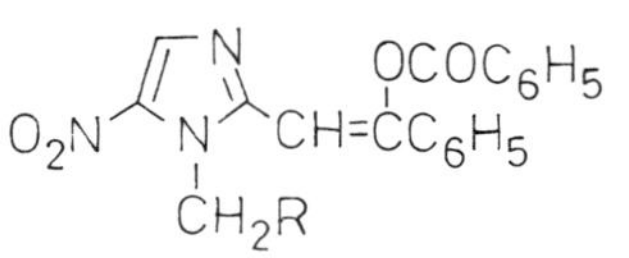

2-222a R = H
2-222b R = C$_6$H$_5$

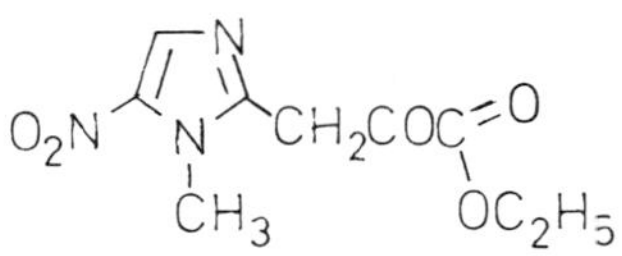

2-222c

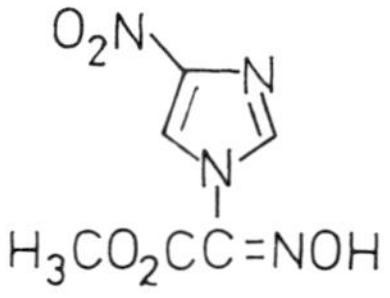

2-223

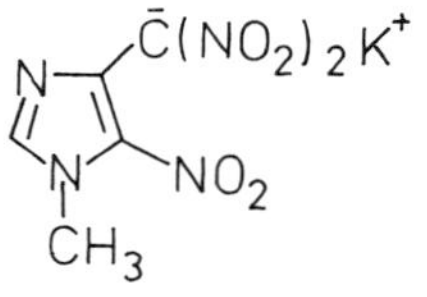

2-224

methyl derivative (**2-224**) is a rarely encountered method of obtaining the salt of a C-H acid.[93] Alkylations of the anion have not been reported.

Amines

Ethyl acetylenedicarboxylate and ethyl ethoxymethylenemalonate condensed with 1-methyl-2-amino-5-nitroimidazole to give the nitroimidazo[1,2-*a*]-pyrimidinones (**2-225** and **2-226**).[281]

When heated in polyphosphoric acid at 110° for 2 h 1(ω-carboxyalkyl)-2-methyl-4-nitro-5-aminoimidazoles cyclized to lactams (**2-227**).[282]

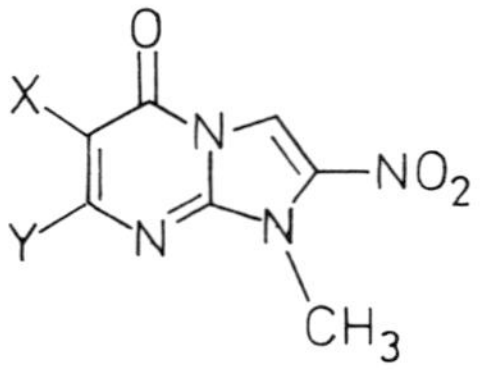

2-225 X = H, Y = CO$_2$C$_2$H$_5$
2-226 X = CO$_2$C$_2$H$_5$, Y = H

Carboximidates

Ethyl 1-methyl-5-nitro-2-imidazolylcarboximidate and acethydrazide condensed to give a hydrazide which cyclized in acetic acid to 3-(1-methyl-5-nitro-2-imidazolyl)-5-methyl-1,2,4-triazole (**2-228**).[283] Related compounds with R = NH$_2$ and X = O,S were also prepared. Anthranilic acid condensed with the carboximidate ester to give a quinazolinone (**2-229**)[284,285] and a benzimidazole derivative (**2-230**) was obtained from *o*-phenylenediamine.[286]

A cyclization between 1-methyl-5-nitroimidazole-2-hydroximoyl chloride and 1,3-cyclohexadiene in the presence of triethylamine at 0° for 3 h gave an isoxazolidine (**2-231**).[287] It seems likely that a nitrile oxide was an intermediate, Eq. (2-23).

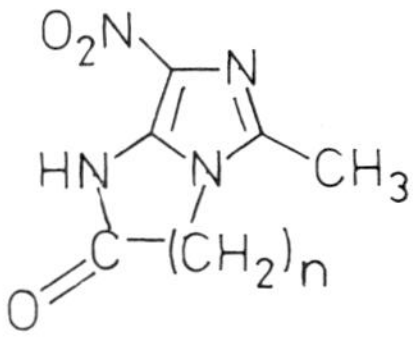

2-227a n = 1
2-227b n = 2
2-227c n = 3

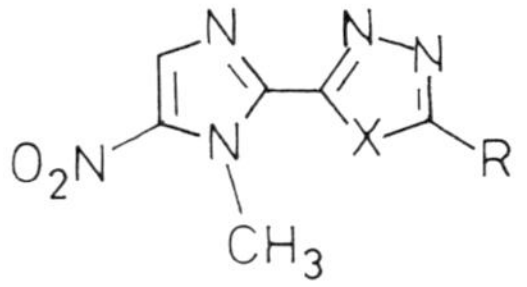

2-228 X = NH, R = CH$_3$

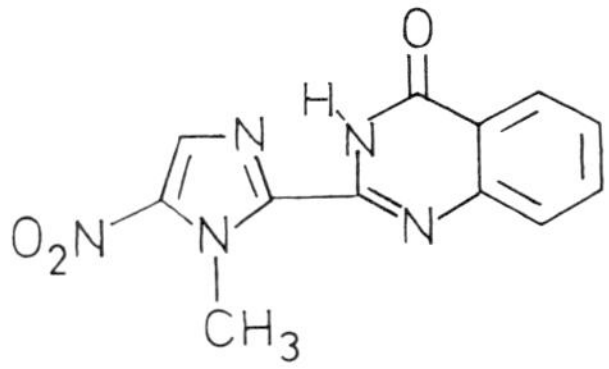

2-229

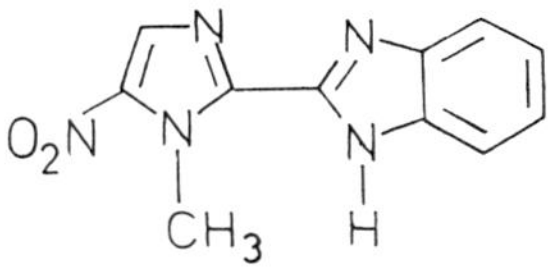

2-230

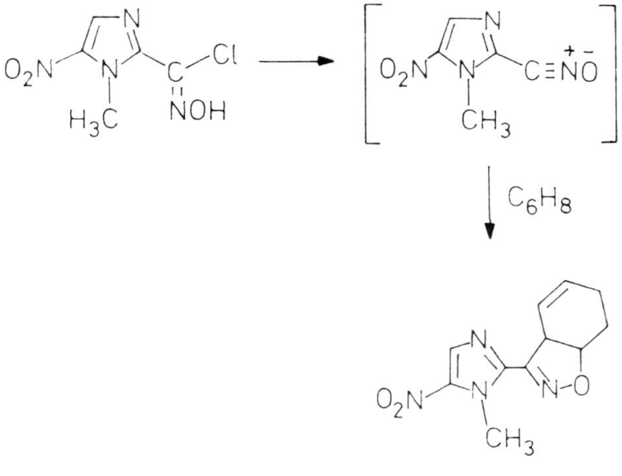

2-231   (2-23)

Aldehyde Derivatives

Oximes and Nitrones

The oxime of 1-alkyl-2-formyl-5-nitroimidazole was obtained by conventional treatment of the aldehyde with hydroxylamine hydrochloride in a mixture of pyridine and ethanol. Treatment with phosphorous oxychloride converted each oxime into a nitrile (2-232) (71% to 88%).[288] The nitrile (2-232a) was also obtained in a high yield directly from the aldehyde by treatment with O,N-bistrifluoroacetylhydroxylamine. Imidazoylthiazoles were obtained by a conversion of the nitrile function to a thiazole ring.[289]

A variety of functional groups have been incorporated in the alkyl portion of oxime-O-ethers (2-233).[290, 291]

Condensation with N-methylhydroxylamine afforded the nitrone (2-234).[292]

A variety of oxime and nitrone derivatives of 1-alkyl-2-nitro-5-formylimidazole are also known.[290-296]

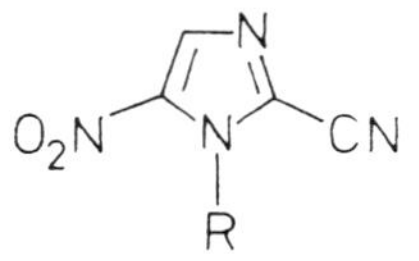

2-232a R = CH_3
2-232b R = C_2H_5
2-232c R = n-C_4H_5
2-232d R = $C_2H_4OCOCH_3$

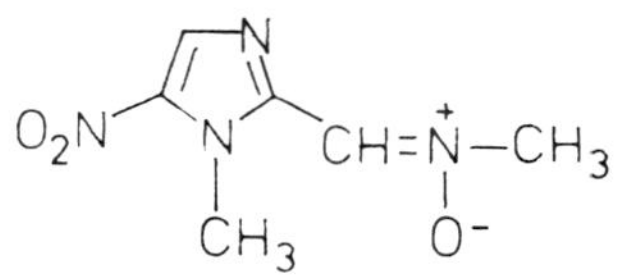

2-234

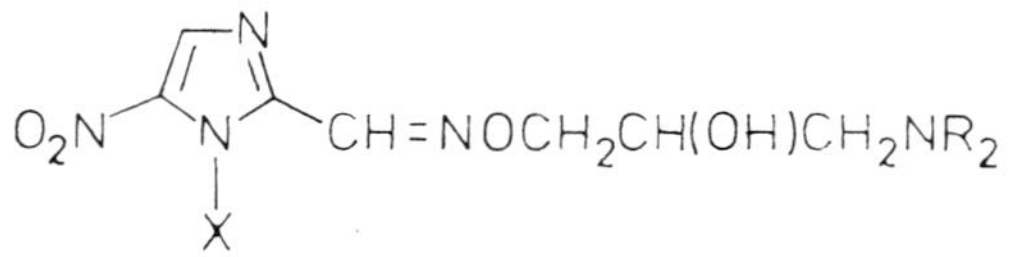

2-233
X = H, CH_3, C_2H_5, CH_2CH_2OH

Hydrazones

Mono and N,N-dialkylhydrazines condensed with 1-methyl-2-formyl-5-nitroimidazole to give expected hydrazones (2-235a–c).[297-304]

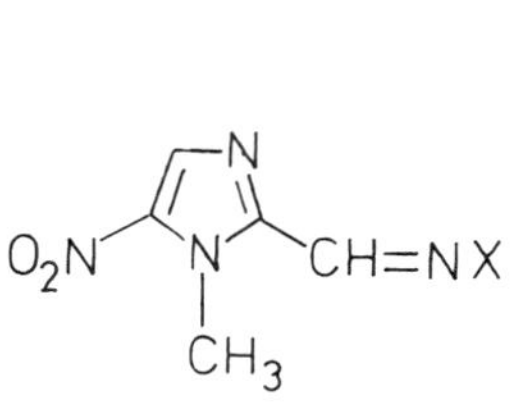

2-235

	X
a	$N(CH_3)_2$
b	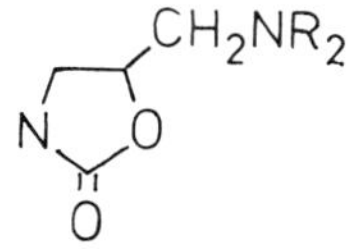 CH_2NR_2 attached to oxazolidinone
c	$NHC(=S)NH_2$

2-235

The dimethylhydrazone (2-235a) was brominated and condensed with malononitrile to give an imidazolylpyrazole (2-236) and the thiosemicarbazone

(**2-235c**) underwent ferric ammonium sulfate oxidative cyclization to 2-amino-5(1-methyl-5-nitro-2-imidazolyl)-1,3,4-thiadiazole (**2-237**).[299-304]

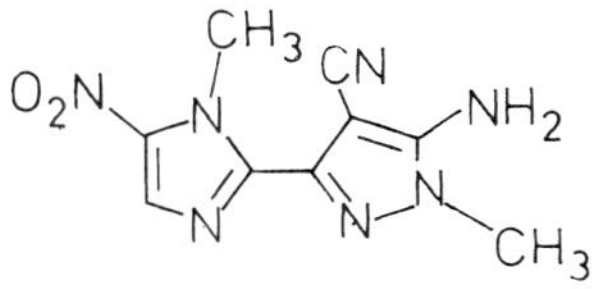

2-236

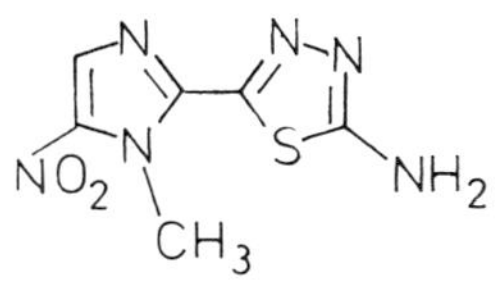

2-237

A variety of hydrazone derivatives of 1-alkyl-2-nitro-5-formylimidazole are known.[294,295,305]

Schiff Bases

Examples of these derivatives, including the compound **2-238**, have been recorded.[294,306]

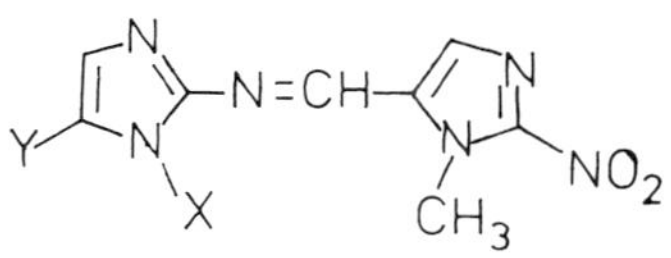

2-238 X = Y = H, alkyl

Vinylimidazoles

A mixture of carbostyril-3-acetic acid and 1-methyl-2-formyl-5-nitroimidazole in acetic acid and acetic anhydride was heated at the reflux temperature for 4 h to give a 3-[2-(1-methyl-5-nitroimidazolyl)vinyl]carbostyril (**2-239**) (64%–77%).[307] Other vinylimidazoles (**2-240a,b**, **2-241a–f**) were obtained by similar reactions.[308-314]

Although β-(1-methyl-2-nitroimidazol-5-yl)acrolein (**2-242a**) could not be made by treating 1-methyl-2-nitro-5-formylimidazole (**2-242e**) with formylmethylenetriphenylphosphorane, the related molecules (**2-242b–d**) (60%–80%) were prepared by the Wittig reaction. The acrolein (**2-242a**) was obtained in poor yield from a base-catalyzed reaction between acetaldehyde and the aldehyde (**2-242e**). The aldehyde (**2-242e**) readily condensed with nitroalkanes to give

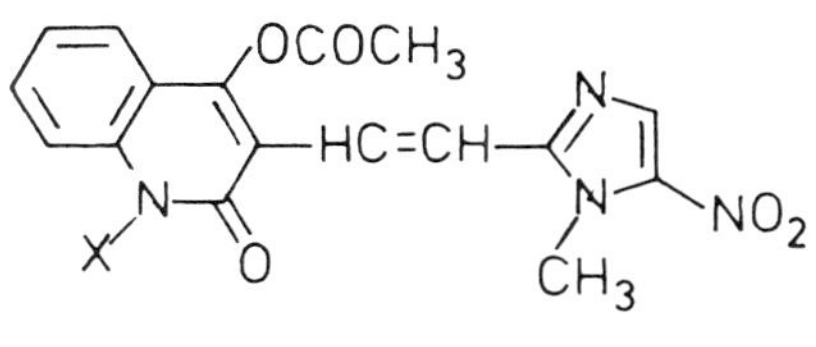

2-239 X = CH_3, C_2H_5

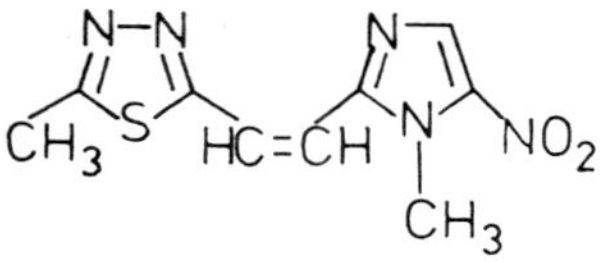

2-240a

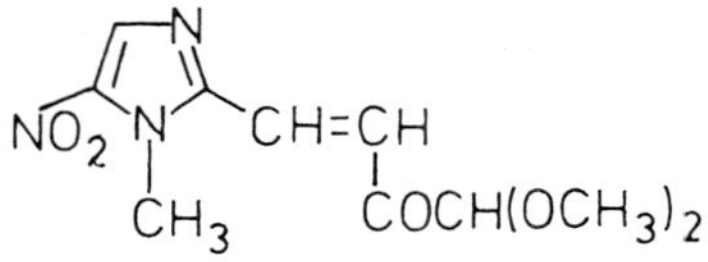

2-240b

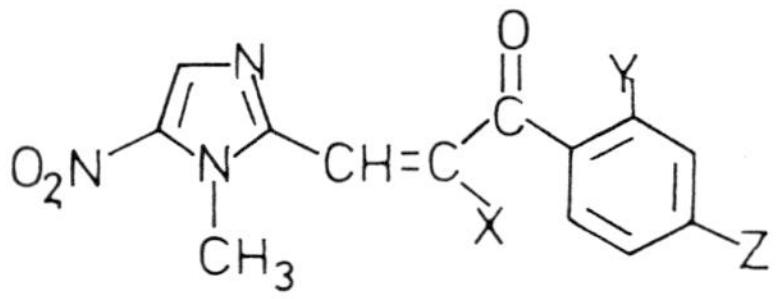

2-241a–f

2-241	-X, Y-	Z	yield%
a	-H, H-	H	8–52
b	-CH$_2$CH$_2$-	OH	24
c	-O-	OAc	33
d	-CH$_2$O-	H	9
e	-S-	H	47
f	-CH$_2$SO-	OCH$_3$	30

nitrovinyl compounds (**2-242f–h**) (9%–33%).[315] Dimethylformamide dicyclohxyl acetal condensed with 1,2-dimethyl-5-nitroimidazole to give an enamine, Eq. (2-24a).[316]

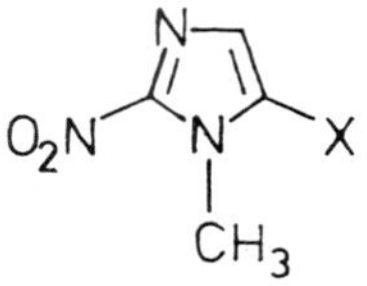

2-242

2-242	X
a	CH=CHCHO
b	CH=CHCO$_2$CH$_3$
c	CH=CHCO$_2$C$_2$H$_5$
d	CH=CHCOCH$_3$
e	CHO
f	CH=CHNO$_2$
g	CH=C(CH$_3$)NO$_2$
h	CH=C(n-C$_4$H$_9$)NO$_2$

A sensitivity of nitroimidazoles to alkaline solutions[183] accounted for the fragmentation of a vinylimidazole in aqueous sodium hydroxide, pH 11–14,

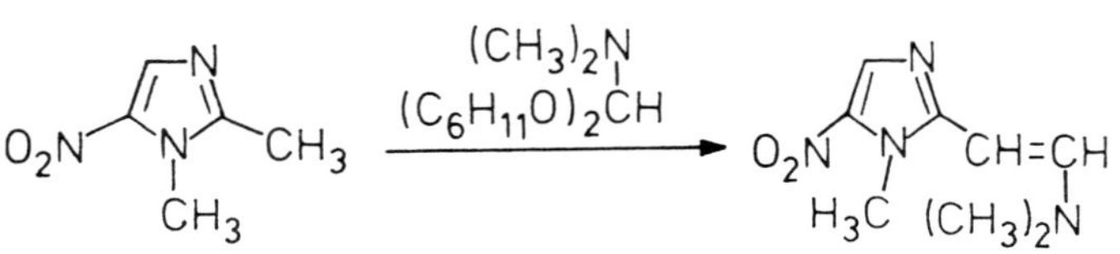

(2-24a)

Eq. (2-24b). Initiation of ring-opening by hydroxide attack at C-4 was proposed.[317]

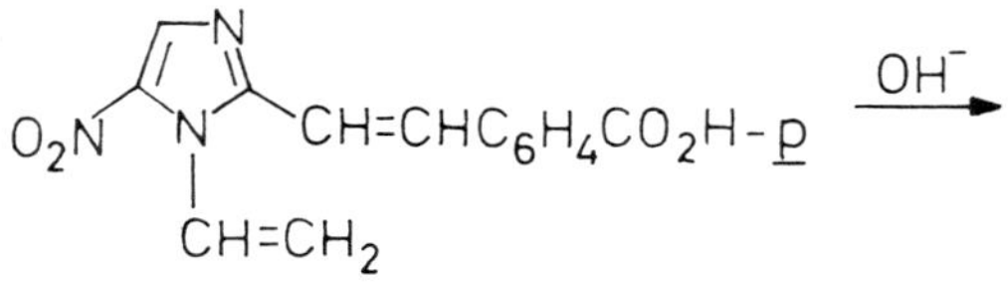

$$O_2N\text{-imidazole-}CH=CHC_6H_4CO_2H\text{-}\underline{p} \xrightarrow{OH^-}$$

$$HC \equiv CH + CH_3CHO + HO_2CCH = CHC_6H_4CO_2H\text{-}p \qquad (2\text{-}24b)$$

α-Hydroxybenzylimidazoles

The formation of 1-methyl-2-[(2-tert-butyl-3-hydroxy-6-methoxyphenyl)-hydroxymethyl]-5-nitroimidazole (**2-243**) was obtained from a reaction between 1-methyl-2-formyl-5-nitroimidazole and 2-tert-butyl-4-methoxyphenol.[318]

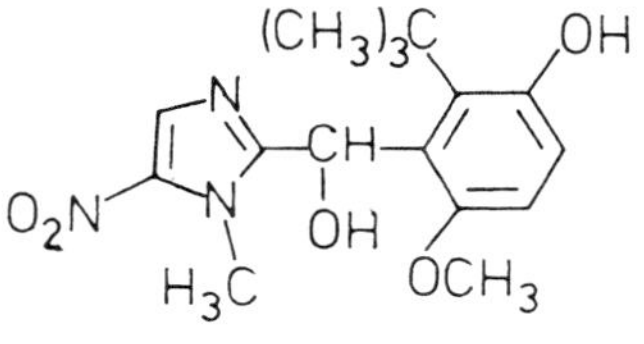

2-243

Replacements of Ring Halogens and Pseudohalogens

Halogen Interchange

A conversion of 4(5)-nitro-5(4)-chloroimidazole to the 5(4)-iodoimidazole was brought about by treatment with potassium iodide in dimethylformamide heated at reflux. Similar conversions were reported for the 1-alkyl and 1,2-dialkyl derivatives.[319-321]

Cyanide

Potassium cyanide in ethanol converted 1,2-dimethyl-4-nitro-5-bromo-imidazole to the expected 5-cyanoimidazole (**2-244a**) which was unstable in the presence of the cyanide anion. Decomposition with no evidence for displacement of bromine by the cyano group was observed when 1,2-dimethyl-4-bromo-5-nitroimidazole was treated with potassium cyanide.[183]

An interesting rearrangement occurred in the conversion of 1-methyl-4-chloro-5-nitroimidazole when treated with a mixture of potassium iodide and potassium cyanide in dimethylformamide to 1-methyl-4-nitro-5-cyanoimidazole (**2-244b**) (52%).[183-185] No reaction occurred in ethanol at 150° in a sealed tube. When the reaction was carried out in the absence of a solvent the same product

(**2-224b**) predominated but a small amount of the expected 1-methyl-4-cyano-5-nitroimidazole (**2-244c**) was also obtained. It is reported that the rearrangement did not occur in the absence of potassium iodide. It was suggested that the reaction proceeded by thermolysis of an intermediate 1,3-dimethyl-4(5)-cyano-5(4)-nitroimidazolium iodide to methyl iodide, the product (**2-244b**), and some of the less stable isomer (**2-244c**).[183]

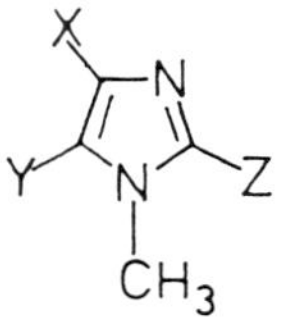

2-244	X	Y	Z
a	NO_2	CN	CH_3
b	NO_2	CN	H
c	CN	NO_2	H

2-244

Amines and Amides

Activation by the nitro substituent accounts for the conversion of 1-methyl-2-bromo-5-nitroimidazole on treatment with piperidine in ethanol at 80° for 1.5 h to 1-methyl-2-piperidino-5-nitroimidazole (60%), Eq. (2-25). A similar conversion of 1-methyl-4-nitro-5-bromoimidazole to 1-methyl-4-nitro-5-piperidinoimidazole (75%) occurred on heating at 80° for 9 h, Eq. (2-25). For comparison, both 1-methyl-2-bromo- and 1-methyl-5-bromoimidazole failed to react with piperidine at 200°.[75]

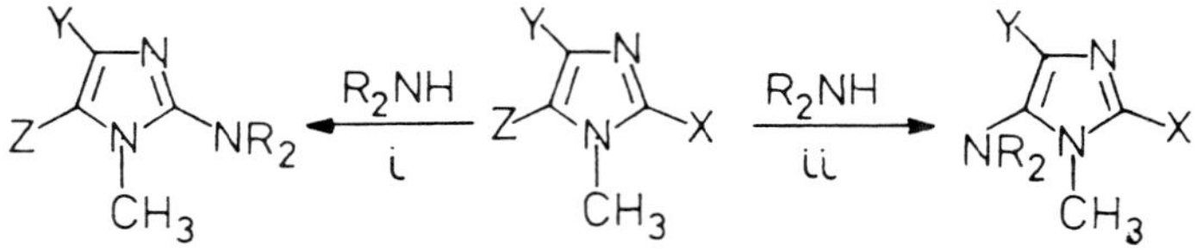

i X = Br, Y = H, Z = NO_2
ii X = H, Y = NO_2, Z = Br
R_2N = ⟨N⟩

(2-25)

Liquid ammonia converted 1-methyl-2-methylsulfonyl-5-nitroimidazole to 1-methyl-2-amino-5-nitroimidazole (**2-245a**) (87%) but with sodamide the major

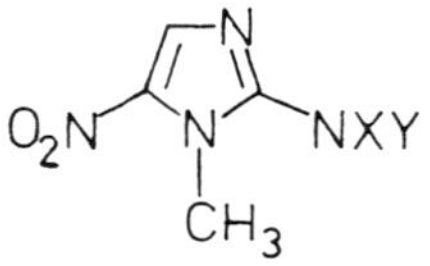

2-245	X	Y
a	H	H
b	CH_3	H
c	CH_3	CH_3
d	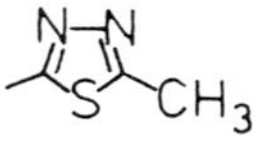 (N-N / S thiadiazole-CH_3)	H
e	p-$O_2NC_6H_4$	H

2-245a–e

product was the sulfone (**2-245f**). Amines also displaced the sulfonyl group but in lower yields: **2-245b** (18%), **2-245c** (25%), **2-245d** (5%), and **2-245e** (11%).[322]

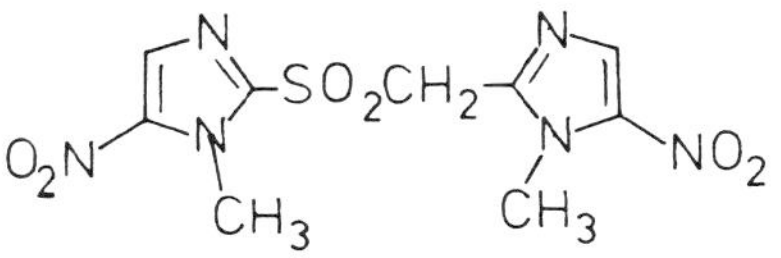

2-245f

Conversions of 4-bromo-5-nitro and 4-nitro-5-bromo derivatives of 1,2-dimethylimidazole to the corresponding 4-amino and 5-amino compounds (**2-246** and **2-247**) were brought about by heating with amines in butanol at 110° for 2–5 h with yields of 57%–82%.[183]

A series of 5-amino-4-nitro-1-methylimidazoles (**2-247c–h**) were obtained from the corresponding 5-chloroimidazole and the appropriate nitrogen compound.[323,324]

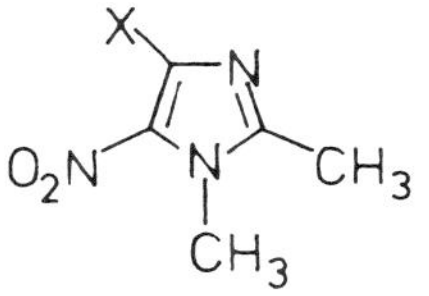

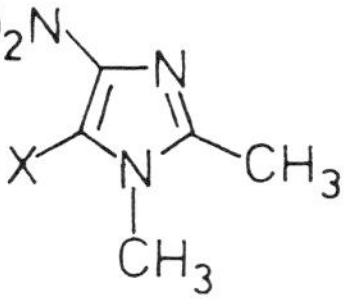

2-246a X = NHCH$_2$CH$_2$OH
2-246b X = N⟨⟩

2-247a X = NHCH$_2$CH$_2$OH
2-247b X = N⟨⟩

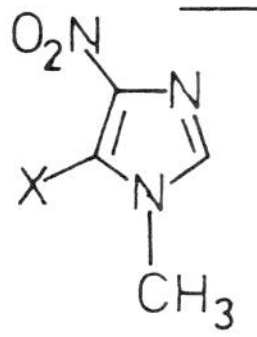

2-247c–h

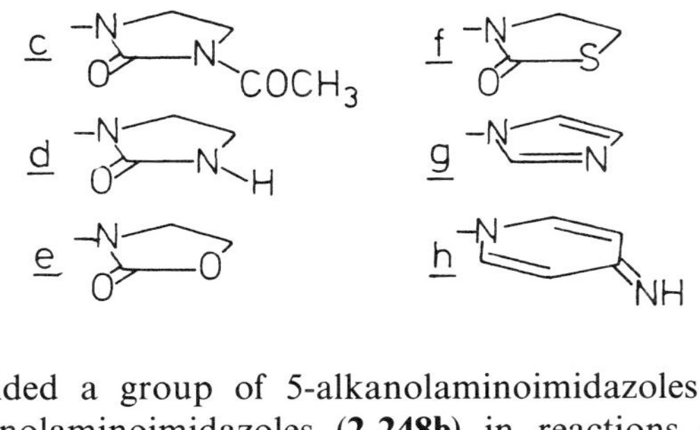

Alkanol amines provided a group of 5-alkanolaminoimidazoles (**2-248a**) (85%–95%) and 4-alkanolaminoimidazoles (**2-248b**) in reactions with the appropriate halide.[325]

Guanidine converted 1-methyl-4-nitro-5-chloroimidazole to the expected 5-guanidinoimidazole (**2-248c**).[326]

A facile replacement of a sulfonyl group occurred in the conversion of 1-methyl-2-methylsulfonyl-5-nitroimidazole in reactions with 1-acyl or 1-heteroaryl-2-oxotetrahyroimidazoles in the presence of a base to 1-(1-methyl-

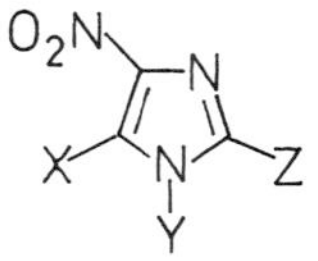

2-248a

X = NH(CH$_2$)$_n$OH, n = X, 3
Y = CH$_3$, C$_2$H$_5$, C$_4$H$_9$
Z = H CH$_3$, C$_2$H$_5$, C$_3$H$_7$, NO$_2$

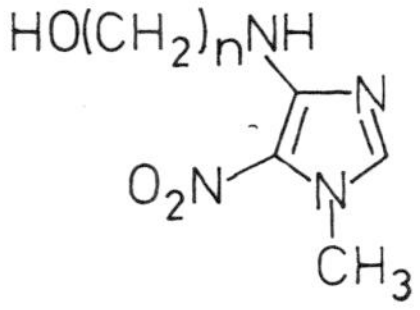

2-248b n = 2, 3

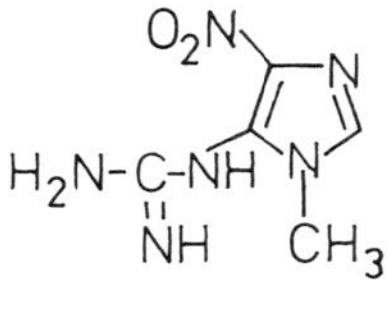

2-248c

5-nitroimidazolyl-2)-2-oxo-3-acyl(or heteroaryl)-tetrahydroimidazoles (**2-249a–h**), Eq. (2-26).[327–331]

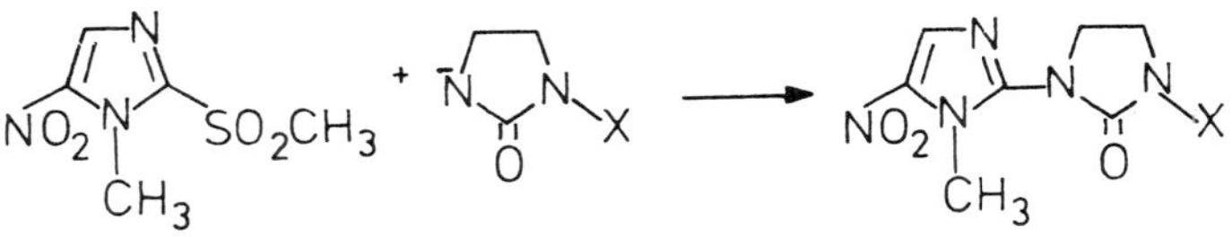

$$\text{(2-26)}$$

2-249

2-249	a(b)	c	d(e)	f	g	h
X	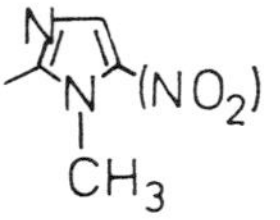 ![pyrrole-NO$_2$ with CH$_3$]	–CO–CH$_3$	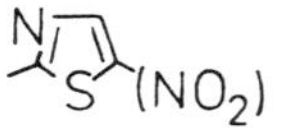 thiazole-NO$_2$	CH$_3$	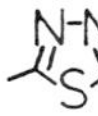 thiadiazole	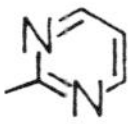 pyrimidine
yield%	54(39)	18	53(23)	10	85	58

Similar reactions afforded products **2-249** (X = SO$_2$R), CSR, and COR (R = alkyl, aryl, dialkylamino, and alkoxy) in yields ranging from 9% to 80%. Analogous products (**2-249**) contained other 1-alkyl substituents in the imidazole ring or additional alkyl substituents in the imidazole and/or the imidazolidone ring.[332] Expected products were obtained when the reaction was extended to the sodium salts of benzimidazolidones, hexahydropyrimidones, triazolidinedione, thiazolidinone, 2-iminothiazolidine, pyrrolidinone, oxazalidinones, and cyclic sulfamides (1,1-dioxo-2,3,4,5-tetrahydro-1,2,5-thiadiazoles).[332–334] The reaction was also extended to the sodium salts of imidazole, pyrrole, indole, benzimidazole, pyrazoles, indazoles, triazole, benzotriazole, and tetrazole.[334]

Chlorine in 1-methyl-2-chloro-5-nitroimidazole was readily displaced by the anion of 1-methanesulfonylimidazolidinone to give product **2-249** (X = SO_2CH_3) (65%), but only trace amounts of the expected isomeric product (**2-250**) were obtained from 1-methyl-2-chloro-4-nitroimidazole (also unreactive to morpholine).[70]

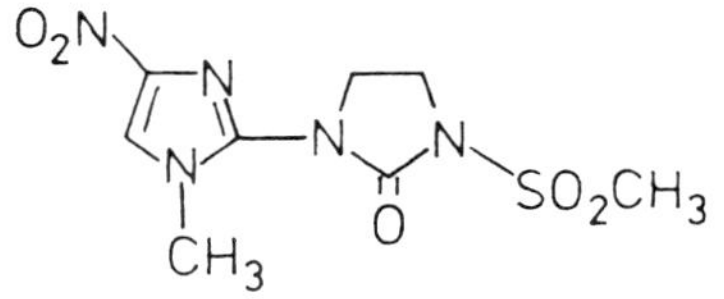

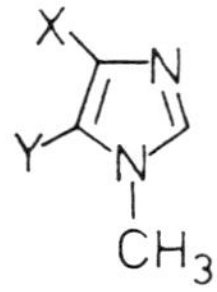

2-250

2-251a X = N͡O, Y = NO_2
2-251b X = NO_2 Y = N͡O

Both 1-methyl-4-iodo-5-nitroimidazole and 1-methyl-4-nitro-5-iodoimidazole reacted with morpholine to give the expected amines (**2-251a,b**) in low yields. The amine products were also obtained from similar reactions between morpholine and the appropriate 4- or 5-chloroimidazole derivative.[70]

When heated in ethanol in the presence of sodium acetate 1-methyl-4-nitro-5-chloroimidazole and o-aminophenol gave 1-methyl-4-nitro-5-o-hydroxyanilinoimidazole (**2-252**).[335]

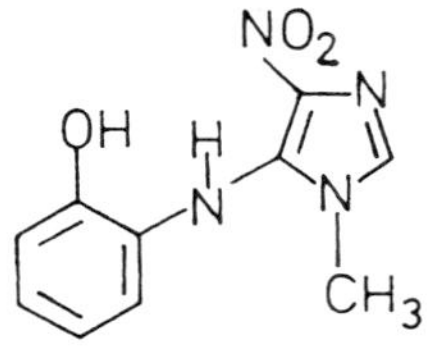

2-252

Nucleophilic substitution at the 2-position in imidazol[1,2-a]pyridine was assisted by the presence of a 3-nitro substituent in the reaction between 2-chloro-3-nitroimidazo[1,2-a]pyridine and dimethylamine in aqueous 1,2-dimethoxyethane to give 2-dimethylamino-3-nitroimidazo[1,2-a]pyridine (80%), Eq. (2-27).[140]

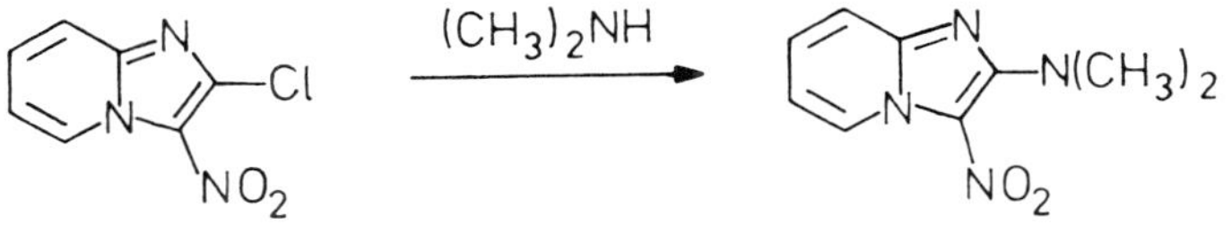

(2-27)

Alkoxides and Phenoxides

In a general procedure a 4-nitro-5-haloimidazole and one or two molar equivalents of sodium alkoxide or hydroxide in an appropriate alcohol or aqueous methanol was heated at reflux temperature for 3–5 h. Thus eight alkoxynitroimidazoles (**2-253a–h**) reported to be the first known examples, and five sodium salts (**2-253i–m**) of unstable hydroxynitroimidazoles, were obtained.[320]

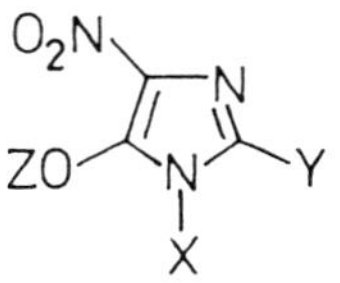

2-253a–m

2-253	X	Y	Z	yield%
a	CH_3	H	CH_3	77–95
b	CH_3	H	C_2H_5	90
c	CH_3	CH_3	CH_3	50
d	CH_2CH_2OH	CH_3	C_2H_5	43
e	CH_3	H	$(CH_3)_2CH$	73
f	H	CH_3	CH_3	87
g	H	CH_3	C_2H_5	79
h	C_2H_5	CH_3	C_2H_5	55
i	CH_3	H	Na	74
j	Na	CH_3	Na	55
k	CH_3	CH_3	Na	86
l	C_2H_5	CH_3	Na	90
m	C_4H_9	C_3H_7	Na	90

In contrast with the formation of the 4-nitro-5-alkoxy compounds (**2-253**), 1,2-dimethyl-4-nitro-5-bromoimidazole did not react with a sodium alkoxide;[183] this discrepancy needs further investigation.

Four equivalents of an alkoxide in the appropriate alcohol at refluxing temperature for ½ h converted 1,2-dimethyl-4-bromo-5-nitroimidazole to its 4-methoxy and 4-ethoxy derivatives (**2-254a**) (62%) and (**2-254b**) (58%).[183] Prolonged heating in the presence of a strong base caused decomposition with the development of an intense red color and the evolution of ammonia. The reaction was suggested as a diagnostic test for the presence of a nitro group at C-5 in an imidazole.

Although further detail was not available from the Chemical Abstract,

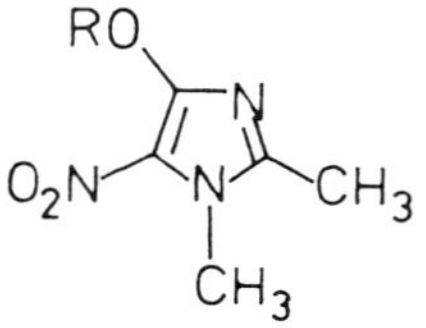

a R = CH_3
b R = C_2H_5
c R = C_6H_5
d R = (quinolinyl)

2-254

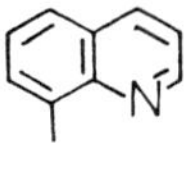

1-methyl-4-phenoxy-5-nitroimidazole (**2-254c**) and 1-methyl-4-(8-quinolyloxy)-5-nitroimidazole (**2-254d**) were obtained from 1-methyl-4-chloro-5-nitroimidazole and the appropriate phenolate.[336]

Mercaptides

Optimal conditions for the preparation of nitromercaptoimidazoles and their sodium salts called for a treatment of a chloro or bromonitroimidazole with sodium sulfide in ethanol and/or water, sometimes with a short heating period to the reflux temperature. The reaction was fast (5–20 min) and the yields were good (77%–73%). Acetic acid liberated the mercaptan from its sodium salt. Longer reaction time, higher temperatures, and air oxidation promoted the formation of sulfides and disulfides. Other procedures utilized thiourea or ammonium sulfide or hydrosulfide, Eq. (2-28, 2-29).[337-339]

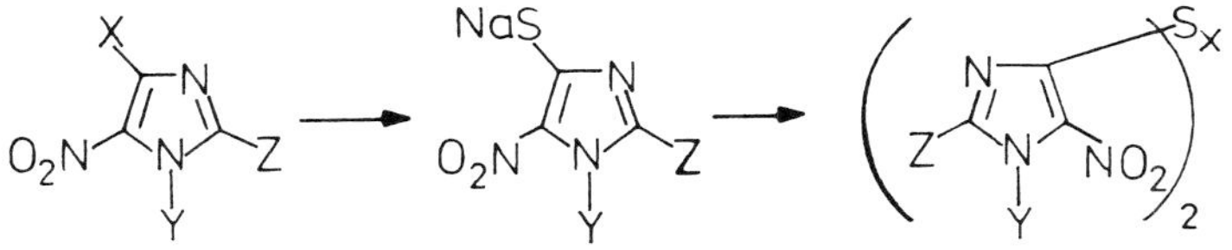

2-255a
X = Cl, Y = CH_3, Z = H, x = 1, 2
X = Cl, Y = C_2H_5, Z = CH_3 (2-28)

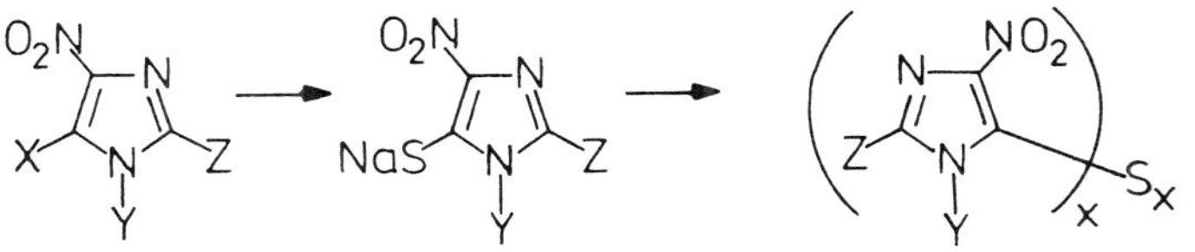

2-255b
X = Cl, Br
Y = H, CH_3, C_2H_5, n-C_3H_7, n-C_4H_9
Z = H, CH_3, C_2H_5, n-C_3H_7
x = 1, 2 (2-29)

The sodium salts (**2-255a, b**) and salts of other mercaptans were readily alkylated.[339] Examples of the arylation of 6-mercaptopurine by alkyl or dialkyl-4(5)-nitro-5(4)-haloimidazoles in ethanol containing sodium hydroxide gave the sulfides **2-256a,b**.[340-345]

A condensation between ethyl 4-(2-thioethoxy)benzoate and 1-methyl-2-bromo-5-nitroimidazole in dimethylformamide containing cuprous oxide heated at reflux temperature in a nitrogen atmosphere for 4 h gave ethyl 4-[2-(1-methyl-5-nitro-2-imidazolyl)-2-thioethoxy]benzoate (**2-257**). The structure of the corresponding acid was confirmed by X-ray crystallographic analysis. This work became the basis for correcting the structure assignments of (a)

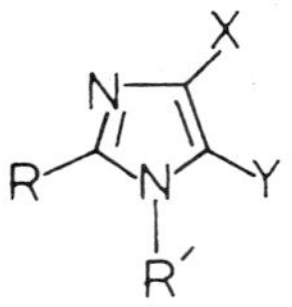

2-256a X = C$_5$H$_3$N$_4$S, Y = NO$_2$
2-256b X = NO$_2$, Y = C$_5$H$_3$N$_4$S

C$_5$H$_3$N$_4$S =

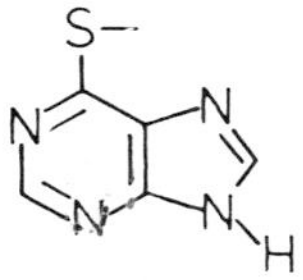

2,4(5)-diiodoimidazole to 4,5-diiodoimidazole and (b) 1-methyl-2-iodo-5-nitroimidazole to 1-methyl-4-iodo-5-nitroimidazole (**2-46**) when it was discovered that **2-46** afforded a position isomer of **2-257**.[71]

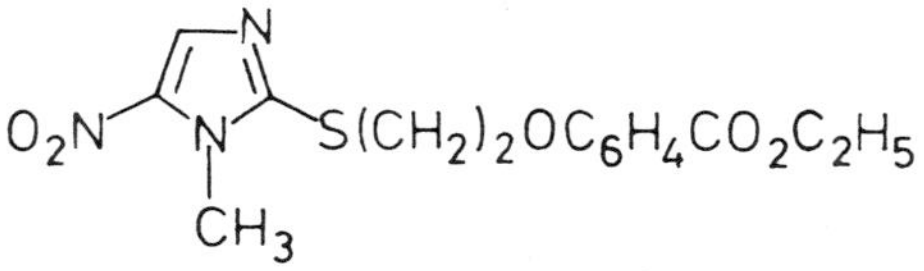

2-257

Potassium sulfide converted halonitroimidazoles to mercaptomono- and dinitroimidazoles obtained as mono- and dipotassium salts (**2-258a–h**) and as the mercaptans **2-258i–l**. The reactions with 2-bromo-4,5-dinitro and 4(5)-iodo-2,5(4)-dinitroimidazoles were sometimes sufficiently vigorous to cause spontaneous combustion.[346] Structures and yields are shown.

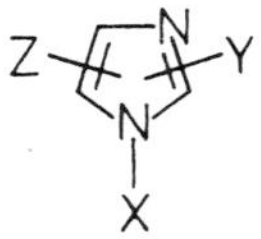

2-258a–l

2-258	X	Y	Z	yield%
a	H	2-SK	4(5)-NO$_2$	70
b	H	4(5)-SK	5(4)-NO$_2$	67
c	H	2-SK	4,5-(NO$_2$)$_2$	89
d	H	4(5)-SK	2,5(4)-(NO$_2$)$_2$	94
e	K	2-SK	4(5)-NO$_2$	88
f	K	4(5)-SK	5(4)-NO$_2$	84
g	K	2-SK	4,5-(NO$_2$)$_2$	99
h	K	4(5)-SK	2,5(4)-(NO$_2$)$_2$	99
i	H	2-SH	4(5)-NO$_2$	81
j	H	4(5)-SH	5(4)-NO$_2$	64
k	H	2-SH	4,5-(NO$_2$)$_2$	80
l	H	4(5)-SH	2,5(4)-(NO$_2$)$_2$	72

Hydrogen sulfide in ammoniacal methanol converted 4(5)-nitro-5(4)-bromoimidazole and its 2-methyl derivative and 1-methyl-4-nitro-5-chloroimidazole to ammonium salts of the corresponding mercaptans (**2-259a–c**).[1k,347] The mercaptans (**2-259a,b**) were alkylated in methanol containing sodium methoxide and the sulfides oxidized to sulfones (**2-259d,e**) by hydrogen peroxide in acetic acid.[347] The ammonium salts of the mercaptoimidazoles (**2-259f**) were converted to sulfides (**2-259g**) by treatment with chloroacetamide.[348,349]

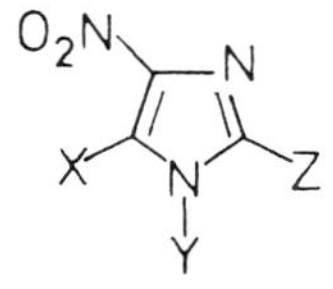

2-259

2-259	X	Y	Z
a	SH	H	H
b	SH	CH_3	H
c	SH	H	CH_3
d	SO_2CH_3	H	H
e	SO_2CH_3	CH_3	H
f	SH	$CH_2Ar‡$	H, CH_3
g	SCH_2CONH_2	$CH_2Ar‡$	H, CH_3

‡Ar = (phenyl with H(NO_2) and H(F)(Cl)(OCH_3) substituents)

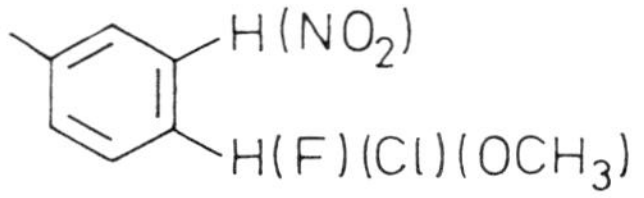

Thiophenol and *p*-substituted derivatives converted 2,4(5)-dibromo-5(4)-nitroimidazole in alcohol containing a sodium alkoxide heated at the reflux temperature for 3 h to 2-bromo-5(4)-nitro-4(5)-imidazolyl-4-(*p*-substituted)-phenyl sulfides (**2-260**).[177]

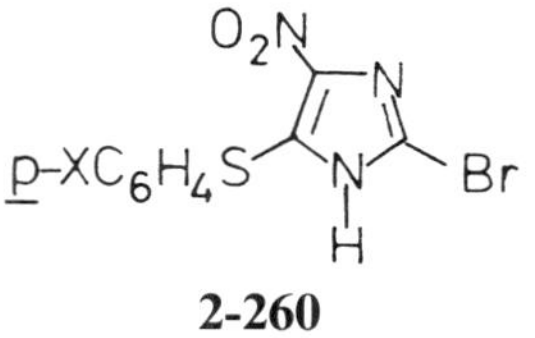

2-260	X	yield%
a	H	41
b	Cl	75
c	NO_2	58
d	CH_3	43

2-260

Replacement of Ring Nitro Groups

A review "Synthesis of Heterocyclic Systems on the Basis of Intramolecular Nucleophilic Substitution of a Nitro Group" appeared in 1982.[350]

The conversion of 2,4(5)-dinitroimidazole to 5-nitro-2,3-dihydroimidazo-[2,1-*b*]oxazoles (**2-151a**) has been noted. Another example of an intramolecular substitution of the nitro group occurred when 1-(2,3,5-tri-O-acetyl-α-D-furanosyl)-2-nitroimidazole (**2-219**) was converted by sodium methoxide in methanol to an $O^2,2'$-anhydronucleoside (**2-261**).[258]

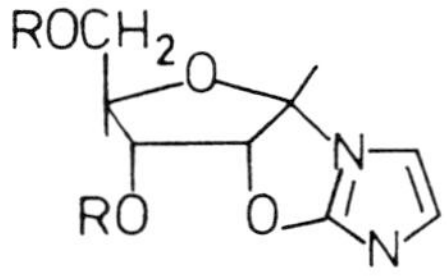

2-261 R = COCH$_3$

With recognition[70] of the probability of the intermediacy of a dihydronitroimidazoxazole, Eq. (2-30), in the assumed conversion of 2,4(5)dinitroimidazole (**2-11**) by treatment with 2-chloroethanol to 4(5)-nitro-5(4)-chloroimidazole (**2-263a**)[94] it was suggested that the product was instead the isomeric 2-chloro-4(5)-nitroimidazole (**2-263b**). The same product, **2-263a** or **b**, was also obtained from the dinitroimidazole (**2-11**) and β-chloropropionitrile.[195] Hydrogen chloride was expected as a byproduct in the reaction. A subsequent replacement of the 2-nitro substituent by chlorine in a reaction between the dinitroimidazole (**2-11**) and hydrogen chloride would lead to the formation of product **2-263b**. This explanation is supported by the formation of chloronitroimidazoles (**2-263c**) from dinitroimidazoles and concentrated hydrochloric acid.[351] In a similar reaction with concentrated hydrochloric acid 2-bromo-4-(5)-nitroimidazole was converted to 2-chloro-4(5)-nitroimidazole.[260]

Other intermolecular replacements of the nitro group are known. Bromine in acetic acid converted 2-nitroimidazole to 2,4,5-tribromoimidazole.[94] At reflux temperature hydrobromic acid converted 4,5-dinitroimidazole to 4(5)-bromo-

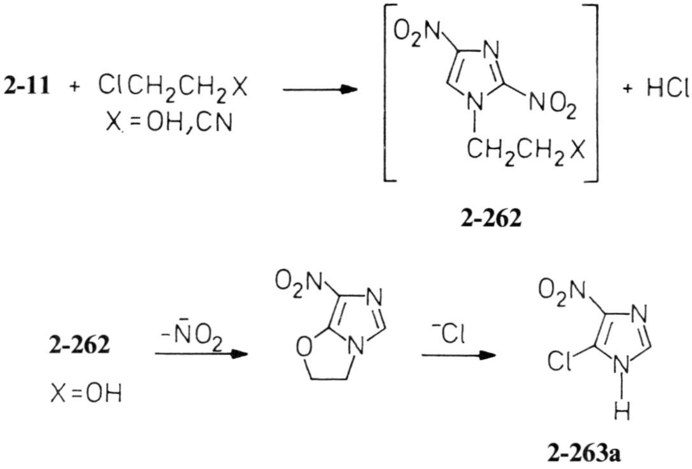

2-11 + ClCH$_2$CH$_2$X ⟶ **2-262** + HCl
X = OH, CN

2-262 $\xrightarrow{-\bar{N}O_2}$ ⟶ $\xrightarrow{-Cl}$ **2-263a**   (2-30)
X = OH

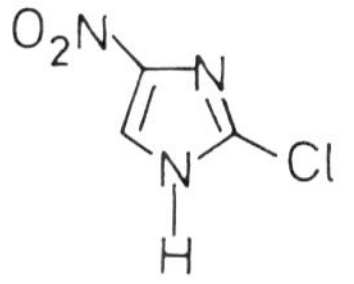

2-263b (structure: O₂N-imidazole-Cl, NH)

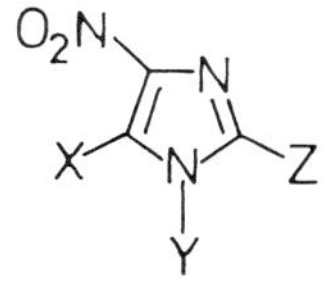

2-263c (structure with X, Y, Z substituents)

X = Y = H, Z = Cl
X = Cl, Y = H, CH₃
Z = H

5(4)-nitroimidazole.[352] Nitrohaloimidazoles (**2-264**) were also obtained from di- and trinitroimidazoles and a phosphorus oxyhalide in dimethylformamide at 70°–100°; thus 1-methyl-2,4-dinitroimidazole converted to 1-methyl-2-nitro-4-chloroimidazole.[352-354]

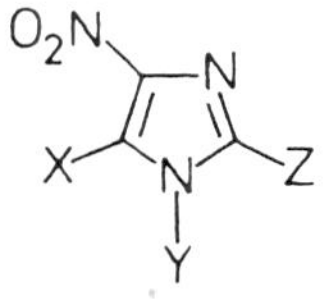

X = Y = H, Z = Cl, Br
X = Cl, Br, Y = Z = H
X = Z = Br, Y = H
X = Cl, Y = H, Z = H, CH₃

2-264

Replacement of the nitro group was reported for the reaction between 1-(2-hydroxyethyl)-2-methyl-5-nitroimidazole and excess 2-aminoethanethiol in water at pH 5.0 and under nitrogen at 37° for 120 h or at 100° for 1.5 h which gave 1-(2-hydroxyethyl)-2-methyl-4-[(2-aminoethyl)thio]imidazole (**2-265c**) (98%). At pH 9.5 a faster but more complex reaction (37°, 24 h) gave the sulfide (**2-265c**) (45%) and 1-(2-hydroxyethyl)-2-methyl-5-[(2-aminoethyl)thio]imidazole (**2-265d**) (22%). The proposed precursors to products **2-265c,d** were the adducts **2-265a,b**, Eq. (2-31).[355]

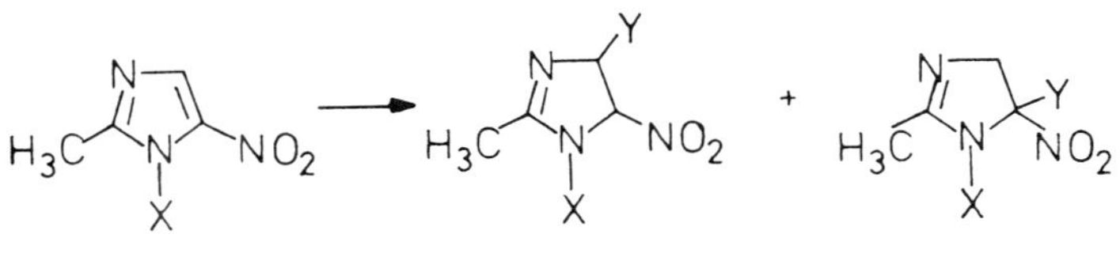

2-265a   **2-265b**

2-265a,b $\xrightarrow{-HNO_2}$ **2-265c** + **2-265d**

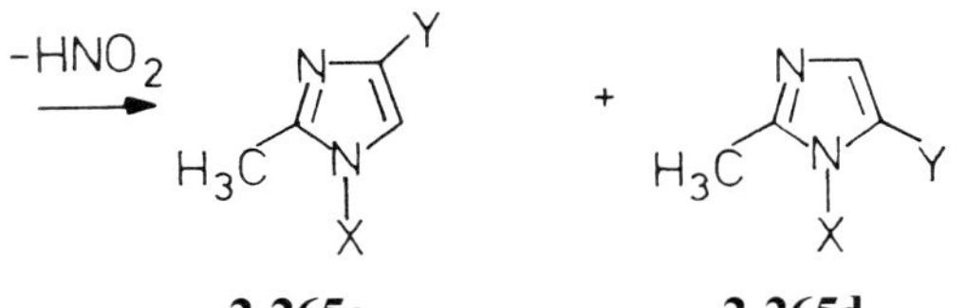

X = HOCH₂CH₂
Y = H₂NCH₂CH₂S

(2-31)

A similar replacement of a nitro group occurred in the reactions between ethyl 3-nitroimidazo[1,2-*a*]pyridine-2-carboxylate or its derivatives and the ethyl thioglycolate anion to produce ethyl 2-carboethoxyimidazo[1,2-*a*]pyridine-3-thioacetates (**2-266**) (14%–60%).[356] The reaction failed when hydrogen replaced the carboethoxy group at C-2.

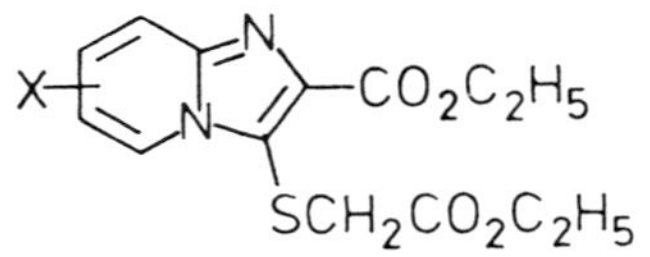

$$X-\text{[imidazo[1,2-a]pyridine]}-CO_2C_2H_5$$
$$SCH_2CO_2C_2H_5$$

2-266 X = H, 5-CH_3, 6-CH_3, 6-Cl, 6-Br, 7-CH_3, 8-CH_3

Photosubstitution of the nitro group in 1-methyl-5-nitroimidazole (the most reactive isomer) with cyano or methoxy proceeded in good yield.[357]

Apparently the heterocycle (**2-267a**) was obtained by an elimination of nitrous acid when the nitroimidazole (**2-252**) was heated to 120° in diethylamine in a sealed tube and rapidly oxidized to a brilliant blue crystalline compound, mp > 310°, insoluble in the usual solvents. The blue radical was reduced by zinc in sulfuric acid to a pink solid, mp 250–255°, which was not characterized because of its easy oxidation to the blue compound, Eq. (2-32).[335]

2-252 $\xrightarrow{120°}_{-HNO_2}$ [structure] $\xrightleftharpoons{O_2, -H}$ [structure]

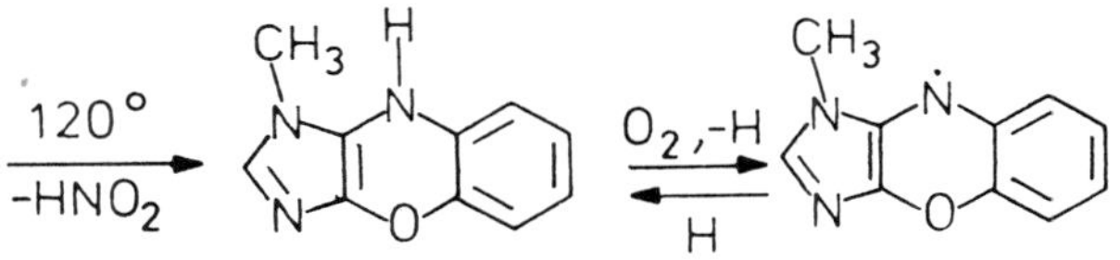

2-267a (2-32)

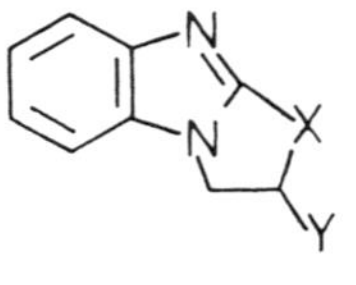

2-267b X = O, Y = CH_2OCH_3
2-267c X = N, Y = H

Conversion of 1-alkyl-2-nitrobenzimidazoles by alkali to 2-benzimidazolones occurred under conditions in which 1-alkyl-2-nitroimidazoles were stable.[101] Treatment with 1,2-epoxy-3-methoxypropane converted 2-nitrobenzimidazole to a benzimidazo[2,1-*b*]oxazoline (**2-267b**). A similar reaction with ethylenimine gave benzimidazo[2,1-*b*]imidazoline (**2-267c**).[358,359]

Treatment with sodium sulfite converted 4,5-dinitroimidazole to the sodium salt of 4(5)-nitro-5(4)-imidazolylsulfonic acid (51%).[359]

Rearrangements

A Curtius rearrangement of 4(5)-nitro-5(4)-imidazolecarboxylic acid azide in ethanol gave ethyl 4(5)-nitro-5(4)-imidazolecarbamate (**2-268a**). Aminolysis with methylamine converted the carbamate to the amine (**2-268b**). The latter was diazotized and treated with amines to give the azo compound (**2-268c**) and the triazene (**2-268d**), and with sodium azide to give the azide (**2-268e**).[360]

Phenyl migration was preferred in the Beckmann rearrangement of the oxime (**2-268f**) in polyphosphoric acid at 100° for 3 h to give the anilide (**2-268g**) (92%).[197]

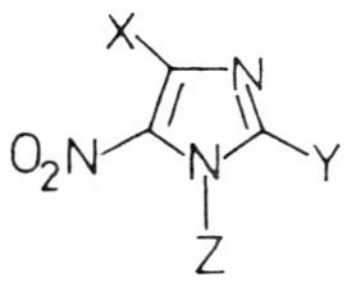

2-268	X	Y	Z
a	$NHCO_2C_2H_5$	H	H
b	NH_2	H	H
c	N=NAr	H	H
d	$N=NNR_2$	H	H
e	N_3	H	H
f	H	$C(C_6H_5)=NOH$	CH_3
g	H	$CONHC_6H_5$	CH_3

Replacements in Side Chains

2-Chloromethylimidazoles

In a series of patents replacement of halogen or pseudohalogen in side chains afforded a variety of derivatives (**2-269**) of 1-methyl-2-substituent-5-

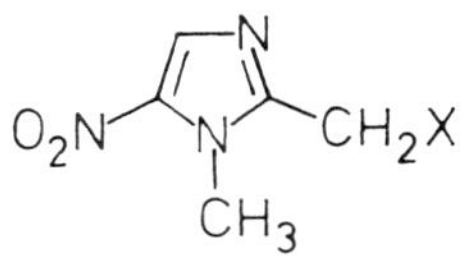

2-269	X
a	$3,4-CH_3(CH_3S)C_6H_3O-$
b	$O(CH_2CH_2)_2N-$
c	$4-H_3CN(CH_2CH_2)_2N-$
d	N_3
e	$p-ClC_6H_4COCH_2S-$
f	C_6H_5S-
g	$CH_3OC(=S)S-$
h	$H_2NC(=NH)S-$
i	CH_3NH-
j	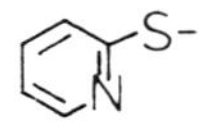 pyridyl-S-
k	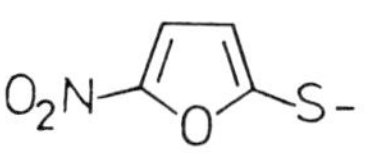 O_2N-furyl-S-
l	C_6H_5CO-
m	$4-H_2NC_6H_4O-$
n	$ROC(=S)S-$
o	$p-O_2NC_6H_4C(=S)O-$
p	thiazoyl
q	pyrimidinyl

nitroimidazoles. Generally 1-methyl-2-chloromethyl-5-nitroimidazole in a solvent, eg, dimethylformamide containing a base such as potassium carbonate, was treated with a nucleophilic reagent. Selected products are shown.[361-376]

Imidazolecarbonyl Chlorides

Amination of 1-alkyl- and 1,2-dialkyl-4-nitroimidazole-5-carbonyl chloride produced imidazolecarboxamides **(2-270)** (38%–80%).[377]

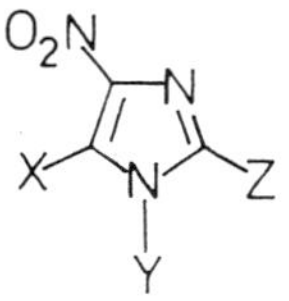

$Z = H, CH_3, C_2H_5, C_3H_7$
$Y = CH_3, C_3H_7, C_4H_9$
$X = CON(CH_2CH_2Cl)_2, CONHC_6H_4NR_2\text{-}p$
$R = CH_2CH_2Cl$

2-270

Esterifications and Ester Reactions

The phosphate ester **(2-271)** was obtained in good yield from 1-β-hydroxyethyl-2-methyl-5-nitroimidazole in phosphoric acid containing phosphorus pentoxide at 105° for 3 h,[378] or by treating the alcohol with β-cyanoethyl diacid phosphate in the presence of dicyclohexylcarbodiimide in pyridine.[379]

Three molar equivalents of benzoyl chloride in the presence of triethylamine in acetonitrile converted 1-β-hydroxyethyl-2-methyl-5-nitroimidazole to 1-(2-benzoyloxyethyl)-5-nitro-2-imidazole-α-phenylethenol benzoate **(2-272a)** in good yield. Three related esters **(2-272b–d)** (55%–67%) were similarly prepared.[380]

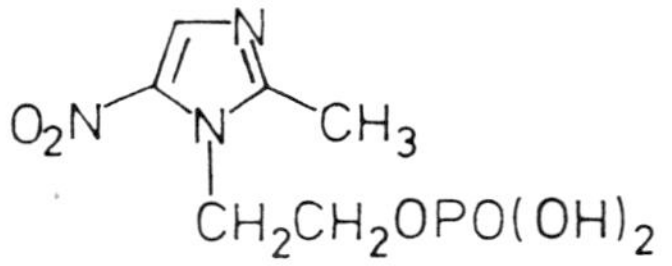

2-271

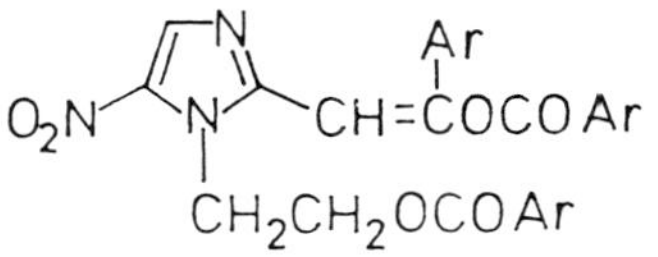

2-272a Ar = C_6H_5
2-272b Ar = $C_6H_5NO_2$-p
2-272c Ar = C_6H_5Cl-p
2-272d Ar = 2-thienyl

The enol–ester **(2-272a)** was hydrolyzed by ethanolic hydrogen chloride to the ketone **(2-273)** but hydrolysis of the esters **(2-272a–d)** with sulfuric acid (1:1) gave the imidazo[2,1-d]-1,4-oxazepines **(2-274a–d)** (52%–79%),[380] unconfirmed structures supported by nmr, mass spectrum, and analytical data.

β,β,β-Trichloroethyl chloroformate converted 1-β-hydroxyethyl-2-nitroimidazole to a carbonate ester **(2-275a)**. The latter gave a carbamate **(2-275b)** on treatment with furfurylamine.[381,382]

Treatment with benzoyl chloride, phenyl chloroformate, or phenyl isocyanate

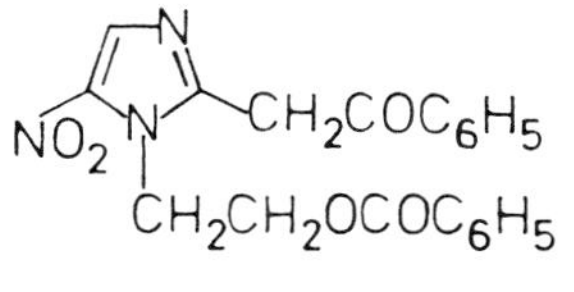

2-273

2-274a–d

2-274	Ar
a	C_6H_5
b	$C_6H_5NO_2$-*p*
c	C_6H_5Cl-*p*
d	2-thienyl

a, $X = OCH_2CCL_3$

b, X =

2-275a,b

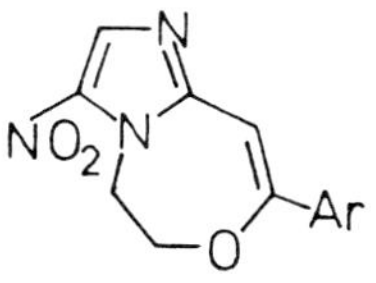

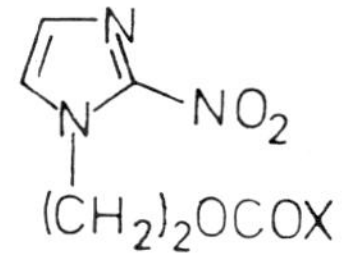

converted 1-methyl-2-nitro-5-(1-hydroxy-1-methyl)ethylimidazole to the benzoate, phenylcarbonate, and phenylcarbamate esters (**2-276a–c**).[383]

a, $X = COC_6H_5$
b, $X = CO_2C_6H_5$
c, $X = CONHC_6H_5$

2-276a–c

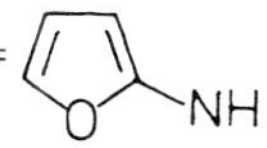

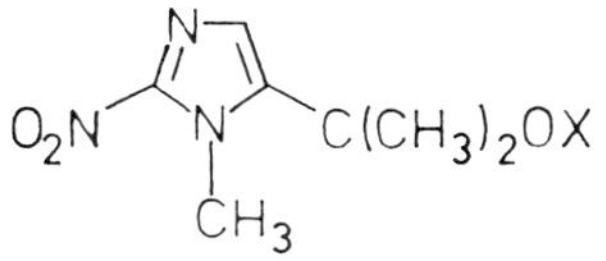

1-ω-Halo(pseudohalo)alkylimidazoles

ω-Halogen and pseudohalogen groups in 1-alkyl substituents have been replaced by treatment with metal sulfides,[384–386] thiocyanates,[387] sulfites,[388, 389] sulfinates,[390] and azides,[391] and with amines,[392] and amides, Eqs. (2-33–2-35).[393]

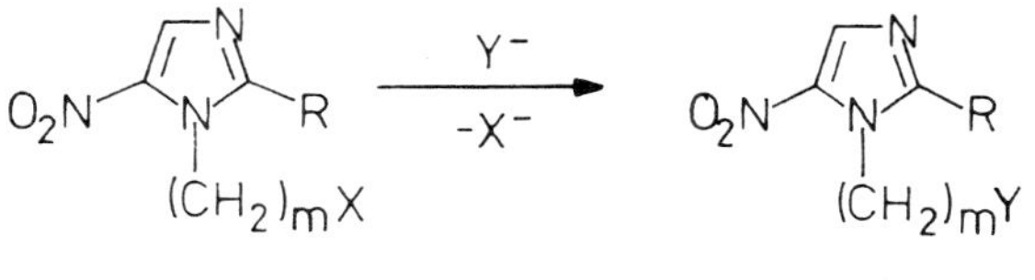

$R = C_nH_{2n+1}$, n = 1-4
$X = Cl, Br, C_7H_7SO_3$, m = 1-4
$Y = SH, SCN, HSO_3, RSO_2, N_3$ (2-33)

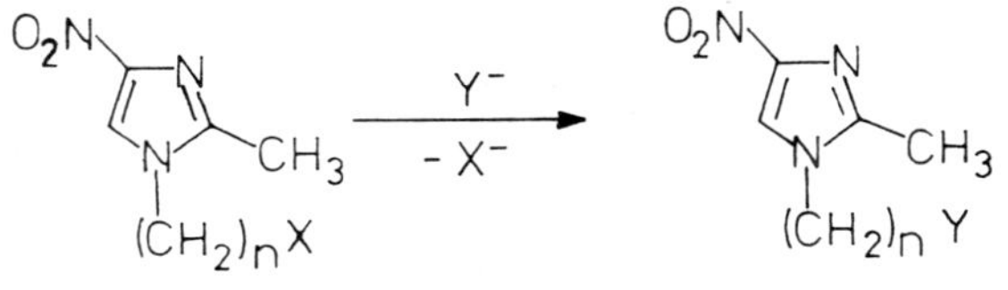

(2-34)

X = Cl, Br, n = 2, 4
Y = [6-thiopurinyl structure]

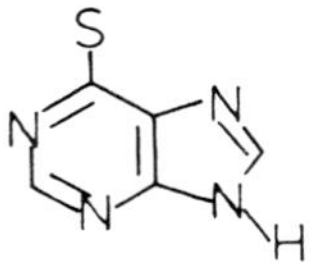

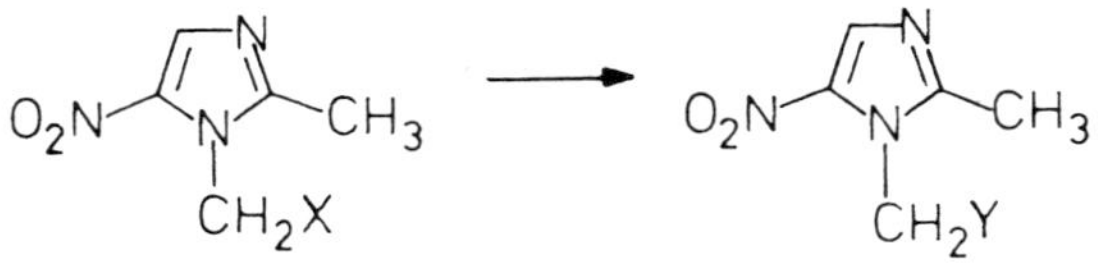

X = CH_2Cl, $CH_2OSO_2C_7H_7$
Y = N(COR)COR'
R,R' = o-C_6H_4, $(CH_2)_{2,3}$, 1,8-naphthalenediyl,
2,3-pyridinediyl, $C(C_6H_5)_2NH$, $C(CH_3)_2NH$,
and other groups
X = CHCH$_2$, Y = CHCH$_2$N(C_2H_5)$_3$
    \ /
     O

(2-35)

Oxidation in Side Chains

Lead tetraacetate and other oxidizing agents converted 1-alkyl-2-hydroxymethyl-5-nitroimidazole to the aldehyde (**2-277a**).[195,271,394-398] Manganese dioxide in chloroform gave yields of 85%–95%.[297,396]

A more imaginative preparation of the aldehyde (**2-277a**) called for an initial condensation between t-butoxymethylenebisdimethylamine and 1,2-dimethyl-5-nitroimidazole to give the dimethylaminovinyl compound (**2-277b**) which was oxidized to the aldehyde (**2-277a**) (90%) by ozone in chloroform at $-30°$ or by a mixture of osmium tetroxide and sodium periodate.[399] The aldehyde (**2-277a**) has been converted, via hydrazone derivatives, to 1,3,4-oxadiazolines,[400] and to 1,3,4-triazoles.[401]

Air or oxygen oxidized 1-methyl-2-isopropyl-5-nitroimidazole in an inert solvent containing potassium $tert$-butoxide at -20–$0°$ to the tertiary alcohol (**2-277c**) (74%).[402]

Oxidation of the nitrofuryl group accompanied nitration of 2-(2-furyl)-imidazo[1,2-a]pyrimidine on treatment with nitric acid (70%) for 2 h at 0°–5° to give products **2-278a** (16%), **2-278b** (20%), and **2-278c** (6%).[403]

A starting material for a variety of useful compounds,[10,404] 1-methyl-2-nitro-

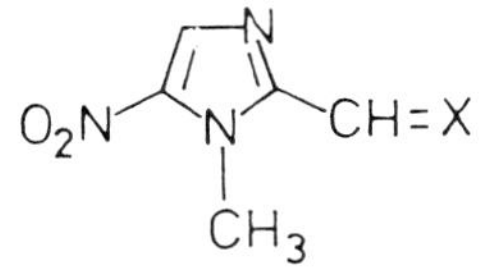

2-277a, X = O
2-277b, X = CHN(CH$_3$)$_2$

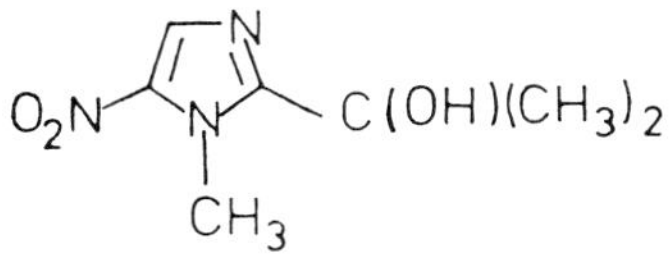

2-277c

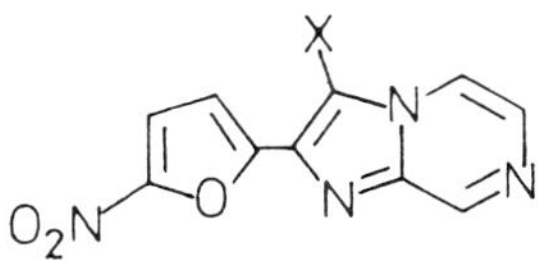

2-278a, X = H
2-278b, X = NO$_2$

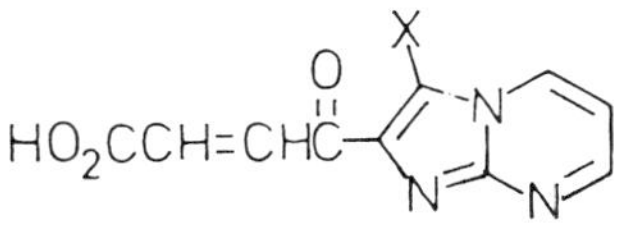

2-278c

5-formylimidazole (**2-279c**) became available by the oxidations of the 2-vinyl precursors (**2-279a** and **2-279b**), Eqs. (2-36–2-38).[110]

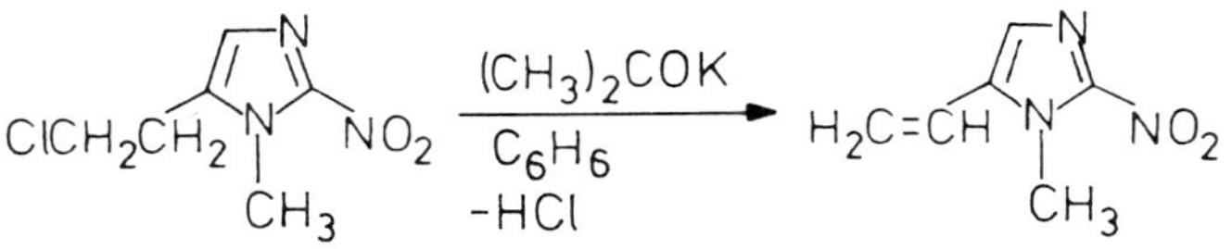

$$\text{2-279a} \quad (2\text{-}36)$$

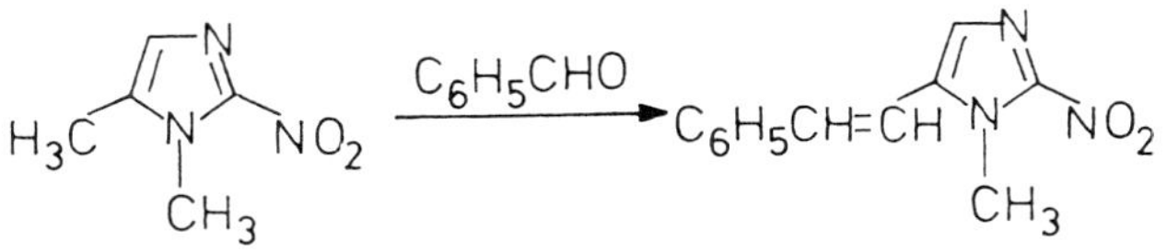

$$\text{2-279b} \quad (2\text{-}37)$$

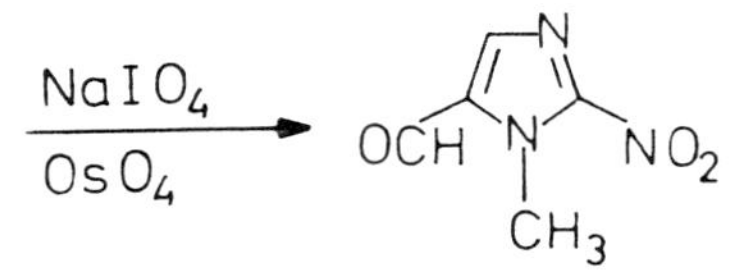

$$\text{2-279c} \quad (2\text{-}38)$$

An unspecified nitroimidazole was effective in quenching a benzo[a]pyrene radical.[405]

Monoperphthalic acid oxidized methyl 1-methyl-5-nitro-2-imidazolyl sulfide to the sulfone.[88]

Chemical Reductions

Radical Anions

Metronidazole in a phosphate buffered solution, pH 7.6, was converted to its radical anion (**2-280**) by an electron transfer from $CO_2^{-\cdot}$. The esr parameters for the intense spectrum are shown in Eq. (2-39).[406] In accord with the basicity, pK 6.1,[407] of this species a different set of esr parameters was obtained at pH 4.9.[406] Attempts to reduce metronidazole to its radical anion by sodium in tetrahydrofuran were unsuccessful as were reactions with $(CH_3)_2CO^{-\cdot}$ generated by the photolysis of acetone in isopropanol containing either sodium hydroxide or diethylamine until water (10% by volume) was added to the system. The radical anion (**2-280**) appeared to be unstable in nonpolar solutions. Radical anions were similarly obtained from 1-(2-hydroxy-3-methoxypropyl)-2-nitroimidazole, 2-methyl-4(5)-nitroimidazole, and 4(5)-nitroimidazole.[406,407]

$$HO(O)N-\text{imidazole}-CH_3 \rightleftharpoons \dot{O}_2N-\text{imidazole}-CH_3 + H^+$$
$$(CH_2)_2OH \quad\quad\quad (CH_2)_2OH$$

2-280 (2-39)

$a_{NO_2}^N = 1.440$ mT
$a^N = 0.035$ mT
$a^N \sim 0.115$ mT
$a_{CH_2}^H$ or $a_{CH_3}^H \sim 0.115$ mT
$g \approx 2.0046$
pH 4.9

$a_{NO_2}^N = 1.565$ mT
$a_1^N = a_2^N = 0.056$ mT
$a_4^H = 0.542$ mT
$a_{CH_3}^H = 0.229$ mT
$a_{CH_2}^H < 0.02$ mT
$g = 2.00445$
pH 7.6

Metmyoglobin adducts of 4(5)-nitro and 2-methyl-4(5)-nitroimidazole were reduced by the dithionate anion, SO_2^-. The primary product was assumed to be the radical anion of the nitroheterocycle portion of the complex.[408]

Quantum chemical calculations by the CNDO/2 method have been made on spin density distribution for anion and dianion radicals of nitro derivatives of pyrrole, pyrazole, imidazole, 1,2,3- and 1,2,4-triazoles. Earlier assignments of hyperfine splitting constants for 3-nitropyrazole, 4(5)-nitroimidazole, and 3-nitro-1,2,4-triazole were corrected.[409]

Radical anions of 2-, 4-, and 5-nitroimidazole derivatives converted oxygen to the superoxide radical anion at pH 8.[410]

Hydroxylamines

Azo and azoxy derivatives produced by reducing misonidazole (**2-281a**) were further reduced to the hydroxylamine (**2-281b**) by *xanthine oxidase* under hypoxic conditions.[411-413] Anaerobic reduction of nitroimidazoles by flavin mononucleotide, or by *xanthine oxidase*, gave aminoimidazole; however, the final step, reduction of a hydroxylamine, was relatively difficult. A reaction scheme involved initial formation of a nitroimidazole radical anion.[414]

2-281a, X = NO$_2$
2-281b, X = NHOH

The presence of a hydroxylamine was detected in the reduction of ^{14}C-misonidazole (2-281a) by zinc dust in an aqueous solution of ammonium chloride. By tracing radioactivity it was demonstrated that an unspecified reduction product combined with calf thymus DNA.[415,416] Intermediate reduction products from nitroimidazoles may be required for their therapeutic and mutagenic activity.

Fragmentation of metronidazole (2-200) (X = OH) resulted from treatment with hypoxanthine as an electron donor in an *in vitro* system containing milk *xanthine oxidase*. Products included N-(2-hydroxyethyl)oxamic acid (2%), N-glycoylethanolamine (3%), N-acetylethanolamine (2%), ethanol amine (15%), ethyl acetate (5%), acetamide (14%), and glycine (8%).[417]

Amines

Generally 4- and 5- aminoimidazoles were too unstable for isolation[1a] and reduction of 4- and 5-nitroimidazoles has been restricted by the opportunity to convert the amines on formation, or shortly thereafter, to derivatives.

Reduction of 4(5)-nitroimidazole by zinc dust in 50% tetrafluoroboric acid was rapid at $-10°$ to $0°$ (nitrogen atmosphere) and gave little evidence of decomposition. Without isolation the amine was diazotized and the diazonium tetrafluoroborate photolyzed to give 4(5)-fluoroimidazole (17%). The procedure afforded 4(5)-fluoro-5(4)-methylimidazole (37%),[418,419] 4-fluoro-1-methyl-imidazole (8%) and 5-fluoro-1-methylimidazole (2%). For comparison diazotization of 2-amino-1-methylimidazole followed by photolysis gave 2-fluoro-1-methylimidazole (48%). Reductions of 4- and 5-nitroimidazoles by sodium amalgam in methanol did not improve the overall conversion to fluorides.[420]

In a structure proof for 2,4(5)-diphenyl-5(4)-nitroimidazole stannous chloride in hydrochloric acid gave the known 2,4(5)-diphenyl-5(4)-aminoimidazole.[120] Stannous chloride also reduced 9-methyl-3-nitro-2-phenylimidazo[1,2-a]-benzimidazole (2-142) to give a complex tin salt (2-282a) from which the amine could not be liberated. Heating in boiling dilute alcohol converted the complex to 2-(α-carboxybenzylamino)-1-methylbenzimidazole (2-282b) (94%). Similar results were obtained with the 9-benzyl derivative.[421] In a third example stannous chloride in hydrochloric acid reduced 1,2-dimethyl-4-isopropyl-5-nitroimidazole to the expected 5-amino derivative. Without isolation it was converted by aqueous sodium hydroxide to valine[$(CH_3)_2CHCH(NH_2)CO_2H$], acetic acid, ammonia, and methyl amine.[49]

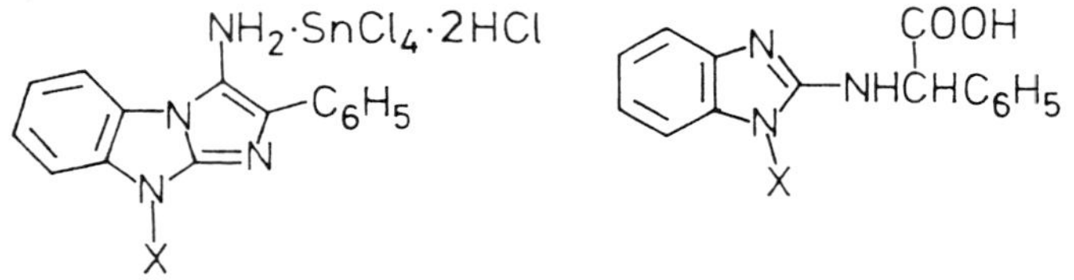

2-282a, X = CH_3, $C_6H_5CH_2$ **2-282b**, X = CH_3, $C_6H_5CH_2$

See **OTHER PHYSICAL PROPERTIES** for reduction of nitroimidazoles to aminoimidazoles by titanium (III) chloride for chromatography.

A surprising stability for 2-methyl-3-amino-6-chloroimidazo[1,2-b]-pyridazine (**2-283**) permitted its isolation (45%) in a reduction of the 3-nitro substituent by treatment with molybdenum (V) chloride in aqueous tetrahydrofuran containing zinc dust heated at reflux temperature until the nitro compound disappeared.[422]

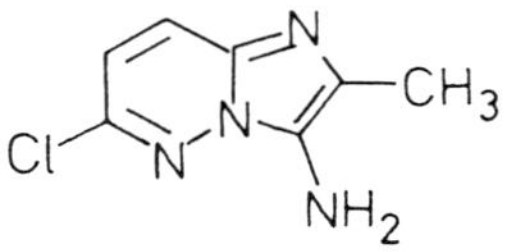

2-283

Sodium borohydride reduced 1-methyl-2-nitro-5-formylimidazole to 1-methyl-2-nitro-5-hydroxymethylimidazole (64%).[404]

Titanium trichloride, known to reduce nitro to amino groups quantitatively (Chapter 1) was developed as a spray applied to paper chromatographs in the analysis of nitroimidazoles. An overspray of a chromogenic reagent gave a sequence of color formations.[18]

In place of an anticipated catalytic (Pd/C) transfer hydrogenation, hydrazine

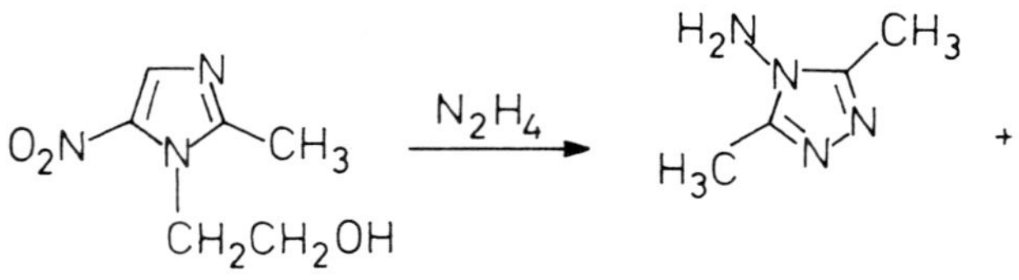

$$(H_2NN=CH)_2 + HOCH_2CH_2\overset{+}{N}H_3\overset{-}{N}O_2 \qquad (2\text{-}40a)$$

in a mixture of tetrahydrofuran and ethanol for 4 h at 55° converted metronidazole **2-200** (X = OH) to a mixture of 3,5-dimethyl-1,2,4-triazole-4-amine (66%), glyoxal dihydrazone (54%), and ethanol ammonium nitrite (47%), Eq. (2-40a). The same reaction occurred in the absence of the metal catalyst.[423]

Catalytic Reduction

The preparation of imidazole amines by the catalytic reduction of nitroimidazoles has also been severely limited by the instability of the amines. In one example hydrogenation over Adams' catalyst converted 2-nitroimidazole to an unstable oil which gave a positive Pauli test.[15] In another example catalytic reduction of 4(5)-nitroimidazole was abandoned when excessive amine decomposition was encountered.[418] It was stated without further detail, that 2-methyl-4(5)-nitroimidazole was reduced to the amine in formic acid over a palladium–charcoal catalyst.[423,424]

An improved stability of 1-alkyl derivatives of 4- and 5-aminoimidazoles apparently accounted for the preparation of the amines (**2-284**) (26%–37%) by reduction of nitroimidazoles over Raney nickel in acetic anhydride at 98°–100°.[425] Slightly higher yields were obtained from the hydrogenation of chloronitroimidazoles.

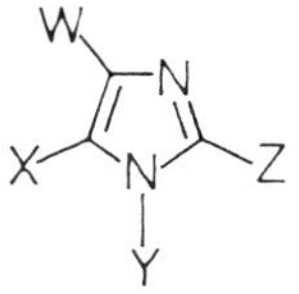

W = AcNH, H
X = H, Ac$_2$N
Y = C$_n$H$_{2n+1}$, n = 1-4
Z = H, C$_n$H$_{2n+1}$, n = 1-3

2-284

Hydrogenation over Raney nickel converted 2-benzoyl- and 2-methylsulfonyl-1-methyl-5-nitroimidazoles to unstable amines (**2-285a,b**) which were immediately converted to amides or thioureas.[426] A similar reduction in acetic anhydride converted 1-ethyl-2-methyl-4-nitro-5-aminoimidazole to the diamine (**2-285c**), isolated as the diacetate or the tetraacetate.[80]

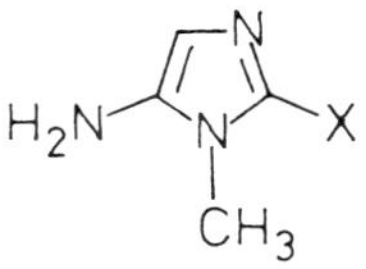

2-285a, X = COC$_6$H$_5$
2-285b, X = SO$_2$CH$_3$

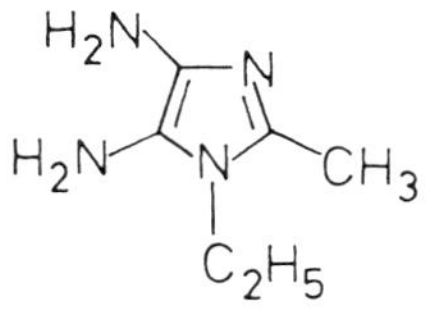

2-285c

Hydrogenation over palladium on charcoal converted 6-chloro-5-nitroimidazol[2,1-b]thiazole (**2-90**) in a mixture of acetic acid and acetic anhydride to the corresponding acetamido compound.[132] Palladium on charcoal also catalyzed the hydrogenation of metronidazole to 1-(2-hydroxyethyl)-2-methyl-5-aminoimidazole. The identification of the amine was determined by uv absorption, mass, and nmr spectra.[427]

Hydrogenation of 5(4)-nitroimidazole-4(5)-sulfonamide in a mixture of acetic acid and acetic anhydride over platinum gave a low yield of 5-acetamido-1-acetylimidazole-4-sulfonamide (8%).[427,428] A similar hydrogenation over 5% Pd/C at 1 atm for 1 h gave the free amine (**2-286**) (80%). When kept below $-30°$ the amine was stable for several weeks, but even mild heating brought about rapid decomposition. On the other hand, the amine in a dilute $(10^{-4}\ M)$ aqueous solution (buffered pH 7) at 37° for three weeks did not lose uv optical density.[429] Other difficulties in the preparation of the amine (**2-286**) by catalytic reduction schemes were attributed to its instability.[347,430-432] Both the 1-methyl and 1-benzyl derivatives (**2-287a–d**) were described as stable.[433]

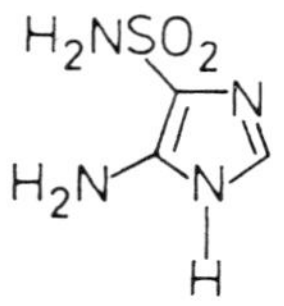

2-286 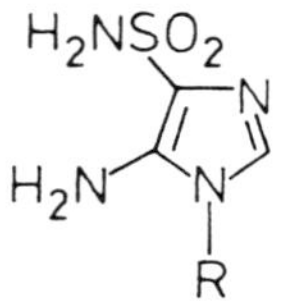   **2-287a,b** 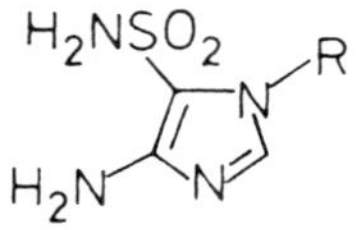   **2-287c,d**

a,c; $R = CH_3$
b,d; $R = C_6H_5CH_2$

Nitronucleosides were reduced to aminonucleosides (**2-288a,b**) by hydrogenation over platinum or palladium in a Parr apparatus.[434]

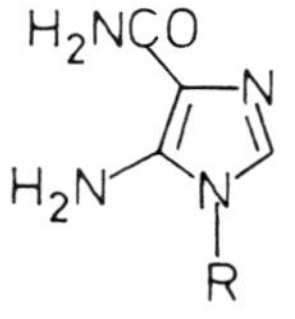

2-288a

$R = \alpha\text{-D-}, \alpha\text{-L-arabinofuranosyl},$
$\beta\text{-D-ribofuranosyl},$
$\beta\text{-D-xylopyranosyl},$
$\alpha\text{-D-}, \alpha\text{-L-arabinopyranosyl}$

2-288b

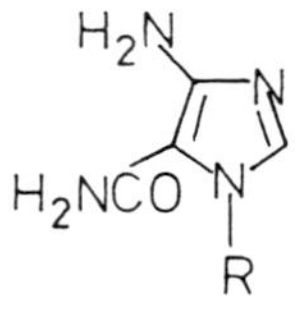

$R = \beta\text{-D-xylopyranosyl},$
$\beta\text{-D-ribopyranosyl},$
$\alpha\text{-D-}, \alpha\text{-L-arabinopyranosyl}$

Hydrogenation of 2-nitro-1-β-D-ribofuranosylimidazole over platinum oxide gave the expected aminoimidazole nucleoside (81%).[435]

Electrochemical and Radiolysis Reduction

One-electron reduction potentials produced by pulse radiolysis[436-438] at pH 7 (E_7^1) and polarographic half-wave potentials[439] of nitroimidazoles have been determined in the collection of data to test the postulate[440] that radiosensitization by a nitro compound was related to its electron affinity. Values for E_7^1 when measured against a quinone couple ranged from -243 mV for 1-methyl-2-nitro-5-formylimidazole, -398 mV for 1-(2-hydroxyethyl)-2-nitroimidazole, -486 mV for 1(2-hydroxyethyl)-2-methyl-5-nitroimidazole to $\leqslant 527$ mV for 4(5)-nitroimidazole. Generally 2-nitroimidazoles showed more positive E_7^1 values and were stronger oxidants than the 5-nitroimidazoles and the 4-nitro analogs were the weakest oxidants.[436] Polarographic half-wave potentials of aqueous nitroimidazoles at pH 7.4 were more positive (150 mV) than E_7^1 values and only qualitatively reflected their redox properties.[441]

Electron affinities of nitroheterocyclic compounds were calculated by HAM/3 and correlated with the E_7^1 values.[442]

Two peaks, -0.7 V to 1.5 V and -1.9 V to -2.4 V, for two one-electron irreversible changes in electrochemical reductions of 2 and 4(5)-nitroimidazoles, 3- and 4-nitropyrazoles, and 3-substituted-1,2,4-triazoles were obtained and confirmed the existence of colorless dianion radicals for the heterocycles.[202,203,443-448] An N-alkyl nitroimidazole in acetonitrile formed a stable yellow–green anion radical by a reversible one-electron change at -1.45 V to -1.75 V. The solution turned red and lost its paramagnetism at the second half-wave potential at -2.5 V to 3.0 V; the second wave was analogous to the second wave for nitrobenzene obtained under identical conditions. A two-step process was proposed, Eq. (2-40b), and supported by esr signals for eight anion radicals and 3 dianion radicals of nitroimidazoles.[447,448] This interpretation has been challenged.[448] A polarographic identification of the position isomers obtained from either nitrating haloimidazoles or N-alkylating halonitroimidazoles has been described.[188]

$$\underset{Y}{\underset{|}{\overset{W}{\underset{X}{\bigwedge}}}}\text{-}Z \xrightarrow{+e^-} \underset{Y}{\underset{|}{\overset{W}{\underset{X}{\bigwedge}}}}\text{-}Z \xrightarrow[4H^+]{3e^-} \text{Reduction Products} \qquad (2\text{-}40\text{b})$$

Identical polarographic $E_{1/2}$ values and identical ultraviolet absorption spectra were obtained for 4(5)-nitroimidazole and its 1-methyl-4-nitro derivative in acid solution.[203]

Imidazoles as Acids

NH Acids

Potentiometric and spectrophotometric determinations of NH acid strengths have been made.[21,75] The introduction of a nitro substituent increased the acidity from pK_{BH}^+ 6.95 for imidazole to pK_a −0.81 and −0.16 for 2- and 4(5)-nitroimidazole.[21] Incremental differences of −7.76, −7.83, −5.13, and −7.11 correlated with the nitro group at positions 2,4,5, and 4(5) for the pK_a values of 12 nitroimidazoles relative to that for the unsubstituted compound. Appreciable errors in their determination were expected for pK_a values so low (independent assignments for 2-nitroimidazole, pK_a −0.20, and for 4(5)-nitroimidazole, pK_a −0.05, have been made).[449] Consequently calculations of tautomer ratios in 4(5)-nitroimidazole systems, which depended on dissociation constants of a compound and one of its N-methyl derivatives, were not always reliable.[21,450] Attempts to apply linear free-energy relationships to correlate substituent effects with pK_a values for nitroimidazole derivatives have been partially successful.[28,449,451]

Nitroimidazoles are also weak acids with pK values of 7.15 and 9.20 for 2- and 4(5)-nitro derivatives; 2,4(5)-dinitroimidazole is a considerably stronger acid with a pK of 2.85.[21] Although a pK value for 2,4,5-trinitroimidazole has not been reported it is known to form a stable ammonium salt.[121,452]

When unsubstituted at nitrogen, nitroimidazoles tend to be stable to alkaline solutions, whereas N-substitution has often facilitated alkaline degradation.[188,257,317]

In an investigation of 16 pairs of isomeric 4- and 5-nitroimidazoles the R_f value (movement on silica gel plates brought about by chloroform and methanol, 97:3 or 97:5) for the 5-nitro isomer (presumably the stronger base)[72] was the larger except in one pair where 1-p-nitrophenyl-2-methyl-4-nitroimidazole and its 5-nitro isomer had the same value, R_f 0.65.[453] A smaller chromatographic distribution coefficient had previously distinguished 4-nitroimidazoles from their 5-nitro isomers.[69]

The ammonium salts of nitro derivatives of imidazoles, pyrazoles, triazoles, and tetrazoles were analyzed by titration with silver nitrate or cetylpyridinium chloride. When a C-5 hydrogen was present titration was feasible only with silver nitrate and when there was no hydrogen then titration was feasible only with cetylpyridinium chloride. Generally the presence of hydrogen in any position promoted the formation of a precipitate on treatment with silver nitrate.[454] The ammonium salt of 2,4(5)-dinitro-5(4)-trinitromethylimidazole decomposed at 135°.

Sodium and/or potassium salts of various mono-, di-, and trinitroimidazoles have been prepared;[455] many have been subjected to kinetic investigations of thermal decomposition.[456] Cobalt (II) and nickel (II) complexes of 4-nitroimidazole were prepared for evaluation of their behavior at high temperatures.[457] Complexes between cobalt chloride and nitroimidazoles were studied spectrophotometrically and calorimetrically, and the electric conductivity of lithium, sodium, potassium, rubidium, cesium, and silver salts of mono- and dinitroimidazoles were determined.[458-463]

A silver salt, $C_3H_2O_2N_3Ag$, was obtained from 2-nitroimidazole (azomycin) and silver nitrate in dilute ammonium hydroxide.[15]

Addition of potassium chloroplatinate to a suspension of 1-β-hydroxyethyl-2-

methyl-5-nitroimidazole (**2-200**) (X = OH) (metronidazole) in water at 50° brought about the formation of a platinum metronidazole complex, *cis*-Pt(metronidazole)$_2$Cl$_2$ (91%). The complex is a radiosensitizer toward hypoxic tumor cells. Its structure was established by X-ray crystallography.[464,465]

CH Acids

Deuterium isotope exchanges at the C-2, C-4, and C-5 positions in imidazole, C- and N-methylimidazoles, and their fluoro and nitro derivatives have been investigated.[29,267,268,466] Base-catalyzed exchanges (D$_2$O, 50°, pD > 5) for imidazole and its C- and N-methyl derivatives occurred at C-2 and more slowly (100°) at C-4 and C-5 positions via ylides (**2-289a–c**). In strong alkaline solution N-H imidazoles exchanged more rapidly at C-4(5) via carbanions (**2-289d–f**). An exchange by the carbanion pathway occurred at the C-5 position in 1-methylimidazole but failed to occur at the C-4 position. An inhibition to carbanion formation at the C-4 position was attributed to an adjacent lone pair (ALP) effect derived from an electrostatic repulsion between lone pairs at N-3 and C-4 positions. At the more complex C-2 position an ALP effect was either very weak or nonexistent. Strong acid promoted an isotope exchange by an electrophilic attack on the heterocycle.

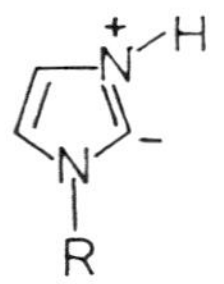

2-289a 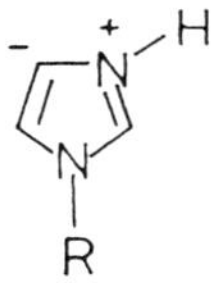   **2-289b** 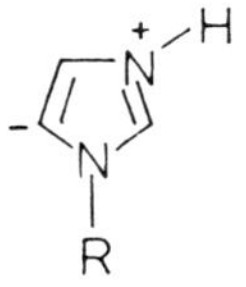   **2-289c**

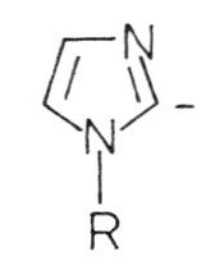

2-289d 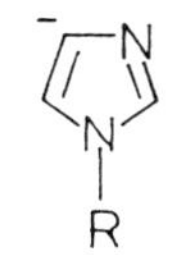   **2-289e** 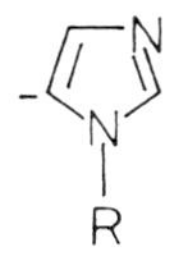   **2-289f**

Base-catalyzed deuterium exchanges (D$_2$O, 50°, 0.2 N NaOD) for nitroimidazoles required a combination of ylide and carbanion intermediates. In 4-nitroimidazoles carbanion exchange was faster at C-5 than at C-2 and in the 2- and 5- nitro derivatives of 1-methylimidazole exchange at C-4 was restricted to the ylide pathway in response to the ALP effect. The faster exchange at C-2 in 1-methyl-5-nitroimidazole than at C-5 in 1-methyl-2-nitroimidazole may have resulted from the extra induction from N-3 but it also afforded the observation that the C-2 position was electronically disturbed by the C-5 nitro group to a greater extent than was the C-5 position disturbed by a C-2 nitro group.[267]

For exchanges by carbanion (C) reactions in 4-X-imidazoles and in 1-methyl-

4-X-imidazoles (X = NO_2, F, H, CH_3, CO_2, C_2H_5, CF_3) values of log $K_{c(5)}$ (where K is a dissociation constant) correlated with σ_o^0 in the former series and with σ_p^0 in the latter. For exchanges by ylide (Y) reactions log $K_{y(2)}$ in both series correlated with σ_m^0. Other exchanges "across" the ring also showed correlations of log K with σ_m^0.[267]

SPECTROSCOPY

Nuclear Magnetic Resonance Spectroscopy

Hydrogen

Chemical shifts (δ) were obtained for 50 simple mono- and dinitroimidazoles in deuterochloroform ($CDCl_3$) and in perdeuterated dimethylsulfoxide (DMSO-d_6). Values ranged from 7.26–8.10 for C-2H, 6.73–8.23 for C-4H, and 6.73–8.73 for C-5H. An inspection of $\Delta\delta$(DMSO-d_6-$CDCl_3$) for C-4H and for C-5H in 5- and 4-nitro-1-alkyl or arylimidazoles revealed that $\Delta\delta > 0.45$ ppm correlated with a 4-nitro derivative and $\Delta\delta < 0.3$ ppm correlated with a 5-nitro derivative. For C-2H $\Delta\delta$ ranged from 0.30–0.75 ppm. An attraction between the positively charged sulfur atom of DMSO and the lone pair of electrons on the substituted nitrogen atom of the imidazole ring was tentatively proposed to produce an inductive deshielding of adjacent C-2 and C-5 protons with a lesser effect on the nonadjacent C-4 protons.[453] Similar results and conclusions came from earlier and simultaneous studies on the six isomeric dimethylnitroimidazoles.[267]

Water induced shifts, $\Delta\delta(H_2O-CDCl_3)$, have also been noted.[267,453]

Structure assignments were based on $\Delta\delta$ values. For 4(5)-nitroimidazole, its 2-methyl derivative, and for 2-chloro-4(5)-nitroimidazoles, the 4-nitro tautomers predominated according to $\Delta\delta$ values of 8.33−7.95 = 0.38, 8.20−7.76 = 0.44, and 8.43−7.87 = 0.56. Methylation of 2,4(5)-dinitroimidazole under acidic, neutral, or basic conditions gave predominantly 1-methyl-2,4-dinitroimidazole ($\Delta\delta = 0.85$ ppm) and a lesser amount of the 5-nitro isomer ($\Delta\delta = 0.15$ ppm).[453]

Assignment of a 1-, 2-, or 3-methyl substituent in a 5-nitroimidazole was based on characteristic signals (δ): 3.99–3.93 (1-methyl), 2.79–2.34 (2-methyl), and 4.13–4.04 (3-methyl).[327a] Methyl and methylene protons in 1-alkyl groups adjacent to a 5-nitro substituent were deshielded by 0.04 ppm to 0.46 ppm relative to signals obtained for the 4-nitro isomers.[453]

Amidinium type resonance stabilization of a protonated imidazole accounted for downfield shifts of 1.26 ppm and 0.52 ppm for signals from C-2H and from C-4H and C-5H. Similar differences between spectra obtained for deuterochloroform and trifluoroacetic acid solutions of the 4-nitro and 5-nitro derivatives of 1-methylimidazole showed downfield shifts of 1.40 ppm and 1.39 ppm for C-2H, 0.51 ppm for C-5H, and 0.45 ppm for C-4H; the 2-bromo and 5-bromo derivatives of 1-methyl-4-nitroimidazole showed downfield shifts of 0.35 ppm for C-5H and 1.24 ppm for C-2H. Comparable results were obtained when sulfuric acid replaced trifluoroacetic acid.[468]

A coupling constant for 4(5)-nitroimidazole in DMSO-d_6 was found to be $J_{2,5(4)} = 1.3$ Hz; $J_{2,4} = 1.1$ Hz and $J_{2,CH^3} = 0.6$ Hz for 1-methyl-5-nitroimidazole, and $J_{2,5} = 1.5$ Hz for 1-methyl-4-nitroimidazole were also obtained. In concentrated sulfuric acid 1-methyl-5-nitroimidazole gave $J_{2,4} = 1.5$ Hz and $J_{3,4} = 3.2$ Hz; 1,2-dimethyl-5-nitroimidazole gave $J_{3,4} = 3.3$ Hz.[466] A dependence of N-1H and N-3H in the conjugate acids of 1-substituted 4- and 5-nitroimidazoles on the relative position of the nitro group was proposed as a basis for assignment to one series or the other.[50]

For 1-methyl-4-nitroimidazole-5-sulfonamide, δ 8.14 (SO_2NH_2) and pK_a 8.1, and for 1-methyl-5-nitroimidazole-4-sulfonamide, δ 7.69 (SO_2NH_2) and pK_a 8.7, were obtained. Additional examples are needed to assess the significance of the observation that the isomer with the sulfonamide signal further downfield had the lower pK_a value.[93]

Other records, sometimes tabulated, of ^1H NMR for nitroimidazoles are available.[91,195,211,214,217,248,280,334,402] Reports with information on selected compounds have also appeared.[39,51,56,63,70,71,95,106,109,110,113,156,183,198,199,255,272,276,322,327,332,333,355,380,467,468]

Carbon

The group of simple mono- and dinitroimidazoles analyzed for ^1H NMR (preceding section) was also analyzed for ^{13}C nmr. The spectra were obtained from solutions in $CDCl_3$, DMSO-d_6, or mixtures of the two at 22.63 mHz. Chemical shifts (δ) were measured downfield from TMS. Solvent dependent chemical shifts, a characteristic of the ^1H spectra, were not encountered. Signals (δ, ppm) for C-2, C-4, and C-5 overlapped considerably and ranged from 81.5 for C-5 in 1-methyl-4-nitro-5-iodoimidazole to 155.7 for C-2 in 4(5)-nitro-5(4)-pyrrolidinomethylimidazole. Isomeric pairs of 4-nitro and 5-nitro-1-alkyl or arylimidazoles were distinguished by two narrow ranges, one at 120–124 ppm attributed to C-5 signals in 4-nitro compounds and the other at 130–133 ppm to C-4 signals in 5-nitro compounds. Both the C-4 and the C-5 signal exhibited one-bond proton coupling of about 200 Hz, but in addition the C-5 signal had extra multiplicity from three-bond coupling with α-protons in 1-substituents. There were a few exceptions to these generalizations, eg, C-4 signal for 1-methyl-2,5-dinitroimidazole came at 129.2 ppm, and the C-5 signals for 2,4-dinitroimidazole and its 1-methyl derivative came at 115.4 and 126.5 ppm.[453,469]

Signal differences for C-4 and C-5 were too insignificant to distinguish derivatives of 1-substituted-2-nitroimidazoles. Both 4(5)-nitroimidazole and its 1-methyl derivative gave chemical shifts at 118.6 and 119.1 ppm each with a large one-bond coupling of about 200 Hz. This was offered as support for the predominance of 1(H)-4-nitroimidazole and its 1-methyl derivative.[453,469]

One-bond proton coupling, larger for C-2 (208 Hz) than for C-4 or C-5 (190 Hz) in imidazole, contributed to the revision in assignment of 2-iodo-4-nitroimidazole to 4-iodo-5-nitroimidazole (predominant form).[70]

A brief account of chemical shift changes brought about by ionization of nitroimidazoles is found in the next section.

Nitrogen

Imidazole, nitroimidazoles, and their anions and cations were investigated in a study extended to the pyrrole counterparts. Imidazole, 4(5)-nitro-(**2-4**), 1-methyl-4-nitro- (**2-14**), 2-methyl-4(5)-nitro- (**2-16**), 4,5-dinitro- (**2-10**), and 2-methyl-4,5-dinitroimidazoles were examined.[470,471]

^{14}N chemical shifts were determined for DMSO solutions and reported in ppm from nitromethane. Signals for the nitrogen atom in the nitro groups of selected compounds **2-4**, **2-14**, **2-16**, and **2-10** were recorded at -16 ± 2 (170), -17 ± 2 (210), -16 ± 2 (340), and -28 ± 2(240) ppm with the ^{14}N linewidths measured at 25° given in parentheses. An additional entry of -28 ± 1 (48) was given for compound **2-10** in acetone. A signal for a one-ring nitrogen atom was observed for each of the six imidazoles examined and for the selected compounds **2-4**, **2-14**, **2-16**, and **2-10** occurred at -202 ± 5 (185), -208 ± 5 (375), -203 ± 5 (390), and -158 (520) with linewidths again in parentheses. The value -158 (520) for 4,5-dinitroimidazole was determined for an acetone solution. The signal intensity corresponded to a one-ring nitrogen atom and it was assigned to the trigonal nitrogen atom because of a closer relationship with the signal for the pyrrole ring nitrogen at $\delta = -229$ ppm ± 2 (137) than with the signal for the pyridine ring nitrogen at $\delta = -68$ ppm. It was suggested that the signal for the second ring nitrogen atom was too broad to be noted.[470]

Chemical shift changes as an imidazole was converted to an anion or a cation were expressed as follows:

$\Delta = \delta$(charged species) $- \delta$(neutral compound). Generally protonation of an unsubstituted heterocycle brought about a diamagnetic shift ($-\Delta$) and the formation of an anion brought about a paramagnetic shift ($+\Delta$).[472-475] For the imidazolium cation, $\Delta = -40$ ppm, and for the imidazole anion, $\Delta = +20$ ppm, for the ring nitrogen atom were observed. For 4(5)-nitroimidazole (**2-4**) a diamagnetic shift for a ring nitrogen atom was not observed; however, $\Delta = -14$ ppm was observed for the nitrogen atom in the nitro group of the cation; paramagnetic shifts, $\Delta = +73$ and $\Delta = +2$, were obtained for a ring nitrogen atom and the nitrogen atom in the nitro group of the anion.[471]

Chemical shift changes (Δ) in the ^{13}C NMR spectra also accompanied ionization of imidazole and its nitro derivatives. Δ Values for the cations were small and ranged from -9.9 ppm to $+1.2$ ppm for C-4 and C-5 in the 4(5)-nitroimidazolium cation and from -1.4 ppm for C-4 in the anion of 4(5)-nitroimidazole to $+15.2$ ppm for C-5 in the anion of 2-methyl-4(5)-nitroimidazole.[471]

A qualitative parallel correlation between chemical shift changes and changes in the ultraviolet spectra for the heterocycle and its charged species was attributed to energy changes in molecular excited states.[471]

Infrared Spectroscopy

A table of infrared (IR) absorption bands (v) for 30 simple mono-, di-, and trinitroimidazoles and their sodium, potassium, or ammonium salts appeared in

1967.[455] The well-known strong asymmetric (v_{as}) and symmetric (v_{sym}) stretching frequencies for the C-nitro group in an uncharged molecule were found in the ranges 1586–1517 and 1408–1320 cm^{-1};[455,468] whereas for the anions the corresponding ranges were found to be 1200–1120 and 1108–950 cm^{-1}. When a 4-trinitromethyl substituent was present the limit for v_{sym} extended to 1633 cm^{-1} for an uncharged molecule. Strong bands ascribed to imidazole ring vibrations were found in the region 1520–1400 cm^{-1}. Imidazole was the richest with six strong bands at 1584, 1550, 1502, 1485, 1454, and 1332 cm^{-1}.[455] Earlier assignments[7] for absorption bands for 4(5)-nitroimidazole (**2-4**) at 1499 and 1384 cm^{-1} and for 2-nitroimidazole (**2-5**) at 1493 and 1375 cm^{-1} to represent the nitro group and bands for **2-5** at 1540 and 1520 cm^{-1} to represent imidazole ring vibration were revised; bands assigned to the nitro group in compound **2-4**: 1561 cm^{-1} and 1344 cm^{-1} (both strong) and in compound **2-5**: 1540, 1518 cm^{-1} (medium) and 1362 cm^{-1} (very strong) and bands for ring vibrations in compound **2-4**: 1516, 1504 cm^{-1} (strong) and 1448 cm^{-1} (very strong) and in compound **2-5**: 1492 cm^{-1} (very strong) and 1423 cm^{-1} (medium).[455]

IR absorption at 750–760 cm^{-1} was characteristic of a 4-nitrobisimidazole whereas a band at 740–760 cm^{-1} was characteristic of a 5-nitrobisimidazole. The bands were sufficiently distinct to permit structure determination.[97] These bands were also noted for C-H vibrations in the 4- and 5- derivatives of 1-alkyl- and 1,2-dialkylimidazoles.[33]

Salt formation required the participation of one nitro group attached to an imidazolone ring, Eq. (2-41). Thus bands at 1561 cm^{-1} and 1344 cm^{-1} for 4(5)-nitroimidazole disappeared on anion formation and new bands for the nitro group in the anion appeared at 1173, 1160, and 950 cm^{-1}. Three skeletal vibrations for the neutral ring at 1516, 1504, and 1448 cm^{-1} were replaced with four bands for the anion at 1550, 1502, 1490, and 1440 cm^{-1}. On the other hand 4,5-dinitroimidazole (two pairs of nitro group bands: 1548 and 1358 cm^{-1}, and 1532 and 1326 cm^{-1}) retained a nitro group (bands at 1528 and 1366 cm^{-1}) in its anionic form. Imidazolone bands were not assigned.[455]

(2-41)

Nitroimidazole structure assignments have had minimal dependence on IR analysis. When IR absorption for nitroimidazoles has been recorded this was often restricted to bands characteristic of nitro groups and the imidazole ring. Selected listings for the routine recordings of absorption characteristic of the nitro group are cited.[15,60,71,95,106,109,110,183,195,214,248,255,315,346,402]

Ultraviolet Spectroscopy

Introduction of the nitro substituent shifted an absorption band at 207–208 nm (log ε 3.70) for un-ionized imidazole to 325 nm (log ε 3.95) for 2-nitro- and

298 nm (log ε 3.86) for 4(5)-nitroimidazole (Table 2-1).[21,72,104] A weaker band for nitroimidazoles in the region 220–260 nm (log ε 3.5 to 3.7) has also been observed.[15,21,476]

N-methylation of 2-nitroimidazole did not shift the absorption from 325 nm but small differences which developed on methylation permitted differentiation between 1-methyl-4-nitroimidazole, 300 nm, and its 5-nitro isomer, 305 nm (Table 2-1). Generally this maximum has occurred at slightly longer wavelengths (7–13 nm) for a 5-nitro-1-alkylimidazole and sometimes distinguished it from its 4-nitro isomer,[72] eg, 1-methyl-4-nitro-5-styrylimidazole, 367 nm, and the 5-nitro-4-styryl isomer, 379 nm.[253,263]

Since the difference was never large and occasionally difficult to detect, isomer differentiation based on hypsochromic shifts (35–45 nm) shown by pairs of 4- and 5-nitroimidazole derivatives was recommended.[72,453] Generally a shift for the more basic 5-nitro isomer was detected in a weaker acid (Table 2-1). A hypsochromic shift for 5-nitroimidazole was brought about by 1 M perchloric acid whereas the 4-nitro isomer required 5 M perchloric acid.[205]

Table 2-1. Principal Absorption for Nitroimidazoles.[21]

W	X	Y	Z	Medium	Species[a]	λ_{max}, mμ	Log ε
H	H	H	NO_2	0.1 M NaOH	CB	372	4.13
				0.5 M H_2SO_4	N	325	3.95
				8.25 M H_2SO_4	CA	298	3.91
NO_2(H)	H(NO_2)	H	H	0.1 M NaOH	CB	350	4.01
				pH 7.38	N	298	3.86
				8.25 M H_2SO_4	CA	264	3.90
H	H	CH_3	NO_2	pH 4.63	N	325	3.93
				5 M H_2SO_4	CA	300	3.89
NO_2	H	CH_3	H	pH 4.63	N	300	3.90
				8.25 M H_2SO_4	CA	266	3.87
H	NO_2	CH_3	H	pH 7.38	N	305	3.81
				2 M H_2SO_4	CA	266	3.70
NO_2(H)	H(NO_2)	H	NO_2	0.1 NaOH	CB	354	4.09
				0.005 M H_2SO_4	N	304	4.05
NO_2	H	CH_3	NO_2	5 M H_2SO_4	N	305	4.06

[a]CB, conjugate base; N, neutral imidazole; CA, conjugate acid.

For those nitroimidazoles unsubstituted at nitrogen an expected bathochromic shift due to the formation of conjugated nitroimidazole anions in alkaline solutions occurred (Table 2-1). For 2-nitroimidazole a small shift, 314 nm to 324 nm, reflected a change in solvent from ethanol (95%) to distilled water; a further shift to 367 nm occurred when the solvent became aqueous sodium hydroxide.[453]

An exceptionally high absorption at 415 nm (log ε 3.71) was observed for 3-nitro-7-methyl-5,6-dihydro-7H-imidazo[1,2-*a*]imidazole (**2-93b**) in distilled water. A maximum at 390 nm (log ε 4.04) in ethanol (95%) was shifted to 340 nm (log ε 3.90) in ethanolic sodium hydroxide, perhaps the result of hydrolytic removal of the 2-amino substituent. Absorption at 374 nm (log ε 4.00) in distilled water and in sodium hydroxide was observed for 1-methyl-2-morphilino-5-nitroimidazole.[453]

UV absorption has been recorded for vinyl-substituted 2-nitroimidazoles,[315] nitrobisimidazoles,[97] nitroimidazo[1,2-*a*]pyridines,[140] glycosylnitroimidazoles,[261] bromonitroimidazoles,[75] and other derivatives.[17,39,51,56,60,63,67,93,106,109,110,199,214,248,255,280,282,327,332,346,355,477]

Mass Spectroscopy

Detailed analyses were made for the spectra obtained from six methylnitroimidazoles (**2-14, 2-15, 2-16, 2-17, 2-183**) ($R = CH_3$) and 2-nitro-4(5)-methylimidazole.[23] Each compound gave a strong signal for the molecular ion and fragmentations associated with aromatic nitro compounds. Interactions with adjacent substituents effected losses of hydroxyl and formyl radicals and water and formaldehyde molecules. Primary fragmentation patterns for each compound are shown, Eqs. (2-42–2-47).

Differences between 2-nitro and 4(5)-nitroimidazoles were noted. A more abundant [M-NO]$^+$ ion was generated from 2-nitromethylimidazoles than

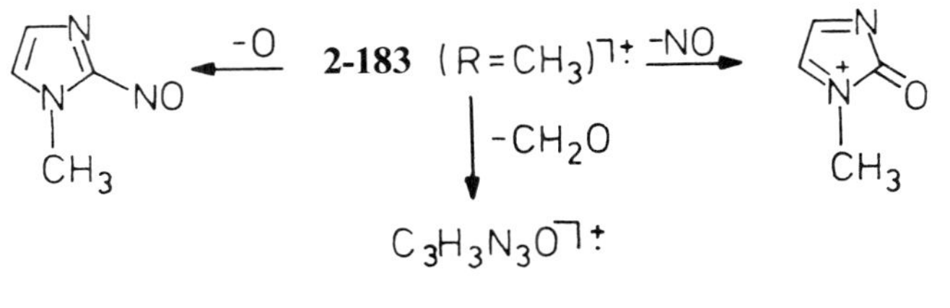

(2-42)

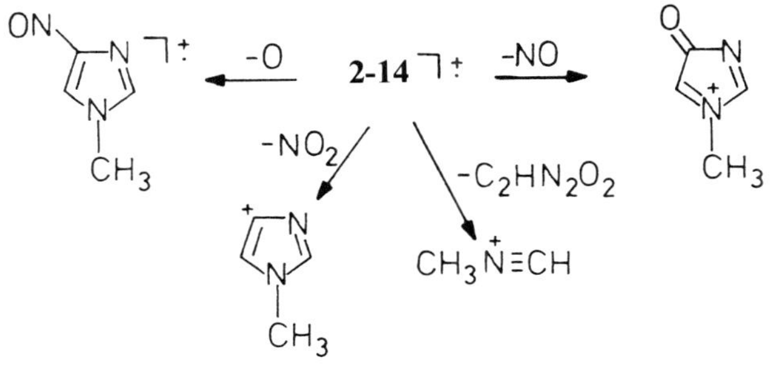

(2-43)

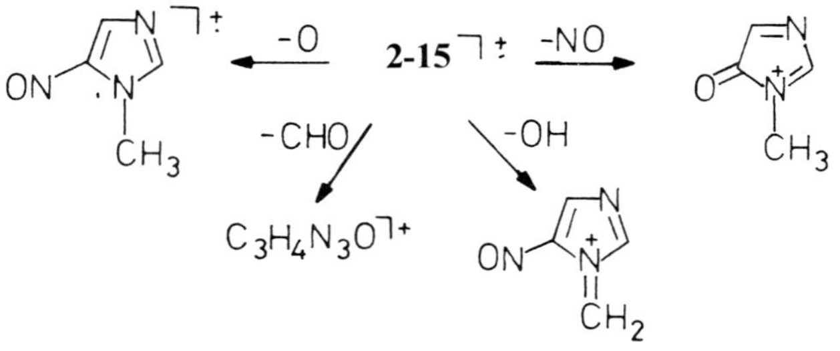

$$\text{(2-44)}$$

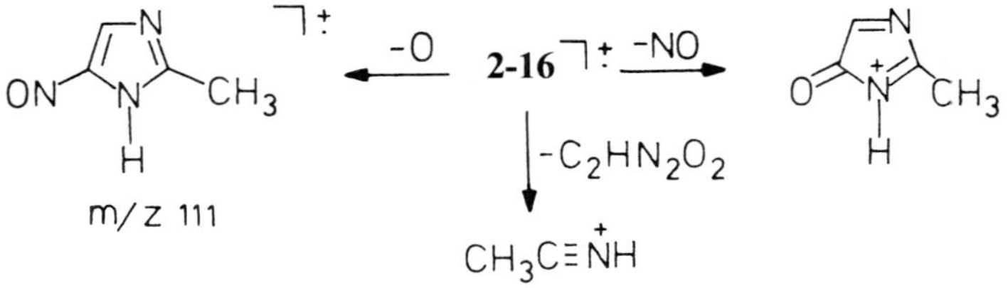

$$\text{(2-45)}$$

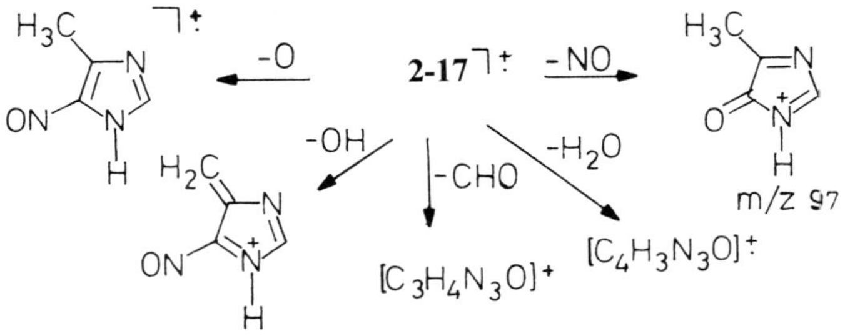

$$\text{(2-46)}$$

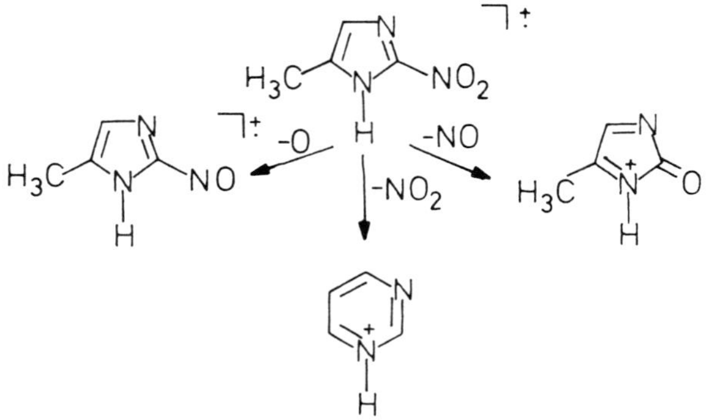

$$\text{(2-47)}$$

from 4(5)-nitro isomers. In addition a primary loss of OH from 1-methyl-2-nitroimidazole (**2-183**) (R = CH$_3$) was not observed; however it was detected in a secondary step.[23]

Analog labelling with ^{13}CH$_3$ and CD$_3$ helped to establish that (1) C and H in formaldehyde came from the methyl group in compound **2-183** (R = CH$_3$); and (2) there was complete retention of ^{13}C and D in the C$_2$H$_4$N$^+$ fragment from compound **2-14**. In a similar way the replacement of >NH with >ND in compound **2-16** afforded $CH_3C\equiv \overset{+}{N}D$, Eq. (2-45).[23] Different secondary fragmentation patterns were needed to accommodate m/z 79 and m/z 52 which were derived from the M^+-NO_2 ion, m/z 81, in Eq. (2-47), but not from the isomeric M^+-NO_2 ion in Eq. (2-45). To meet the requirement a ring expanded protonated pyrimidine was proposed for m/z 81 in Eq. (2-47). Earlier reports[478-482] that HCN was derived from the C-2 and N-3 positions in an imidazole ring system were confirmed and compatible with the ring expansion.[23]

It was concluded from the study of these six isomers of methylnitroimidazole that sufficient detail was available in their mass spectra to distinguish them from each other. The following observations tend to moderate the conclusion.

Mass spectra for 26 nitroimidazoles were collected in an evaluation of methods whereby 1-substituted-4-nitroimidazoles were differentiated from their 5-nitro isomers.[453] Generally a more intense M^+-NO_2 signal from a 1-substituted-5-nitroimidazole (10%–800% relative to M^+ at 100%) than from the 4-nitro isomer (negligible signals) was obtained Eq. (2-48).[195,453] That neither 2,4-dinitro- nor 2,5-dinitro-1-methylimidazole (**2-166**, **2-167**) gave a significant M^+-NO_2 signal was exceptional but not unique. For comparison the M^+-NO_2 signal relative to M^+(100) was 100 for 4(5)-nitroimidazole, 80 for 1-methyl-5-nitroimidazole, 20 for 1-methyl-4-nitroimidazole, and 26 for 1-methyl-2-nitroimidazole.[23,453] A reverse pattern of intensity for M^+-NO_2 signals was noted for 1-methyl-4-chloro-5-nitroimidazole (10) and the 4-nitro-5-chloro isomer (55). Generally a participation by the neighboring ring nitrogen atom and the possibility for transfer of an α-hydrogen from a 1-alkyl substituent to the C-5 position facilitated the loss of a C-5 nitro group, Eq. (2-48).[453]

$$\left[\underset{\underset{CHR_2}{|}}{O_2N-\text{imidazole}}\right]^{+\cdot} \xrightarrow{-NO_2} \underset{\underset{CHR_2}{|}}{\text{imidazole}^+} \longrightarrow \underset{\underset{CR_2}{\|}}{\text{imidazole}^+} \qquad (2\text{-}48)$$

A signal for loss of hydroxyl was characteristic of the 1-alkyl-5-nitroimidazoles and negligible for the 4-nitro isomers.[195,453] The usual explanation of hydrogen abstraction from a neighboring group applied, Eq. (2-49).[453] An exception was found in the negligible signals obtained for both 4- and 5-nitro-1,2-dimethylimidazole.

$$\left[\underset{\underset{HCR_2}{|}}{O_2N-\text{imidazole}}\right]^{+\cdot} \xrightarrow{-OH} \underset{\underset{CR_2}{\|}}{ON-\text{imidazole}^+} \qquad (2\text{-}49)$$

Loss of nitric oxide following rearrangement of a nitro group to a nitrite function generally gave signals of very weak to medium intensity for both 4- and 5-nitro-1-substituted imidazoles, Eq. (2-50); however, there were exceptional examples where a distinctly stronger signal was detected from one of two isomers: 1-β- hydroxyethyl-2-methyl-5-nitroimidazole(5) and the 4-nitro isomer (88). A similar observation differentiated between compound **2-241**, X = SO_2CH_3 (50) and its 4-nitro isomer (5).[453] A pattern was not discernible.

$$ImNO_2 \rceil^{+\cdot} \longrightarrow ImONO\rceil^{+\cdot} \xrightarrow{-NO} ImO^+ \qquad (2\text{-}50)$$

A signal M^+-O for loss of oxygen from a nitro group was consistently weak and nondiscriminating between 4- and 5-nitroimidazole isomers.[195,453] It was also detected for 1-β-cyanoethyl-2-nitroimidazole.[195]

An interesting fragmentation of certain 1-alkyl-2-heterocyclyl-5-nitroimidazoles required the loss of the 1-alkyl group together with an oxygen atom of the nitro group to give a 2-heterocyclyl-4(5)-nitrosoimidazole radical cation [compare m/z 111 in Eq. (2-45)].[483,484] Since 1-β-hydroxyethyl-2-methyl-5-nitroimidazole did not show a fragment m/z 111 this was not helpful in distinguishing the compound from its 4-nitro isomer. Aziridinone ions from 4-nitroimidazoles and cyclopropenones from 5-nitroimidazoles have been noted.[484]

Photoelectron Spectroscopy

The PE spectra were recorded for imidazole and its derivatives: 1-methyl, 2-methyl-, 4(5)-nitro, 2-methyl-4(5)-nitro, 1,2-dimethyl-5-nitro, 1-ethyl-2-methyl-5-nitro, 1-β-bromoethyl-2-methyl-5-nitro, and 1-β-hydroxyethyl-2-methyl-5-nitro.[485]

X-Ray Spectroscopy

In an investigation on the structure–activity relationship of 5-nitroimidazoles as anti-protozoa and anti-bacteria agents the structure of 1-β-hydroxyethyl-2-methyl-5-nitroimidazole (metronidazole) was determined. A distinguishing feature was not detected for correlation with either the pharmacological or the physico-chemical properties of the compound.[486]

An X-ray crystallographic analysis of ethyl 4-[2-(1-methyl-5-nitro-2-imidazolyl)-2-thioethoxy]benzoate (**2-257**) was carried out to determine the structure. Similar X-ray analytical assistance contributed to the structure determinations of 2,3-dihydro-5-nitroimidazo[2,1-*b*]oxazole (**2-151a**),[487,488] 2-methyl-4(5)-nitroimidazole (**2-16**),[489] and various derivatives of imidazo-[1,2-*a*]pyridine (**2-98–2-100**).[143] An almost localized double bond, C-2 to N-3, and the unambiguous location of the mobile proton at N-1 in compound **2-16** were revealed.[489]

Electron Spin Resonance Spectroscopy

Tetrahydrofuran solution of 11 nitro aromatic compounds and nitromethane were irradiated with a 200-W high pressure mercury lamp (PEK 210) to produce spin-adduct free radicals; the hyperfine splitting constants a^N were determined in situ by electron spin resonance (esr) spectroscopy. A proportionality was found between the a^N values obtained by photolysis with those obtained by other workers by radiolysis. Since photolysis is a more accessible technique than radiolysis it was suggested that effects of molecular structure on the electron affinity of anion radicals of radiosensitizers can be more easily obtained by photolysis. Five of the nitro compounds, including 2-methyl-5-nitroimidazole (the only nitroimidazole in the study) failed to give an esr spectrum on photolysis.[490]

OTHER PHYSICAL PROPERTIES

An inhibition to bulk polymerization of methyl methacrylate by 1,2-dimethyl-5-nitroimidazole, and by other nitro compounds, was correlated with the polarographic half-wave reduction potential.[491]

Luminescence of derivatives of imidazo[1,2-b]-1,2,4-triazine (**2-290**) became more intense with electron-donating substituents and less intense with electron-withdrawing substituents, eg, 3-nitro.[492]

$$W, X = CH_3, C_6H_5$$
$$Y = CH_3, C_6H_4, C_6H_4R\text{-}p, H,$$
$$Z = H, CH_3, C_6H_5, Br, NO_2$$

2-290

Fluorescence of the imidazo[1,2-a]pyridine and pyrimidine ring systems disappeared when a ring nitro substituent was present.[493]

Thermomicroanalysis revealed a eutectic mixture of the 1-β-bromoethyl and the 1-β-iodoethyl derivatives of 2-methyl-4(5)-nitroimidazole (**2-16**). Additional eutectic mixtures for pairs of 1-substituted derivatives of the imidazole (**2-16**) were obtained when the substituents were: (a) $CH_2CH_2OCH_2CH_2Cl$ and CH_2CH_2Cl; (b) $CH_2CH_2OCH_2CH_2Cl$ and CH_2CH_2Br; and (c) CH_2CH_2Cl and CH_2CH_2I. The pair of 1-β-chloroethyl and 1-β-bromoethyl derivatives of the imidazole (**2-16**) were isomorphous and each derivative gave eutectic mixtures of the same eutectic temperature with acetanilide and phenacetin.[494]

Antistatic filaments of polyamides, polyolefins, polystyrenes, polyvinyl chloride, and polyvinylidene chloride were obtained by including 0.005%–5% of an imidazole (**2-291**) in the polymerization recipe.[495]

Photographic developers for use in high-temperature, high-speed processing

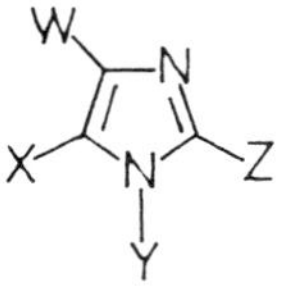

W = H, CO$_2$H, C$_6$H$_5$, CH$_3$, NO$_2$
X = C$_6$H$_5$, H, CO$_2$H, COCH$_3$, NH$_2$
Y = H, CH$_3$, CH$_2$CH$_2$OH, SO$_3$H,
CH$_2$CH$_2$NH$_2$, C$_6$H$_5$, C$_8$H$_{17}$
Z = C$_{11}$H$_{23}$, CH$_3$, C$_{17}$H$_{35}$, C$_6$H$_5$, SH,
cyclo-C$_6$H$_{11}$, C$_2$H$_5$, Br, C$_{15}$H$_{31}$

2-291

of black and white films have included 5-nitroimidazole in their composition.[496] It has also been included in infectious developers for lithographic printing plates.[497]

Nitroimidazoles were included in the compounds investigated for the development of an empirical linear relationship, $D' = (F - 0.26)/0.55$, between detonation velocity D' and a factor F that depended on the chemical composition and structure of the high explosive.[498]

The corrosion of steel by hydrofluoric acid was inhibited by the presence of 2-methyl-4-nitroimidazole.[499]

Reactions between metronidazole (**2-200**) (X = OH) and sulfhydryl groups in cysteine and cysteamine were promoted by the presence of ferrous or cupric ions. A similar reaction with glutathione in the presence of the ferrous ion led to reduction of 1-(3-methoxy-2-hydroxy)propyl-2-nitroimidazole (misonidazole) but metronidazole (**2-200**) (X = OH) was reduced slowly if at all. The reactions were inhibited by the presence of zinc ions.[500]

A determination of the presence of cysteine was based on a reaction with 1-methyl-4-nitro-5-chloroimidazole in an ammonia buffer (pH 10) for 1 h at 50°. A 1:1 complex gave an absorption maximum at 410 nm. Beer's law was obeyed for 10^{-5} to 10^{-3} M cysteine.[501]

Azathioprine (**2-292**) was reduced by zinc in ammonium hydroxide (5%) saturated with ammonium chloride to give a purple–red product. This color reaction was used to determine the presence of azathioprine in mixtures with 6-mercaptopurine and 5-chloro-1-methyl-4-nitroimidazole.[502]

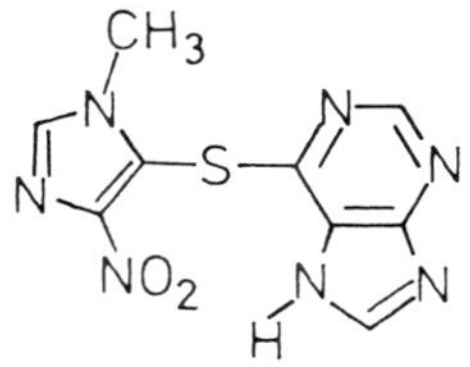

2-292 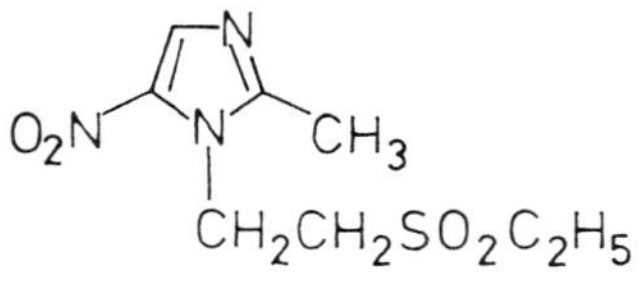   **2-293**

A colorimetric method for the determination of 1-(2-sulfonylethyl)ethyl-2-methyl-5-nitroimidazole (**2-293**) (tinidazole) consisted of a reaction with a mixture of Na$_2$Fe(CN)$_5$NO, K$_3$Fe(CN)$_6$, and sodium hydroxide and the measure of absorption at 460 nm. A gravimetric method called for weighing a precipitate obtained in a reaction between tinidazole and molybdophosphoric acid.[503]

An acid dye method for the determination of metronidazole (**2-200**) (X = OH) involved the use of bromothymol blue at pH 4.4 to form a complex which was extracted in chloroform and gave a λ_{max} of 440 nm.[504]

For detection and determination of 1,2-dimethyl-5-aminoimidazole (obtained by reduction of the corresponding nitroimidazole) the diazotized amine was coupled with N-(1-naphthyl)ethylenediamine to give a product with λ_{max} 495 nm (solvent not specified).[505]

Chromatograms of 14 nitroimidazole derivatives were sprayed with titanium (III) chloride in acetic acid and heated at 80° for 20 min to reduce the nitro to an amino group. The plates were then sprayed with one of the following chromogenic reagents: *p*-dimethylaminobenzaldehyde (5%) in ethanolic hydrogen chloride (Ehrlich's reagent); diazotized sulfanilic acid[506,507] (Pauli's reagent); or ninhydrin (0.2%) in acetone to develop characteristic colors.[508] Liquid chromatographic assays for ornidazole[509] and misonidazole [510] have been developed.

The electric conductivity of metallic salts of nitro and halonitro derivatives of imidazole in dimethylformamide has been determined. Dissociation and association constants were also reported for salts of Li, Na, K, Rb, Cs, and Ag.[511] Radiation-induced hydroxylation of thymine to thymine glycol was promoted by the presence of various nitroimidazoles, *p*-nitroacetophenone, and 5-nitro-2-furoic acid, although overall conversion of thymine was somewhat depressed.[512]

BIOCHEMICAL PROPERTIES

After 2-nitroimidazole (**2-5**) (azomycin), an antibiotic, was identified as the active principle in an extract of Streptomyces 6670 which inhibited the growth of the protozoan, *Trichomonas vaginalis*,[513] other nitroimidazole derivatives were prepared and investigated for anti-protozoan properties. Two promising compounds were found. For treatment of *Trichomonas vaginalis* infections in man 1-β-hydroxyethyl-2-methyl-5-nitroimidazole (**2-200**) (X = OH) (metronidazole) was selected.[514] The more toxic 1,2-dimethyl-5-nitroimidazole (**2-20**) (dimetridazole) was effective in treating blackhead in turkeys and dysentery in swine. Metronidazole was also found to be effective in the treatment of amoebiasis and giardiasis.[515,516] It is also one of the nitroimidazoles effective in the treatment of infections caused by obligate anaerobic bacteria[517-519] and the treatment of Crohn's disease.[520]

Numerous nitroimidazole derivatives have been prepared and investigated for therapeutic and prophylactic applications.[106,107,305,376] Drugs developed for human medicine included 1-(2-sulfonylethyl)ethyl-2-methyl-5-nitroimidazole (**2-293**) (tinidazole), 1-β-morpholinoethyl-5-nitroimidazole (**2-294**) (nimorazole), and 1-(3-chloro-2-hydroxy)propyl-2-methyl-5-nitroimidazole (**2-295**) (ornidazole). Amongst those developed for veterinary medicine were

1-methyl-2-carbamoyloxymethyl-5-nitroimidazole (**2-296**) (ronidazole) and 1-methyl-2-isopropyl-5-nitroimidazole (**2-297**) (ipronidazole).[521]

$$\text{2-294 } X = CH_2N(CH_2CH_2)_2O, Y = H$$
$$\text{2-295 } X = CH(OH)CH_2Cl, Y = CH_3$$
$$\text{2-296 } X = CH_3, Y = CH_2OCONH_2$$
$$\text{2-297 } X = CH_3, Y = CH(CH_3)_2$$
$$\text{2-298 } X = CH_3, Y = CH_2OC_6H_4SCH_3\text{-}p$$

More recently, attention was given to metronidazole and to 1-(3-methoxy-2-hydroxy)propyl-2-nitroimidazole (**2-281a**) (misonidazole) as radiosensitizers to increase the effect of therapeutic radiation with X-rays.[521]

Numerous nitroimidazoles are known to be mutagenic and/or carcinogenic. Metronidazole, one of the most thoroughly tested, was mutagenic on bacteria but did not show mutagenic activity in mammalian cells in vitro. It was carcinogenic in mice and rats but carcinogenic effects have not been observed in man. Although an active reduction product has not been isolated, due to instability, there is strong evidence that nitroimidazoles are effective as drugs and cause genetic damage only when reduced.[521]

Reduction of the nitro group in a nitroimidazole was found only in a minor pathway in mammalian metabolism. Metronidazole metabolites in man include 1-β-hydroxyethyl-2-hydroxymethyl-5-nitroimidazole (a potent mutagenic compound), 2-methyl-5-nitroimidazole-1-ylacetic acid, 1-β-hydroxyethyl-5-nitroimidazole-2-carboxylic acid, and some glucuronides. The carcinogen acetamide, a bacterial metabolite, was not found in germ-free animals.[521]

Many reviews cover the biochemical and allied properties of:

1. nitroimidazoles: general,[522] biochemistry,[523] toxicity,[524,525] mutagenicity,[521] radiosensitization,[526-528] action on microorganisms,[529] anaerobic chemotherapy,[530,531] steroid analogs,[532] cancer chemotherapy,[533,534] metabolism,[535] antiparasitic agents,[536] and correlation between radiosensitization and polarographic reduction.[537]
2. metronidazole (**2-200**) (X = OH): history,[538] conference proceedings,[539] drug therapy,[540] pharmacology,[541-543] biology,[544] antimicrobial action,[545] treatment of anaerobic infections,[546-549] toxicity,[550,551] and carcinogenicity.[552]
3. misonidazole (**2-281a**): radiosensitization.[553-555]
4. ronidazole (**2-296**).[556]
5. tinidazole (**2-293**).[557,558]
6. 1-methyl-2-p-thiomethylphenoxymethyl-5-nitroimidazole (**2-298**).[559]

REFERENCES

1. (a) Fargher, R. G., Pyman, F. L., *J. Chem. Soc.*, **1919**, *115*, 217, 1015; (b) Grant, R. L., Pyman, F. L., *ibid.*, **1921**, *119*, 1893; (c) Balaban, I. E., Pyman, F. L., *ibid.*, **1922**, *121*, 947; **1924**, *125*, 1564; (d) Pyman, F. L., *ibid.*, **1916**, *109*, 186; **1922**, *121*, 2616; (e) Light, L., Pyman, F. L., *ibid.*, **1922**,

121, 2626; (f) Pyman, F. L., Timmis, G. M., *ibid.*, **1923**, *123*, 494; (g) Lamb, I. D., Pyman, F. L., *ibid.*, **1924**, *125*, 706; (h) Hazeldine, C. E., Pyman, F. L., Winchester, J., *ibid.*, **1924**, *125*, 1431; (i) Pyman, F. L., Stanley, E., *ibid.*, **1924**, *125*, 2484; (j) Forsyth, W. G., Pyman, F. L., *ibid.*, **1925**, *127*, 573; (k) Bhagwat, V. K., Pyman, F. L., *ibid.*, **1925**, *127*, 1832; (m) Forsyth, R., Nimkar, V. K., Pyman, F. L., *ibid.*, **1926**, 800; (n) Forsyth, R., Pyman, F. L., *ibid.*, **1926**, 800; (o) Forsyth, R., Pyman, F. L., *ibid.*, **1930**, 397; (p) Pyman, F. L., Timmis, L. B., *J. Soc. Dyers Colorists*, **1922**, *38*, 269; (q) Fargher, R. G., *J. Chem. Soc.*, **1920**, *117*, 668; (r) Fargher, R. G., *ibid.*, **1921**, *119*, 158; (s) Grindley, R., Pyman, F. L., *ibid.*, **1927**, 3128.
2. Schultzen, H., *Z. Physiol. Chem.*, **1867**, 614.
3. Wohl, A., Marckwald, W., *Ber.*, **1889**, *22*, 1353.
4. Rung, F., Behrend, M., *J. Liebig's Ann. Chem.*, **1892**, *271*, 28.
5. Behrend, M., Schitz, J., *ibid.*, **1893**, *277*, 310, 338.
6. Windaus, A., *Ber.*, **1901**, *42*, 758.
7. Maeda, K., Osato, T., Umezawa, H., *J. Antibiot. Ser. A*, **1953**, *6*, 182.
8. Hofmann, K., "Imidazole and Its Derivatives. Part I," Interscience Publishers, Inc., New York, **1953**, pp. 127–136.
9. Schofield, K., Grimmett, M. R., Keene, B. R. T., "The Azoles," Cambridge University Press, Cambridge, **1976**, pp. 18, 22, 24, 60–68, 80, 133, 138, 141, 147, 168, 232–235, 344–350, 367, 368.
10. Cavalleri, B., "Nitroimidazole Chemistry, Synthetic Methods," in "Nitroimidazoles: Chemistry, Pharmacology, and Clinical Application," ed. A. Breccia, B. Cavalleri, and G. E. Adams, Vol. 42 in NATO Adv. Study Inst. Ser., Ser. A, Plenum Press, New York, **1982**, pp. 9–34.
11. Breccia, A., "Chemical Properties and Reaction Mechanisms of Nitroimidazoles," *ibid.*, pp. 35–49.
12. Suwinski, J., Salwinska, E., *Wiad. Chem.*, **1982**, *36*, 721.
13. Grimmett, M. R., "Advances in Imidazole Chemistry," *Advances in Heterocyclic Chemistry*, **1970**, *12*, 141–183.
14. *Ibid.*, **1980**, *27*, 242–326.
15. Nakamura, S., *Phar. Bull.* (Japan), **1955**, *3*, 379; *Chem. Abstr.*, **1956**, *50*, 15897f.
16. Novikov, S. S., Khmel'nitskii, L. I., Lebedev, O. V., Sevast'yanova, V. V., Epeshina, L. V., *Khim. Geterotsikl. Soedin*, **1970**, 503; Eng. 465.
17. Austin, M. W., Blackborow, J. R., Ridd, J. H., Smith, B. V., *J. Chem. Soc.*, **1965**, 1051.
18. Stambaugh, J. E., Manthei, R. W., *J. Chromatog.*, **1967**, *31*, 128.
19. Butler, K., Howes, H. L., Lynch, J. E., Pirie, D. K., *J. Med. Chem.*, **1967**, *10*, 891.
20. Grimson, A., Ridd, J. H., Smith, B. V., *J. Chem. Soc.*, **1960**, 1352.
21. Gallo, G. G., Pasqualucci, C. R., Radaelli, P., Lancini, G. C., *J. Org. Chem.*, **1964**, *29*, 862.
22. Sklarz, W. A., Epstein, A. D., U.S. 3,631,060; *Chem. Abstr.*, **1972**, *76*, 99662w.
23. Luijten, W. C. M. M., van Thuijl, J., *Org. Mass Spectrom.*, **1981**, *16*, 199.
24. *Ibid.*, **1982**, *17*, 299.
25. *Ibid.*, 304.
26. Sharnin, G. P., Fassakhov, R. Kh., Orlov, P. P., U.S.S.R. 458,853; *Chem. Abstr.*, **1975**, *82*, 156316m.
27. Fassakhov, R. Kh., Sharnin, G. P., Sabirzyanov, R. G., Naumov, V. N., U.S.S.R. SU 1,011,644; *Chem. Abstr.*, **1983**, *99*, 122453q.
28. Storm, C. B., Freeman, C. M., Butcher, R. J., Turner, A. H., Rowan, N. S., Johnson, J. O., Sinn, E., *Inorg. Chem.*, **1983**, *22*, 678.
29. Takeuchi, Y., Yeh, H. J. C., Kirk, K. L., Cohen, L. A., *J. Org. Chem.*, **1978**, *43*, 3565.
30. Wetzler, M., Dockner, T., Ger. Offen. 2,645,172; *Chem. Abstr.*, **1978**, *88*, 190837r.
31. Grenda, V. I., Sklarz, W. A., Lindberg, G. W., Epstein, A. D., Ger. Offen. 1,953,999; *Chem. Abstr.*, **1970**, *73*, 25472a.
32. Miller, M. W., Howes, H. L., Jr., Kasubick, R. V., Englich, A. R., *J. Med. Chem.*, **1970**, *13*, 849.
33. Cosar, C., Crisan, C., Horclois, R., Jacob, R. R. M., Robert, J., Tchelitcheff, S., Vaupre, R., *Arzneim. Forsch.*, **1966**, *16*, 23.
34. Brit. 837,838; *Chem. Abstr.* **1960**, *54*, 24804g.
35. Allsebrook, W. E., Guilland, J. M., Story, L. F., *J. Chem. Soc.*, **1942**, 232.
36. Chan, E., Putt, S. R., Showalter, H. D. H., Baker, D. C., *J. Org. Chem.*, **1982**, *47*, 3457.
37. Spaening, H., Dockner, T., Frank, A., Ger. Offen. 2,208,924, *Chem. Abstr.*, **1973**, *79*, 137157g.
38. Brit. 1,119,636; *Chem. Abstr.* **1968**, *69*, 106705y.
39. Nagarajan, K., Arya, V. P., Shenoy, S. J., Shah, R. K., Goud, A. N., Bhat, G. A., *Indian J. Chem.*, **1977**, *15B*, 629.
40. Tautz, W., Teitel, S., Brossi, A., *J. Med. Chem.*, **1973**, *16*, 705.

41. Kochergin, P. M., Tsyganova, A. M., Bilinova, L. S., Shlikhunova, V. S., *Khim. Geterotsikl. Soedin.*, **1965**, *1*, 875; Eng. 594.
42. Kochergin, P. M., Tsyganova, A. M., U.S.S.R. 164,289; *Chem. Abstr.*, **1964**, *61*, 16075b.
43. Zmojdzin, A., Utecht, E., Florczak, K., Harwazinski, L., Stelmachowski, W., Pol. 92,065; *Chem. Abstr.*, **1979**, *90*, 87460e.
44. *Ibid.*, *Chem. Abstr.*, **1978**, *89*, 109494c.
45. U.S. 4,209,631; *Chem. Abstr.*, **1980**, *93*, 1863547s.
46. Pol. 79,434; *Chem. Abstr.*, **1977**, *86*, 5458r.
47. Pol. 79,417; *Chem. Abstr.*, **1977**, *86*, 5459s.
48. Sawa, N., Okamura, S., Hoda, M., *Nippon Kagaku Zasshi*, **1968**, *89*, 1239; *Chem. Abstr.*, **1969**, *70*, 77865p.
49. Matsura, T., Banba, A., Ogura, K., *Tetrahedron*, **1971**, *27*, 1211.
50. Cox, J. S. S., Fitzmaurice, C., Katritzky, A. R., Tiddy, G. J. T., *J. Chem. Soc. B.*, **1967**, 1251.
51. Bochwic, B., Frankowski, A., Kuswik, G., Seliga, C., *Pol. J. Chem.*, **1981**, *55*, 1055.
52. Neth. Appl. 6,413,815; *Chem. Abstr.*, **1965**, *63*, 18096h.
53. Christensen, B. G., Hoff, D. R., Fr. Demande 2,070,695; *Chem. Abstr.*, **1972**, *77*, 34525e.
54. Iddon, B., Lim, B. L., *J. Chem. Soc., Perkin Trans. I*, **1983**, 271.
55. Wu, D. C. J., Cheer, C. J., Panzica, R. P., Abushanab, E., *J. Org. Chem.*, **1982**, *47*, 2661.
56. Grimmett, M. R., Hartshorn, S. R., Schofield, K., Weston, J. B., *J. Chem. Soc., Perkin Trans. II*, **1972**, 1654.
57. Neth. Appl. 6,609,552; *Chem. Abstr.*, **1967**, *67*, 54123u.
58. Neth. Appl. 6,609,553; *Chem. Abstr.*, **1967**, *67*, 11487.
59. Haugwitz, R. D., Narayanan, V. L., U.S. 3,380,871; *Chem. Abstr.*, **1975**, *83*, 79247f.
60. Julia, M., Tam, H. D., *Bull. Soc. Chim. Fr.*, **1971**, 1303.
61. Ochiai, E., Utahasi, K., *J. Pharm. Soc. Japan*, **1940**, *60*, 312; *Chem. Abstr.*, **1941**, *35*, 458.
62. Ellis, G. P., Epstein, C., Fitzmaurice, C., Goldberg, L., Lord, G. H., *J. Pharm. Pharmacol.*, **1967**, *19*, 102.
63. Amato, J. S., Grenda, V. J., Liu, T. M. H., Grabowski, E. J. J., *J. Heterocycl. Chem.*, **1979**, *16*, 1153.
64. Sarett, J. H., Hoff, D. R., Henry, D. W., Brit. 1,215,858; *Chem. Abstr.*, **1971**, *74*, 141792z.
65. Iradyan, M. A., Torosyan, A. G., Mirzoyan, A. G., Aroyan, A. A., *Khim. Geterotsikl. Soedin.*, **1977**, 1384; Eng. 1110.
66. Melloni, P., Metelli, R., Bassini, D. F., Confalonieri, C., Logemann, W., de Carneri, I., *Arzneim.-Forsch.*, **1975**, *25*, 9.
67. Remers, W. A., Gibs, G. J., Weiss, M. J., *J. Heterocycl. Chem.*, **1969**, *6*, 835.
68. Kochergin, P. M., Tsyganova, A. M., Viktorova, L. M., Peresleni, E. M., *Khim. Geterotsikl. Soedin.*, Sb. 1: Azotsoderszhashchie Geterotsikly, **1967**, 126; *Chem. Abstr.*, **1969**, 71, 13052j.
69. Novikov, S. S., Khmel'nitskii, L. I., Lebedev, O. V., Epeshina, L. V., Sevost'yanova, V. V., *Khim. Geterotsikl. Soedin.*, **1970**, 664; Eng. 614.
70. Sudarsanan, V., Nagarajan, K., George, T., Shenoy, S. J., Iyer, V. V., Kaulgud, A. P., *Indian J. Chem.*, **1982**, *21B*, 1022.
71. Dickens, J. P., Dyer, R. L., Hamill, B. J., Harrow, T. A., Bible, R. H., Jr., Finnegan, P. M., Henrick, K., Ouston, P. G., *J. Org. Chem.*, **1981**, *46*, 1781.
72. Hoffer, M., Toome, V., Brossi, A., *J. Heterocycl. Chem.*, **1966**, *3*, 454.
73. Hoffer, M., U.S. 3,341,548; *Chem. Abstr.*, **1968**, *68*, 105198c.
74. Neth. Appl. 6,606,853; *Chem. Abstr.*, **1968**, *68*, 21933a.
75. Barlin, G. B., *J. Chem. Soc. B*, **1967**, 641.
76. Sarasin, J., Wegmann, E., *Helv. Chim. Acta*, **1924**, *7*, 713, 720.
77. Mann, F. G., Porter, J. W. G., *J. Chem. Soc.*, **1945**, 751.
78. Block, M. S., Towne, E. B., U.S. 3,213,080; *Chem. Abstr.*, **1966**, *64*, 2200d.
79. Wolski, T., Golkiewicz, W., Kaczorek, M., Kiszcak, W., *Acta Pol. Pharm.*, **1977**, *34*, 601; *Chem. Abstr.* **1978**, *89*, 43233h.
80. Kochergin, P. M., *Khim. Geterosikl. Soedin.*, Akad. Nauk Latv. SSR **1965**, 761, 765; *Chem. Abstr.* **1966**, *64*, 9709e, 9709g.
81. Kochergin, P. M., U.S.S.R. 143,401; *Chem. Abstr.*, **1962**, *57*, 9859b.
82. Trout, G. E., Levy, P. R., *Rec. Trav. Chim.*, **1965**, *84*, 1257.
83. Hoffer, M. E., MacDonald, A., Jr., U.S. 3,652,579; *Chem. Abstr.*, **1972**, *76*, 153741m.
84. *Ibid.*, Can. 914,196; *Chem. Abstr.*, **1974**, *81*, 3943t.
85. Nagarajan, K., Arya, V. P., George, T., Bhat, G. A., Kulkarni, Y. S., Shenoy, S. J., Rao, M. K., *Indian J. Chem.*, **1982**, *21B*, 949.

References

86. Tweit, R. C., Muir, R. D., Ziecina, S., *J. Med. Chem.*, **1977**, *20*, 1697.
87. Tweit, R. C., Kreider, E. M., Muir, R. D., *J. Med. Chem.*, **1973**, *16*, 1161.
88. Anjaneyulu, B., Maller, R. K., Nagarajan, K., *J. Label. Compound, Radiopharm.*, **1983**, *20*, 951.
89. Belgodere, E., Bossio, R., Chimichi, S., Marcaccini, S., Pepino, R., *Heterocycles*, **1983**, *20*, 2019.
90. Neth. Appl. 6,609,553; *Chem. Abstr.*, **1967**, *67*, 11487y.
91. Miyano, M., Smith, J. N., *J. Heterocyclic Chem.*, **1982**, *19*, 659.
92. Lebedev, O. V., Epishina, L. V., Sevost'yanova, V. V., Novakova, T. S., Khmel'nitskii, L. I., Novikov, S. S., U.S.S.R. 184,868; *Chem. Abstr.*, **1967**, *67*, 90809f.
93. Novikov, S. S., Khmel'nitskii, L. I., Novikova, T. S., Lebedev, O. V., Epeshina, L. V., *Khim. Geterotsikl. Soedin.*, **1970**, *6*, 669; Eng. 619.
94. Lancini, G. C., Maggi, N., Sensi, P., *Farmaco Ed. Sci.*, **1963**, *18*, 390; *Chem. Abstr.*, **1963**, *59*, 10032.
95. Agrawal, K. C., Bears, K. B., Sehgal, R. K., Brown, J. N., Rist, P. E., Rupp, W. D., *J. Med. Chem.*, **1979**, *22*, 583.
96. Sarett, L., Hoff, D. R., Henry, D. W., Belg. 660,836; *Chem. Abstr.*, **1965**, *63*, 18097d.
97. Melloni, P., Dradi, E., Logemann, W., de Carneri, I., Trane, F., *J. Med. Chem.*, **1972**, *15*, 926.
98. Lehmstedt, K., Rolker, H., *Ber.*, **1943**, *76*, 879 (earlier references cited).
99. Lancini, G. C., Lazzari, E., *Experientia*, **1965**, *21*, 83.
100. Lebedev, O. V., Epishina, L. V., Sevost'yanova, V. V., Novikova, T. S., Khmel'nitskii, L. I., Novikov, S. S., Prikhod'ko, A. S., U.S.S.R. 177,897; *Chem. Abstr.*, **1966**, *64*, 19630d.
101. Beaman, A. G., Tautz, W., Gabriel, T., Keller, O., Toome, V., Duchinsky, R., *Antimicrobial Agents Chemotherapy*, **1966**, 469.
102. Beaman, A. G., Tautz, W., Gabriel, T., Duchinsky, R., *J. Am. Chem. Soc.*, **1965**, *87*, 389.
103. Neth. Appl. 6,514,946; *Chem. Abstr.*, **1966**, *65*, 13725b.
104. Neth. Appl. 6,510,485; *Chem. Abstr.*, **1966**, *65*, 724c.
105. Lancini, G. C., Lazzari, E., Pallanza, K., *Farmaco Ed. Sci.*, **1966**, *21*, 278; *Chem. Abstr.*, **1966**, *65*, 700f.
106. Cavalleri, B., Volpe, G., Arioli, V., Lancini, G., *J. Med. Chem.*, **1977**, *20*, 1522.
107. *Ibid.*, *Arzneim.-Forsch.*, **1977**, *27*, 1391.
108. Sartori, G., Lancini, G. C., Cavalleri, B., *J. Labelled Compd. Radiopharm.*, **1978**, *15*, 673.
109. Cavalleri, B., Volpe, G., Arioli, V., Pizzocheri, F., Diena, A., *J. Med. Chem.*, **1978**, *21*, 781.
110. Cavalleri, G., Ballotta, R., Lancini, G. C., *J. Heterocycl. Chem.*, **1972**, *9*, 979.
111. Lancini, G. C., Lazzari, E., Arioli, V., Bellani, P., *J. Med. Chem.*, **1969**, *12*, 775.
112. Beaman, A. G., Tautz, W., Gabriel, T., Keller, O., Loome, V., Duschinsky, R., *Antimicrobial Agents Chemotherapy*, **1965**, 469.
113. Davis, D. P., Kirk, K. L., Cohen, L. A., *J. Heterocycl. Chem.*, **1982**, *19*, 253.
114. Povstyanoi, M. V., Idzikovskii, V. A., Kruglenko, V. P., *Khim. Geterotsikl. Soedin*, **1981**, 1284; *Chem. Abstr.*, **1982**, *96*, 6677e.
115. Milligan, B., *J. Org. Chem.*, **1983**, *48*, 1495.
116. Tsuruta, T., Oda, R., *J. Chem. Soc. Japan*, **1949**, *70*, 65; *Chem. Abstr.*, **1951**, *45*, 6592i.
117. Tertov, B. A., Burykin, V. V., Morkovnik, A. S., U.S.S.R. 437, 763; *Chem. Abstr.*, **1974**, *81*, 169542m.
118. Martin, J., Johnson, F., U.S. 3,838,064; *Chem. Abstr.*, **1974**, *81*, 120625.
119. Cherkofsky, S. C., U.S. 4,199,592; *Chem. Abstr.*, **1980**, *93*, 18635q; Eur. Pat. Appl., 8755; *Chem. Abstr.*, **1980**, *93*, 114527c.
120. Ruccia, M., Cusmano, S., *Atti Accad. Sci. Lettere Arti, Palermo*, **1959–1960**, *20*, 31; *Chem. Abstr.*, **1963**, *58*, 5661a.
121. Cady, H. H., Coburn, M. D., Harris, B. W., Rogers, R. N., *Report* **1977**, LA-6802-MS; *Energy Res. Abstr.*, **1977**, *2*, 52829; *Chem. Abstr.*, **1978**, *88*, 152497x.
122. Coburn, M., U.S. 4,028,154; *Chem. Abstr.*, **1978**, *88*, 9198v.
123. Fr. Demande 2,436,780; *Chem. Abstr.*, **1981**, *94*, 4012u.
124. Al-Jobour, N. H., *J. Iraqi, Chem. Soc.*, **1980**, *5*, 37; *Chem. Abstr.*, **1982**, *96*, 16254f.
125. Coburn, M. D., *J. Heterocycl. Chem.*, **1970**, *7*, 455.
126. Lebedev, O. V., Epishina, L. V., Sevost'yanova, V. V., Novikova, T. S., Khmel'nitskii, L. I., Novikov, S. S., U.S.S.R. 177,896; *Chem. Abstr.*, **1966**, *64*, 19630d.
127. Pozharskii, A. F., Filipskikh, T. P., Zvezdina, E. A., *Tezisy Vses. Soveshch. Khim. Nitrosoedinenii*, *5th*, **1974**, 57; *Chem. Abstr.*, **1977**, *87*, 135191e.
128. Lancini, G. C., Kluepfel, D., Lazzari, E., Sartori, G., *Biochim. Biophys, Acta*, **1966**, *130*, 37.
129. Seki, Y., Nakamura, T., Okami, Y., *Biochem. J.*, **1970**, *67*, 389.
130. Lancini, G. C., Lazzari, E., Sartori, G., *J. Antibiotics*, **1968**, *6*, 387.

131. Lancini, G., Sensi, P., *Fr.* 1,528,151; *Chem. Abstr.*, **1969**, *71*, 38958x.
132. Paolini, J. P., Lendvay, L. J., *J. Med. Chem.*, **1969**, *12*, 1031.
133. Pyl, T., Bülling, L., Wünsch, K.-H., Beyer, H., *J. Liebig's Ann. Chem.*, **1961**, *643*, 153.
134. Pentimalli, L., Guerra, A. M., *Gazz. Chim. Ital.*, **1967**, *97*, 1286.
135. Pyl, T., Wünsch, K. H., Beyer, H., *J. Liebig's Ann. Chem.*, **1962**, *657*, 108.
136. Miller, L. F., Bambury, R. E., *J. Org. Chem.*, **1973**, *38*, 1955.
137. Paolini, J. P., Robins, R. K., *J. Org. Chem.*, **1965**, *30*, 4085.
138. Saldabol, N. O., Popelis, J., Mazeika, I., *Zh. Org. Khim.*, **1983**, *19*, 1558; *Chem. Abstr.*, **1983**, *99*, 175663q.
139. Andreasson, E., Newton, C. G., Ollis, D. W., Rees, C. W., Smith, D. I., Wright, D. E., *J. Chem. Soc., Chem. Commun.*, **1983**, 816.
140. Paudler, W. W., Blewitt, H. L., *Tetrahedron*, **1965**, *21*, 353.
141. *Ibid.*, *J. Org. Chem.*, **1965**, *30*, 4081.
142. Hand, E. S., Paudler, W. W., *J. Org. Chem.*, **1978**, *43*, 2900.
143. Teulade, J.-C., Escale, R., Rossi, J. C., Chapat, J.-P., Grassy, G., Payard, M., *Austr. J. Chem.*, **1982**, *35*, 1761.
144. Teulade, J.-C., Escale, R., Grassy, G., Girard, J.-P., Chapat, J.-P., *Bull. Soc. Chim. Fr.*, **1979**, Pt. 2, 529.
145. Teulade, J.-C., Grassy, G., Gerard, J.-P., Chabat, J.-P., de Bouchberg, M. S., *Eur. J. Med. Chem.—Chim. Ther.*, **1978**, *13*, 271.
146. Winkelmann, E., Raether, W., Hartung, H., Wagner, W.-H., *Arzneim. Forsch.*, **1977**, *27*, 82.
147. Gol'dfarb, Ya. L., Kondakova, M. S., *J. Gen. Chem. (USSR)*, **1940**, *10*, 1055; *Chem. Abstr.*, **1941**, *35*, 4020.
148. Kondakova, M. S., Gol'dfarb, Ya. L., *Izvest Akad. Nauk S.S.S.R., Otdel Khim. Nauk*, **1946**, 523; *Chem. Abstr.*, **1948**, *42*, 6364.
149. Saldabol, N. O., Liepin'sh, É. É., Popelis, Yu. Yu., Gavar, R. A., Baumane, L. Kh., Birgele, I. S., *Zhur. Org. Khim.*, **1979**, *15*, 2534, Eng. 2292.
150. Saldabol, N. O., Popelis, Yu. Yu., Aleekseeva, L. N., Yalnskaya, A. K., Moskaleva, N. D., *Khim.—Farm. Zh.*, **1977**, *11*, 64; *Chem. Abstr.*, **1977**, *87*, 848798f.
151. Saldabol, N. O., Hillers, S. A., *Tezisy Vses. Soveshch. Khim. Nitrosoedinenii, 5th*, **1974**, 30; *Chem. Abstr.*, **1977**, *87*, 39370s.
152. Saldabol, N. O., Popelis, Yu. Yu., *Vses. Nauchn. Konf. Khim. Teknol. Furanovykh Soedin. [Tezisy Dokl.] 3rd*, **1978**, 85; *Chem. Abstr.*, **1980**, *92*, 215139w.
153. Saldabol, N. O., Popelis, Yu. Yu., *Khim. Geterotsikl. Soedin.*, **1977**, 556; Eng. p. 451.
154. Glover, E. E., Peck, L. W., *J. Chem. Soc., Perkin I*, **1980**, 959.
155. Glover, E. E., Peck, L. W., Daughty, D. G., *J. Chem. Soc., Perkin I*, **1979**, 1833.
156. Satoh, K., Miyasaka, T., *Heterocycles*, **1978**, *10*, 269.
157. Fabio, P. F., Lanzilotti, A. E., Lang, S. A., *J. Labelled Compd. Radiopharm.*, **1978**, 407.
158. Kozuka, H., Koyama, M., Okitsu, T., *Chem. Pharm. Bull. (Tokyo)*, **1982**, *30*, 941.
159. Brunner, H., Leins, H., *Ber.*, **1897**, *30*, 2584.
160. Cacace, F., Masironi, R., *Ann. Chim. (Rome)*, **1957**, *47*, 366.
161. Marquardt, P., Müller-Erbeling, I., *Ger.* 859,470; *Chem. Abstr.*, **1953**, *47*, 11237.
162. Duesel, B. G., Bermann, H., Schachter, R. J., *J. Amer. Pharm. Assoc.*, **1954**, *43*, 619.
163. Serchi, G., Sancio, L., Bichi, C., *Il Farmico (Pavia)*, *Ed. Sci.*, **1955**, *10*, 733.
164. Morozowich, W., Bope, F. W., *J. Amer. Pharm. Assoc.*, **1958**, *47*, 173.
165. Jones, J. W., Robins, R. K., *J. Amer. Chem. Soc.*, **1960**, *82*, 3773.
166. Blitz, H., Sauer, J., *Ber.*, **1931**, *64*, 752 (earlier references cited).
167. Simonov, A. M., Anisimova, V. A., *Khim. Geterotsikl. Soedin.*, **1968**, *4*, 1102; Eng. p. 801.
168. Anisimova, V. A., Simonov, A. M., *Khim. Geterotsikl. Soedin.*, **1975**, 258; Eng. p. 222.
169. Strokin, Yu. V., Priimenko, B. A., Sheinkman, A. K., Klyuev, N. A., *Khim. Geterotsikl. Soedin.*, **1979**, 1404; Eng. p. 1131.
170. Eichler, E., Rooney, C. S., Williams, H. W. R., *J. Heterocycl. Chem.*, **1983**, *20*, 419.
171. Dunwell, D. W., Evans, D., *J. Chem. Soc., Perkin Trans I*, **1973**, 1588.
172. Nair, M. D., Sudarsanam, V., Desai, J. A., *Indian J. Chem.*, **1982**, *21B*, 1030.
173. Sehgal, K., Agrawal, K. C., *J. Heterocycl. Chem.*, **1979**, *16*, 1499.
174. Sehgal, R. K., Webb, M. W., Agrawal, K. C., *J. Med. Chem.* **1981**, *24*, 601.
175. Klykov, M. A., Povstyanoi, M. V., Kochergin, P. M., *Khim. Geterotsikl. Soedin.*, **1979**, 1542; Eng. p. 1237.
176. Povstyanoi, M. V., Klykov, M. A., Klyuev, N. A., *ibid.*, **1981**, 837; Eng. p. 622.
177. Koehler, F., *J. Prakt. Chem.*, **1963**, *21*, 50.

178. Kochergin, P. M., Tsyganova, A. M., Shlikhunova, V. S., *U.S.S.R.* 192,212; *Chem. Abstr.*, **1968**, *69*, 19150v.
179. Kochergin, P. M., Tsyganova, A. M., Shlikhunova, V. S., *U.S.S.R.* 223,097; *Chem. Abstr.*, **1969**, *70*, 28919a.
180. Iradyan, M. A., Iradyan, N. S., *Sint. Geterotsikl. Soedin.*, **1981**, *12*, 11; *Chem. Abstr.*, **1981**, *95*, 150540b.
181. Šunjić, V., Fajdiga, T., Japelj, M., Reims, P., Jerman, P., *Ger. Offen.* 1,954,705; *Chem. Abstr.*, **1970**, *73*, 25477f.
182. Kochergin, P. M., Tsyganova, A. M., Shlikhunova, V. S., *Khim.-Farm. Zh.*, **1968**, *2*, 22; *Chem. Abstr.*, **1969**, *70*, 37714g.
183. Šunjić, V., Fajdiga, T., Japelj, M., Reims, P., *J. Heterocycl. Chem.*, **1969**, *6*, 53.
184. Baddiley, J., Buchanan, J. G., Hardy, F. E., Stewart, J., *J. Chem. Soc.*, **1959**, 2893.
185. Taylor, E. C., Loeffler, P. K., *J. Am. Chem. Soc.*, **1960**, *82*, 3147.
186. Newton, C. G., Ollis, W. D., Podmore, M. L., Wright, D. E., *J. Chem. Soc., Perkin Trans. I*, **1984**, 63.
187. Newton, C. G., Ollis, W. D., Wright, D. E., *J. Chem. Soc., Perkin Trans. I*, **1984**, 69.
188. Dumanović, D., Maksimović, R., Ćirić, J., Jeremić, D., *Talanta*, **1975**, *22*, 811.
189. Iradyan, M. A., Iradyan, N. S., Avetyan, Sh. A., *Arm. Khim. Zh.*, **1978**, *31*, 435; *Chem. Abstr.*, **1978**, *89*, 197409g.
190. Bakulev, V. A., Mokrushin, V. S., Pushkareva, Z. V., *Khim. Geterotsikl. Soedin.*, **1982**, 704; Eng. p. 539.
191. *Neth. Appl.* 6,413,814; *Chem. Abstr.*, **1965**, *63*, 18097b.
192. *Belg.* 639,469; *Chem. Abstr.*, **1965**, *62*, 9142.
193. Winklermann, E., Kroha, H., *Ger. Offen.* 2,425,292; *Chem. Abstr.*, **1976**, *85*, 123919.
194. *Neth. Appl.* 6,413,814; *Chem. Abstr.*, **1965**, *63*, 18097.
195. Sehgal, R. K., Agrawal, K. C., *J. Heterocycl. Chem.*, **1979**, *16*, 871.
196. Parthasarathy, P. C., Desai, H. K., Saindane, M. T., *Indian. J. Chem.*, **1983**, *22B*, 157.
197. Nair, M. D., Sudarsanam, V., Desai, J. A., *Indian J. Chem.*, **1982**, *21B*, 1027.
198. Grabowski, E. J. J., Liu, T. M. H., Salce, L., Schoenenwaldt, E. F., *J. Med. Chem.*, **1974**, *17*, 547.
199. Kajfež, F., Kohbah, D., Oklobdžija, M., Fajdiga, T., Slamnik, M., Šunjić, V., *Croat. Chem. Acta*, **1967**, *39*, 199; *Chem. Abstr.*, **1968**, *68*, 49513n.
200. Kajfež, F., Šunjić, V., Kolbah, D., Fajdiga, T., Oklobdžija, H., *J. Med. Chem.*, **1968**, *11*, 167.
201. Grimson, A., Smith, B. V., *Chem. Ind. (London)*, **1956**, 983.
202. Laviron, E., *Bull. Soc. Chim. France*, **1963**, 2840.
203. Lavirori, E., *Compt. Rend.*, **1962**, *255*, 2603.
204. *Ibid., Abhandl. Deut. Akad. Wiss. Berlin, Kl. Chem., Geol. Biol.*, **1964**, 63; *Chem. Abstr.*, **1965**, *62*, 8674f.
205. Grimson, A., Ridd, J. H., Smith, B. V., *J. Chem. Soc.*, **1960**, 1352, 1357.
206. Ridd, J. H., Smith, B. V., *J. Chem. Soc.*, **1960**, 1363.
207. Beak, P., Lee, J., McKinnie, B. G., *J. Org. Chem.*, **1978**, *43*, 1367.
208. Finotto, M., *Fr. Demande* 2,263,241; *Chem. Abstr.*, **1976**, *84*, 150632f.
209. Kajfež, F., Šunjić, V., Caplar, V., *Ger. Offen.* 2,620,316; *Chem. Abstr.*, **1977**, *86*, 898783d.
210. Frank, A., Karn, H., Dockner, T., *Ger. Offen.* 2,424,280; *Chem. Abstr.*, **1976**, *84*, 17357d.
211. Nagarajan, K., Sudarsanam, V., Shenoy, S. J., Rama Roa, K., *Indian J. Chem.*, **1982**, *21B*, 997.
212. Beaman, A. G., Duschinsky, R., Tautz, W. P., *U.S.* 3,391,156; *Chem. Abstr.*, **1968**, *69*, 96718p.
213. Beaman, A. G., Tautz, W. P., *U.S.* 3,793,317; *Chem. Abstr.*, **1974**, *80*, 95956e.
214. Klötzer, W., Baldinger, H., Karpitschka, E. M., Knoflach, J., *Synthesis*, **1982**, 592.
215. Khan, M. A., Freitas, A. C. C., *Monatash.*, **1981**, *112*, 675.
216. *Indian* 147, 422; *Chem. Abstr.*, **1981**, *95*, 43117x.
217. Huang, B. S., Lauzon, M. J., Parham, J. C., *J. Heterocycl. Chem.*, **1979**, *16*, 811.
218. Beaman, A. G., Duschinsky, R., Tautz, W. P., *U.S.* 3,255,201; *Chem. Abstr.*, **1966**, *65*, 13724a.
219. Miyake, T., Takeda, K., Tada, K., *Ger. Offen.* 3,026,410; *Chem. Abstr.*, **1981**, *95*, 62203p.
220. Baker, D. C., Putt, S. R., *J. Amer. Chem. Soc.*, **1979**, *101*, 6127.
221. Kajfež, F., Blažević, N., Šunjić, V., *Farm. Glas.*, **1969**, *25*, 49; *Chem. Abstr.*, **1969**, *71*, 70534s.
222. Reddy, K. C., Kumar, K. A., Srimannarayana, G., *Indian J. Pharm. Sci.*, **1982**, *44*, 6.
223. Šunjić, V., Fajdiga, T., Blažević, N., Kajfež, F., *Acta Pharm. Jugoslav.*, **1969**, *19*, 65; *Chem. Abstr.*, **1970**, *72*, 66866e.
224. Valles, P., *Swiss* 520,090; *Chem. Abstr.*, **1972**, *77*, 101609a.
225. Jacob, R. M., Régnier, G. L., Crisan, C., *U.S.* 2,944,061; *Chem. Abstr.*, **1961**, *55*, 1657g.
226. Iradyan, M. A., Torosyan, A. G., Mirzoyan, R. G., Badalyants, I. P., Isaakya, A. S., Manu-

charyan, D. Sh., Dayan, M. Kh., Sakanyan, G. S., Dzhagatspanyan, I. A., et al., *Khim.-Farm. Zh.*, **1977**, *11*, 42; *Chem. Abstr.*, **1978**, *88*, 22759y.
227. *Japan, Kokai* 7,451,276; *Chem. Abstr.*, **1976**, *85*, 21359u.
228. Beaman, A. G., Tautz, W. P., *U.S.* 3,803,165; *Chem. Abstr.*, **1974**, *81*, 13511q.
229. Suwinski, J., Salwinski, E., Waters, J., Widel, M., *Acta Pol. Pharm.*, **1978**, *35*, 529; *Chem. Abstr.*, **1979**, *91*, 107932m.
230. *Span.* 305,073; *Chem. Abstr.*, **1965**, *63*, 2981b.
231. *Span.* 467,734; *Chem. Abstr.*, **1979**, *91*, 5225r.
232. Dockner, T., Frank, A., *Ger. Offen.* 2,827,796; *Chem. Abstr.*, **1980**, *92*, 146775c.
233. Arya, V. P., David, J., Honkan, V., Shenoy, S., *Indian J. Chem.*, **1977**, *15B*, 141.
234. Vogel, H. H., Strickler, R., Oppenlaender, K., Baur, R., *Ger. Offen.* 2,948,884; *Chem. Abstr.*, **1981**, *95*, 132884n.
235. *Japan. Kokai* 76,143,680; *Chem. Abstr.*, **1978**, *88*, 121175m.
236. Frank, A., Karn, H., Spaenig, H., *Ger. Offen.* 2,359,625; *Chem. Abstr.*, **1975**, *83*, 147480d.
237. *Fr. Demande* 2,260,996; *Chem. Abstr.*, **1976**, *84*, 150631e.
238. Suwinski, J., Rajca, A., Waters, J., Widel, M., *Acta Pol. Pharm.*, **1980**, *37*, 59; *Chem. Abstr.*, **1981**, *94*, 30649b.
239. Mroczkiewicz, A., *Acta. Pol. Pharm.*, **1981**, *38*, 559; *Chem. Abstr.*, **1982**, *97*, 72287z.
240. Hoffer, M., Grunberg, E., *J. Med. Chem.*, **1974**, *17*, 1019.
241. Heeres, J., Mostmans, J. H., Maes, B., Backx, L. J. J., *Eur. J. Med. Chem.—Chim. Ther.*, **1976**, *11*, 237.
242. Heeres, J., Mostmans, J. H., Maes, R., *Ger. Offen.* 2,429,755; *Chem. Abstr.*, **1975**, *82*, 156309m.
243. Iradyan, M. A., Torosyan, A. G., Agababyan, R. V., Aroyan, A. A., *Arm. Khim. Zh.*, **1977**, *30*, 756; *Chem. Abstr.*, **1978**, *88*, 152498y.
244. Povstyanoi, M. V., Klykov, M. A., Klyuev, N. A., *Khim. Geterotsikl. Soedin.*, **1981**, 833; Eng. p. 622.
245. Lee, W. W., *U.S. Pat. Appl.* US 180,373; *Chem. Abstr.*, **1982**, *97*, 55811y.
246. *Neth. Appl.* 7,403,438; *Chem. Abstr.*, **1976**, *85*, 46675g.
247. Makovec, F., Senin, P., Rovati, L., *Brit. UK Pat. Appl.* 2,022,073; *Chem. Abstr.*, **1980**, *93*, 95294f.
248. Crnic, Z., Gluncic, B., *Croat. Chem. Acta*, **1981**, *54*, 217; *Chem. Abstr.*, **1981**, *95*, 187151b.
249. Herz, J. E., Torres, R. S., *Org. Prep. and Proc. Internat.*, **1975**, *7*, 211.
250. Miller, M. W., *Fr.* 1,535,023; *Chem. Abstr.*, **1969**, *71*, 38959y.
251. Šunjić, V., Kalbah, D., Kajfež, F., Blažević, N., *J. Med. Chem.*, **1968**, *11*, 1264.
252. Giraldi, P. N., Mariotti, V., Nannini, G., Tosolini, G. P., Dradi, E., Logemann, W., de Carneri, I., Minti, G., *Arzneim-Forsch.*, **1970**, *20*, 52.
253. Giraldi, P. N., Mariotti, V., de Carneri, I., *J. Med. Chem.*, **1968**, *11*, 66.
254. Ross, W. J., Todd, A., *J. Med. Chem.*, **1973**, *16*, 863.
255. Caplar, V., Šunjić, V., Kajfež, F., *J. Heterocycl. Chem.*, **1974**, *11*, 1055.
256. Berg, S. S., Sharp, B. W., *Eur. J. Med. Chem.*, **1975**, *10*, 171.
257. Rousseau, R. J., Robins, R. K., Townsend, L. B., *J. Heterocycl. Chem.*, **1967**, *4*, 311.
258. Prisbe, E. J., Verheyden, J. P. H., Moffatt, J. G., *J. Org. Chem.*, **1978**, *43*, 4784.
259. Sakaguchi, M., Larroquette, C. A., Agrawal, K. C., *J. Med. Chem.*, **1983**, *26*, 20.
260. Barascut, J. L., Tanby, C., Imbach, J. L., *J. Carbohydr., Nucleosides, Nucleotides*, **1974**, *1*, 77.
261. Guglielmi, H., Vergin, H., *J. Liebigs Ann. Chem.*, **1972**, *761*, 67.
262. Chavis, C., Grodenic, F., Imbach, J.-L., *Eur. J. Med. Chem.—Chim. Ther.*, **1979**, *14*, 123.
263. Baddiley, J., Buchanan, J. G., Hardy, F. E., Stewart, J., *J. Chem. Soc.*, **1959**, 2893.
264. Chavis, C., Dumont, F., Grodenic, F., Imbach, J. L., *J. Carbohydr., Nucleosides, Nucleotides*, **1981**, *8*, 507.
265. Nesvadba, H., Reinshagen, H., *Swiss* 597,197; *Chem. Abstr.*, **1978**, *89*, 43421t.
266. Bruckner, G., Gorog, K., Nemessany, Z., Havasi, M., Andriska, V., Raskay, B., Marosvolgyi, S., Pinter, Z., Gribovszki, P., et al., *Austrian* 342,616; *Chem. Abstr.*, **1978**, *89*, 43426y.
267. Rav-Acha, C., Cohen, L. A., *J. Org. Chem.*, **1981**, *46*, 4717.
268. Takeuchi, Y., Kirk, K. L., Cohen, L. A., *J. Org. Chem.*, **1978**, *43*, 3570.
269. Zatsepina, N. N., Tupitsyn, I. F., Belyashova, A. I., Kirova, A. V., Konyakhina, E. Ya., *Khim. Geterotsikl. Soedin.*, **1977**, 1196; Eng. p. 963.
270. Zatsepina, N. N., Tupitsyn, I. F., Kane, A. A., Sudakova, G. N., *Khim. Geterotsikl. Soedin.*, **1977**, 1192; Eng. p. 959.
271. Henry, D. W., Wollf, D. R., *Belg.* 661,262; *Chem. Abstr.*, **1966**, *64*, 2093g.
272. Albright, J. D., Shepherd, R. V., *J. Heterocycl. Chem.*, **1973**, *10*, 899.
273. Asata, G., Berkelhammer, G., *J. Med. Chem.*, **1972**, *15*, 1086.

274. Shimada, K., Kuriyama, S., Kanazawa, T., Satoh, M., *Yakugaku Zasshi*, **1971**, *91*, 231; *Chem. Abstr.*, **1971**, *74*, 141633.
275. Bastiannsen, L. A. M., Macco, A. A., Godefroi, E. F., *J. Chem. Soc., Chem. Commun.*, **1974**, 36.
276. Maccao, A. A., Godefroi, F. F., Dronen, J. J. M., *J. Org. Chem.*, **1975**, *40*, 252.
277. Godefroi, E. F., Geenen, J., van Klingeren, B., Van Wijngaarden, L. J., *J. Med. Chem.*, **1975**, *18*, 530.
278. *U.S.* 3,652,555; *Chem. Abstr.*, **1972**, *77*, 34517d.
279. Hagen, H., Kohler, R. D., *Ger. Offen.* 2,827,351; *Chem. Abstr.*, **1980**, *92*, 181191f.
280. Elitropi, G., Panto, E., Tricerri, S., *J. Heterocycl. Chem.*, **1979**, *16*, 1545.
281. Nair, M. D., Sudarsanam, V., Desai, J. A., *Indian J. Chem.*, **1982**, *21B*, 1030.
282. Šunjić, V., Fajdiga, T., Japelj, M., *J. Heterocycl. Chem.*, **1970**, *7*, 211.
283. Asato, G., Berkelhammer, G., Gastrock, W. H., *U.S.* 4,026,903; *Chem. Abstr.*, **1977**, *87*, 102330w.
284. Nesvadba, H., Reinshagen, H., *Belg.* 841,669, *Chem. Abstr.*, **1977**, *87*, 135380r.
285. *Ibid., Ger. Offen.* 2,619,110, *Chem. Abstr.*, **1977**, *86*, 72696e.
286. Rufer, C., Kessler, H. J., Schroeder, E., *Progr. Antimicrob. Anticancer Chemother. Proc. Int. Congr. Chemother.*, *6th*, **1969**, *1*, 145; *Chem. Abstr.*, **1971**, *74*, 141632x.
287. Mrozik, H. H., Kulsa, P., *U.S.* 4,144,345; *Chem. Abstr.*, **1979**, *91*, 20505a.
288. Rufer, C., Kessler, H. J., Schröder, E., *Chim. Ther.*, **1971**, *6*, 286.
289. Neville, M. C., Verge, J. P., *J. Med. Chem.*, **1977**, *20*, 946.
290. *Japan Kokai* 7,859,660; *Chem. Abstr.*, **1978**, *89*, 146903f.
291. Gebert, U., Raether, W., *Ger. Offen.*, 2,651,084; *Chem. Abstr.*, **1979**, *90*, 152184j.
292. Kulsa, P., Rooney, C. S., *U.S.* 4,010,176; *Chem. Abstr.*, **1977**, *87*, 23279h.
293. Bambury, R. E., Kim, H. K., *Ger. Offen.* 2,017,023; *Chem. Abstr.*, **1971**, *74*, 3626x.
294. Cavalleri, B., Volpe, G., Ripamonti, A., Arioli, V., *Arzneim.-Forsch.*, **1977**, *27*, 1131.
295. Cavalleri, B., Ballotta, R., Arioli, V., *Arzneim.-Forsch.*, **1975**, *25*, 338.
296. Cavalleri, B., Volpe, G., Arioli, V., Lancini, G. C., *Arzneim.-Forsch.*, **1977**, *27*, 1391.
297. Rufer, C., Kessler, H.-J., Schröder, E., *J. Med. Chem.*, **1971**, *14*, 94.
298. Kessler, H.-J., Rufer, C., Schwarz, K., *Eur. J. Med. Chem.*, **1976**, *11*, 19.
299. Verge, J. P., Neville, M. C., Friedman, H., *U.S.* 3,980,794; *Chem. Abstr.*, **1977**, *86*, 29817h.
300. *Ibid., Ger. Offen.*, 2,431,775; *Chem. Abstr.*, **1975**, *83*, 10079.
301. Berkelhammer, G., Asato, G., *Science*, **1968**, *162*, 1146.
302. *Ibid., U.S.* 3,991,200; *Chem. Abstr.*, **1977**, *86*, 106597r.
303. *Ibid., U.S.* 3,740,434; *Chem. Abstr.*, **1973**, *79*, 66365.
304. Remers, W. A., Gibs, G. J., Weiss, J. M., *U.S.* 3,790,590; *Chem. Abstr.*, **1974**, *80*, 95964f.
305. Cavalleri, B., Volpe, G., Pallanza, R., *Arzneim.-Forsch.*, **1975**, *25*, 148.
306. Malabarba, A., Berti, M., DePaoli, A., Cavalleri, B., *Il Farmaco, Ed. Sci.*, **1979**, *34*, 105.
307. Shridhar, D. R., Sastry, C. V. R., Mehotra, A. K., Nagarajan, R., Lai, B., Bhopale, K. K., *Indian J. Chem.*, **1980**, *19B*, 59.
308. Dockner, T., Fleig, H., Hagen, H., Kohlmann, F. W., *U.S.* 4,218,460; *Chem. Abstr.*, **1981**, *94*, 30764k.
309. Fleig, H., Hagen, H., Dockner, T., Kohlmann, F. W., *Ger. Offen.* 2,544,460; *Chem. Abstr.*, **1977**, *87*, 53316k.
310. *Belg.* 877,171; *Chem. Abstr.*, **1980**, *92*, 128922u.
311. Rufer, C., Kessler, H.-J., Schröder, E., Damerius, A., *Chim. Thér.*, **1972**, *7*, 5.
312. *Ibid.*, **1973**, *8*, 567.
313. Rufer, C., Kessler, H.-J., Schröder, E., *J. Med. Chem.*, **1975**, *18*, 253.
314. Bradamante, S., Colombo, A., Vittadini, G., *J. Heterocyclic Chem.*, **1981**, *18*, 1399.
315. Cavalleri, B., Volpe, G., Arioli, V., *J. Med. Chem.*, **1977**, *20*, 656.
316. Albright, J. D., Shepherd, R. G., *J. Heterocyclic Chem.*, **1973**, *10*, 899.
317. Cockerill, A. F., Harden, R. C., Mallen, D. N. B., *J. Chem. Soc., Perkin Trans. II*, **1972**, 1428.
318. Bononi, L. J., *Belg.* BE 892,911; *Chem. Abstr.*, **1983**, *98*, 16688q.
319. Kochergin, P. M., Klykov, M. A., Ordzhonikidze, S., *U.S.S.R.* 230,825; *Chem. Abstr.*, **1969**, *70*, 77964v.
320. Kochergin, P. M., Tsyganova, A. M., Shlikhunova, V. S., Klykov, M. A., *Khim. Geterotsikl. Soedin.*, **1971**, *7*, 689; Eng. p. 648.
321. Kochergin, P., Tsyganova, A. M., Shlikhunova, V. S., *U.S.S.R.* 232,271; *Chem. Abstr.*, **1969**, *70*, 96793e.
322. Sudarsanam, V., Nagarajan, K., Arya, V. P., Kaulgud, A. P., Shenoy, S. J., Shah, R. K., *Indian J. Chem.*, **1982**, *21B*, 989.
323. Arya, V. P., Nagarajam, K., Shenoy, S. J., *Indian J. Chem.*, **1982**, *21B*, 1115.

324. Naragarajam, K., Arya, V. P., Thomas, G., *Indian* 143,968; *Chem. Abstr.*, **1980**, *92*, 128919y.
325. Gireva, R. N., Reznichenko, L. A., Kochergin, P. M., *Khim.-Farm. Zh.*, **1977**, *11*, 38; *Chem. Abstr.*, **1978**, *88*, 22758x.
326. Carbon, J. A. *J. Org. Chem.*, **1961**, *26*, 455.
327. Iivespää, A. O., Jarumilinta, R., *Indian J. Chem.*, **1982**, *21B*, 923.
328. *Israeli* 44,282; *Chem. Abstr.*, **1980**, *92*, 163965g.
329. Nagarajan, K., Arya, V. P., Thomas, G., *Can.* 1,039,732, *Chem. Abstr.*, **1979**, *90*, 87456h.
330. *Ibid.*, *Australian* 492,051, *Chem. Abstr.*, **1978**, *89*, 179996s.
331. *Ibid.*, *S. African* 7,405,157, *Chem. Abstr.*, **1978**, *88*, 50855v.
332. Nagarajan, K., Arya, V. P., George, T., Sudarsanam, V., Shah, R. K., Goud, A. N., Shenoy, S. J., Honkan, V., Kulkarni, Y. S., Rao, M. K., *Indian J. Chem.*, **1982**, *21B*, 928.
333. Arya, V. P., Nagarajan, K., Shenoy, S. J., *Indian J. Chem.*, **1982**, *21B*, 941.
334. Nagarajan, K., Arya, V. P., Shah, R. K., Shenoy, S. J., Bhat, G. A., *Indian J. Chem.*, **1982**, *21B*, 945.
335. Berg, S. S., Petrov, V., *J. Chem. Soc.*, **1952**, 784.
336. Shimada, K., Kuriyama, S., Kanozawa, T., Satoh, M., Toyoshima, S., *Yakugaku Zasshi*, **1971**, *91*, 221; *Chem. Abstr.*, **1974**, *74*, 141631w.
337. Kochergin, P. M., *Khim. Geterotsikl. Soedin.*, **1966**, 749; Eng. p. 576.
338. Kochergin, P. M., Bashkir, E. A., *Khim. Geterotsikl. Soedin.*, **1966**, 754; Eng. p. 581.
339. *Ibid.*, 762; Eng. p. 588.
340. Kochergin, P. M., Shmidt, I. S., *Khim. Geterotsikl. Soedin. Sb.1: Azotsoderzhashchie Geterotsikily*, **1967**, 130; *Chem. Abstr.*, **1969**, *70*, 87758v.
341. Kochergin, P. M., Karsunskü, V. S., Shlikhunova, V. S., *U.S.S.R.* 384,822; *Chem. Abstr.*, **1973**, *79*, 105297m.
342. Kochergin, P. M., Shmidt, I. S., *U.S.S.R.* 172,814; *Chem. Abstr.*, **1966**, *64*, 741c.
343. Kochergin, P. M., Shmidt, I. S., *U.S.S.R.* 173,381; *Chem. Abstr.*, **1966**, *64*, 2109b.
344. Kochergin, P. M., *U.S.S.R.* 158,280; *Chem. Abstr.* **1960**, *60*, 9284f.
345. Wolski, T., *Pol. J. Pharmacol. Pharm.*, **1978**, *30*, 593; *Chem. Abstr.*, **1979**, *90*, 204020z.
346. Nurgatin, V. V., Sharnin, G. P., Nurgatina, R. B., Ginzburg, B. M., *Khim. Geterotsikl. Soedin.*, **1982**, 812; Eng. p. 616.
347. Bennett, L. L., Jr., Baker, H. T., *J. Am. Chem. Soc.*, **1957**, *79*, 2188.
348. Iradyan, M. A., Stepanyan, G. M., Alvazyan, A. Kh., Mirzoyan, V. S., Avetyan, Sh. A., Isaakyan, Z. S., Manucharyan, D. Sh., Dayan, M. Kh., Garibdzhanyan, B. T., *Khim.-Farm. Zh.*, **1981**, *15*, 40; *Chem. Abstr.*, **1981**, *95*, 24915q.
349. Mroczkiewicz, A., *Acta Pol. Pharm.*, **1981**, *38*, 379; *Chem. Abstr.*, **1982**, *96*, 104146a.
350. Migachev, G. I., Danilenko, V. A., *Khim. Geterotsikl. Soedin.*, **1982**, 867; Eng. p. 649.
351. Sharnin, G. P., Fassakhov, R. L., Eneikina, T. A., *U.S.S.R.* 565,913; *Chem. Abstr.*, **1977**, *87*, 184505t.
352. *Ibid.*, *Khim. Geterotsikl. Soedin.*, **1977**, 1666; Eng. p. 1332.
353. Sharnin, G. P., Fassakhov, R. K., Orlov, P. P., Eneikina, T. A., *U.S.S.R.* 479,767; *Chem. Abstr.*, **1976**, *84*, 17343w.
354. *Ibid.*, *Khim. Geterotsikl. Soedin.*, **1977**, 653; Eng. p. 529.
355. Goldman, P., Wuest, J. D., *J. Amer. Chem. Soc.*, **1981**, *103*, 6224.
356. Teulade, J.-C., Grassy, G., Escale, R., Chapat, J.-P., *J. Org. Chem.*, **1981**, *46*, 1026.
357. Oldenhof, C., Cornelisse, J., *Recl. Trav. Chim. Pays-Bas*, **1978**, *97*, 35.
358. Giupta, R. P., Larroquette, C. A., Agrawal, K. C., *J. Med. Chem.*, **1982**, *25*, 1342.
359. Mokrushin, V. S., Belyaev, N. A., Kolobov, M. Yu., Fedotov, A. N., *Khim. Geterotsikl. Soedin.*, **1983**, 808; *Chem. Abstr.*, **1983**, *99*, 139847u.
360. Mokrushin, V. S., Selezneva, I. S., Pospelova, T. A., Usova, V. K., Malinskaya, S. M., Anoshina, G. M., Zubova, T. E., Pushkareva, Z. V., *Khim.-Farm. Zh.*, **1982**, *16*, 303; *Chem. Abstr.*, **1982**, *97*, 6218b.
361. Winkelmann, E., Raether, W., *Arzneim-Forsch.*, **1978**, *28*, 739.
362. *Ibid.*, *Ger. Offen.* 2,515,515; *Chem. Abstr.*, **1977**, *86*, 55448h.
363. *Ibid.*, *Ger. Offen.* 2,515,522; *Chem. Abstr.*, **1977**, *86*, 72642j.
364. *Ibid.*, *Ger. Offen.* 2,516,366; *Chem. Abstr.*, **1977**, *86*, 55450c.
365. *Ibid.*, *Ger. Offen.* 2,531,303; *Chem. Abstr.*, **1977**, *86*, 140051t.
366. *Ibid.*, *Ger. Offen.* 2,625,195; *Chem. Abstr.*, **1978**, *88*, 121183n.
367. *Ibid.*, *Ger. Offen.* 2,650,659; *Chem. Abstr.*, **1978**, *89*, 109479b.
368. *Ibid.*, *Ger. Offen.* 2,605,222; *Chem. Abstr.*, **1978**, *89*, 109480v.
369. Winkelmann, E., Raether, W., Sinharay, A., *Arzneim.-Forsch.*, **1978**, *28*, 351.

370. *Neth. Appl.* 7,513,132; *Chem. Abstr.*, **1977**, *86*, 72640g.
371. *Neth. Appl.* 7,611,848; *Chem. Abstr.*, **1978**, *89*, 109482x.
372. Winkelmann, E., Raether, W., *Ger. Offen.* 2,522,176; *Chem. Abstr.*, **1977**, *86*, 72654q.
373. *Ibid.*, 2,521,358; *Chem. Abstr.*, **1977**, *86*, 89822r.
374. Austrian 351,021; *Chem. Abstr.*, **1979**, *91*, 107981b.
375. *Fr. Demande* 2,374,904; *Chem. Abstr.*, **1979**, *91*, 20499b.
376. Winkelmann, E., Raether, W., Gebert, U., *Arzneim.-Forsch.*, **1978**, *28*, 1682.
377. Gireva, R. N., Aleshina, G. A., Reznichenko, L. A., Zyukina, G. V., Mal'tseva, L. F., Kochergin, P. M., *Khim.-Farm. Zh.*, **1976**, *10*, 48; *Chem. Abstr.*, **1977**, *86*, 106470u.
378. Postescu, I. D., Mustea, I. V., *Rom.* 64,528; *Chem. Abstr.*, **1980**, *92*, 128918x.
379. Cho, M. J., Biermacher, J. J., *S. African* 1,700,282; *Chem. Abstr.*, **1980**, *93*, 173749z.
380. Nair, M. D., Desai, J. H., *Indian J. Chem.*, **1980**, *19B*, 338.
381. Beaman, A. G., Tautz, W. P., *U.S.* 3,865,823; *Chem. Abstr.*, **1970**, *82*, 170944w.
382. *Ibid.*, *U.S.* 3,803,165; *Chem. Abstr.*, **1974**, *81*, 13511q.
383. Cavalleri, B., Volpe, G., *Ger. Offen.* 2,700,346; *Chem. Abstr.*, **1977**, *87*, 184501q.
384. Kajfež, F., Šunjić, V., Šunjić, V., *Swiss* 582,678; *Chem. Abstr.*, **1977**, *86*, 121336y.
385. *Ibid.*, *Swiss* 605,811; *Chem. Abstr.*, **1979**, *90*, 38926c.
386. Rashid, H., Kobah, D., Ghani, A., *Bangladesh Pharm. J.*, **1976**, *5*, 17; *Chem. Abstr.*, **1977**, *86*, 89760u.
387. Schlager, L. H., Austrian 337,696; *Chem. Abstr.*, **1978**, *88*, 6884e.
388. Kajfež, F., Šunjić, V., Šunjić, V., *Swiss* 605,817; *Chem. Abstr.*, **1979**, *90*, 38922y.
389. *Ibid.*, *Swiss* 605,816; *Chem. Abstr.*, **1979**, *90*, 38924a.
390. *Ibid.*, *Swiss* 605,818; *Chem. Abstr.*, **1979**, *90*, 38923z.
391. Schlager, L. H., Austrian 345,277; *Chem. Abstr.*, **1979**, *90*, 23055u.
392. Smithen, C. E., *Ger. Offen.* 2,836,073; *Chem. Abstr.*, **1979**, *90*, 186950w.
393. Cusic, J. W., Levon, E. F., *U.S.* 3,997,572; *Chem. Abstr.*, **1977**, *86*, 140049y.
394. Shridhar, D. R., Lal, B., Vaidya, N. K., Bhopale, K. K., Tripathi, H. N., *Indian J. Chem.*, **1979**, *18B*, 251.
395. Kulsa, P., Rooney, C. S., *U.S.* 3,915,978; *Chem. Abstr.*, **1976**, *84*, 59471s.
396. Rufer, C., Kessler, H.-J., Schröder, E., *J. Med. Chem.*, **1971**, *14*, 94.
397. Bambury, R. E., Lutz, C. M., Miller, L. F., Kim, H. H., Ritter, H. W., *J. Med. Chem.*, **1973**, *16*, 566.
398. Bambury, R. E., Kim, H. K., *U.S.* 3,583,985; *Chem. Abstr.*, **1971**, *74*, 3626x.
399. Rufer, C., Schwarz, K., Winterfildt, E., *J. Liebig's Ann. Chem.*, **1975**, 1465.
400. Wu, M. T., *J. Heterocyclic Chem.*, **1972**, *9*, 31.
401. Berger, H., Gall, R., Stack, K., Voemel, W., Hoffmann, R., *Ger. Offen.* 2,261,693; *Chem. Abstr.*, **1974**, *81*, 120674s.
402. Hofheinz, W., *Ger. Offen.* 2,613,345; *Chem. Abstr.*, **1977**, *86*, 55445e.
403. Saldabol, N., Popelis, Yu. Yu., Liepin'sh, E. E., *Khim. Geterotsikl. Soedin.*, **1978**, 1566; Eng. p. 1279.
404. Cavalleri, B., Ballotta, R., Arioli, V., Lancini, G., *J. Med. Chem.*, **1973**, *16*, 557.
405. Caspary, W., Cohen, B., Lesko, S., Ts'O, P. O. P., *Biochemistry*, **1973**, *12*, 2649.
406. Ayscough, P. B., Elliott, A. J., Salmon, G. A., *J. Chem. Soc., Faraday Trans. I*, **1978**, *74*, 511.
407. Wardman, P., *Int. J. Radiation Biol.*, **1975**, *28*, 585.
408. Eaton, D. R., Wilkins, R. G., *J. Biol. Chem.*, **1978**, *253*, 908.
409. Larina, L. I., Dubnikov, V. M., Lur'e, F. S., Vakul'skaya, T. I., Vitkovskaya, N. M., Lopyrev, V. A., Voronkov, M. G., *Zh. Strukt. Khim.*, **1980**, *21*, 203; *Chem. Abstr.*, **1981**, *94*, 46331c.
410. Wardman, P., Clarke, E. D., *NATO Adv. Study Inst. Ser., Ser. C*, **1978**, C50 (Tech. Appl. Fast Rea.), 535; *Chem. Abstr.*, **1980**, *92*, 93741e.
411. Josephy, D. P., Palcic, B., Skarsgaard, L. D., *Biochem. Pharmacol.*, **1981**, *30*, 849.
412. *Ibid.*, *Radiation Sensitizers: Their Use Clin. Manage. Cancer [Proc. Conf.]*, **1979**, 61; *Chem. Abstr.*, **1981**, *95*, 54647w.
413. Wardman, P., Anderson, R. F., Clarke, E. D., Jones, N. R., Minchinton, A. I., Patel, K. B., Stratford, M. R. L., Watts, M. E., *Int. J. Radiat. Oncol., Biol. Phys.*, **1982**, *8*, 777.
414. Clarke, E. D., Wardman, P., Goulding, K. H., *Biochem. Pharmacol.*, **1980**, *29*, 2684.
415. Varghese, A. J., Whitmore, G. F., *Radiat. Sensitizers: Their Use Clin. Manage. Cancer, [Proc. Conf.]*, **1979**, 57; *Chem. Abstr.*, **1981**, *95*, 17915m.
416. *Ibid.*, *Cancer Res.*, **1983**, *43*, 78.
417. Chrystal, E. J. T., Koch, R. L., Goldman, P., *Mol. Pharmacol.*, **1980**, *18*, 105.
418. Kirk, K. L., Cohen, L. A., *J. Org. Chem.*, **1973**, *38*, 3647.

419. Varghese, A. J., Whitmore, G. F., *Cancer Clin. Trials*, **1980**, *3*, 43.
420. Fabra, F., Gálvez, C., González, A., Viladoms, P., Vilarasa, J., *J. Heterocycl. Chem.*, **1978**, *15*, 1227.
421. Simonov, A. M., Anisimova, V. A., Koshchienko, Yu. V., *Khim. Geterotsikl. Soedin.*, **1969**, 184; Eng. p. 140.
422. Polanc, S., Stanovnik, B., Tisler, M., *Synthesis*, **1980**, 129.
423. Goldman, P., Ramos, S. M., Wuest, J. D., *J. Org. Chem.*, **1984**, *49*, 932.
424. Entwistle, I. D., Jackson, A. E., Johnstone, R. A. W., Telford, R. P., *J. Chem. Soc., Perkin I*, **1977**, 443.
425. Kochergin, P. M., Klykov, M. A., Korsunski, V. S., Povstyanoi, M. V., *Ukr. Khim. Zh. (Russ. Ed.)*, **1975**, *41*, 1293; *Chem. Abstr.*, **1976**, *84*, 74183t.
426. Sudarsanam, V., Nagarajan, K., Gokhale, N. G., *Indian J. Chem.*, **1982**, *21B*, 1087.
427. Sullivan, C. E., Tally, F. P., Goldin, B. R., Vouros, P., *Biochem. Pharmacol.*, **1982**, *31*, 2689.
428. Robinson, B., Zubair, M. U., *Bull. Chem. Soc. Japan*, **1977**, *50*, 561.
429. Huang, B.-S., Chello, P. L., Yip, L., Parham, J. C., *J. Med. Chem.*, **1980**, 575.
430. Barnes, G. R., Pyman, F. L., *J. Chem. Soc.*, **1927**, 2711.
431. Fisher, M. H., Nickolson, W. H., Stuart, S. R., *Can. J. Chem.*, **1961**, *39*, 501.
432. Robinson, B., Zubair, M. U., *Bull. Chem. Soc. Jpn.*, **1977**, *50*, 561.
433. Huang, B.-S., Parham, J. C., *J. Org. Chem.*, **1979**, *44*, 4046.
434. Guglielmi, H., Jung, A., *Hoppe-Syeler's Z. Physiol. Chem.*, **1977**, *358*, 1463.
435. Revankar, G. R., Robins, R. K., *Nucleic Acid Chem.*, **1978**, *1*, 207.
436. Wardman, P., Clarke, E. D., *J. Chem. Soc., Faraday Trans. 1*, **1976**, *72*, 1377.
437. Whillans, D. W., Whitmore, G. F., *Radiat. Res.*, **1981**, *86*, 311.
438. Sjöberg, L., Eriksen, T. E., Mustea, I., Révész, L., *Radiochem. Radioanal. Letters*, **1977**, *29*, 19.
439. Ruddock, G. W., Greenstock, C. L., *Biochim. et Biophys. Acta*, **1977**, *496*, 197.
440. Adams, G. E., Dewey, D. L., *Biochem. and Biophys. Research Commun.*, **1963**, *12*, 473.
441. O'Neill, P., *Anal. Proc. (London)*, **1980**, *17*, 282.
442. Sjoberg, L., Eriksen, T. E., *Radiochem. Radioanal. Lett.*, **1978**, *35*, 275.
443. Lopyrev, V. A., Larina, L. I., Rakhmatulina, T. N., Shibanova, E. F., Vakul'skaya, T. I., Voronkov, M. G., *Dokl. Akad. Nauk SSSR*, **1978**, *242*, 142; Eng. p. 770.
444. Kargin, Yu. M., Latypova, V. Z., Fassakhov, R. Kh., Arkhipov, A. I., Eneikina, T. A., Sharnin, G. P., *Zh. Obshch. Khim.*, **1979**, *49*, 2139; *Chem. Abstr.*, **1980**, *92*, 12817j.
445. Kargin, Yu. M., Latypova, V. Z., Arkhipov, A. I., Fassakhov, R. Kh., Eneikina, T. A., Sharnin, G. P., *Zh. Obshch. Khim.*, **1980**, *50*, 2582; *Chem. Abstr.*, **1981**, *94*, 111337a.
446. Kargin, Yu. M., Latypova, V. Z., Arkhipov, A. I., *Deposited Doc.*, **1980**, SPSTL 839 Khp-D80; *Chem. Abstr.*, **1982**, *97*, 215240s.
447. Vakul'skaya, T. I., Larina, L. I., Nefedova, O. B., Pétukhov, L. P., Voronkov, M. G., Polyrev, V. A., *Khim. Geterotsikl. Soedin.*, **1979**, 1398; Eng. p. 1127.
448. Roffia, S., Gottardi, C., Vianello, E., *J. Electroanal. Chem.*, **1982**, *142*, 263.
449. Blažević, N., Kajfež, F., Šunjić, V., *J. Heterocycl. Chem.*, **1970**, *7*, 227.
450. Paiva, A. C. M., Juliano, L., Boschov, P., *J. Am. Chem. Soc.*, **1976**, *98*, 7645.
451. Tomasik, P., Zalewski, R., Chodzinski, J., *Chem. Zvesti*, **1979**, *33*, 105.
452. Coburn, M., *U.S.* 4,028,154; *Chem. Abstr.*, **1978**, *88*, 9198v.
453. Nagarajan, K., Sudarsanam, V., Parthasarathy, P. C., Arya, V. P., Shenoy, S. J., *Indian J. Chem.*, **1982**, *21B*, 1006.
454. Selig, W., *Mikrochim. Acta*, **1981**, *2*, 251.
455. Epishina, L. V., Slovetskii, V. L., Osipov, V. G., Lebedev, O. V., Khmel'nitskii, L. I., Sevost'yanova, V. V., Novikova, T. S., *Khim. Geterotsikl. Soedin.*, **1967**, 716; Eng. p. 570.
456. Sharnin, G. P., Khavbirov, R. A., Nurgatin, V. V., Khmel'nitskii, L. I., Lebedev, O. V., Prikhod'ko, A. S., *Izv. Akad. Nauk SSSR, Ser. Khim.*, **1977**, 2711; Eng. p. 2506.
457. Batyr, D. G., Marchenko, G. N., Nurgatin, V. V., Korsakov, A. G., Nikol'skaya, V. F., Baranova, G. S., Zubareva, V. E., *Izv. Akad. Nauk Mold. SSR, Ser. Biol. Khim. Nauk*, **1980**, 68; *Chem. Abstr.*, **1980**, *93*, 160459e.
458. Dulova, V. I., Brezhe, A. L., Myagchenko, A. P., Artyukhova, E. P., Molchanova, I. R., *Vopr. Khim. Khim. Teknol.*, **1980**, *59*, 72; *Chem. Abstr.*, **1981**, *94*, 146082r.
459. Barabanov, V. P., Tret'yakova, A. Ya., Sharnin, G. P., Fassakhov, R. Kh., Eneikina, T. A., *Zh. Obshch. Khim.*, **1980**, *50*, 2318; *Chem. Abstr.*, **1981**, *94*, 83447t.
460. Breccia, A., Balducci, R., Stagni, G., *Int. J. Radiat. Oncol. Biol. Phys.*, **1982**, *8*, 423.
461. Norris, A. R., Buncel, E., Taylor, S. E., *J. Inorg. Biochem.*, **1982**, *16*, 279.
462. Suwinski, J., Salwinska, E., *Pol. J. Chem.*, **1981**, *55*, 2325; *Chem. Abstr.*, **1983**, *99*, 175141t.

463. Suwinski, J., Salwinska, E., Watras, J., Widel, M., *Zesz. Nauk Politech. Slask.*, *Chem.*, **1981**, *678*, 99; *Chem. Abstr.*, **1982**, *96*, 122108u.
464. Bales, J. R., Coulson, C. J., Gilmour, D. W., Mazid, M. A., Neidle, S., Kuroda, R., Peart, B. J., Ramsden, C. A., Sadler, P. J., *J. Chem. Soc., Chem. Commun.*, **1983**, 432.
465. Callaghan, V., Goodgame, D. M. L., Tooze, R. P., *Inorg. Chim. Acta*, **1983**, *78*, L1.
466. Staab, H. A., Irngartiner, H., Mannschreck, A., Wu, M. Th., *Justus Liebegs Ann. Chem.*, **1966**, *695*, 55.
467. McKillop, A., Wright, D. E., Podmore, M. L., Chambers, R. K., *Tetrahedron*, **1983**, *39*, 3797.
468. Stambaugh, J. E., Manthei, R. W., *Can. Spectrosc.*, **1968**, *13*, 134.
469. Barlin, G. B., Batterham, T. J., *J. Chem. Soc. (B)*, **1967**, 516.
470. Lippmaa, E., Mägi, M., Novikov, S. S., Khmel'nitskii, L. I., Prihodko, A. S., Lebedev, O. V., Epishina, L. V., *Org. Magn. Reson.*, **1972**, *4*, 153.
471. *Ibid.*, 197.
472. Pugmire, R. J., Grant, D. M., *J. Am. Chem. Soc.*, **1968**, *90*, 697, 4232.
473. Adam, W., Grimson, A., Rodriguez, G., *J. Chem. Phys.*, **1969**, *50*, 645.
474. Mathias, A., Gil, V. M. S., *Tetrahedron Lett.*, **1965**, 3163.
475. Baldeschwieler, J. D., Randall, E. W., *Proc. Chem. Soc.*, **1961**, 303.
476. Epishina, L. V., Slovetskii, V. I., Lebedev, O. V., Filippova, L. I., Khmel'mitskii, L. I., Sevost'yanova, V. V., Novikova, T. S., *Khim. Geterotsikl. Soedin., Sb. 1: Azotsoderzhashchei Geterotsikly*, **1967**, 102; *Chem. Abstr.*, **1969**, *71*, 17276h.
477. Joshi, N. G., Saojii, D. G., *Indian J. Pharm. Sci.*, **1979**, *41*, 226.
478. Bowie, J. H., Cooks, R. G., Lawesson, S.-O., Schroll, G., *Austr. J. Chem.*, **1967**, *20*, 1613.
479. Bowie, J. H., Donaghue, P. F., Rodda, H. J., Simons, B. K., *Tetrahedron*, **1968**, *24*, 3965.
480. Hodges, R., Grimmett, M. R., *Austr. J. Chem.*, **1968**, *21*, 1085.
481. Klebe, K. J., van Houte, J. J., van Thuijl, J., *Org. Mass Spectrom.*, **1972**, *6*, 1363.
482. van Thuijl, J., Klege, K. J., van Houte, J., *Org. Mass Spectrom.*, **1973**, *7*, 1165.
483. Van Lear, G. E., *Org. Mass Spectrom.*, **1972**, *6*, 1117.
484. Haskins, N. J., Ford, G. C., Waddell, K. A., Dickens, J. P., Dyer, R. L., Hamill, B. J., Harrow, T. A., *Biomed. Mass Spectrom.*, **1981**, *8*, 351.
485. Kajfež, F., Klasinc, L., Šunjić, V., *J. Heterocycl. Chem.*, **1979**, *16*, 529.
486. Blaton, N. M., Peeters, O. M., DeRanter, C. J., *Acta Cryst.*, **1979**, *B35*, 2465.
487. Brown, J. N., Rist, P. E., Agrawal, K., *Cryst. Struct. Commun.*, **1979**, *8*, 761.
488. Agrawal, K. C., Bears, K. B., Sehgal, R. K., Brown, J. N., Rist, P. E., Rupp, W. D., *J. Med. Chem.*, **1979**, *22*, 583.
489. Kalman, A., Van Meurs, F., Toth, J., *Cryst. Struct. Commun.*, **1980**, *9*, 709.
490. Sancier, K. M., *Radiat. Res.*, **1980**, *81*, 487.
491. Stankovic, R., Jovanovic, S., Kosanovic, D., *Glas. Hem. Drus. Beograd.*, **1979**, *44*, 581; *Chem. Abstr.*, **1980**, *92*, 94766d.
492. Povstyanoi, M. V., Kruglenko, V. P., Gachkovskii, V. F., *Izv. Vyssh. Uchebn. Zaved., Khim. Khim. Teknol.*, **1979**, *22*, 23; *Chem. Abstr.*, **1979**, *90*, 185903w.
493. Rackham, D. M., *Appl. Spectrosc.*, **1979**, *33*, 561.
494. Šunjić, V., Kajfež, F., Slamnik, M., Kolbah, D., *Bull. Sci., Cons. Acad. RSF Yougoslvie*, **1967**, *12*, 61; *Chem. Abstr.*, **1969**, *70*, 23473r.
495. Fr. 1,564,308; *Chem. Abstr.*, **1969**, *71*, 92608w.
496. Shibaoka, H., Ger. Offen. 2,455,002; *Chem. Abstr.*, **1976**, *84*, 67789s.
497. Sakazume, Y., Shirasaki, J., *Japan Kokai* 7,763,335; *Chem. Abstr.*, **1978**, *88*, 1978.
498. Rothstein, L. R., Petersen, R., *Propellants Explos.*, **1979**, *4*, 56; *Chem. Abstr.*, **1979**, *91*, 177511x.
499. Ujma, J., *Ochr. Koroz.*, **1980**, *23*, 163; *Chem. Abstr.*, **1981**, *94*, 107293r.
500. Rahnemann, D., Basaga, H., Dunlop, J. R., Searle, A. J. F., Wilson, R. L., *Br. J. Cancer Suppl.*, **1978**, *37*, 16.
501. Klinek, J., Wolski, T., Wawrzycki, S., Wronska, J., *Chem. Anal. [Warsaw]*, **1978**, *23*, 317; *Chem. Abstr.*, **1978**, *89*, 52930k.
502. Czarnecki, W., *Acta Pol. Pharm.*, **1977**, *34*, 515; *Chem. Abstr.*, **1978**, *88*, 158549n.
503. Patel, R. B., Patel, A. A., Gandhi, T. P., Patel, P. R., Patel, V. C., Manakiwala, S. C., *Indian Drugs*, **1980**, *18*, 76; *Chem. Abstr.*, **1981**, *94*, 162825r.
504. Bhatkar, R. G., Chodankar, S. K., *Indian J. Pharm. Sci.*, **1980**, *42*, 127.
505. Colvin, L. B., Silvaramakrishnan, R., Couch, J. R., *Chemist-Analyst*, **1963**, *52*, 9.
506. Grimmett, M. R., Richards, E. L., *J. Chromatogr.*, **1965**, *20*, 171.
507. Klimova, V. A., Dubinskii, R. A., *Izv. Akad. Nauk SSSR, Ser. Khim.*, **1974**, 640; Eng. p. 640.
508. Gattavecchia, E., Tonelli, D., *J. Chromatogr.*, **1980**, *193*, 340.

509. Merdjan, H., Bonnat, C., Singlas, E., Diquet, B., *J. Chromatrogr.*, **1983**, *273*, 475.
510. Hubbard, R. W., Beierle, F. A., *J. Chromatogr.*, **1982**, *232*, 443.
511. Barabanov, V. P., Tret'yakova, A. Ya., Sharnin, G. P., Fassakhov, R. Kh., Eneikina, T. A., *Zh. Obshch. Khim.*, **1980**, *50*, 2318; *Chem. Abstr.*, **1981**, *94*, 83447r.
512. Wada, T., Ide, H., Nishimoto, S., Kagiya, T., *Chem. Lett.*, **1982**, 1041; Sundararajan R., Jain, S. R., *Combust. Flame*, **1982**, *45*, 47; *Chem. Abstr.*, **1982**, *96*, 220118s.
513. Jolles, G. E., *Int. Congr. Congr. Ser.-Excerpta Med.*, **1977**, 438 (Metronidazole), 3; *Chem. Abstr.*, **1978**, *89*, 84428u.
514. Cosar, C., Julou, L., *Ann. Inst. Pasteur*, **1959**, *96*, 238.
515. O'Holohan, D. R., Hugoe-Matthews, J., *Ann. Trop. Med. Parasitol.*, **1972**, *66*, 181.
516. Jokipii, L., Jokipii, A. M. M., *J. Infect. Dis.*, **1979**, *140*, 984.
517. Perera, M., Chipping, P. M., Noone, P., *J. Antimicrob. Chemother.*, **1980**, *6*, 105.
518. Christensson, B., Heström, S. A., Ursing, B., *Scand. J. Infect. Dis.*, **1979**, *11*, 69.
519. Tally, F. P., Goldin, B. R., Sullivan, N., Johnston, J., Gorbach, S. L., *Antimicrobial Agents Chemother.*, **1978**, *13*, 460.
520. Ursing, B., *Int. Congr. Ser.-Excerpta Med.*, **1977**, *438* (Metronidazole), 415–421.
521. Voogd, C. E., *Mutat. Res.*, **1981**, *86*, 243.
522. Thin, R. N., *Br. J. Vener. Dis.*, **1978**, *54*, 69.
523. Edwards, D. I., Knox, R. J., Rowley, D. A., Skolimowski, I. M., Knight, R. C., *Janssen Res. Found. Ser.*, **1980**, *2*, 673; *Chem. Abstr.*, **1981**, *94*, 115098w.
524. *Ibid.*, *NATO Adv. Study Insti. Ser., Ser. A*, **1982**, 105.
525. Edwards, D. I., *Br. J. Vener. Dis.*, **1980**, *56*, 285.
526. Adams, G. E., Fowler, J. F., *Modif. Radiosensitivity Biol. Syst., Proc. Advis. Group Meet.*, **1975**, **1976**, 103; *Chem. Abstr.*, **1977**, *87*, 78105d.
527. Smithen, C. E., Hardy, C. R., *NATO Adv. Study Inst. Ser., Ser. A*, **1982**, *43*, 1; *Chem. Abstr.*, **1982**, *97*, 19860e.
528. Workman, P., *NATO Adv. Study Inst. Ser., Ser. A*, **1982**, *43*, 143; *Chem. Abstr.*, **1982**, *97*, 19863h.
529. Mueller, M., *Scand. J. Infect. Dis. Suppl.*, **1981**, *26*, 31.
530. Simmen, H. P., *Schweiz. Apoth.-Ztg.*, **1981**, 227.
531. Tally, F. P., Goldin, B., Sullivan, N. E., *Scand. J. Infect. Dis. Suppl.*, **1981**, *26*, 46.
532. Van Lier, J. E., *Cytotoxic Estrogens Horm. Recept. Tumors*, [Proc. Workshop], **1979**, **1980**, 207; *Chem. Abstr.*, **1981**, *94*, 58510y.
533. Edwards, D. I., Knight, R. C., Kantor, I., *Curr. Chemother., Proc. Int. Congr. Chemother., 10th*, **1977**, **1978**, 714; *Chem. Abstr.*, **1978**, *89*, 190820a.
534. Stratford, I. J., Adams, G. E., *NATO Adv. Study Inst. Ser., Ser. A*, **1982**, *42*, 67; *Chem. Abstr.*, **1982**, *97*, 33021p.
535. Schwartz, D. E., Hofheinz, W., *ibid.*, 91; *Chem. Abstr.*, **1982**, *97*, 33022q.
536. De Carneri, I., *ibid.*, 115; *Chem. Abstr.*, **1982**, *97*, 33023r.
537. Leach, S. C., Weaver, R. D., Kinoshita, K., Lee, W. W., *J. Electroanal. Chem.*, **1981**, *129*, 213.
538. Baines, E. J., *J. Antimicrob. Chemother.*, **1978**, *4* (Suppl. C), 97.
539. Finegold, S. M., McFadzean, J. A., Roe, J. C., editors, *International Congress Series-Excerpta Medica*, **1978**, *438*, (methronidazole), 1–436; *Chem. Abstr.*, **1978**, *89*, 53557f.
540. Goldman, P., *N. Engl. J. Med.*, **1980**, *303*, 1212.
541. Tally, F. P., Sullivan, C. E., *Pharmacotherapy*, **1981**, *1*, 28.
542. Offerhaus, L., *Ned. Tijdschr. Geneeksd.*, **1977**, *121*, 1550; *Chem. Abstr.*, **1978**, *88*, 68808s.
543. Wearley, L. L., Anthony, G. D., *Anal. Profiles Drug Subst.*, **1976**, *5*, 327.
544. Goldman, P., *Johns Hopkins Med. J.*, **1980**, *147*, 1.
545. Edwards, D. I., *J. Antimicrob. Chemother.*, **1979**, *5*, 499.
546. Brogden, R. N., Heel, R. C., Speight, T. M., Avery, G. S., *Drugs*, **1978**, *16*, 387.
547. Graber, H., *Ther. Hung.*, **1977**, *25*, 109; *Chem. Abstr.*, **1978**, *88*, 98794b.
548. Playle, A. C., *Med. Actual.*, **1977**, *13*, 147.
549. Mueller, M., *Int. Congr. Symp. Ser.-R. Soc. Mod.*, **1979**, *18* (metronidazole), 223; *Chem. Abstr.*, **1980**, *92*, 158254v.
550. Roe, F. J. C., *J. Antimicrob. Chemother.*, **1977**, *3*, 205.
551. Foster, J. L., *Int. J. Radiat. Oncol., Biol. Phys.*, **1978**, *4*, 153.
552. Goldman, P., Ingelfinger, J. A., Friedman, P. A., *Cold Spring Harbor Conf. Cell Proliferation*, **1977**, *4*, 465; *Chem. Abstr.*, **1978**, *89*, 5t.
553. Plowman, P. N., *Trends Pharmacol. Sci.*, **1981**, *2*, 8.
554. Fowler, J. F., Adams, G. E., *High-Energy Photons Electron. Clin. Appl. Cancer Manage., Proc. Int. Symp.*, **1975**, **1976**, 309; *Chem. Abstr.*, **1978**, *88*, 46964n.

555. Denekamp, J., Fowler, J. F., *Int. J. Radiat. Oncol., Biol. Phys.*, **1978**, *4*, 143.
556. Bogan, J. A., *Med. Actual.*, **1977**, *13*, 510.
557. Naguchi, Y., Tanaka, T., *Drugs*, **1978**, *15*, (Suppl. 1), 10; *Chem. Abstr.*, **1978**, *89*, 140012t.
558. Coulter, J. R., Turner, J. V., *Mutat. Res.*, **1978**, *57*, 97.
559. Winkelmann, E., Raether, W., *Curr. Chemother. Infect. Dis., Proc. Int. Congr. Chemother, 11th*, **1979**, **1980**, *2*, 969; *Chem. Abstr.*, **1980**, *93*, 106558y.

BIBLIOGRAPHY

Agrawal, K. C., Sakaguchi, M., Nitroimidazole Radiosensitizers and Composition Thereof, *PCT. Int. Appl.* WO 83 02,774; *Chem. Abstr.*, **1984**, *100*, 6514a.
Ahmed, I., Adams, G. E., Stratford, I. J., Gibson, D., Compounds Useful in Radiotherapy or Chemotherapy, *Eur. Pat. Appl.* EP 95,906; *Chem. Abstr.*, **1984**, *100*, 139107s.
Aicardi, G., Cantelli-Forti, G., Guerra, M. C., Barbaro, A. M., Biagi, G. L., Electroreduction, Mutagenicity and Antimicrobial Activity of 5-Nitroimidazole Derivatives, *Fortschr. Onkol.*, **1983**, 300.
Aicardi, G., Forti, G. C., Guerra, M. C., Electroreduction, Mutagenicity and Antimicrobial Activity of 5-Nitroimidazole Derivatives, *Dev. Oncol.*, **1983**, *15 (Control Tumour Growth Its Biol. Bases)*, 300.
Al-Badr, A. A., Carbon-13 Nuclear Magnetic Resonance Spectroscopy of Some Biologically Active Imidazoles, *Spectrom. Lett.*, **1983**, *16*, 613.
Adreasson, E., Newton, C. G., Ollis, W. D., Rees, C. W., Smith, D. I., Wright, D. E., Unusual Synthesis of Stable Pyridinium Dinitromethylides, *J. Chem. Soc., Chem. Commun.*, **1983**, 816.
Alcade, E., Manos, L., Valls, N., Elguero, J., Synthesis of Derivatives of 5-Nitroimidazole That Are Potentially Active in Chemotherapy, *J. Heterocyclic Chem.*, **1984**, *21*, 1647.
Arya, V. P., Nagarajan, K., Shenoy, S. J., Nitroimidazoles. Part XV. 1-Methyl-5-nitro-2-oxy(mercapto)Imidazoles, 1-Methyl-5-nitroimidazole-2-methanol(Carboxaldehyde and Glyoxalic Ester) Derivatives and 1-Substituted Alkyl-2- Methyl-5- and -4-Nitroimidazoles, *Indian J. Chem., Sect. B*, **1982**, *21B*, 1078.
Arya, V. P., Nagarajan, K., Shenoy, S. J., Nitroimidazoles. Part XVI. Some 1-Methyl-4-nitro-5-substituted Imidazoles, *Indian J. Chem., Sect. B*, **1982**, *21B*, 1115.
Baker, D. C., Putt, S. R., Showalter, H. D. H., Studies Related to the Total Synthesis of Pentostatin. Approaches To the Synthesis of (8R)-3,6,7,8-Tetrahydroimidazo (4,5-d)(1,3)Diazepin-8-01 and N-3 Alkyl Congeners., *J. Heterocycl. Chem.*, **1983**, *20*, 629.
Barety, D., Resibois, B., Vergoten, G., Moschetto, Y., Electrochemical Behaviour of Nitroimidazole Derivatives in Dimethyl Sulfoxide, *J. Electroanal. Chem. Interfacial Electrochem.*, **1984**, *162*, 335.
Bhatia, S. C., Shanbhag, V. D., Electron-Capture Gas Chromatographic Assays

of 5-Nitroimidazole Class of Antimicrobials in Blood, *J. Chromatogr.*, **1984**, *305*, 325.

Biagi, G. L., Barbaro, A. M., Guerra, M. C., Cantelli Forti, G., Aicardi, G., Borea, P. A., Quantitative Relationship Between Structure and Mutagenic Activity in Series of 5-Nitroimidazoles, *Teratog., Carcinog., Mutagen.*, **1983**, *3*, 429.

Born, J. L., Smith, B. R., The Synthesis of Tritium-Labeled Misonidazole, *J. Labelled Compd. Radiopharm.*, **1983**, *20*, 429.

Brown, D. M., Dionet, C., Brown, J. M., Inhibition of X-Ray-Induced Potentially Lethal Damage (PLD) Repair in Aerobic Plateau-Phase Chinese Hamster Cells by Misonidazole, *Radiat. Res.*, **1984**, *97*, 162.

Brown, D. M., Cohen, M. S., Sagerman, R. H., Gonzales-Mendez, R., Hahn, G. M., Brown, J. M., Influence of Heat on the Intracellular Uptake and Radiosensitization of 2-Nitroimidazole Hypoxic Cell Sensitizers in Vitro, *Cancer Res.*, **1983**, *43*, 3138.

Bundgaard, H., Larsen, C., Thorbek, P., Prodrugs as Drug Delivery Systems. XXVI. Preparation and Enzymic Hydrolysis of Various Water-Soluble Amino Acid Esters of Misonidazole, *Int. J. Phar.*, **1984**, *18*, 67.

Carlson, J. A., Antibacterial and Antiprotozoal 1-Methyl-5-nitro-2-(2-Phenylvinyl) Imidazoles, *U.S.* US 4, 423, 046; *Chem. Abstr.*, **1984**, *100*, 132577m.

Catalan, J., Perez, P., Elguero, J., Substituent Effects on Proton Affinities: Through Bonds or Through Space Mechanism, *Heterocycles*, **1983**, *20*, 1717.

Chasseaud, L. F., Henrick, K., Matthews, R. W., Scott, P. W., Wood, S. G., Metabolic Ring Hydroxylation of Tinidazole Involving a Novel Nitro-Group Migration: X-ray Structures of Tinidazole and the $NH_4(+)$ Salt of Its Ring Hydroxylated Metabolite, *J. Chem. Soc., Chem. Commun.*, **1984**, 491.

Chen, B. C., Von Philipsborn, W., Nagarajan, K., Nitrogen-15 NMR Spectroscopy. Part X. Nitrogen-15 NMR Spectra of Azoles with Two Heteroatoms, *Helv. Chim. Acta*, **1983**, *66*, 1537.

Chowdary, K. P. R., Kumar, K. T. R., A New Spectrophotometric Method for the Estimation of Metronidazole and Metronidazole Benzoate, *Indian J. Pharm. Sci.*, **1983**, *45*, 182.

Cornea, A. F., Ionescu, D., 5-Amino-1H-Imidazole-4-Carboxamides and –4-Carboxylate Esters, *Rom.* RO 80,161; *Chem. Abstr.*, **1984**, *100*, 103346e.

De, A. U., Sengupta, C., Pal, D., Mandal, A., Chatterjee, J., Partition Coefficients of Some Metronidazole and Furacin Analogs in Relation to Biological Activity, *Indian J. Pharm. Sci.*, **1983**, *45*, 123.

Declerck, P. J., De Ranter, C. J., Volckaert, G., Base Specific Interaction of Reductively Activated Nitroimidazoles with DNA, *FEBS Lett.*, **1983**, *164*, 145.

Dereu, N., Welter, A., Ghyczy, M., 2,4(5)-Disubstituted-5(4)-Nitroimidazoles, *Ger. Offen.* DE 3, 247, 646; *Chem. Abstr.*, **1985**, *102*, 24618h.

Dodali, V. A., Zubareva, T. M., Litvinenko, L. M., Simanenko, Y. S., Mechanism of Nucleophilic Catalysis in Acyl Transfer Reactions in Protoinert Media. Effect of an Intermediate Product Anion, *Zh. Org. Khim.*, **1983**, *19*, 1468; *Chem. Abstr.*, **1983**, *99*, 157518s.

Dumanovic, D., Ciric, J., Kosanovic, D., Jeremic, D., The Possibilities of Polarography in the Chemistry of Nitroazoles, *Collect. Czech. Chem. Commun.*, **1984**, *49*, 1342.
Eichler, E., Rooney, C. S., Williams, H. W. R., Concerning the Preparations of Some Pyridylimidazothiazole Derivatives, *J. Heterocycl. Chem.*, **1983**, *20*, 419.
Fassakhov, R. K., Sharnin, G. P., Sabirzyanov, R. G., Naumov, V. N., 1-Alkyl-4-Nitroimidazoles, *U.S.S.R.* SU 1,011,644; *Chem. Abstr.*, **1983**, *99*, 122453q.
Garcia, M. L. S., Smith, J. A. S., Bavin, P. M. G., Ganellin, C. R., Nitrogen-14 and Deuterium Quadrapole Double Resonance in Substituted Imidazoles, *J. Chem. Soc., Perkin Trans. 2*, **1983**, *(9)*, 1391.
Gilmour, D. W., Sadler, P. J., Metal Complexes of Nitro-Substituted Pyrazoles, Imidazoles, and Isothiazoles, *Brit. UK Pat. Appl.* GB 2,122,194; *Chem. Abstr.*, **1984**, *100*, 203599h.
Goldman, P., Ramos, S. M., Wuest, J. D., Reactions of Nitroimidazoles with Hydrazine, *J. Org. Chem.*, **1984**, *49*, 932.
Hinchliffe, M., McNally, N. J., Stratford, M. R. L., The Effect of Radiosensitizers on the Pharmacokinetics of Melphalon and Cyclophosphamide in the Mouse, *Br. J. Cancer*, **1983**, *48*, 375.
Hirst, D. G., Hazlehurst, J. L., Brown, J. M., Sensitization of Normal and Malignant Tissues to Cyclophosphamide by Nitroimidazoles with Different Partition Coefficients, *Br. J. Cancer*, **1984**, *49*, 33.
Hofheinz, W., 2-Nitroimidazoles, *Brit. UK Pat. Appl.* GB 2,110,211; *Chem. Abstr.*, **1983**, *99*, 175763x.
Holmwood, G., Regel, E., Jaeger, G., Buechel, K. H., Plempel, M., Antimycotic Substituted 1-Hydroxyalkylazolyl Derivatives, *Ger. Offen.* DE 3,202,613; *Chem. Abstr.*, **1983**, *99*, 200501v.
Holmwood, G., Regel, E., Jaeger, G., Buechel, K. H., Luerssen, K., Frohberger, P. E., Brandes, W., Paul, V., Substituted 1-Hydroxyalkyl-Azolyl Derivatives and their Use as Fungicides and Plant Growth Regulators, *Ger. Offen.* DE 3,202,601; *Chem. Abstr.*, **1983**, *99*, 194972f.
Iivespää, A. O., Jarumilinta, R., Nitroimidazoles. Part III. 1-[1-Methyl-5-Nitroimidazol-2-yl]-2-oxo-3-acyl/Hetaryltetrahydroimidazoles, *Indian J. Chem., Sect. B*, **1982**, *21B*, 923.
Jensen, J. C., Gugler, R., Sensitive High-Performance Liquid Chromatographic Method for the Determination of Metronidazole and Metabolites, *J. Chromatogr.*, **1983**, *277*, 381.
Jensen, J. C., Gugler, R., Single- and Multiple-Dose Metronidazole Kinetics, *Clin. Pharmacol. Ther.*, **1983**, *34*, 481.
Johnson, C. R., Shepherd, R. E., Mössbauer Study of Imidazole and N-Heterocyclic Complexes of Pentacyanoiron(II) and -Iron(III), *Inorg. Chem.*, **1983**, *22*, 3506.
Kamalapurkar, O. S., Chudasama, J. J., Colorimetric Estimation of Tinidazole in Pharmaceutical Dosage Forms, *East. Pharm.*, **1983**, *26*, 207.
Kjoeller, L. I., Structure of Axomycin (2-Nitroimidazole), $C_3H_3N_3O_2$, at 105 K., *Acta Crystallogr., Sect. C: Cryst. Struct. Commun.*, **1984**, *C40*, 285.

Knox, R. J., Knight, R. C., Edwards, D. I., Studies on the Action of Nitroimidazole Drugs. The Products of Nitroimidazole Reduction, *Biochem. Pharmacol.*, **1983**, *32*, 2149.
Kruglenko, V. P., Povstyanoi, M. V., Klyuev, N. A., Condensed Imidazo-1,2,4-azines.9. Imidazo(1,2-b)-1,2,4-triazine Derivatives in Electrophilic Substitution Reactions, *Khim. Geterotskl. Soedin*, **1984**, 413; *Chem. Abstr.*, **1984**, *101*, 55069p.
Lokajicek, M., Kozubed, S., Prokes, K., Cumulative Effect of Various Types of Ionizing Particles During Fractional Irradiation, *Soveshch Ispolz. Nov. Yad.-Fiz. Metodov Resheniya Nauchno-Tekh. Narodnokhoz. Zadach (Dokl.)*, 4th 1981, 376; *Chem. Abstr.*, **1984**, *100*, 205759j.
McClelland, R. A., Fuller, J. R., Seaman, N. E., Rauth, A. M., Battistella, R., 2-Hydroxylaminoimidazoles—Unstable Intermediates in the Reduction of 2-Nitroimidazoles, *Biochem. Pharmacol.*, **1984**, *33*, 303.
McFadzean, J. A., Combating Industrial Waste or Decay with Nitroimidazole Compounds, *Ger. Offen.* DE 3,230,887; *Chem. Abstr.*, **1983**, *99*, 1816b.
McKillop, A., Wright, D. E., Podmore, M. L., Chambers, R. K., 4- And 5-Nitroimidazoles: Carbon-13 NMR Assignment of Structure, *Tetrahedron*, **1983**, *39*, 3797.
Michaels, H. B., Peterson, E. C., Epp, E. R., Effects of Modifiers of the Yield of Hydroxyl Radicals on the Radiosensitivity of Mammalian Cells at Ultrahigh Dose Rates, *Radiat. Res.*, **1983**, *95*, 620.
Middleton, R. W., Monney, H., Parrick, J. K., N-Methylation of Heterocycles with Dimethylformamide Dimethyl Acetal, *Synthesis*, **1984**, 740.
Miyake, T., Takeda, K., Tada, K., Basic Imidazolylmethyl-Styrene Compound, Its Polymer, and Its Use as an Ion Exchange Resin, *U.S.* US 4,430,445; *Chem. Abstr.*, **1984**, *100*, 193229w.
Moellgaard, B., Hoelgaard, A., Vehicle Effect on Topical Drug Delivery. I. Influence of Glycolysis and Drug Concentration on Skin Transport, *Acta Pharm. Suec.*, **1984**, *20*, 433.
Mokrushin, V. S., Belyaev, N. A., Kolobov, M. Y., Fedotov, A. N., Reactions of 4,5-Dinitroimidazole and 4(5)-Nitroimidazole-5(4)-sulfonic Acid with Nucleophiles, *Khim. Geterotsikl. Soedin.*, **1983**, *(6)*, 808; *Chem. Abstr.*, **1983**, *99*, 139847u.
Muir, R. D., Use of Metronidazole in Oil Recovery, *U.S.* US 4,395,341; *Chem. Abstr.*, **1983**, *99*, 125378e.
Nagarajan, K., Arya, V. P., George, T., Bhat, G. A., Kulkarni, Y. S., Shenoy, S. J., Rao, M. K., Nitroimidazoles. Part VII. 1-(1-Alkyl-5-nitroimidazol-2-yl)aza-(diaza,oxaza)cycloalkanes, *Indian J. Chem., Sect. B*, **1982**, *21B*, 949.
Nagarajan, K., Shenoy, S. J., Nitroimidazoles: Part XX—Reactions of 2,4-Dinitroimidazole with 2-Haloethanols, 3-Chloropropionitrile and Propylene Oxide. *Indian J. Chem., Sect. B*, **1984**, *23(B)*, 363.
Nagarajan, K., Sudarsanam, V., Parthasarathy, P. C., Arya, V. P., Shenoy, S. J., Nitroimidazoles. Part X. Spectral Studies on Isomeric 1-Substituted 4- and 5-Nitroimidazoles and Some 2-Nitroimidazoles, *Indian J. Chem., Sect. B*, **1982**, *21B*, 1006.

Nair, M. D., Desai, J. A., Nitroimidazoles: Part XIII -(1-Methyl-5-nitro-2-imidazolyl-methyl Aryl Ketones and Derived Heterocycles, *Indian J. Chem., Sect. B*, **1984**, *23(B)*, 480.
Nair, M. D., Nagarajan, K., Nitroimidazoles as Chemotherapeutic Agents, *Prog. Drug. Res.*, **1983**, *27*, 163.
Newton, C. G., Ollis, W. D., Podmore, M. L., Wright, D. E., Cyclic Meso-Ionic Compounds. Part 21. The Examination of Nitro Derivatives of Meso-Ionic Heterocycles as Potential Pharmaceuticals, *J. Chem. Soc., Perkin Trans. 1*, **1984**, 63.
Newton, C. G., Ollis, W. D., Wright, D. E., Cyclic Meso-Ionic Compounds. Part 22. Meso-Ionic Derivatives of the Imidazo[1,2-a]pyridinium Dinitromethylides, *J. Chem. Soc., Perkin Trans. 1*, **1964**, 69.
Nishimoto, S., Ide, H., Wada, T., Kagiya, T., Radiation-induced by Hydroxylation of Thymine Promoted by Electron-Affinic Compounds, *Int. J. Radiat. Biol. Relat. Stud. Phys., Chem. Med.*, **1983**, *44*, 585.
Nothenberg, M. S., Korolkovsa, A., Spectrometric Analysis of Benznidazole, *Rev. Farm. Bioquim. Univ. Sao Paulo*, **1983**, *19*, 1; *Chem. Abstr.*, **1984**, *100*, 91271y.
Odintsova, S. P., Gonikberg, E. M., Increase of Radiosensitivity of DNA under Effect of Reducible Nitroimidazoles Nitrofurans, *Radiobiologiya*, **1983**, *23*, 291; *Chem. Abstr.*, **1983**, *99*, 49618r.
Oeckl, S., Schmitt, H. G., Brandes, W., Alkylene (cycloalkylene) Bisheterocyclylbiguanide, *Eur. Pat. Appl.* EP 96,280; *Chem. Abstr.*, **1984**, *100*, 174835z.
Omura, M., Radiosensitization of Hypoxic Cancer Cells, *Nippon Sanka Fujinka Gakkai Zasshi*, **1983**, *35*, 1972; *Chem. Abstr.*, **1984**, *100*, 19830a.
Parthasarathy, P. C., Desai, H. K., Saindane, M. T., Nitroimidazoles. Part XVIII. Mannich Reactions of Azomycin -2-Nitro-4,5-bisaminoalkylimidazoles, *Indian J. Chem., Sect. B*, **1983**, *22B*, 157.
Raleigh, J. A., Liu, S. F., Reductive Fragmentation of 2-Nitroimidazoles in the Presence of Nitroreductases-Glyoxal Formation from Misonidazole, *Biochem. Pharmacol.*, **1983**, *32*, 1444.
Rossignol, J. F., Maisonneuve, H., Cho, Y. W., Nitroimidazoles in the Treatment of Trichomoniasis, Giardiasis, and Amebiasis, *Int. J. Clin. Pharmacol., Ther. Toxicol.*, **1984**, *22*, 63.
Rusczak, J., Pakula, R., Rudnicki, A., Magielka, S., Szczepanik, H., Grzeszkiewicz, A., Spychala, S., Imidazole Derivatives, *Pol. PL* 115,196; *Chem. Abstr.*, **1983**, *99*, 70729n.
Saldabol, M. O., Popelis, Y. Y., Mazheika, I. B., Effect of a Mixture of Nitric and Trifluoroacetic Acids on 2-Phenylimidazo(1,2-Alpha)pyrimidine, *Zh. Org. Khim.*, **1983**, *19*, 1558; *Chem. Abstr.*, **1983**, *99*, 175663q.
Sanghavi, N. M., Sathe, V. H., Padki, M. M., 9-Chloroacridine as a Reagent for the Analysis of Drugs Having Amine Groups, *Indian Drugs*, **1983**, *20*, 341.
Schrautemeier, B., Boehme, H., Boeger, P., *In vitro* Studies on Pathways and Regulation of Electron Transport to Nitrogenase with a Cell-Free Extract from Heterocysts of Anabaena Variabilis, *Arch. Microbiol.*, **1984**, *137*, 14.
Schumann, H., Scheithauer, S., Friese, J., Imidazole Derivatives, *Ger. (East)* DD

159,638; *Chem. Abstr.*, **1983**, *99*, 122455s.
Shubin, V. E., Kuropteva, Z. V., EPR Study of Nitrogen Oxide (NO) Evolution During Reduction of Nitrofurans and Nitroimidazoles. I. Hemoglobin Solutions, *Stud. Biophys.*, **1983**, *97*, 157; *Chem. Abstr.*, **1984**, *100*, 61312h.
Sitzmann, M. E., Gilligan, W. H., Derivatives of Energetic Orthoformates, *U.S. Pat. Appl.* US 467,715; *Chem. Abstr.*, **1984**, *100*, 24077q.
Skwarski, D., Sobiak, S., Synthesis of Some 1-Phenacyl-4-Nitroimidazole Derivatives, *Polish J. Chem.*, **1983**, *57*, 551.
Smith, B. R., Born, J. L., Garcia, D. J., Influence of Hypoxia on the Metabolism and Excretion of Misonidazole by the Isolated Perfused Rat Liver—A Model System, *Biochem. Pharmacol.*, **1983**, *32*, 1609.
Souney, P. F., Colucci, R. D., Mariani, G., Campbell, D., Compatibility of Magnesium Sulfate Solutions with Various Antibiotics during Simulated Y-Site Injection, *Am. J. Hosp. Pharm.*, **1984**, *41*, 323.
Sudarsanam, V., Nagarajan, K., Arya, V. P., Kaulgud, A. P., Shenoy, S. J., Shah, R. K., Nitroimidazoles. Part VIII. 2-Amino-1-methyl-5-nitroimidazoles and Derivatives, *Indian J. Chem., Sect. B*, **1982**, *21B*, 989.
Sudarsanam, V., Nagarajan, K., George, T., Shenoy, S. J., Iyer, V. V., Kaulgud, A. P., Nitroimidazoles. Part XI. Some Halo- Nitro- and Dinitroimidazoles, *Indian J. Chem., Sect. B*, **1982**, *21B*, 1022.
Sudarsanam, V., Nagarajan, K., Gokhale, N. G., Nitroimidazoles. Part XVII. 5-Aminoimidazoles, *Indian J. Chem., Sect. B*, **1982**, *21B*, 1087.
Sutherland, R. M., Siemann, D. W., Conroy, P. J., Potentials and Limitations for the Use of Radiation Sensitizers of Resistant Hypoxic Cells in Tumors, *Adrenal Endocr. Tumors Child.*, **1984**, 19.
Suwinski, J., Salwinska, E., Watras, J., Widel, M., Nitroimidazoles. Part V. Chloronitroimidazoles from Dinitroimidazoles. A Reinvestigation, *Pol. J. Chem.*, **1982**, *56*, 1261.
Suwinski, J., Salwinska, E., Nitroimidazoles. Part III. Ionization Constants and Tautomerism of Imidazoles. A Reinvestigation, *Pol. J. Chem.*, **1981**, *55*, 2525.
Thind, P. S., Gandhi, J. S., Analytical Applications of Ion-Exchange Materials. V. Chromatographic Separation of Drugs on Ferric Phosphate Papers, *J. Liq. Chromatogr.*, **1983**, *6*, 1153.
Thorbeck, P., Bundgaard, H., Larsen, C., Ester of Metronidazole with N,N-Dimethylglycine and its Acid Addition Salt, *Eur. Pat. Appl.* EP 96,870; *Chem. Abstr.*, **1984**, *100*, 121076w.
Tonelli, D., Gattavecchia, E., Breccia, A., Simultaneous Determination of Some Radiosensitizing and Chemotherapeutic Drugs in Plasma by Thin-Layer Chromatography, *J. Chromatogr.*, **1983**, *275*, 223.
Vajtner, Z., Determination of Some Organic Intermediates and Products by Electroanalytical Methods. III. Determination of Chlorine and Bromine by Potentiometric Titration, *Kem. Ind.*, **1983**, *32*, 461; *Chem. Abstr.*, **1984**, *100*, 61145f.
Vecchia, L. D., Dellureficio, J., Kisis, B., Vlattas, I., Imidazole Systems. I. Imidazo [2,1-b][1,3,5]Benzothiadiazepine Derivatives as Potential Antipsycho-

tic Agents, *J. Heterocycl. Chem.*, **1983**, *20*, 1287.
Vinge, E., Andersson, K. E., Ando, G., Lunell, E., Biological Availability and Pharmacokinetics of Tinidazole after Single and Repeated Doses, *Scand. J. Infect. Dis.*, **1983**, *15*, 391.
Walker, H., Cullen, B., Radiosensitization of Cells *in vitro* with Misonidazole: Dependence on Endogenous Sulfhydryl, *Br. J. Radiol.*, **1983**, *56*, 871.
Wolf, A. D., Rorer, M. P., Herbicidal Sulfonamides, *Eur. Pat. Appl.* EP 83,975; *Chem. Abstr.*, **1983**, *99*, 175812n.
Wolff, D., Hertel, O., Voges, D., Bis(aminocyclohexyl)dialkylmethanes, *Ger. Offen.* DE 3,226,889x; *Chem. Abstr.*, **1984**, *100*, 174333j.
Workman, P., Development of Nitroimidazoles, *Dev. Oncol.*, **1983**, 15 (*Control Tumour Growth Its Biol. Bases*), 166.
Workman, P., Development of Nitroimidazoles, *Fortschr. Onkol.*, **1983**, *10*, 166.
Yeung, T. C., Sudlow, G., Koch, R. L., Goldman, P., Reduction of Nitroheterocyclic Compounds by Mammalian Tissue *in vivo*, *Biochem. Pharmacol.*, **1983**, *32*, 2249.
Zheng, K., Shen, D., Ni, Z., Chen, L., Dai, Z., Ma, Z., Studies on Antimalarials. X. Synthesis and Antimalarial Effect of 2,4-Diamino-6-N^1,N^2-D-Disubstituted Hydrazinoquinazoline Derivatives, *Yaoxue Xuebao*, **1983**, *18*, 673; *Chem. Abstr.*, **1984**, *100*, 121002u.
Color Photography, *Jpn. Kokai Tokkyo Koho* JP 57,132,145 (82,132,145); *Chem. Abstr.*, **1983**, *99*, 203531r.
Thermosensitive Diazo Recording Material, *Jpn. Kokai Tokkyo Koho* JP 59 02,888 [84 02,888]; *Chem. Abstr.*, **1984**, *100*, 183257q.

Chapter 3

Nitropyrazoles

INTRODUCTION

Nitropyrazole, pyrazoline, and pyrazolidine derivatives of mono- and polycyclic systems have been briefly mentioned in reviews of the heterocycles.[1-5] In this review the literature has been covered through *Chemical Abstracts*, **1983**, *98*. There is a bibliographic addendum to cover *Chemical Abstracts 99* and *100*.

PREPARATIONS OF MONOCYCLIC PYRAZOLES

Nitration and Nitrolysis

Pyrazole, Alkylpyrazoles, and Arylpyrazoles

Nitration of pyrazole (**3-1**) by fuming nitric acid in "anhydrosulfuric" acid to give a nitropyrazole, mp 162°, was first reported in 1893.[6] In a 1955 modification[7-9] pyrazole was treated with a small amount of concentrated sulfuric acid with cooling, then with a mixture (1:1) of fuming sulfuric and fuming nitric acids, and heated at 110° for 40 h after an initial vigorous reaction had subsided; only 4-nitropyrazole (**3-2**) (80%), mp 161°–162°, was detected.[6-11] A spectrophotometric examination showed the reaction to be quantitative.[12] Initial formation of 1-nitropyrazole (**3-3**) may not have been competitive with protonation; however, this unstable compound was obtained at lower acidities[12] and gave in the cold an acid-catalyzed rearrangement to the 4-nitro isomer (**3-2**).[13,14] Loss of the nitro group through transfer to another substrate has been noted.[14,15]

Over the range 90%–99% sulfuric acid, the nitration of pyrazole was a second-order reaction and proceeded from the conjugate acid of the heterocycle.[12,16,17] A partial rate factor for the pyrazolium ion (4-position) was determined to be 2.1×10^{-10}.[12] Since 1,2-dimethylpyrazolium cation (from a methylation of 1-methylpyrazole (**3-4**)) was nitrated faster than pyrazole in 98.7% sulfuric acid whereas at a lower acidity in 76% sulfuric acid the order was reversed, it was concluded that pyrazole was nitrated as the free base at lower acidities.[11] Nitration of 1,3,5-trimethylpyrazole over the range 73%–98% sulfuric acid occurred on the conjugated acid.[18] MO calculations were in agreement with the observed orientation of substitution in strongly acidic media.[19]

3-1 X = Y = Z = H
3-2 X = Y = H, Z = NO$_2$
3-3 X = NO$_2$, Y = Z = H
3-4 X = CH$_3$, Y = Z = H
3-5 X = CH$_3$, Y = H
   Z = NO$_2$

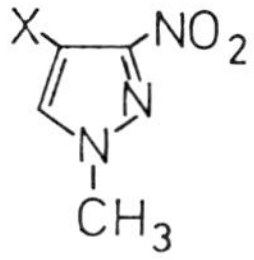

3-6a X = NO$_2$
3-6b X = H

A fivefold excess of fuming nitric acid in 80% sulfuric acid at 25° for 2 days failed to react with 1-methylpyrazole, but conversion to the 4-nitro derivative (**3-5**) (52%) and also to the 3,4-dinitro derivative (**3-6a**) (13%) was completed at 100° after 18 h. The 3-nitropyrazole (**3-6b**) was detected in the reaction mixture by an nmr analysis, but it could not be isolated from the mixture. The dinitro compound (**3-6a**) was also obtained by a nitration of 1-methyl-3-nitropyrazole (**3-6b**), but the 4-nitro isomer (**3-5**) resisted nitration.[20]

In similar reactions mixed-acid nitration afforded 4-nitro derivatives (**3-7**)–(**3-11**) of alkylpyrazoles: 3(5)-methyl-,[10,21] 3,5-dimethyl-,[10,22,23] 1,5-dimethyl-,[24] 1,3-dimethyl-,[25] 3(5)-*tert*-butyl-,[26] and 3(5)-mono-, di-, and trideuteromethyl-.[27,28] Nitration of 1,4-dimethylpyrazole in sulfuric acid gave the 3-nitro derivative (**3-12**) (40%) and small amounts of the 5-nitro isomer (**3-13**).[29]

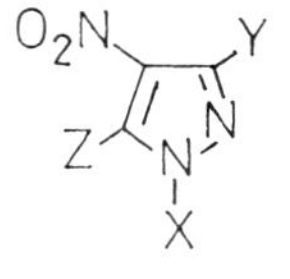

3-7 X = Z = H, Y = CH$_3$
3-8 X = H, Y = Z = CH$_3$
3-9 X = Z = CH$_3$, Y = H
3-10 X = Y = CH$_3$, Z = H
3-11 X = Z = H, Y = C(CH$_3$)$_3$

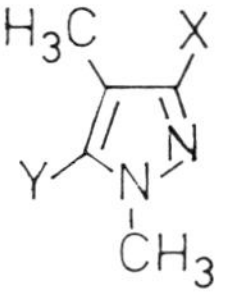

3-12 X = NO$_2$, Y = H
3-13 X = H, Y = NO$_2$

Four 3-perfluoroalkylpyrazoles gave 4-nitro derivatives (**3-14a–d**) in good yields as shown on treatment with a mixture of nitric and sulfuric acids at 25° for 12 h. Similar nitration was unsuccessful with 1,3-dimethyl-5-perfluoroheptylpyrazole, 3,5-diperfluoroheptylpyrazole, and the 1-methyl derivative of the latter compound. Apparently the reaction was inhibited by perfluoroalkyl substituents at the 5- and 3,5-positions.[30]

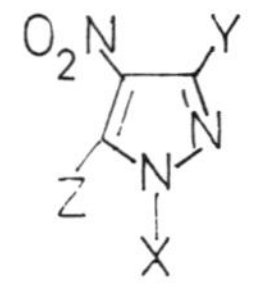

3-14a X = H, Y = C_3F_7, Z = CH_3(88%)
3-14b X = H, Y = C_5F_{11}, Z = CH_3(75%)
3-14c X = H, Y = C_7F_{15}, Z = CH_3(93%)
3-14d X = Z = CH_3, Y = C_7F_{15}(92%)

Methylene-1,1′-dipyrazole was nitrated at 25° (22 h) to give the mono- (**3-15a**) (30%) and the dinitro derivative (**3-15b**) (60%).[31]

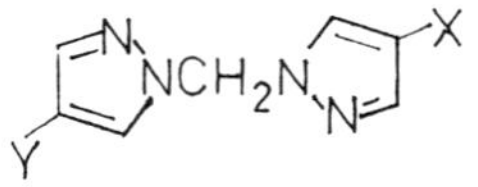

3-15a X = NO_2, Y = H
3-15b X = Y = NO_2

Nitration of phenylpyrazoles generally gave first a *p*-nitrophenyl derivative, then a 4-nitro-X-(4-nitrophenyl)pyrazole (X = 1,3,5).[32,33] A solution of 1-phenylpyrazole in sulfuric acid (d 1.84) was slowly combined with a mixture of sulfuric acid (d 1.84) and nitric acid (d 1.42) and kept at 12° for 30 min to produce 1-(4-nitrophenyl)pyrazole (**3-16a**) (86%); a similar reaction at 22° for 16 h gave 1-(4-nitrophenyl)-4-nitropyrazole (**3-16b**) (76%) and at 100° for 30 min gave 1-(2,4-dinitrophenyl)-4-nitropyrazole (**3-16c**) (83%).[32–39] Although other 1-, and/or 3-, and/or 5-arylpyrazoles gave similar results,[32,34,40,41] attempts to nitrate 4-phenyl and 4-picrylpyrazole were unsuccessful under conditions (0°) whereby 1-methyl-4-phenylpyrazole gave the 2′,4′-dinitro derivative and 1-methyl-4-(4-tolyl)pyrazole gave the 3-nitropyrazole derivative.[38] 1-(3-Nitrophenyl)pyrazole nitrated at the pyrazole 4-position.[32]

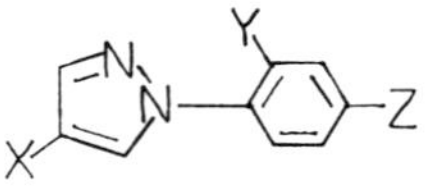

3-16a X = Y = H, Z = NO_2
3-16b X = Z = NO_2, Y = H
3-16c X = Y = Z = NO_2
3-23 X = NO_2, Y = Z = H

A mixed-acid treatment converted 1-picrylpyrazole to its 4-nitro derivative (70%),[42] and 1-methyl-4-picrylpyrazole to its 3,5-dinitro derivative (**3-17**) (55%).[43] In refluxing nitric acid (90%), 1-methyl-4-picrylpyrazole was converted to 1-methyl-3-nitro-4-picrylpyrazole (**3-18**) (55%). Similar treatment converted 1-methyl-4-(2,4-dinitrophenyl)pyrazole to a mixture of the 3-nitro- (**3-19**), 5-nitro- (**3-20**), and 3,5-dinitro derivative (**3-21**) (29:13:58); however, the reaction in nitric acid (70%) gave a product ratio of 24:17:0 with recovery (59%) of the starting material (**3-18**). Both product mixtures were completely converted to the tetranitro compound (**3-21**) in a mixture of concentrated nitric and sulfuric acids at reflux.[43] A kinetic analysis resolved the observation that the ratio (**3-19**)/(**3-20**) increased with time by establishing that compound (**3-20**) slowly disappeared (it was shown that compound (**3-20**) was not converted to compound (**3-19**)). It was then postulated that mononitration at the 5-position in 1-methyl-4-picrylpyrazole also occurred but that it decomposed at a faster rate.[44,45]

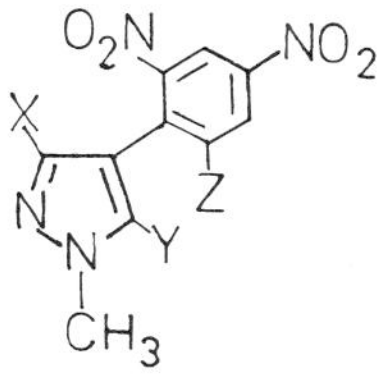

3-17 X = Y = Z = NO$_2$
3-18 X = Z = NO$_2$, Y = H
3-19 X = NO$_2$, Y = Z = H
3-20 X = Z = H, Y = NO$_2$
3-21 X = Y = NO$_2$, Z = H

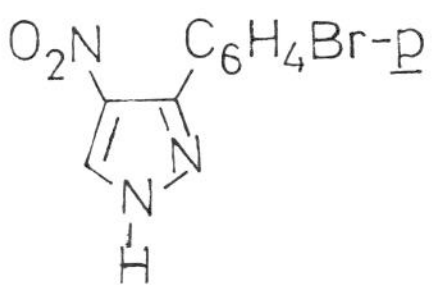

3-22

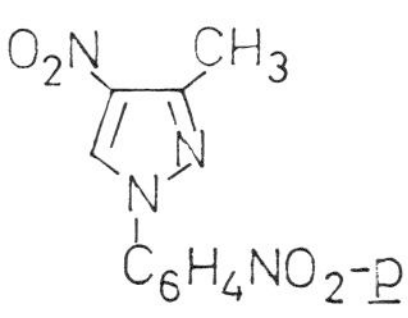

3-24

The ease of nitrating 3-*p*-bromophenylpyrazole in nitric acid (d 1.5) at $-10°$ to $-15°$ for 1 h to give the 4-nitro derivative (**3-22**) (44%)[46] illustrated the generalization that an arylpyrazole as a free base nitrated in the heterocyclic ring whereas at higher acidity it nitrated as the conjugate acid in the phenyl ring.[32,40] In another example nitronium tetrafluoroborate in sulfolan converted 1-phenylpyrazole to its 4-nitro derivative (**3-23**) without competitive formation of 1-(4-nitrophenyl) pyrazole.[36] It was reported that 1-*p*-nitrophenyl-3-methyl-pyrazole mononitrated at 5° in the heterocyclic ring to give compound (**3-24**).[37]

Nitration by nitric acid (100%) in sulfuric acid (98%) converted thienylpyrazoles to nitrothienylpyrazoles without competitive substitution in the pyrazole ring. At 0° for 1 h 1-(2-thienyl)pyrazole (**3-25**) gave a mixture (90%) of the 4'-nitro and 5'-nitro derivative (7:93) and 1-(3-thienyl)pyrazole (**3-26**) gave a mixture (87%) of the 2'-nitro, 4'-nitro, and 5'-nitro derivative (11:3:86).[47] Similar reactions at 0° for 20 and 45 min converted 2'- and 3'-thienyl derivatives (**3-27a, 3-28a**) of 1-methylpyrazole to mixtures (each 52%) of mono- and dinitrothienylpyrazoles (**3-27b–e, 3-28b–f**) in the ratios shown.[48] Nitration in sulfuric acid also gave the 5'-nitro derivatives (**3-29** to **3-31**) of 3-(2-thienyl)pyrazole, 3-(3-thienyl)pyrazole, and 3-(2-furyl)pyrazole,[49] and it afforded the dinitrobipyrazolyl (**3-32**) (48%) from 4-nitro-5(3)-(1-pyrazolyl)pyrazole.[50] Nitric acid in trifluoroacetic acid also nitrated thienylpyrazoles in the thiophene ring.[47,48]

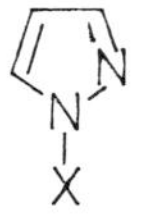

3-25 X = 2-thienyl
3-26 X = 3-thienyl

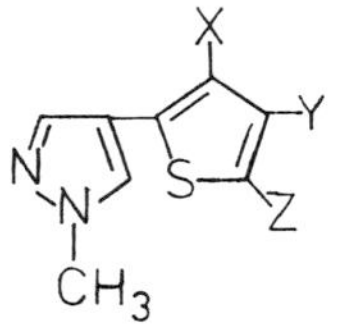

3-27a X = Y = Z = H
3-27b X = NO$_2$, Y = Z = H (4%)
3-27c X = Z = H, Y = NO$_2$ (1%)
3-27d X = Y = H, Z = NO$_2$ (25%)
3-27e X = Z = NO$_2$, Y = H (70%)

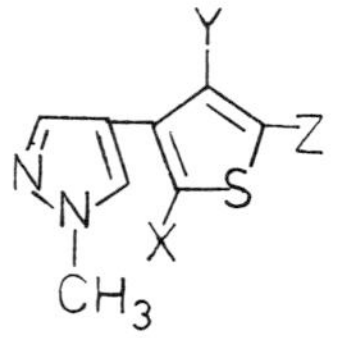

3-28a X = Y = Z = H
3-28b X = NO$_2$, Y = Z = H (34%)
3-28c X = Z = H, Y = NO$_2$ (1%)
3-28d X = Y = H, Z = NO$_2$ (8%)
3-28e X = Y = NO$_2$, Z = H (14%)
3-28f X = Z = NO$_2$, Y = H (44%)

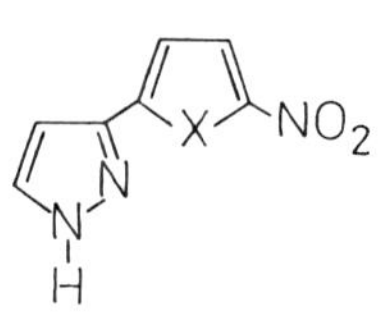

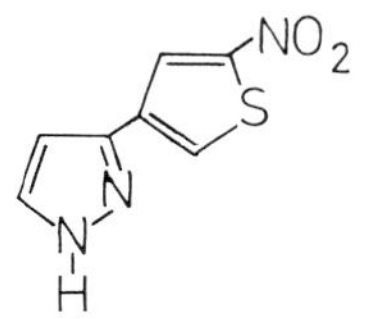

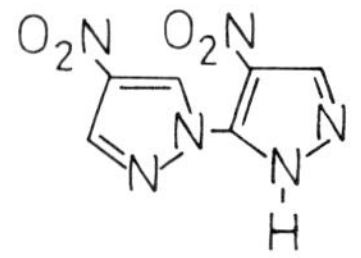

3-29 X = S
3-31 X = O

3-30

3-32

One (**3-33**) of the few known dinitropyrazoles was obtained by the nitration of 5(3)-nitro-3(5)-(3-pyridyl)pyrazole. As expected, 3(5)-(3-pyridyl)pyrazole was nitrated to give 4-nitro-3(5)-(3-pyridyl)pyrazole (**3-34a**), which failed to give further nitration;[51,52] nitro compound (**3-34b**) was obtained from 3(5)-methyl-5(3)-(3-pyridyl)pyrazole.[23]

A direct conversion of 1,5-dimethylpyrazole to the 3,4-dinitro derivative (**3-35**) has been reported.[53]

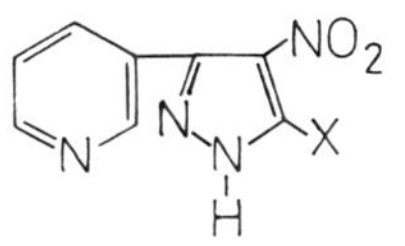

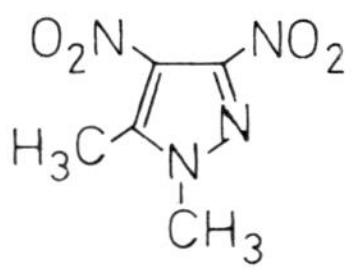

3-33 X = NO$_2$
3-34a X = H
3-34b X = CH$_3$

3-35

Three 3-perfluoroalkyl-5-phenylpyrazoles and 1,5-diphenyl-3-perfluoroheptylpyrazole were substituted in 4- and/or p-positions to give products (**3-36a–d**) in good yields as shown on treatment with a mixture of nitric and sulfuric acids at 25° for 12 h. In 1,3-diphenyl-5-perfluoroheptylpyrazole and 1-phenyl-3,5-diperfluoroheptylpyrazole, p-phenyl but not 4-pyrazolyl nitration occurred.[30] The formation of 1,5-di-p-nitrophenyl-3-perfluoroheptyl-4-nitropyrazole (**3-36d**) contrasted with initial nitration in a phenyl ring in 1-phenylpyrazole.[35]

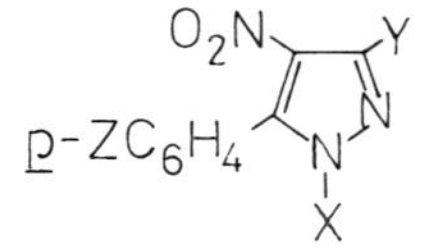

3-36a X = H, Y = C$_5$F$_{11}$, Z = NO$_2$ (85%)
3-36b X = H, Y = C$_7$F$_{15}$, Z = NO$_2$ (90%)
3-36c X = CH$_3$, Y = C$_7$F$_{15}$, Z = NO$_2$ (87%)
3-36d X = p-O$_2$NC$_6$H$_4$, Y = C$_7$F$_{15}$, Z = NO$_2$ (82%)

Fuming nitric acid and acetic anhydride in acetic acid converted pyrazole, on gentle heating, to 1-nitropyrazole (**3-3**) (70%); ice-cold concentrated sulfuric acid isomerized the nitramine (**3-3**) to 4-nitropyrazole (**3-2**) (80%). Similarly,

1-nitro derivatives were obtained from 3,5-dimethyl-, 3,4,5-trimethyl-, 3-methyl-4-nitro-, 3,5-dimethyl-4-nitro-, and 3,5-diethylpyrazoles and rearranged to the 4-nitro isomer when the position was not blocked by another substituent.[13,14] In one instance dinitration occurred to give 1,4-dinitro-3,5-dimethylpyrazole.[13] Nitric acid in acetic anhydride (acetyl nitrate) converted 3(5)-phenylpyrazole to a mixture of 1-acetyl-3(5)-(4-nitrophenyl)pyrazole (**3-37**) and 1-nitro-3(5)-phenylpyrazole (**3-38**); both products reacted with sulfuric acid to give 3(5)-(4-nitrophenyl)pyrazole (**3-39**), but an isomerization (**3-38**) → (**3-40**) was not detected.[54]

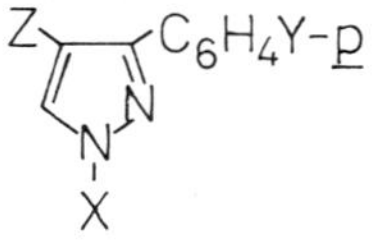

3-37 X = COCH$_3$, Y = NO$_2$, Z = H
3-38 X = NO$_2$, Y = Z = H
3-39 X = Z = H, Y = NO$_2$
3-40 X = Y = H, Z = NO$_2$

Nitration of 1-substituted pyrazoles with nitric acid in acetic anhydride gave 4-nitro derivatives.[33,36,40] Thus 1-phenylpyrazole and its 3-methyl, 5-methyl, and 3,5-dimethyl derivatives gave the 4-nitropyrazoles (**3-41**) to (**3-44**).[36,40] A small amount of 1-(4-nitrophenyl)-3-methylpyrazole was also obtained with its isomer (**3-42**).[37] Similar reactions produced other derivatives of 4-nitropyrazole: 1,5-diphenyl-3-methyl- (**3-45**),[33] 1-(2-tolyl)- (**3-46**),[40] and 1-(2,6-dimethylphenyl)- (**3-47**).[40] An excess of nitric acid led to the formation of 1-(2,6-dimethyl-3-nitrophenyl)-4-nitropyrazole (**3-48**). Eratic kinetic behavior of the 1-arylpyrazoles and poor yields (25%–80%) of products precluded an assignment of mechanism. The coformation of 1-phenyl-3-nitro-5-methylpyrazole (**3-49**) with its isomer (**3-43**) required an uncommon nitration at a pyrazole 3-position.[40]

Nitration of 3(5)-(pyrazol-1-yl)-4-nitropyrazole gave the trinitro derivative (**3-51**); the desired dinitro product (**3-50**) was not detected.[50]

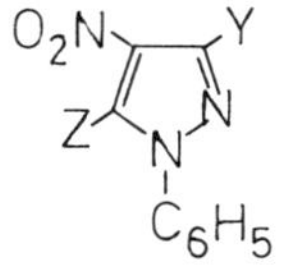

3-41 Y = Z = H
3-42 Y = CH$_3$, Z = H
3-43 Y = H, Z = CH$_3$
3-44 Y = Z = CH$_3$

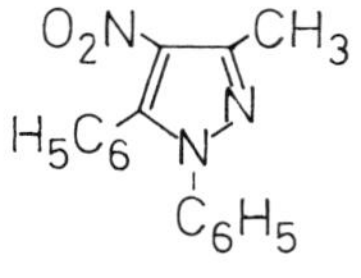

3-45

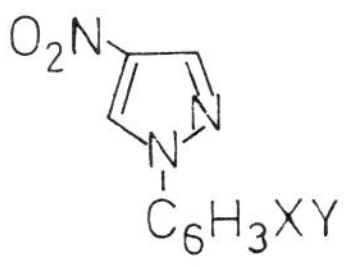

3-46 X = 2'-CH₃ Y = H
3-47 XY = 2',6'-(CH₃)₂

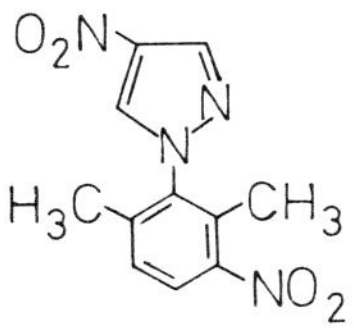

3-48

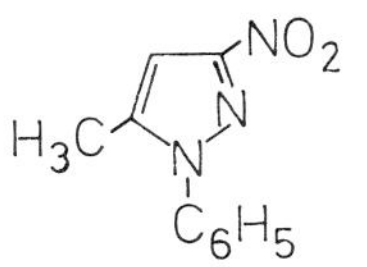

3-49

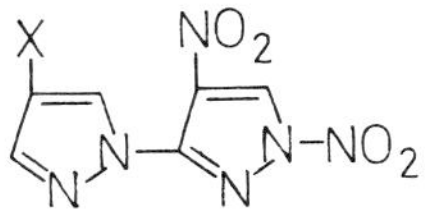

3-50 X = H
3-51 X = NO₂

Nitration of 1,4-disubstituted pyrazoles with nitric acid in acetic anhydride gave product mixtures that contained low yields of 3-nitropyrazole derivatives as the main product: 1,4-dimethyl-3-nitropyrazole (**3-52**) (22%);[29,55] 1-(4-nitrophenyl)-3-nitro-4-methylpyrazole (**3-53**) (35%);[29] 1-methyl-3-nitro-4-ethylpyrazole (**3-54**);[55] and 1-phenyl-3-nitro-4-ethylpyrazole (**3-55**).[55] The nitration of 1-methyl-4-phenylpyrazole gave a mixture of six compounds (**3-56a–f**).[38] The prominence of *ortho* nitration in the phenyl ring (**3-56a,c,e**) contrasted with other known nitrations of phenylpyrazoles where exclusive *para* substitution was detected. It was suggested that an unprotonated pyrazole ring was involved in the nitration with acetyl nitrate[29,55] and that the formation of products (**3-56b,d**) in equal amounts resulted from comparable participation from the phenyl and pyrazolyl groups in the electrophilic substitution.[38]

Nitration of a triacetate ester of 1-(2,4-dinitrophenyl)-3-ribofuranosylpyrazole with cupric nitrate in acetic anhydride gave the 4-nitro derivative (**3-57**) (93%). The same reagent converted the pyrazole (**3-58**) to its 1-nitro derivative (**3-59**) (87%).[56]

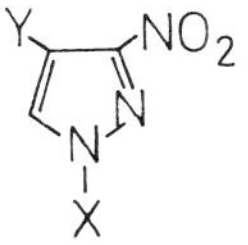

3-52 X = Y = CH₃
3-53 X = *p*-O₂NC₆H₄, Y = CH₃
3-54 X = CH₃, Y = C₂H₅
3-55 X = C₆H₅, Y = C₂H₅

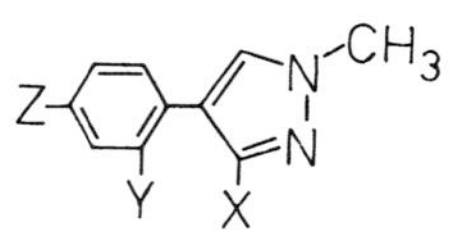

3-56a X = H, Y = Z = NO$_2$
3-56b X = NO$_2$, Y = Z = H
3-56c X = Z = H, Y = NO$_2$
3-56d X = Y = H, Z = NO$_2$
3-56e X = Y = NO$_2$, Z = H
3-56f X = Z = NO$_2$, Y = H

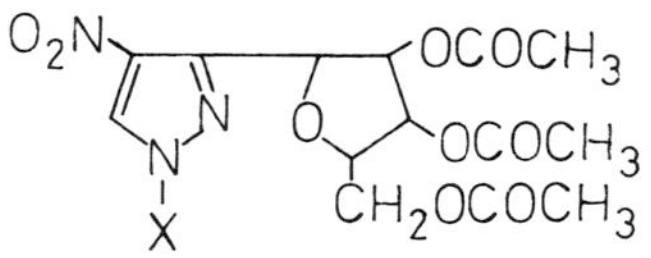

3-57 X = C$_6$H$_3$(NO$_2$)$_2$-2,4
3-58 X = H
3-59 X = NO$_2$

Pyrazolyl Halides

Nitrolysis of a 4-bromopyrazole to give a 4-nitropyrazole may accompany nitration at the 3- (first) and the 5-positions.[25] Structural features of the pyrazole and reaction conditions affect the predominant formation of 4-nitro-[13,57,58] or 3,5-dinitro-4-bromopyrazoles.[45]

A mixture of nitric acid and oleum at 65° for 2 h converted 4-bromo-1-methylpyrazole-5-carboxylic acid to the 3-nitro derivative (43%) and 3,5-dinitro-4-bromo-1-methylpyrazole (51%). Only the latter product (69%) was detected from the reaction at 65° after 4 h or at 75° after 2 h. Nitration of 4-chloro-1-methylpyrazole-5-carboxylic acid gave similar results. In these reactions replacement of a 4-halo substituent was not observed. Nitrodecarboxylation was characteristic of 4-halo-1-methyl-3-nitropyrazole-5-carboxylic acids but not of the 5(3)-nitropyrazole-3(5)-carboxylic acid isomers.[25]

Nitric acid in sulfuric acid at 0° converted both 1,5-diphenyl-3-methylpyrazole and its 4-bromo derivative to the corresponding 5-p-nitrophenylpyrazole (**3-60a,b**) (60%, 45%). Nitration at 0°–15° for 3 h gave 1,5-di-p-nitrophenylpyrazoles (**3-61a,b**) (73%, 90%) and at 100° for 30 min gave the 4-nitropyrazole (**3-62**) (30%, 27%).[33]

Similarly, nitration converted 5-chloropyrazoles to 4-nitro derivatives (**3-63**),[57–64] and nitrolysis converted 4-bromo- to 4-nitropyrazoles and (**3-2**), (**3-6**), (**3-7**), (**3-64**), and (**3-65**).[65,66] Nitration gave the p-nitrophenyl derivative (**3-66**) from 3-phenyl-4-bromo-pyrazole.[65]

An interesting conversion of 1-methyl-4-bromopyrazole to its 3,5-dinitro derivative (**3-67**) (26%) was brought about by mixed acid at reflux temperature

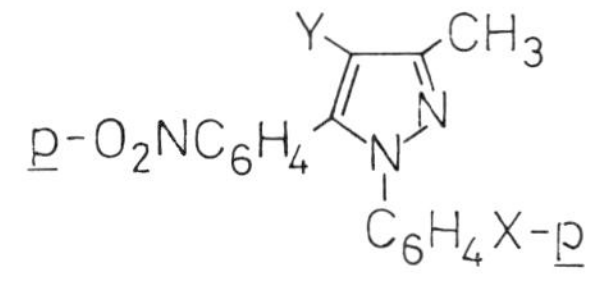

3-60a X = Y = H
3-60b X = H, Y = Br
3-61a X = NO$_2$, Y = H
3-61b X = NO$_2$, Y = Br
3-62 X = Y = NO$_2$

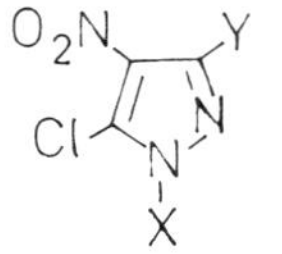

X = H, alkyl, aryl, acyl, carbamoyl, etc.
Y = H, C$_{1-7}$ alkyl, p-O$_2$NC$_6$H$_4$

3-63

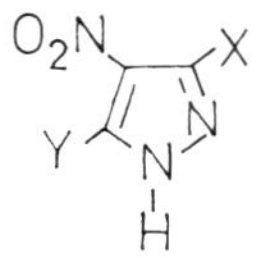

3-64
X = C$_2$H$_5$, Y = H

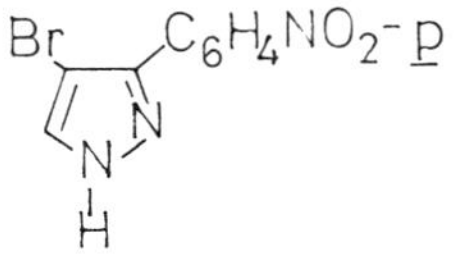

3-66

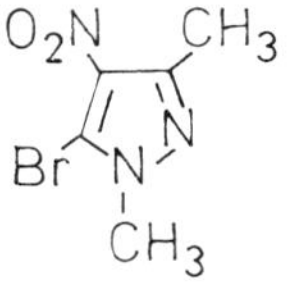

3-65

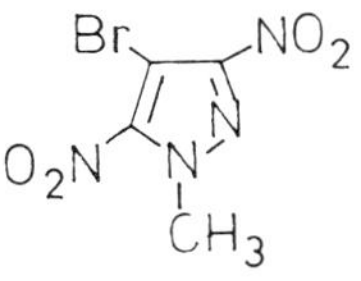

3-67

for 2 h. Competitive nitrolysis to 1-methyl-4-nitropyrazole (**3-5**) was not detected.[45]

A preparation of 3-fluoro-4-nitropyrazole from 3,4-difluoropyrazole by an "acetylation-nitration-deacetylation sequence" was claimed.[67]

Pyrazolecarboxylic Acids and Derivatives

A mixture of nitric and sulfuric acids at 90° for 3 h converted 3-methylpyrazole-5-carboxylic acid to its 4-nitro derivative (**3-68a**) (43%);[68] the N-methylamide, methyl, and isopropyl ester derivatives of the acid gave the corresponding 4-nitropyrazoles (**3-68b–d**).[23] In similar reactions 1-ethyl-3-methylpyrazole-5-carboxylic acid and 1-ethyl-5-methylpyrazole-3-carboxylic acid were converted to their 4-nitro derivatives (**3-69a,b**);[69,70] 3-n-propylpyrazole-5-carboxylic acid and its ethyl ester gave the 4-nitropyrazoles (**3-68e**)[71] and (**3-68f**).[23] In contrast

1-phenylpyrazole-3-carboxylic acid at $-5°$ gave 1-*p*-nitrophenylpyrazole-3-carboxylic acid (**3-70**).[37]

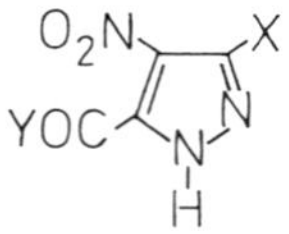

3-68a X = CH$_3$, Y = OH
3-68b X = CH$_3$, Y = NHCH$_3$
3-68c X = CH$_3$, Y = OCH$_3$
3-68d X = CH$_3$, Y = OCH(CH$_3$)$_2$
3-68e X = *n*-C$_3$H$_7$, Y = OH
3-68f X = *n*-C$_3$H$_7$, Y = OC$_2$H$_5$

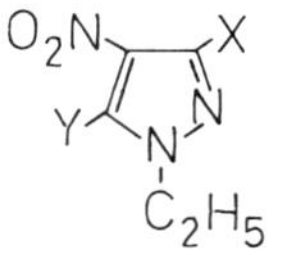

3-69a X = CH$_3$ Y = CO$_2$H
3-69b X = CO$_2$H Y = CH$_3$

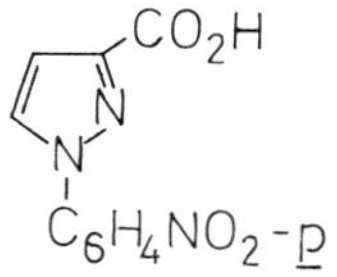

3-70

Fuming nitric acid and 3,5-dimethyl-1-guanylpyrazole nitrate in equimolar amounts failed to react at 70° for 2 h, but an 8:1 mixture at 100° gave 3,5-dimethyl-4-nitropyrazole (**3-8**) (50%), which was also produced (76%) in a reaction with mixed acid at 40°.[72] The 4-nitro compound was obtained in related reactions with 1-N-benzoylguanyl-3,5-dimethylpyrazole and with 3,6-bis(3,5-dimethyl-1-pyrazolyl)-1,2-dihydrotetrazine; the former reaction also produced N-benzoylurea, Eq. (3-1).[72,73]

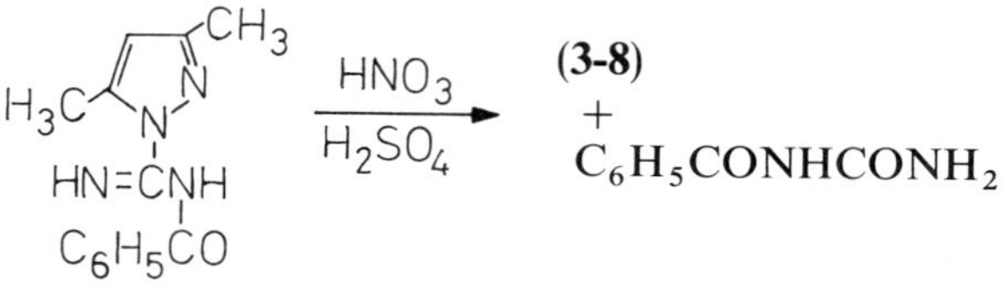

(3-1)

Hydroxy- and Alkoxypyrazoles

Enolic tautomers for 1-substituted 3-hydroxypyrazoles in nonpolar media and in the solid state and coexistence with comparable amounts of the oxotautomers in aqueous solution were determined by spectroscopic and basicity evidence. Under most conditions in nonpolar media 1-phenyl- and 1-methylpyrazolin-5-ones preferred the Δ^3 tautomeric form and in aqueous solution the NH (90%) and OH tautomers (10%) were in equilibrium.[74,75]

Similar evidence showed that 1-phenyl-3-hydroxy-4-nitro-5-methylpyrazole (**3-71**) and the corresponding 1-*p*-nitrophenyl dinitro derivative preferred the enolic form in the solid state and in aqueous solution. In the solid state 1-phenyl-3-methyl-4-nitro-5-hydroxypyrazole and the corresponding 1-*p*-nitrophenyl dinitro derivative preferred the enolic form, whereas in neutral and acidic solutions there was not a clear distinction between the predominancy of NH (**3-73**) or OH tautomers.

Calculations by the Hückel MO method showed the pyrazolin-5-one tautomers to be energetically favored by electron-donating substituents at 3- and 4-positions, whereas an electron-withdrawing substituent—eg, 3-nitro—favored the enol tautomer.[76] The results agreed with ir spectroscopic results.[77]

Amyl nitrite in acetone at 20° for 3 days converted 1-phenyl-3-hydroxy-5-methylpyrazole to the 4-nitro derivative (**3-71**) (66%).[74] Similar reactions afforded 3-methyl-4-nitro-Δ^3-pyrazolin-5-one (**3-72**) and its 1-phenyl derivative (**3-73**) but gave the oximes (**3-74**) and (**3-75**) from 1,3-diphenyl- and 1-*p*-nitrophenyl-3-methyl-Δ^3-pyrazolin-5-one (**3-76**). The latter product (**3-75**) was also obtained from the pyrazolone (**3-76**) by treatment with amyl nitrite and sodium ethoxide.[78] Initial electrophilic attack at the 4-position to give a nitrosopyrazolone tautomer of the oxime (**3-75**) can be presumed, Eq. (3-2). In certain examples nitrosopyrazolones were oxidized to nitro compounds (**3-71**) to (**3-73**). Comparable reactivity was not found in 1-*p*-nitrophenyl-3-methylpyrazole, since it was not converted by amyl nitrite under the conditions cited.[78]

Nitric acid in sulfuric acid at 20° for 48 h converted 1-phenyl-3-hydroxy-5-methylpyrazole to the *p*,4-dinitro derivative or its keto tautomer (**3-77**) (67%). Similarly 1-*p*-nitrophenyl-2,5-dimethyl-4-nitro-Δ^4-pyrazolin-3-one (**3-78**) (57%) was obtained. Nitric acid at 20° for 12 h mononitrated 1-phenyl-3-methoxy-5-methylpyrazole to its 4-nitro derivative (**3-79**) (61%), whereas mixed acid at 50°

3-71 X = H
3-77 X = NO$_2$

3-72 X = H
3-73 X = C$_6$H$_5$

3-74 X = Y = C$_6$H$_5$
3-75 X = *p*-O$_2$NC$_6$H$_4$, Y = CH$_3$

3-76

(3-2)

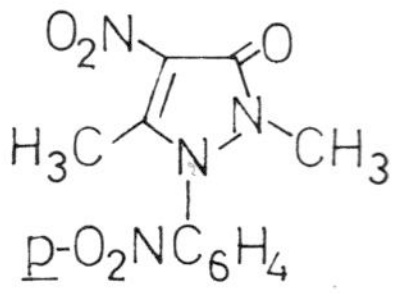

3-78

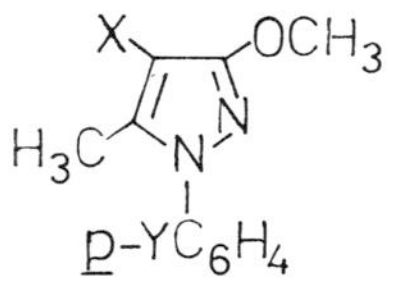

3-79 X = NO$_2$, Y = H
3-80 X = H, Y = NO$_2$
3-81 X = Y = NO$_2$

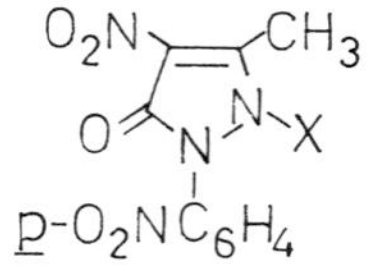

3-82 X = CH$_3$
3-83 X = H

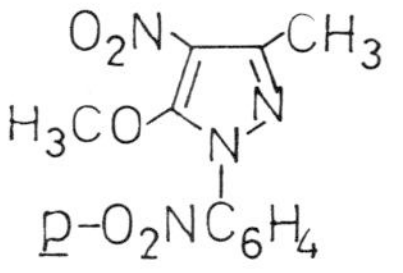

3-84

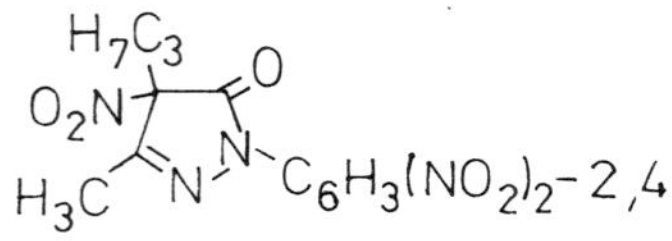

3-85

for 3 h gave the *p*-nitrophenyl derivative (**3-80**) (51%) and mixed acid at 40° for 24 h gave the dinitro derivative (**3-81**) (85%).[74]

Mixed-acid treatment also produced 1-*p*-nitrophenyl-2,3-dimethyl-4-nitro-Δ^3-pyrazolin-5-one (**3-82**). At 20° for 10 h it converted 1-*p*-nitrophenyl-3-methyl-Δ^3-pyrazolin-5-one to its 4-nitro derivative (**3-83**) (90%) (picrolonic acid) also obtained from 1-phenyl-3-methyl-Δ^3-pyrazolin-5-one at 20° for 12 h. Nitric acid (d 1.42) at 40° for 3 h converted 1-phenyl-3-methyl-5-methoxypyrazole to its 4-nitro derivative (70%), whereas the 4,*p*-dinitro derivative (**3-84**) was produced by mixed acid at 20° for 2 h.[79]

These nitrations of 1-phenyl-3-hydroxy-4-nitro-5-methylpyrazole and its N- and O-methyl derivatives to give 1-*p*-nitrophenylpyrazoles and of 1-*p*-nitrophenyl-3-hydroxy-5-methylpyrazole and its N- and O-methyl derivatives to give 4-nitropyrazoles changed from nitration of the free bases at lower acidity to nitration of the conjugate acids at higher acidity.[74] A similar conclusion was reached for nitration of 1-*p*-nitrophenyl-3-methyl-5-hydroxypyrazole and its N- and O-methyl derivatives at the 4-position and of 1-phenyl-3-methyl-4-nitro-5-hydroxypyrazole and its N- and O-methyl derivatives at the *para*-position of the 1-phenyl substituent.[79]

Fuming nitric acid and 1-phenyl-2,3-dimethyl-4-*n*-propyl-Δ^3-pyrazolin-5-one (propylphenazone) gave at least 27 products including the trinitro compound (**3-85**).[80]

Mixed acid nitrated antipyrine to give a p,4-dinitro derivative (**3-86**).[81] Dinitrogen tetroxide oxidized dipyrafene (**3-87**) to 1,2-diphenyl-3,4,5-trioxopyrazolidine isolated as the hydrate (**3-88a**).[82]

A mixture of concentrated nitric and sulfuric acids at 0° for a few minutes converted 1,2-dihydro-2,5-dimethylpyrazol-3(3H)-one to its 4-nitro derivative (**3-88b**) (43%).[83,84]

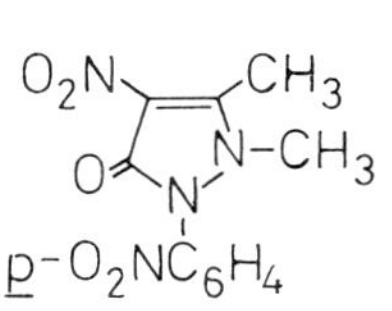

3-86

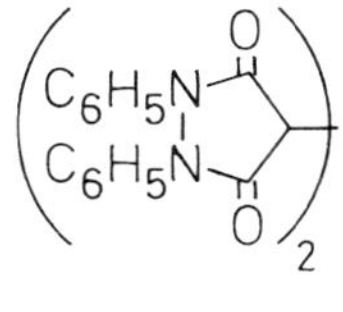

3-87

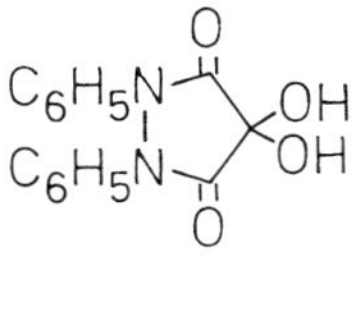

3-88a

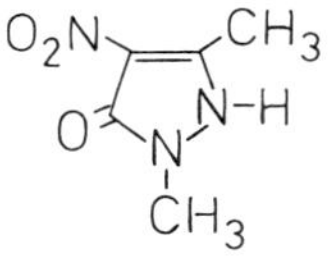

3-88b

Amino- and Amidopyrazoles

Mixed acid at 0° for 2 h nitrated 3(5)-acetamidopyrazole to give the 4-nitro derivative (**3-89a**) (70%), which was quantitatively hydrolyzed by hydrochloric acid to the amine (**3-89b**). A similar procedure afforded 3(5)-amino-4-nitro-5(3)-methylpyrazole (**3-89c**) (90%).[85-88] At 25° for 3 h mixed acid converted 1-methyl-5-aminopyrazole to its 4-nitro derivative (**3-90a**) (23%); the corresponding 1-phenyl-4-nitro-5-amino-pyrazole (**3-90b**) (92%) was obtained from 1-phenyl-5-aminopyrazole and nitric acid in acetic anhydride at 0° for 5 min.[89]

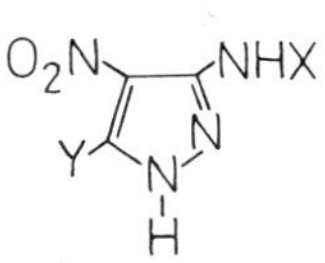

3-89a X = COCH$_3$, Y = H
3-89b X = Y = H
3-89c X = H, Y = CH$_3$

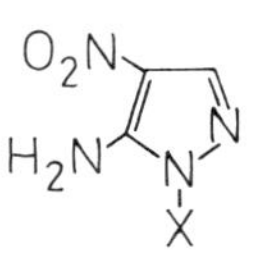

3-90a X = CH$_3$
3-90b X = C$_6$H$_5$

Pyrazole-N-Oxides

In a kinetic study 1-methylpyrazole-2-oxide was nitrated by mixed acid at 25° over the range 66%–76% sulfuric acid as the free base. At higher acid strength the unsatisfactory kinetics obtained was attributed to dinitration. Mononitration in 66% sulfuric acid at 25° for 12 h gave the 5-nitro derivative (**3-91**) (90%). Dinitration in 87% sulfuric acid at 25° for 75 min gave the dinitro derivative

(**3-92**) (78%), also obtained (16% yield) from the mononitro compound (**3-91**) on nitration in 88% sulfuric acid at 25° for 12 min. The latter reaction also gave small amounts of 1-methyl-5-nitropyrazole (**3-93**) and 1-methyl-3,5-dinitropyrazole (**3-94**).[90]

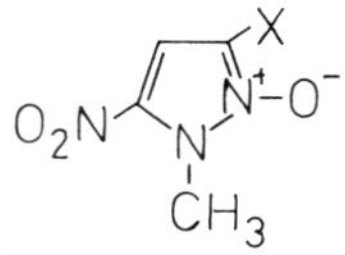

3-91 X = H
3-92 X = NO_2

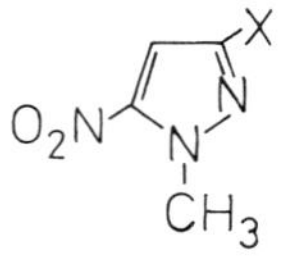

3-93 X = H
3-94 X = NO_2

Nitrosyl chloride and 4-bromo-3,5,5-trimethylpyrazole-1,2-dioxide (**3-95a**) in a sealed glass tube at 25° for 24 h gave 3-chloro-4-nitro-3,5,5-trimethyl-pyrazoline-1,2-dioxide (**3-95b**) (71%).[91]

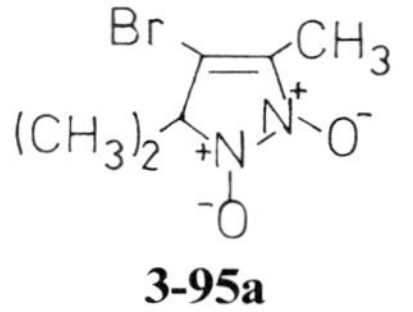

3-95a

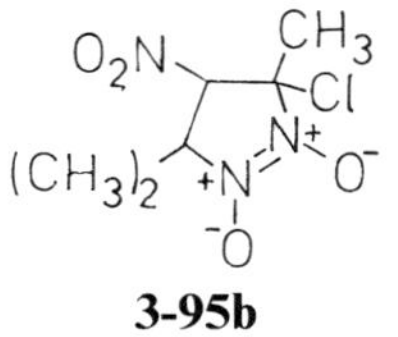

3-95b

Silylpyrazoles

Mixed acid converted 3(5)-trimethylsilylpyrazole to the 4-nitro derivative (**3-96**) (94%), also obtained from 3(5),4-bistrimethylsilylpyrazole by similar treatment.[92]

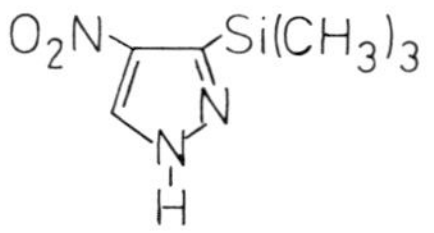

3-96

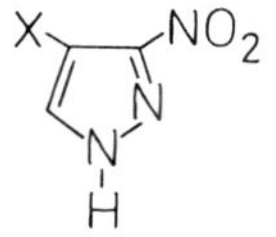

3-97a X = H
3-97b X = CH_3
3-97c X = C_2H_5
3-97d X = C_6H_5
3-97e X = NO_2

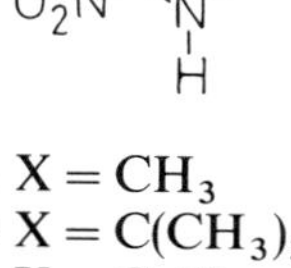

3-98a X = CH_3
3-98b X = $C(CH_3)_3$
3-98c X = C_6H_5
3-98d X = p-$O_2NC_6H_4$
3-98e X = NO_2

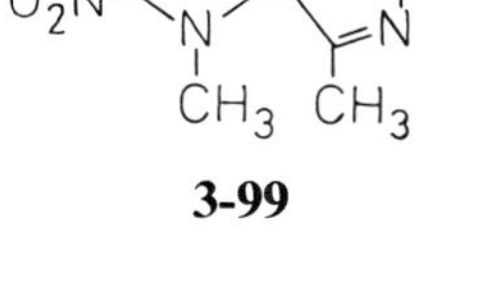

3-99

Rearrangements of 1-Nitropyrazoles

In benzonitrile at 180° for 2 h 1-nitropyrazole (**3-3**) rearranged to 3(5)-nitropyrazole (**3-97a**) (86%).[14,26,93-99] Isomerization of a 4-substituted 1-nitropyrazole in benzonitrile or anisole at 120°–190° for 2.5–5 h gave lower yields of 4-methyl-, 4-ethyl-, 4-phenyl-, and 4-nitro-3(5)-nitropyrazoles (**3-97b–e**) along with considerable decomposition. A 3-substituted-1-nitropyrazole in anisole or benzonitrile at 130°–145° for 1.5–110 h gave excellent yields of 3(5)-methyl-, 3(5)-*tert*-butyl-, 3(5)-phenyl-, 3(5)-*p*-nitrophenyl-5(3)-nitropyrazoles, and 3,5-dinitropyrazole (**3-98a–e**).[26] A nearly quantitative conversion of 3-methyl-1-nitropyrazole (neat) at 145° gave exclusively 3(5)-methyl-5(3)-nitropyrazole (**3-98a**); 5-methyl-1-nitropyrazole gave 3(5)-methyl-4-nitropyrazole (**3-7**) (93%) and the isomer (**3-98a**) (7%).[14,95] In an example of isomerization of a 3,4-disubstituted derivative of 1-nitropyrazole, 3(5)-methyl-4-(1-methyl-5-nitro-2-imidazolyl)-5(3)-nitropyrazole (**3-99**) (22%) was obtained from 1-nitro-3-methyl-4-(1-methyl-5-nitro-2-imidazolyl)pyrazole.[99]

This thermally induced migration of a nitro group from a ring nitrogen to a ring carbon atom, the Habraken reaction, appears to be characteristic of N-nitroazoles.[14] The reaction gave first-order kinetics and was not affected by acids, bases, scavengers for free radicals, or the nitronium cation. Replacement of 3(5)H by D had no effect on the rate of the intramolecular change. Solvents and substituents had little effect on activation parameters: $\Delta H^{\ddagger} = 30$–36 kcal/mol; $\Delta S^{\ddagger} = 2 \pm 5$ e.u. Since in benzene the reaction gave only trace quantities of corresponding 1-phenylpyrazoles, a homolytic mechanism appeared unlikely; a heterolytic reaction was also judged unlikely, since the reaction could be achieved in the vapor phase. A rate-determining [1,5] shift of the nitro group to give a 3H-pyrazole that subsequently isomerized to a 3(5)-nitropyrazole was found compatible with all experimental data. A reverse reaction was not observed. Some denitration competed with the rearrangement of 4-substituted 1-nitropyrazoles and was attributed to steric hindrance.[94]

(3-3)

In a new method for the preparation of heterocyclic N-nitro compounds a heterocycle bearing "positive" halogen was treated with a complex (1:1) between trimethyl phosphite and silver nitrate. Thus 3,5-dimethylpyrazole and bromine at 25° in the presence of aqueous sodium carbonate gave the *gem*-dibromide (**3-100a**) (70%), in turn converted to 1-nitro-3,5-dimethyl-4-bromopyrazole (**3-100b**) (61%) on treatment with the trimethyl phosphite/silver nitrate complex. The following explanation was offered, Eq. (3-3).[100]

Pyrazole Diazonium Compounds and Nitrites

A suspension of 1-methyl-5-aminopyrazole-4-carboxamide in water was treated with a molar excess of sodium nitrite and hydrochloric acid at 25°. Once foaming subsided the mixture was heated at 100° for 1 h to give 1-methyl-5-nitropyrazole-4-carboxamide (**3-101a**) (78%).[101,102] When hydrochloric acid was added to the aqueous solution of the amine and sodium nitrite at 0°–5° and the mixture was stirred at 0° for 30 min, 1-methyl-5-nitrosaminopyrazole-4-carboxylic acid (**3-101b**) (49%) was obtained. The suggestion[101] that the latter compound (**3-101b**) could be converted to the nitro compound (**3-101a**) by treatment with sodium nitrite and acid at the higher temperature was inconsistent with the formation of 1-methyl-4-hydroxypyrazolo[3,4-d]-v-triazine (**3-102**) (85%) when the mixture prepared at 0°–5° was heated at 70° for 1 h.[101] 3(5)-Nitropyrazole (**3-101c**) (63%) was similarly prepared from 3(5)-aminopyrazole.[102] These preparations appear to be related to other isolated examples of the formation of aromatic nitro compounds from diazonium salts by treatment with sodium nitrite (in the absence of copper salts these are not Sandmeyer reactions).[103,104]

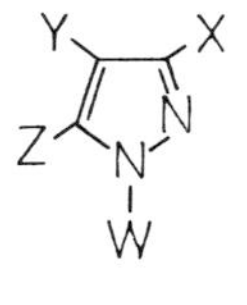

3-101a W = CH$_3$, X = H, Y = CONH$_2$, Z = NO$_2$
3-101b W = CH$_3$, X = H, Y = CO$_2$H, Z = NHNO
3-101c W = Y = Z = H, X = NO$_2$
3-101d W = CH$_3$, X = NO$_2$, Y = CO$_2$CH$_3$, Z = H
3-101e W = Z = H, X = NO$_2$, Y = CN

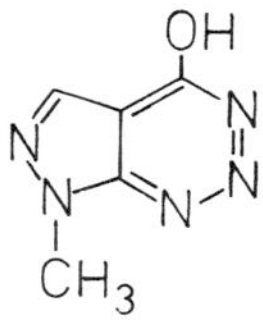

3-102

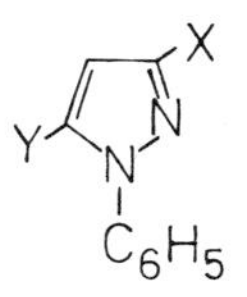

3-103a X = NO$_2$, Y = H
3-103b X = H, Y = NO$_2$

Sandmeyer conversions to the nitropyrazoles (**3-101d,e**) were realized when methyl 1-methyl-3-aminopyrazole-4-carboxylate was treated with sodium nitrite and copper in aqueous fluoroboric acid[105] and when 3(5)-aminopyrazole-4-carbonitrile was diazotized and treated with a mixture of sodium nitrite and copper.[106,107]

Oxidation of Amino- and Nitrosopyrazoles

Peroxytrifluoroacetic acid in methylene chloride oxidized the 3- and 5-amino derivatives of 1-phenylpyrazole at reflux temperature for 2 h to the 3-nitro and 5-nitropyrazoles (**3-103a,b**) (47%, 21%). In a similar reaction 1-methyl-5-aminopyrazole gave 1-methyl-5-nitropyrazole-2-oxide (**104a**) (10%).[108]

Concentrated nitric acid (d 1.39) converted 3(5)-nitrosopyrazole at 100° for 30 min, then at 120° for 5 min to 3(5)-nitropyrazole (**3-104b**) (92%), also obtained (83%) from 3(5)-nitroso-4-trimethylsilylpyrazole by similar treatment; 4-nitrosopyrazole gave 4-nitropyrazole (**3-2**) (73%).[109] Nitrous acid nitrosated and oxidized 1-hydroxy-4-alkylpyrazole-2-oxides to the corresponding 4-nitropyrazolenine-1,2-dioxides (**3-104c**).[110-112] Other examples of the nitration of a pyrazole derivative by treatment with nitrous acid are known.[7,113]

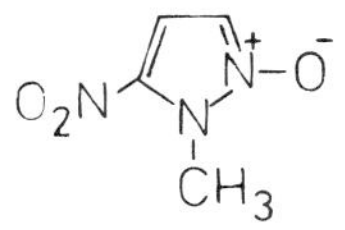

3-104a 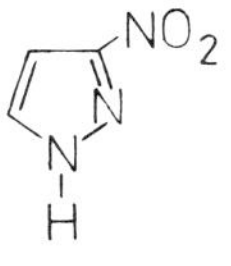   **3-104b** 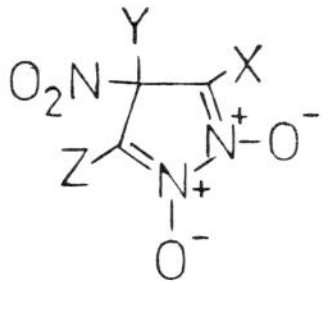   **3-104c**

Aromatization of Nitropyrazolines and Nitropyrazolidines

Nitropyrazolines have been obtained as adducts from nitroolefins and diazoalkanes, Eqs. (3-4), (3-5),[114,121] and from pyrazolines by nitration, Eqs. (3-6), (3-7).[122,123] An unisolated nitropyrazoline was presumed to be an unstable intermediate adduct from diazofluorene and nitroethylene for the formation of a nitropropane (97%), Eq. (3-8).[124]

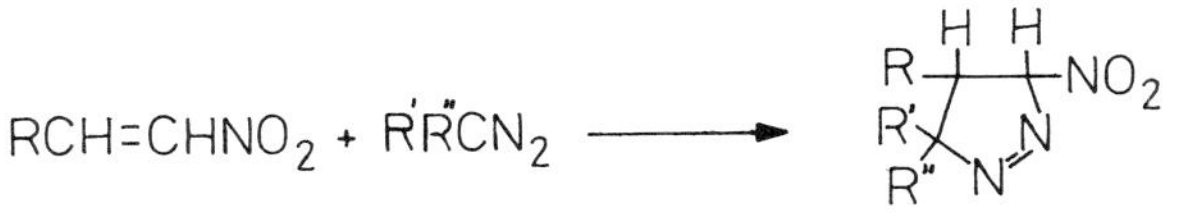

$$RCH=CHNO_2 + RR'CN_2 \longrightarrow$$

3-105   (3-4)

$$C_6H_5\underset{NO_2}{C}=CHNO_2 + RR'CN_2 \xrightarrow{0°}$$

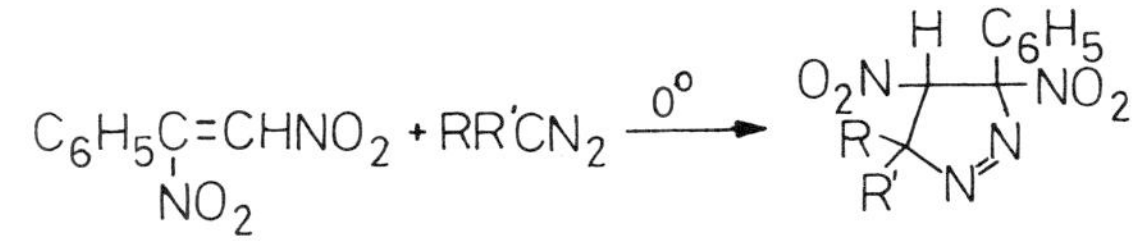

3-106   (3-5)

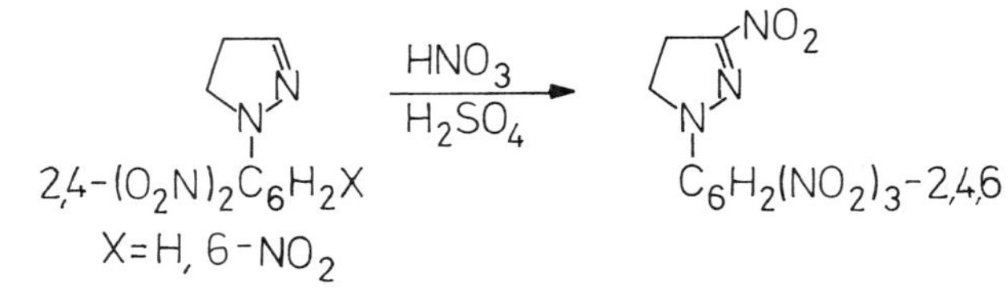

$$2,4\text{-}(O_2N)_2C_6H_2X \xrightarrow[H_2SO_4]{HNO_3} C_6H_2(NO_2)_3\text{-}2,4,6$$
$$X = H, 6\text{-}NO_2 \tag{3-6}$$

$$\textbf{3-106} \xrightarrow[20°]{HNO_3} \text{[pyrazoline product]} \tag{3-7}$$

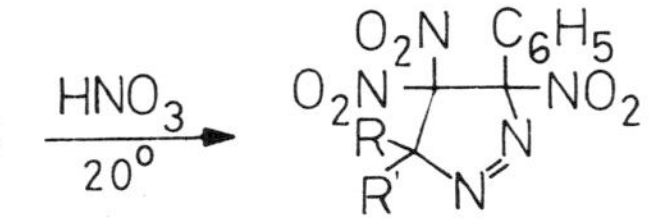

$$\text{fluorenyl-}N_2 + CH_2=CHNO_2 \rightarrow [\text{pyrazoline intermediate}] \rightarrow O_2N\text{-cyclopropane-}C_{12}H_8 \tag{3-8}$$

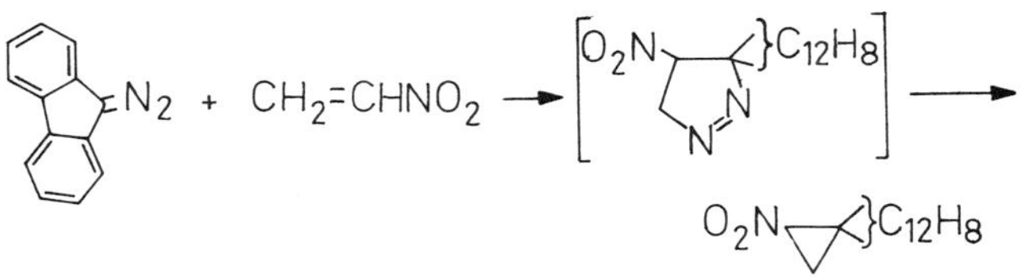

Although treatment with acids or bases converted nitropyrazolines (**3-105**) to pyrazoles by an elimination of nitrous acid,[114–118] the presence of a labile hydrogen atom at the 4-position afforded a base-catalyzed conversion of 3,5,5-triphenyl-3,4-dinitropyrazoline to 1,1,2-triphenyl-3-alkoxycyclopropene, Eq. (3-9).[115] An attractive oxidation of certain nitropyrazolines (**3-105**) by manganese dioxide gave nitropyrazoles (**3-107a,b**) quantitatively; however, similar attempts to obtain the nitropyrazoles (**3-107c,d**) were unsuccessful.[125]

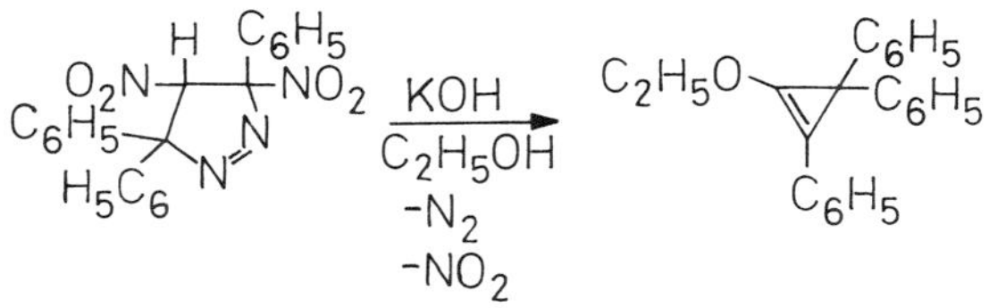

$$\text{pyrazoline} \xrightarrow[\substack{-N_2 \\ -NO_2}]{KOH/C_2H_5OH} \text{cyclopropene product} \tag{3-9}$$

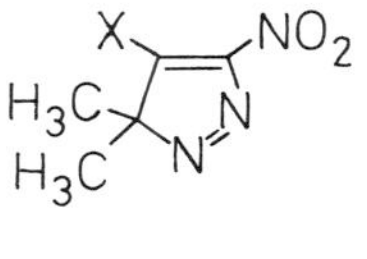

3-107a X = C$_6$H$_5$ **3-108a** X = NO$_2$
3-107b X = CN   **3-108b** X = C$_6$H$_5$
3-107c X = H
3-107d X = CH$_3$

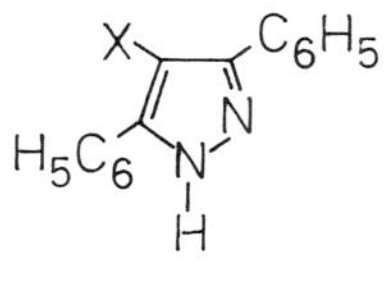

On formation at 25° the dinitropyrazoline (**3-106**) (R = C$_6$H$_5$, R' = H) gave 3,5-diphenyl-4-nitropyrazole (**3-108a**) (86%), but the pyrazoline (**3-106**) (R = R' = C$_6$H$_5$) at 25° gave 3,4,5-triphenylpyrazole (**3-108b**) (88%) by migration of a phenyl group.[119,121]

The formation of 3,5-diphenyl-4-nitropyrazole (**3-108a**) (67%) from cis- or trans-1,2-diphenyl-1,2-dinitroethene and diazomethane in ether at 25° for 24 h required an elimination of nitrous acid and a migration of phenyl from C-4 to C-5.[120]

In an interesting extension of the treatment of a Δ^2-pyrazoline with mixed acid, 1-picryl-Δ^2-pyrazoline (**3-109a**) was converted to 1-picryl-3-nitropyrazole (**3-109b**) (80%).[122] This method of preparation of 3-nitropyrazoles has not been further developed.

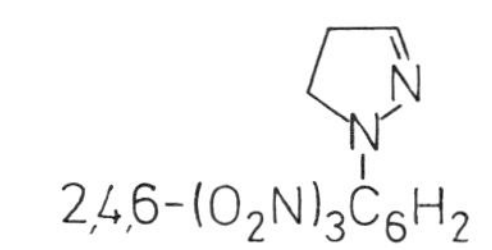

$$2,4,6-(O_2N)_3C_6H_2 \quad \text{(3-109a)} \qquad C_6H_2(NO_2)_3-2,4,6 \quad \text{(3-109b)}$$

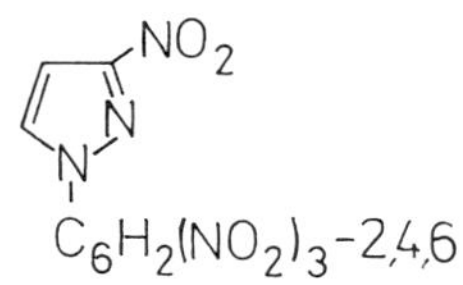

Diazomethane and 2-nitrochloroethene in ether at 20° for 12 h gave 3(5)-nitropyrazole (**3-97a**) (65%); a similar reaction gave ethyl 3(5)-nitropyrazole-5(3)-carboxylate (**3-110**) (43%), Eq. (3-10).[126] Whether an elimination of hydrogen chloride preceded or followed cycloaddition was not determined; the presumed instability of unknown nitroacetylene precluded its direct examination in cycloaddition.

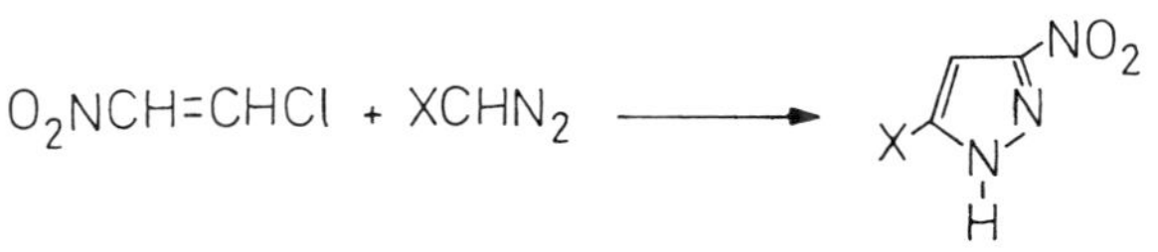

$$O_2NCH=CHCl + XCHN_2 \longrightarrow$$

3-97a X = H
3-110 X = $CO_2C_2H_5$ \hspace{2em} (3-10)

A preparation of 3-nitropyrazole derivatives from 2,2-dinitroethanol and diazocarbonyl compounds[127] appears to be a related reaction in which eliminations of water and nitrous acid are required.

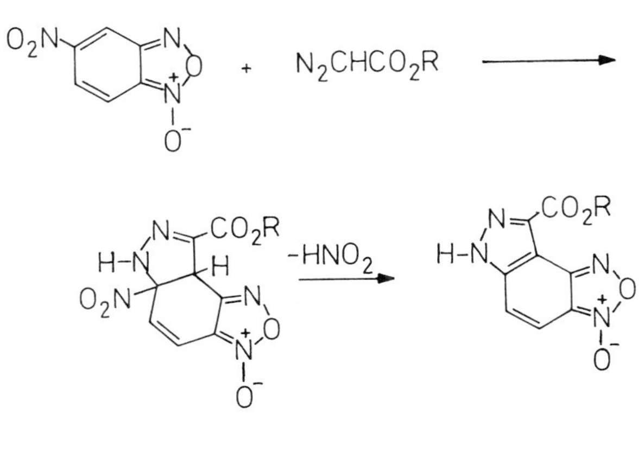

$$+ \; N_2CHCO_2R \longrightarrow$$

$$\xrightarrow{-HNO_2}$$

(3-11)

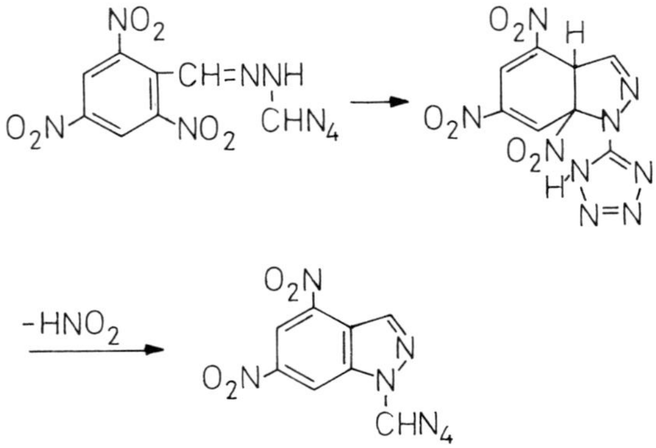

It was assumed that intermediate nitropyrazolines were produced in reactions between diazoacetates and 5-nitrobenzofuroxan. Loss of nitrous acid then accounted for the formation of a pyrazole, Eq. (3-11).[128]

Cyclization and elimination of nitrous acid accounted for the formation of 1-tetrazoyl-4,6-dinitroindazole via an intermediate nitropyrazoline, Eq. (3-12).[129]

A Michael reaction between a hydrazone and β-nitrostyrene followed by cyclization provided an attractive preparation of nitropyrazolidines and, by air oxidation, nitropyrazolines and nitropyrazoles. Thus a mixture of propionaldehyde phenylhydrazone and β-nitrostyrene at 25° for 48 h gave the nitropyrazolidine (**3-111**) (57%); by air oxidation, brought about by stirring (**3-111**) in chloroform at 25° for 48 h, the nitropyrazoline (**3-113**) (quantitative) was obtained; and finally, oxidation of (**3-112**) on storage in air gave the nitropyrazole (**3-114**) (apparently quantitative), Eq. (3-13). An intermediate isomerization (**3-112 → 3-113**) was detected by nmr spectroscopic analysis. In a similar reaction 1,5-diphenyl-3-isopropyl-4-nitropyrazolidine was obtained.[130]

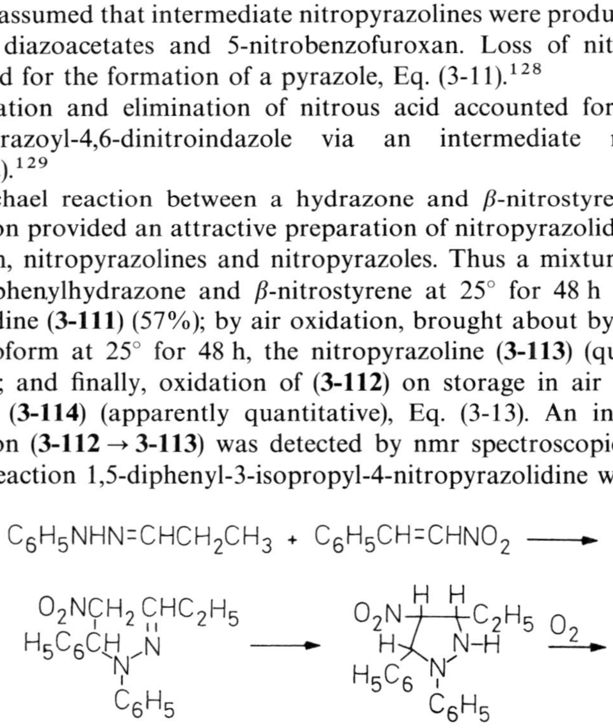

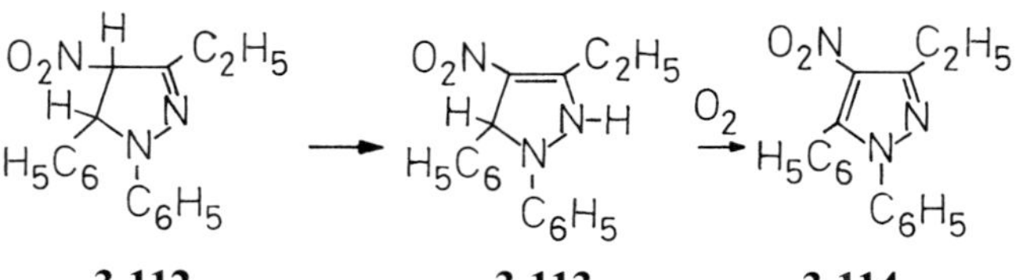

Bicyclic nitropyrazolidines have been prepared from dipolar pyrazolidone-azomethinimines and nitroolefins, Eq. (3-14). They have not been converted to pyrazolines and pyrazoles.[131–133]

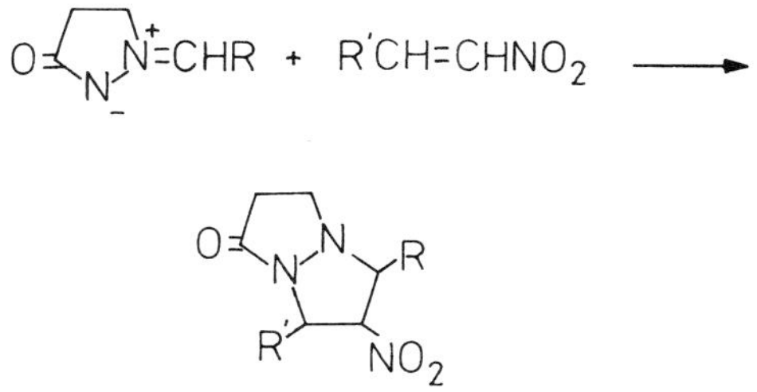

$$O \overset{+}{\underset{N^-}{\diagdown}} N=CHR + R'CH=CHNO_2 \longrightarrow$$

$$O \underset{R'}{\diagdown} \underset{NO_2}{\diagdown} R$$

(3-14)

Hydrazines combined with either 1,3-dihydroxy or 1,3-diamino derivatives of 2-nitro- or 2,2-dinitropropane to give 4-nitro- or 4,4-dinitropyrazolidines (**3-115a,b**).[131–133] The pyrazolidine (**3-115b**) at 25° slowly converted to the corresponding nitropyrazole (**3-116**).[134–138]

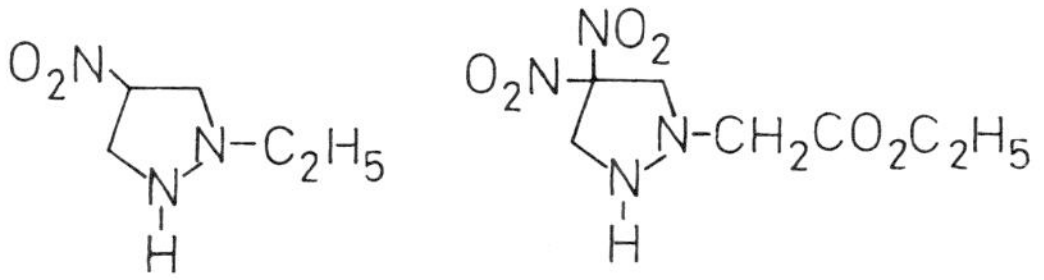

3-115a   **3-115b**

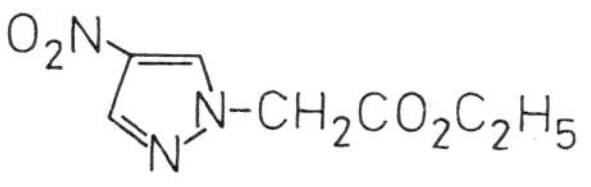

3-116

RING FORMATIONS

Nitrilimines and Nitroethenes

Low yields (as shown) of 4-nitro-1,3-disubstituted pyrazoles were obtained from 2-morpholino-1-nitroethene and hydrazonoyl halides in the presence of an amine, Eq. (3-15).[139] Presumably nitrilimines were intermediates

Bromonitroformaldehyde *p*-nitrophenylhydrazone and sodioacetoacetic ester in ethanol at 80° for 10 h produced 1-*p*-nitrophenyl-3-nitro-4-acetylpyrazolin-5-one (26%), Eq. (3-16). The corresponding 1-*p*-amidosulfonylphenylpyrazole derivative (44%) was similarly prepared.[140]

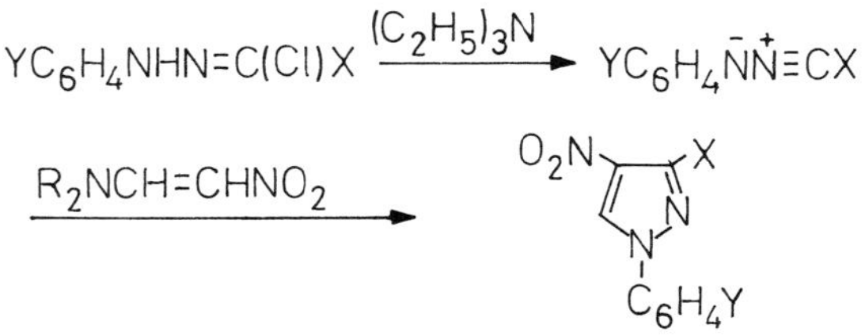

$$YC_6H_4NHN=C(Cl)X \xrightarrow{(C_2H_5)_3N} YC_6H_4\bar{N}\overset{+}{N}\equiv CX$$

$$\underset{R_2NCH=CHNO_2}{\xrightarrow{\hspace{2cm}}}$$

[pyrazole structure with O_2N, X, and C_6H_4Y] (3-15)

X	Y	Yield,%
C_6H_5	H	5
$CO_2C_2H_5$	H	20
$CO_2C_2H_5$	p-NO$_2$	6

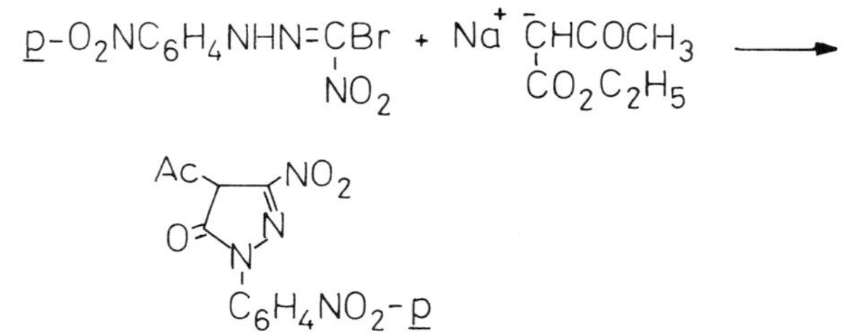

$$\underset{NO_2}{p\text{-}O_2NC_6H_4NHN=CBr} + \underset{CO_2C_2H_5}{Na^+ \bar{C}HCOCH_3} \longrightarrow$$

[pyrazolone structure with Ac, NO_2, and $C_6H_4NO_2$-p] (3-16)

Nitrodicarbonyl Compounds and Hydrazines

A mixture of *o*-chlorophenylhydrazine and sodionitromalonaldehyde was heated in ethanol for 2 h to give 1-*o*-chlorophenyl-4-nitropyrazole (**3-117**) (65%), Eq. (3-17).[141] Similarly the *m*- and *p*-chloro isomers (**3-117**) and 1-(2,4-dichlorophenyl)-4-nitropyrazole (60%) were prepared. Methylhydrazine afforded 1-methyl-4-nitropyrazole (**3-5**) (65%).[142]

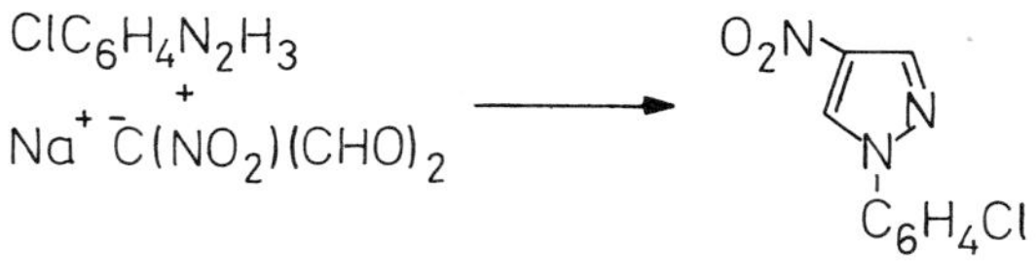

$$\underset{Na^+ \bar{C}(NO_2)(CHO)_2}{ClC_6H_4N_2H_3} \longrightarrow$$

[4-nitropyrazole structure with O_2N and C_6H_4Cl]

3-117 (3-17)

An older reaction is probably related. In alkali the diacetate ester of the dioxime of sodionitromalonaldehyde gave 4-nitro-5-hydroxypyrazole, Eq. (3-18). Treatment with phosphorus oxychloride converted the product to 4-nitro-5-chloropyrazole.[143]

A report without experimental directions claimed the conversion of chromium(III) [or aluminum(III)] tris-(3-nitro-2,4-pentanedionate) to 3,5-

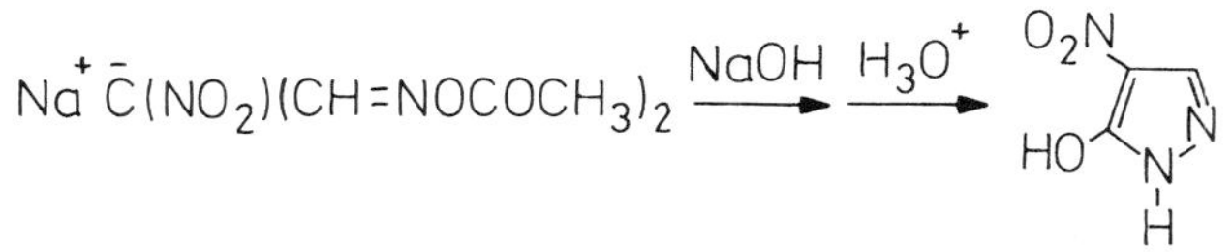

$$\text{Na}^+\bar{\text{C}}(\text{NO}_2)(\text{CH}=\text{NOCOCH}_3)_2 \xrightarrow{\text{NaOH}} \xrightarrow{\text{H}_3\text{O}^+} \underset{\underset{H}{\overset{}{\text{HO}}}}{\overset{\text{O}_2\text{N}}{\bigg\langle}}$$

(3-18)

dimethyl-4-nitropyrazole (**3-8**), Eq. (3-19).[144] The intermediate formation of nitroacetylacetone was suggested.

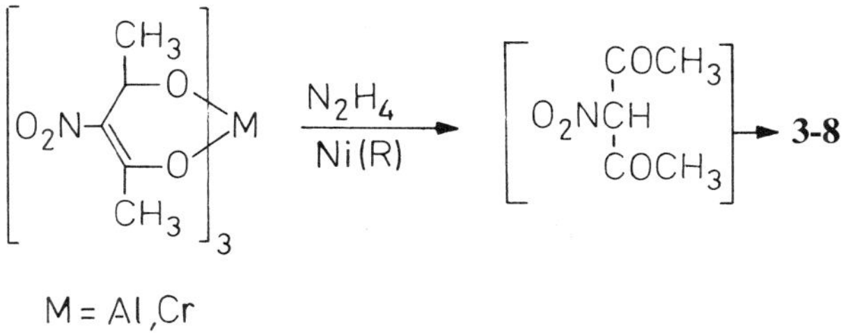

$$\left[\underset{\text{CH}_3}{\overset{\text{CH}_3}{\text{O}_2\text{N}}}\begin{array}{c}\text{O}\\\text{M}\\\text{O}\end{array}\right]_3 \xrightarrow{\underset{\text{Ni(R)}}{\text{N}_2\text{H}_4}} \left[\underset{\text{COCH}_3}{\overset{\text{COCH}_3}{\text{O}_2\text{NCH}}}\right] \rightarrow \textbf{3-8}$$

M = Al, Cr

(3-19)

Nitroketenaminals and Hydrazines

A mixture of β,β-bisbenzylamino-α-nitroacrylonitrile and hydrazine in methanol at 60° for 16 h gave 3-benzylamino-4-nitro-5-aminopyrazole (**3-118**) (80%), Eq. (3-20).[145] Methylhydrazine gave the 1-methyl derivative of (**3-118**),[145] and the cyclic aminal (**3-119**) gave the 4-nitropyrazole derivative (**3-120**).[146]

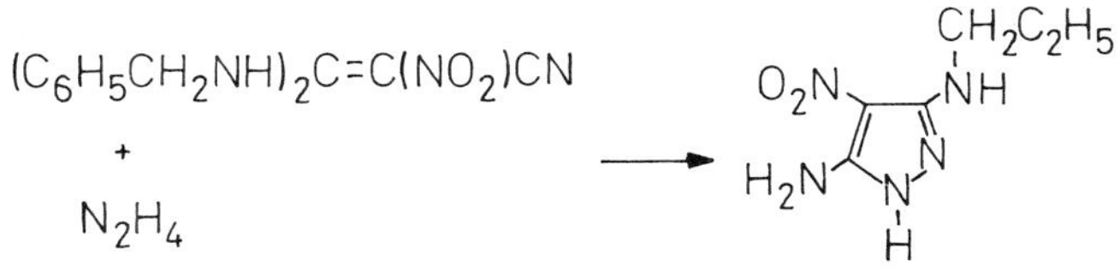

$$(\text{C}_6\text{H}_5\text{CH}_2\text{NH})_2\text{C}=\text{C}(\text{NO}_2)\text{CN}$$
$$+$$
$$\text{N}_2\text{H}_4$$

3-118

(3-20)

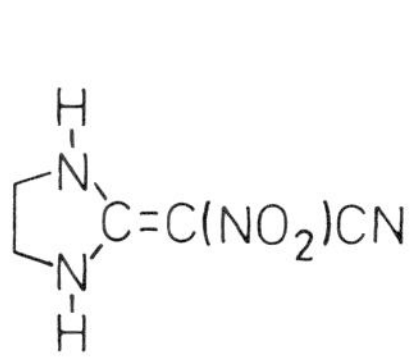

3-119 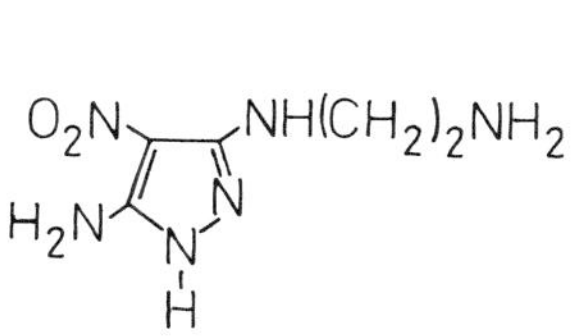   **3-120**

β,β-Bisamino-α-nitrothioacrylic acid amides, methyl iodide, and hydrazine gave the 4-nitropyrazoles (**3-121a–e**) (40%–95%), Eq. (3-21).[147]

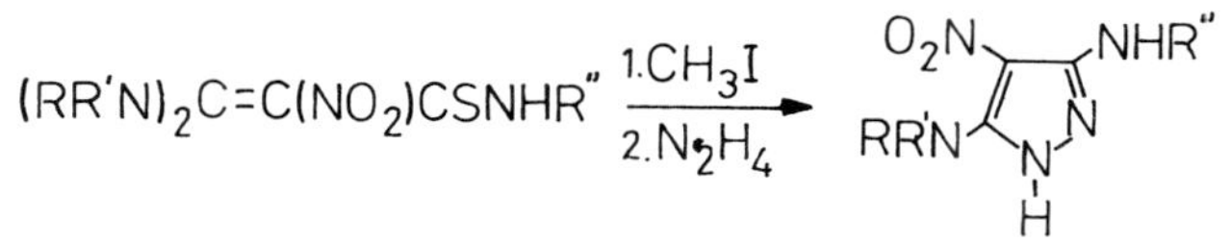

$$(RR'N)_2C=C(NO_2)CSNHR'' \xrightarrow[2.N_2H_4]{1.CH_3I} \underset{\textbf{3-121}}{\text{[pyrazole with } O_2N, NHR'', RR'N\text{]}}$$ (3-21)

3-121	R	R'	R"	Yield%
a	CH_3	CH_3	C_6H_5	60
b	CH_3	H	C_6H_5	95
c	CH_3	CH_3	CH_3	73
d	CH_3	CH_3	$CH_2CH=CH_2$	40
e	$(CH_2)_4$		C_6H_5	61

α-Nitro-β-Hydrazinoacrylates

On formation, α-nitro-β-hydrazinoacrylate esters cyclized to 4-nitropyrazolin-3-ones, Eqs. (3-22), (3-23).[148,149] In one procedure the unisolated intermediate acrylate ester was provided by a reaction between 2,4-dinitrophenylhydrazine and 1-benzylideneamino-2-(2-ethoxycarbonyl-2-nitromethylidene)imidazolidine. The latter was conveniently obtained from ethyl nitroacetate and 1-benzyl-

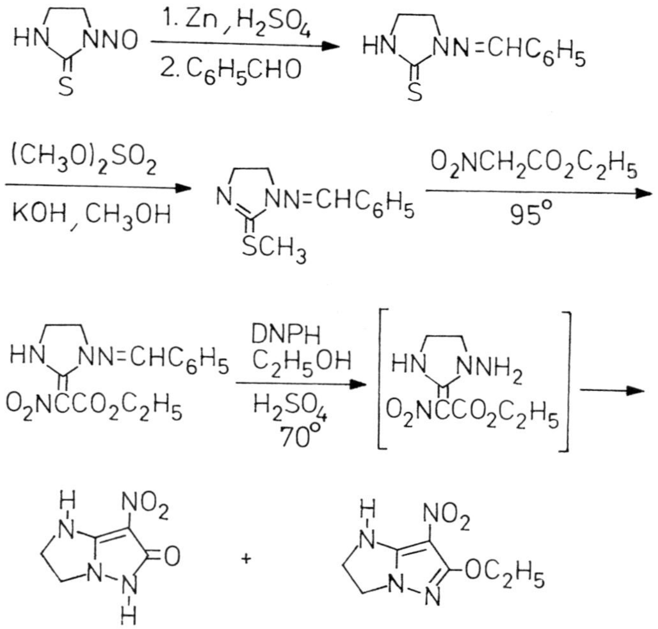

3-122 55%   +   **3-123** 14%   (3-22)

ideneamino-2-methylthioimidazoline. Cyclization converted the acrylates to 2,3-dihydro-1H-imidazolo[1,2-b]pyrazoles (**3-122**) and (**3-123**).[148]

In another procedure ethyl α-nitro-β-hydrazinoacrylate condensed with a hydrazine to give mono- and disubstitued derivatives (**3-124a–e**) (70%) of 4-nitro-pyrazolin-3-ones, Eq. (3-23).[149]

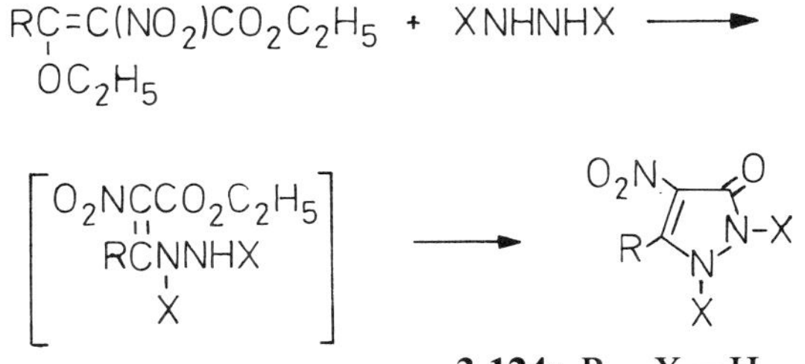

$$RC=C(NO_2)CO_2C_2H_5 + XNHNHX \longrightarrow$$
$$\underset{OC_2H_5}{}$$

$$\begin{bmatrix} O_2NCCO_2C_2H_5 \\ \underset{\|}{R\overset{}{C}NNHX} \\ X \end{bmatrix} \longrightarrow$$

3-124a R = X = H
3-124b R = H, X = CH$_3$
3-124c R = H, X = C$_6$H$_5$
3-124d R = CH$_3$, X = H
3-124e R = CH$_3$, X = C$_6$H$_5$ (3-23)

Cyclization of 1,3-Dinitro-1,3-bisphenylhydrazono-2-propanones

On heating 2 h in methanol, 1,3-dinitro-1,3-bisphenylhydrazono-2-propanone gave 1-phenyl-3-nitro-4-hydroxy-5-phenylazopyrazole (**3-125**) (93%) Eq. (3-24).[150]

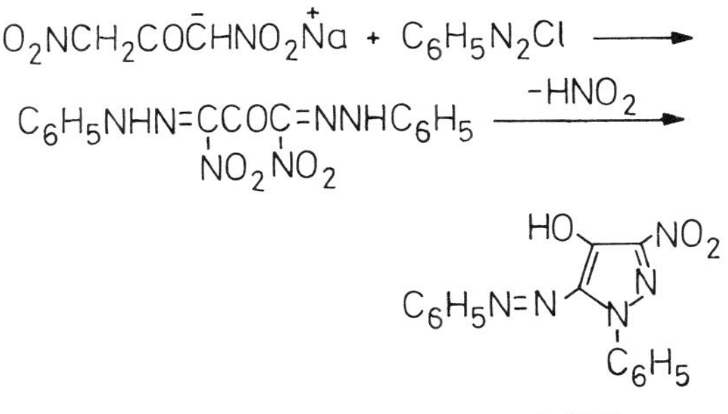

$$O_2NCH_2COCHNO_2^- \overset{+}{N}a + C_6H_5N_2Cl \longrightarrow$$

$$C_6H_5NHN=\underset{NO_2}{C}CO\underset{NO_2}{C}=NNHC_6H_5 \xrightarrow{-HNO_2}$$

3-125   (3-24)

RING TRANSFORMATIONS

Pyrimidines

Hydrazine in ethanol at 25° for 45 min converted 4-methoxy-5-nitropyrimidine to 3(5)-amino-4-nitropyrazole (**3-126**) (60%).[151] A suggested mechanism[151] is shown in Eq. (3-25).[152–154]

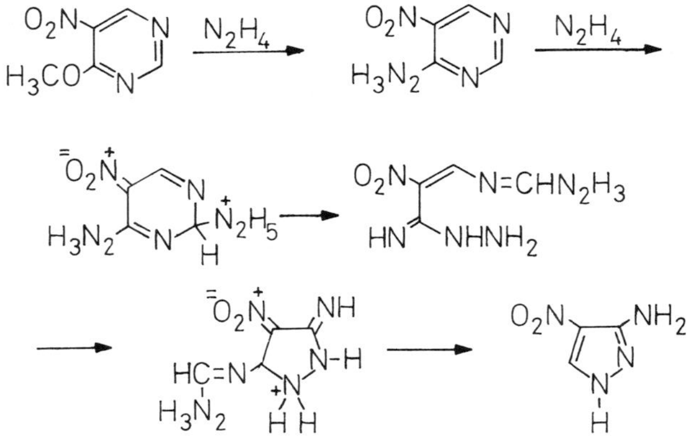

[Scheme for Eq. (3-25) producing **3-126**]

In a related reaction hydrazine in isopropyl alcohol at reflux temperature for 2 h converted 1,3,6-trimethyl-5-nitrouracil to 4-nitro-5-methylpyrazolin-3-one (**3-127**) (36%); methylhydrazine afforded the N-methyl derivative (**3-88b**) (55%), Eq. (3-26).[84,155]

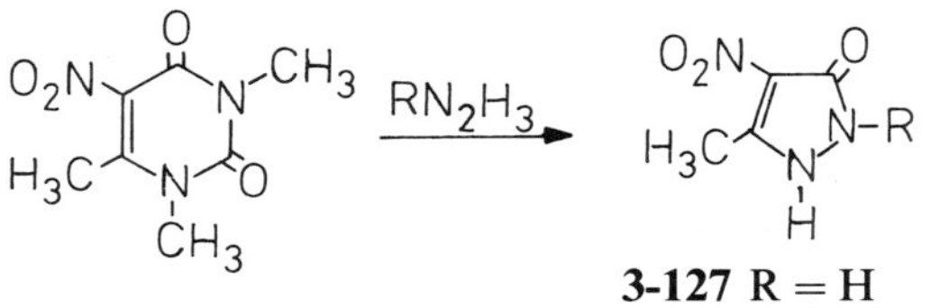

[Scheme for Eq. (3-26)]

3-127 R = H
3-88b R = CH_3 (3-26)

Hydrazine cleavage of 7-substituted amino-2,3-dihydro-5-methyl-6-nitroimidazo[1,2-a]pyrimidines (**3-128**) and similar dihydropyrimidopyrimidines (**3-129**) gave a variety of substituted nitropyrazoles (**3-130**) to (**3-132**), Eq. (3-28) to (3-30). In a suggested explanation of the conversions an initial π complex between a hydrazine and the pyrimidine (**3-128**) or (**3-129**) was proposed, Eq. (3-27).[156]

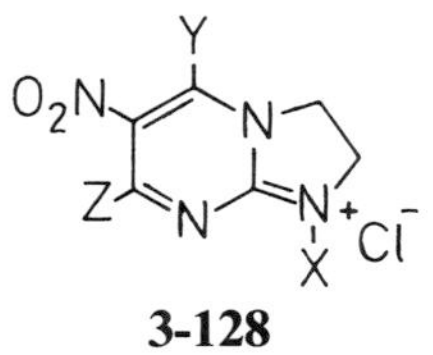

3-128 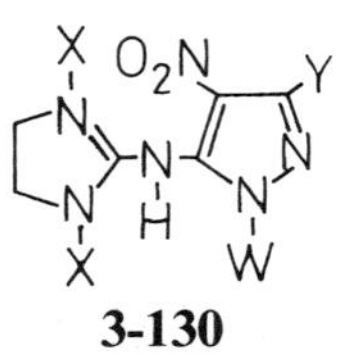   **3-130** 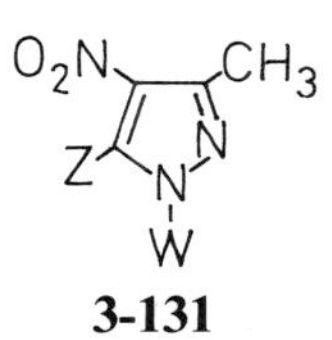   **3-131**

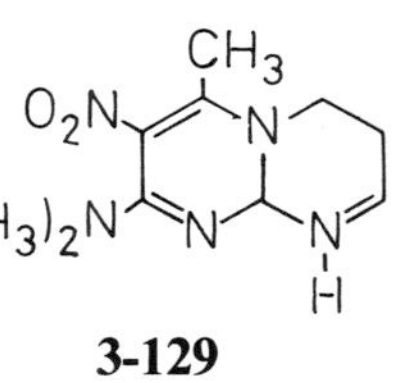

3-129 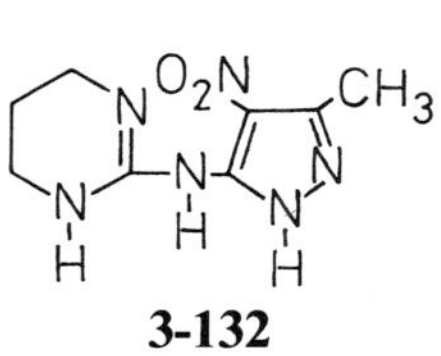   **3-132**

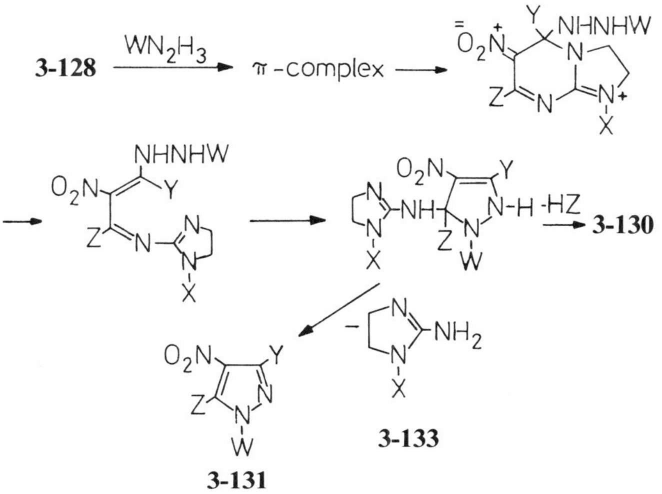

Ring Transformations

[Scheme showing conversion of 3-128 via WN₂H₃ to π-complex, then through intermediates to 3-130, 3-131, and 3-133]

(3-27)

$$\textbf{3-128} + N_2H_2 \xrightarrow{\text{cold}} HZ + \textbf{3-130} \ (60-90\%)$$

a X=H; Y=CH₃; Z=N(CH₃)₂, pyrrolidinyl, piperidyl, morpholino
b X=Y=CH₃, Z=N(CH₃)₂
c X=Y=H, Z=N(CH₃)₂

(3-28)

$$\textbf{3-128} + N_2H_2 \xrightarrow[80\%]{(CH_3)_2CHOH} \textbf{3-131}\ (50\%) + \textbf{3-133}$$

a X=H, Y=CH₃, Z=N(CH₃)₂, pyrrolidinyl, piperidyl, morpholino
b X=H, Y=CH₃, Z=NHR(R=CH₃, C₂H₅, C₆H₅CH₂, CH₂CH₂OH)

(3-29)

$$\textbf{3-128} + WN_2H_3 \longrightarrow HZ + \textbf{3-130}$$

X=H; Y=CH₃; Z=N(CH₃)₂; W=CH₃, C₆H₅

(3-30)

Pyrrolidines

Oxidation of nicotine (**3-134a**) with concentrated nitric acid gave mainly nicotinic acid and 3(5)-nitro-5(3)-(3-pyridyl)pyrazole (**3-135**) (5%); addition of sodium nitrite to the reaction mixture raised the yield to 9%, and the addition of urea (a scavenger for nitrosating agents) to the reaction mixture reduced the yield to 0.8%. Nitric acid failed to convert either cotinine (**3-134b**) (a minor product of the reaction) or nornicotine (**3-134c**) to the pyrazole (**3-135**). Nicotine with ^{15}N in the pyrrole ring gave the pyrazole (**3-135**) with ^{15}N enrichment only at the 1(2)-position. An explanation for the reaction has been offered, Eq. (3-31).[157]

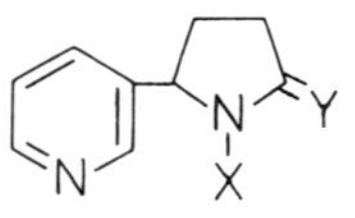

3-134a X = CH$_3$, Y = H, H
3-134b X = CH$_3$, Y = O
3-134c X = H, Y = H, H

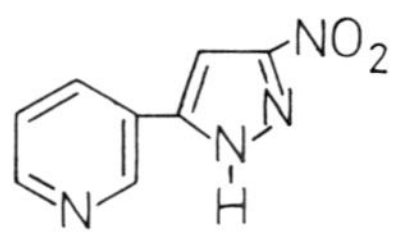

3-135

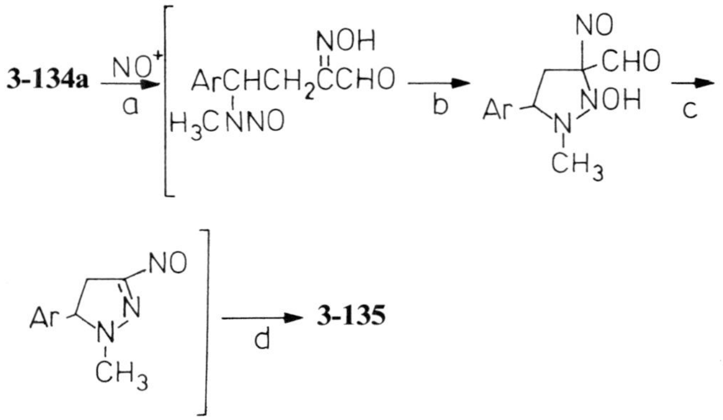

$$3\text{-}134a \xrightarrow{NO^+}_{a} \left[\underset{H_3C\,NNO}{ArCHCH_2\overset{NOH}{\overset{\|}{C}}CHO} \xrightarrow{b} Ar\underset{\underset{CH_3}{|}}{\overset{NO}{\underset{N}{\diagdown}}}\overset{CHO}{\underset{NOH}{\diagup}} \xrightarrow{c} \right.$$

$$\left. Ar\underset{\underset{CH_3}{|}}{\overset{}{\underset{N}{\diagdown}}}\overset{NO}{\underset{N}{\diagup}} \right] \xrightarrow{d} \text{3-135}$$

(3-31)

Step *a* was seen as an example of (1) nitrous acid degradation of a tertiary amine to a nitrosamine and an aldehyde and (2) nitrosation of an active methylene unit with isomerization to an oxime. The cyclization, step *b*, required an unprecedented addition to a nitrosamine. Oxidation, decarboxylation, and dehydration occurred in step *c*. Aromatization and demethylation in step *d* then accounted for the formation of the pyrazole (**3-135**). Although oxidative demethylation during nitration of N,N-dimethylaniline was cited to support step *d*, this is not a common conversion of tertiary amines brought about by nitric acid and has no precedent in the chemistry of pyrazolines or pyrazoles. It seems unlikely that the explanation, Eq. (3-31), is adequate for this unique reaction.

Isoxazoles

In the presence of potassium hydroxide methyl- and phenylhydrazine converted 4-nitro-5-methylisoxazole to the 1-methyl and 1-phenyl derivatives (**3-136a,b**) of

3-methyl-4-nitro-5-aminopyrazole.[158] An explanation for the reaction recognized initiation by base abstraction of the C-3 proton, ring opening, and condensation with the hydrazine molecule, Eq. (3-32).[159]

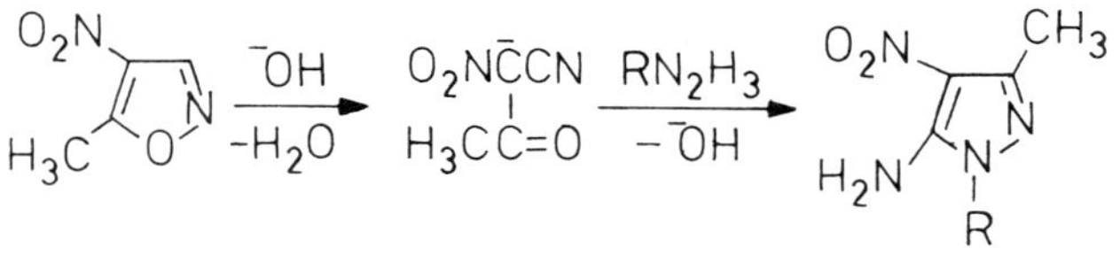

$$\text{3-136a } R = CH_3$$
$$\text{3-136b } R = C_6H_5 \quad (3\text{-}32)$$

Alkylhydrazines converted 3,5-dimethyl-4-nitroisoxazole to 1-alkyl-3,5-dimethyl-4-nitropyrazole (**3-137**). It was shown that a strong electron-withdrawing substituent at the C-4 position was required for the reaction to proceed.[160-162] A similar reaction converted the bisisoxazolyl (**3-138a**) to the bispyrazolyl (**3-138b**).[163] Initial attack by the hydrazine reagent at the C-5 isoxazole position, ring opening, and cyclization with an elimination of hydroxylamine were the proposed steps to account for these conversions, Eq. (3-33).[159]

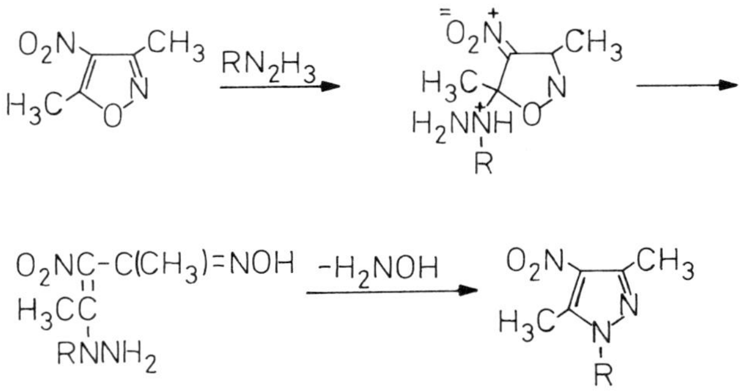

$$(3\text{-}33)$$

3-138a **3-138b**

Pyridines

Hydrazine and 3-nitro-4-chloroquinoline gave 4-nitro-5-(2-aminophenyl)-pyrazole (**3-139b**) (83%).[164] The intermediacy of 3-nitro-4-hydrazinoquinoline (**3-139a**) was presumed.

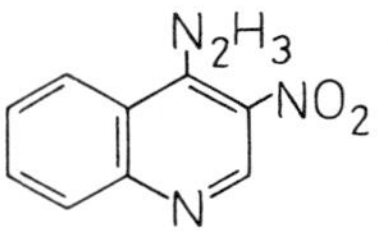

3-139a

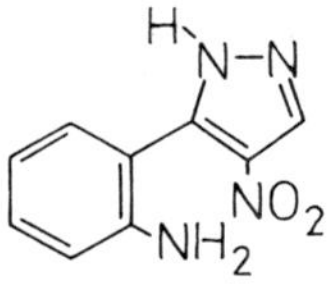

3-139b

Pyridazines

Thermolysis of 2-phenyl-4-hydroxy-5-nitro-3(2H)pyridazinone in dimethylsulfoxide for 1 h at 150° gave 1-phenyl-4-nitropyrazole (**3-139c**) (80%), Eq. (3-34). Similar thermolyses afforded 4-nitropyrazole and its 1-tosyl and 1-methyl derivatives.[165] An explanation for the reactions was not given. This extraordinary ring contraction is apparently related to an older reaction whereby, hydrochloric acid at 170° converted 2-phenyl-4-hydroxy-6-methyl-3(2H)pyridazinone to 1-phenyl-3-methylpyrazole-5-carboxylic acid, Eq. (3-35).[166]

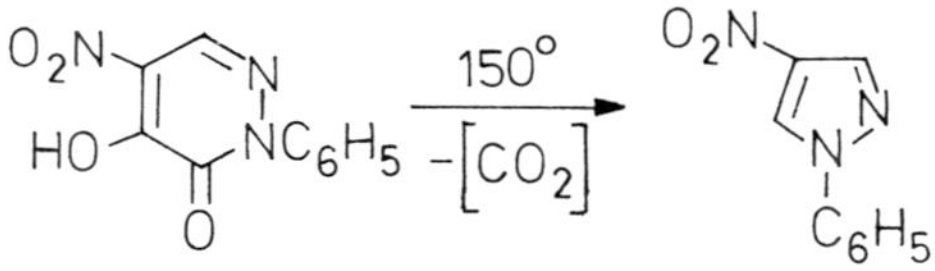

3-139c (3-34)

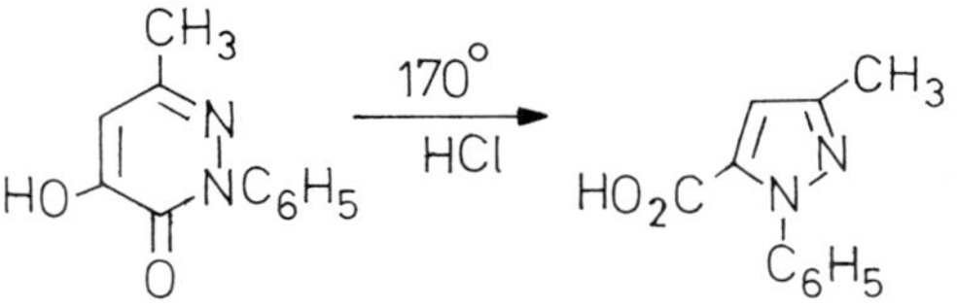

(3-35)

POLYCYCLIC NITROPYRAZOLES

Indazoles

A mixture of nitric and acetic acids and acetic anhydride at 40° for a few minutes converted indazole to 3-nitroindazole (**3-140**) (55%) and 3,5-dinitroindazole (**3-141**) (20%). A similar nitration converted 5-nitroindazole to the dinitro derivative (**3-141**) (42%) and 2,5-dinitroindazole (**3-142**) (51%); from 6-nitroindazole only 3,6-dinitroindazole (**3-143**) (97%) was obtained; 4-nitro- and 7-nitroindazole gave 2,4-dinitroindazole (**3-144**) (95%) and 2,7-dinitroindazole (**3-145**) (85%). Each N-nitro compound (**3-142**, **3-144**, and **3-145**) on heating in anisole was readily converted to the corresponding 3-nitro derivative (**3-141**,

3-146, and **3-147**). Compound (**3-142**) was not obtained pure, since it disproportionated in solution; after heating a neat sample at 184° for 1.5 h the presence of the isomer (**3-141**) was noted. At lower temperatures 2,6-dinitro- and 2,3-dinitroindazole (**3-148**) (75%) and (**3-149**) (30%) were obtained on nitration of 6-nitro- and 3-nitroindazole; the former thermolyzed to the isomer (**3-143**), but the latter decomposed on heating. Initial nitration at the 2-position was facilitated by weaker acid media and at positions in the benzene ring by stronger acid media.[167]

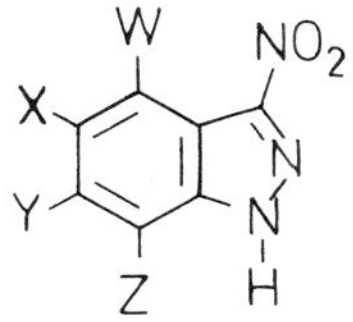

3-140 W = X = Y = Z = H
3-141 W = Y = Z = H, X = NO_2
3-143 W = X = Z = H, Y = NO_2
3-146 W = NO_2, X = Y = Z = H
3-147 W = X = Y = H, Z = NO_2

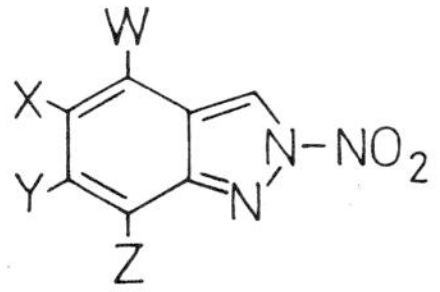

3-142 W = Y = Z = H, X = NO_2
3-144 W = NO_2, X = Y = Z = H
3-145 W = X = Y = H, Z = NO_2
3-148 W = X = Z = H, Y = NO_2

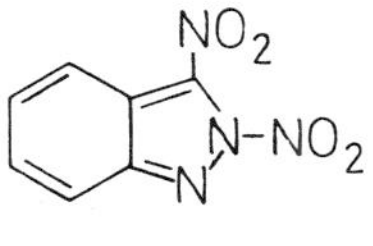

3-149

A mixture of concentrated sulfuric and nitric acids converted 5,7-dinitro- (at 20° for 48 h), 5,6-dinitro- (same conditions), and 4,6-dinitroindazole (at 70° for 3 h) to 3,5,7-, 3,5,6-, and 3,4,6-trinitroindazoles (**3-150**), (**3-151**) (55%), and (**3-152**) (28%). Mixed acid further nitrated compound (**3-151**) (at 80° for 4 h) to 2,3,5,6-tetranitroindazole (**3-153**) (84%).[168]

Nitric acid in acetic anhydride converted 5,6-dinitro- and 4,6-dinitroindazole (both at −5° with immediate quenching by ice) to 2,5,6- and 2,4,6-trinitroindazoles (**3-154**) (55%) and (**3-155**) (53%). The same reagent converted compounds (**3-151**) and (**3-152**) (both at 0°–15° for 15 h) to the tetranitro derivative (**3-153**) (58%) and 2,3,4,6-tetranitroindazole (**3-156**) (65%); 3,6-dinitroindazole gave 2,3,6-trinitroindazole (**3-157**) (57%).[168]

Nitration of 5- and 6-nitroindazoles gave 2-nitro derivatives that were thermolyzed to 3,5- and 3,6-dinitroindazoles.[169] There is a report that nitration of 6-nitroindazole gave 5,6-dinitroindazole.[170]

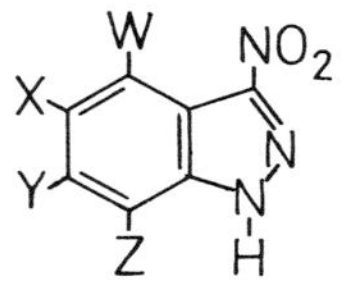

3-150 W = Y = H, X = Z = NO_2
3-151 W = Z = H, X = Y = NO_2
3-152 W = Y = NO_2, X = Z = H

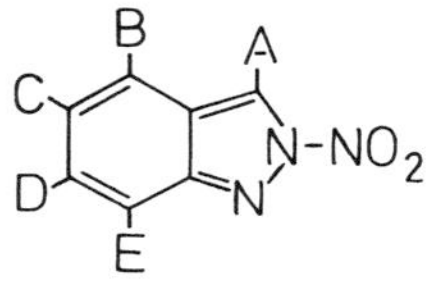

3-153 A = C = D = NO_2, B = E = H
3-154 A = B = E = H, C = D = NO_2
3-155 A = C = E = H, B = D = NO_2
3-156 A = B = D = NO_2, C = E = H
3-157 A = D = NO_2, B = C = E = H

Pyrazolo[1,5-a]pyridines

Pyrazolo[1,5-*a*]pyridine in a mixture of nitric and sulfuric acids at 0° for 1 h gave the 3-nitro derivative (**3-158a**) (78%);[171] A similar reaction with an excess of nitric acid at 20° for 1 h gave 3,4-dinitropyrazolo[1,5-*a*]pyridine (**3-158b**) (53%) (previously obtained from the unsubstituted heterocycle in a mixture of concentrated nitric and sulfuric acids at reflux temperature).[172] Peroxyacetic acid oxidized the 3-nitroso analog [from nitrosation (60%) at 15°] to the nitro compound (**3-158a**) (90%).[171]

Fuming nitric acid in acetic anhydride at 0° for 1 h converted pyrazolo-[1,5-*a*]pyridine to the 3-nitro derivative (**3-158a**) (60%) and to di(3-pyrazolo-[1,5-*a*]pyridyl)oxido-ammonium nitrate (**3-159**), a structure consistent with elemental analysis and analysis of mass, ir, and nmr spectra.[171]

Concentrated nitric acid at 25° for 1 h converted 2-hydroxypyrazolo-[1,5-*a*]pyridine to the 3-nitro derivative (**3-158c**) (48%).[173] Nitrolysis of 3-acyl-2-alkyl-pyrazolo[1,5-*a*]-pyridines gave the corresponding 3-nitro compounds (**3-158d**).[174]

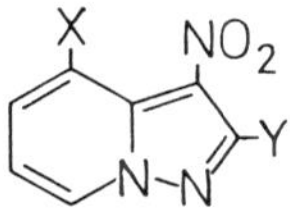

3-158a X = Y = H
3-158b X = NO$_2$, Y = H
3-158c X = H, Y = OH
3-158d X = H, Y = R

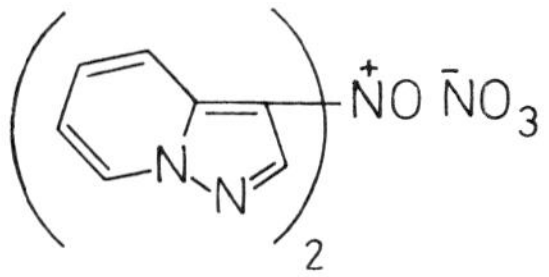

3-159

Pyrazolo[4,3-b]pyridines

A mixture of nitric and sulfuric acid at 110° for 3 h converted 5-methyl-1H-pyrazolo[4,3-b]pyridine to the 3-nitro derivative (**3-160a**) (98%); similar treatment converted 1H-pyrazolo[4,3-b]pyridin-5(4H)-one to the 3,6-dinitro derivative (**3-160b**) (73%).[175]

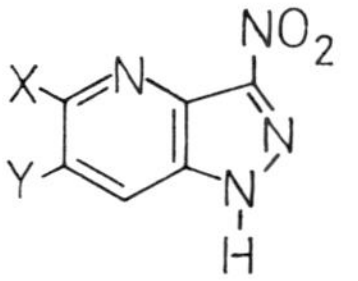

3-160a X = CH$_3$, Y = H
3-160b X = OH, Y = NO$_2$

Pyrazolo[3,4-b]pyridines

Nitration in sulfuric acid at 100° for 2 h gave the 3-nitro derivative (**3-161**) (91%) of 1-methyl-4-hydroxy-1H-pyrazolo[3,4-b]pyridine-5-carboxylic acid.[176]

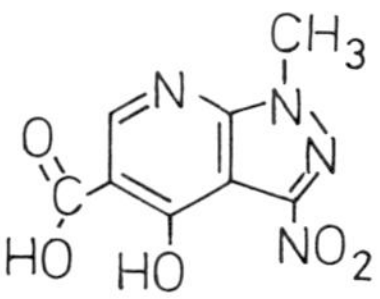

3-161

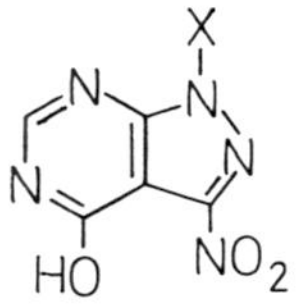

3-162a X = CH$_3$
3-162b X = 2,4(NO$_2$)$_2$-C$_6$H$_3$
3-162c X = 3-NO$_2$-4-Cl-C$_6$H$_3$
3-162d X = 3,5-(O$_2$N)$_2$C$_6$H$_2$CH$_3$-4
3-162e X = cyclo-C$_6$H$_{11}$

Pyrazolo[3,4-d]pyrimidines

The 3-nitro derivative (**3-162a**) (46%) of 1-methyl-4-hydroxypyrazolo[3,4-d]-pyrimidine was prepared by treatment with a mixture of concentrated nitric and sulfuric acids at 100° for 2 h. Similar preparations afforded 1-(2,4-dinitrophenyl)-, 1-(4-chloro-3-nitrophenyl)-, and 1-(3,5-dinitro-4-methylphenyl)-4-hydroxy-3-nitropyrazolo[3,4-d]pyrimidines (**3-162b–d**) (90%, 40%, 41%) from the corresponding unnitrated 1-arylpyrazolopyridinone.[176]

Nitric acid in acetic anhydride converted 1-cyclohexyl-4-hydroxypyrazolo[3,4-d]pyrimidine at a temperature under 40° for 24 h to the 3-nitro derivative (**3-162e**) (84%).[176]

Pyrazolo[1,5-a]pyrimidines

A mixture of concentrated sulfuric acid and nitric acid (90%) at 0°–5° for 30 min converted pyrazolo[1,5-a]pyrimidine to the 3-nitro derivative (**3-163a**) (44%). The heterocycle was nitrated to the 6-nitro derivative (**3-163b**) (43%) with nitric acid in acetic anhydride. Each reaction was regiospecific. The results contrasted with the nitrations of pyrazolo[1,5-a]pyridine where nitric acid in acetic anhydride attacked the pyrazole ring.[177]

Complementary analyses of approximate molecular orbital calculations and of the variation of coupling constant patterns before and after protonation identified the majority species in strongly acid media as the 1-protonated derivative (**3-163c**) of pyrazolo[1,5-a]pyrimidine. The coupling constant $J_{2,3}$ changed from 2.0 for the heterocycle in chloroform-d to 3.0 for the protonated derivative in trifluoroacetic acid-d, whereas $J_{5,6}$ and $J_{6,7}$ remained constant. Calculations strongly favored electrophilic substitution at position 3 followed by position 6. Nitration in mixed acid gave predominantly the 3-nitro derivative (**3-163a**), but nitration by nitric acid in acetic anhydride (acetyl nitrate) gave the 6-nitro derivative (**3-163b**). This was rationalized by an addition-elimination mechanism in which an acetyl nitrate adduct to the 6–7 bond was followed by an elimination of acetic acid, Eq. (3-36).[177]

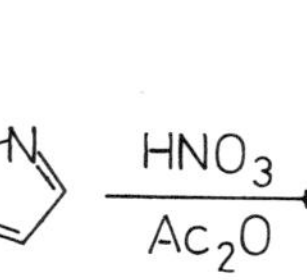

3-163a X = NO$_2$, Y = H
3-163b X = H, Y = NO$_2$

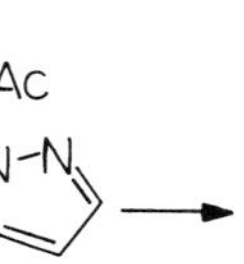

3-163c

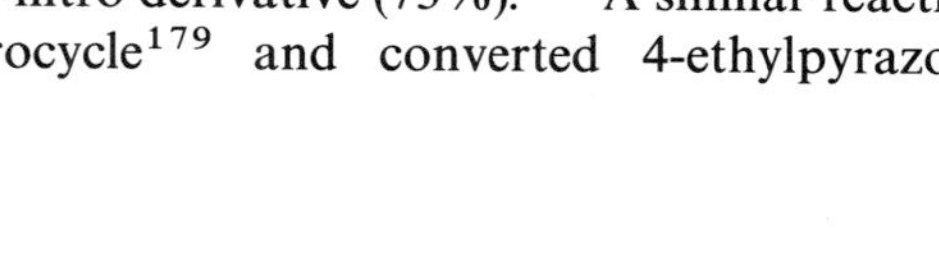

(3-36)

The mixed-acid treatment converted 5-methyl-7-ethylpyrazolo[1,5-a]pyrimidine to its 3-nitro derivative (73%).[178] A similar reaction gave the 5,7-dimethyl-3-nitroheterocycle[179] and converted 4-ethylpyrazolo[1,5-a]pyrimidin-7-one and its 6-carboethoxy derivative to the 3-nitro derivatives (**3-164a**) (38%) and (**3-164b**) (71%). Diethyl ethoxymethylenemalonate in acetic acid at reflux temperature for 7 h converted 3-amino-4-nitropyrazole (**3-89b**) to 3-nitro-6-carbethoxy-7-hydroxypyrazolo[1,5-*a*]pyrimidine (**3-164c**) (61%).[180]

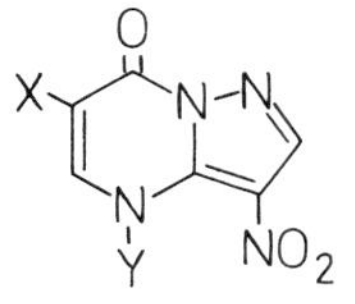

3-164a X = H, Y = C$_2$H$_5$
3-164b X = CO$_2$C$_2$H$_5$, Y = C$_2$H$_5$
3-164c X = CO$_2$C$_2$H$_5$, Y = H

Condensations between 2-alkyl-3-ethoxyacroleins and 4-nitro-5(3)-aminopyrazole or its 3(5)-methyl derivative gave a group of 12 mono- and dialkyl derivatives of 2-nitropyrazolo[1,5-*a*]pyrimidine (**3-165a–l**) (67%–97%), Eq. (3-37); condensations with 5-formyl-3,4-dihydro-2*H*-pyran afforded 4 mono- and dialkyl derivatives (**3-166a–d**) (56%–90%), Eq. (3-38).[86]

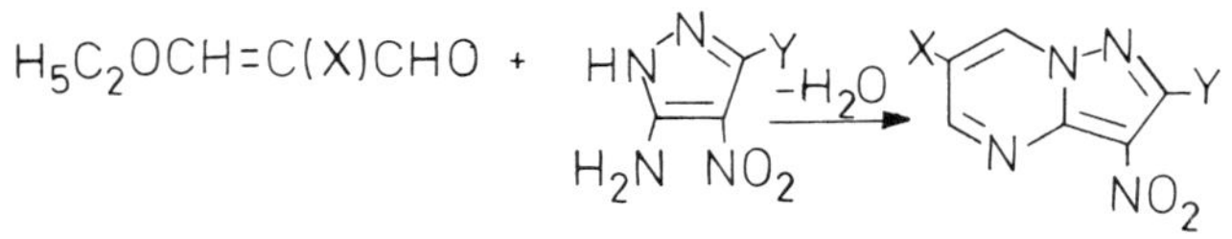

3-165(a–l) (3-37)

X,Y = CH$_3$, H; *n*-C$_3$H$_7$, H; *i*-C$_3$H$_7$, H; *n*-C$_4$H$_9$, H;
n-C$_5$H$_{11}$, H; CH$_3$, CH$_3$; C$_2$H$_5$, CH$_3$; *n*-C$_3$H$_7$,
CH$_3$; *i*-C$_3$H$_7$, CH$_3$; *n*-C$_4$H$_9$, CH$_3$; *n*-C$_5$H$_{11}$, CH$_3$

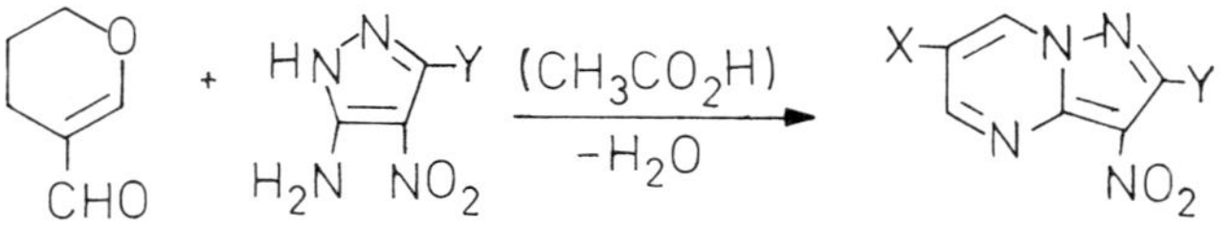

3-166a–d (3-38)

X,Y = (CH$_2$)$_3$OH, H; (CH$_2$)$_3$OH, CH$_3$; (CH$_2$)$_3$OCOCH$_3$,
H; (CH$_2$)$_3$OCOCH$_3$, CH$_3$

In a similar condensation diethyl malonate and 3-amino-4-nitropyrazole gave 3-nitro-5,7-dihydroxypyrazolo[1,5-*a*]pyrimidine (**3-166e**);[87] ethyl ethoxymethylenecyanoacetate afforded ethyl 3-nitro-7-aminopyrazolo[1,5-*a*]pyrimidine-6-carboxylate (**3-166f**) (13%).[88]

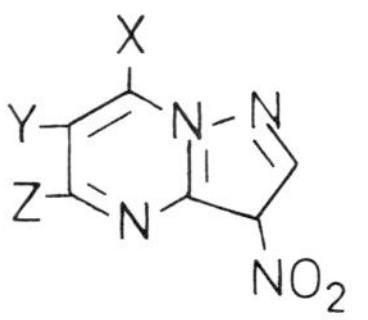

3-166e X = Z = OH
3-166f X = NH$_2$, Y = CO$_2$C$_2$H$_5$
Z = H

4H-Pyrazolo[1,5-a]benzimidazoles

In a mixed acid 4H-pyrazolo[1,5-a]benzimidazole gave the trinitro derivative (**3-167a**) (35%).[181]

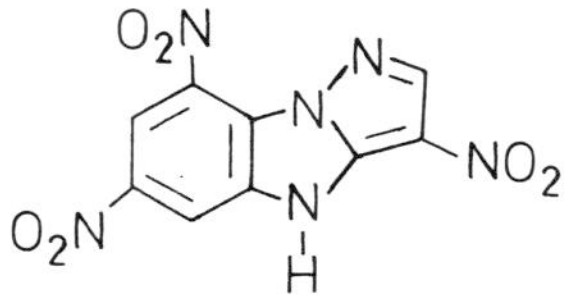

3-167a 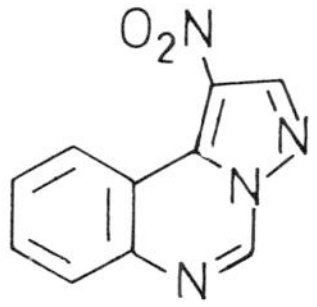   **3-167b**

Pyrazolo[1,5-c]quinazolines

Ethyl orthoformate converted 4-nitro-5-(2-aminophenyl)pyrazole (**3-139b**) to 1-nitropyrazolo[1,5-c]quinazoline (**3-167b**) (92%).[164]

Pyrazolo[1,5-a]-s-triazines

A patent reported the preparation of 3-nitro-5,7-dioxo-6-isopropyl-4H,6H-pyrazolo[1,5-a]-s-triazine from a reaction (assumed) between 3(5)-amino-4-nitropyrazole and N,N-dichlorocarbonylisopropylamine, Eq. (3-39).[182,183] In another patent 4-nitropyrazolo[1,5-a]-s-triazines are listed.[184]

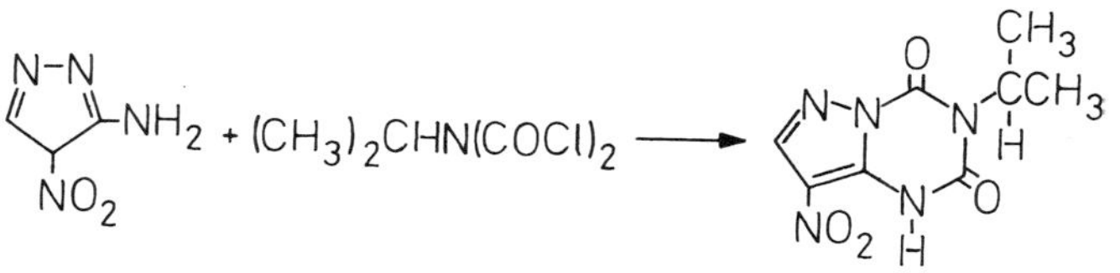

(3-39)

2H,5H-Pyrazolo[4,3-c]pyrazoles

Dehydrochlorination of 1,2-dichloroglyoxal-bis-phenylhydrazone resulted in the formation of the oxalodinitrile-bis-phenylimine (**3-168a**) which dimerized to 2,5-diphenyl-3,6-diphenylazo-2H,5H-pyrazolo[4,3-c]pyrazole (**3-168b**) (X = C$_6$H$_5$N=N). Nitric acid (100%) at reflux temperature for 10 min nitrolyzed the latter to 2,5-diphenyl-3,6-dinitro-2H,5H-pyrazolo[4,3-c]pyrazole (**3-168b**) (X = NO$_2$) (81%), Eq. (3-40).[185]

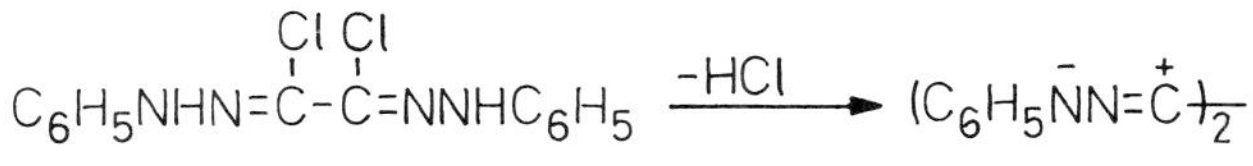

$$C_6H_5NHN=\overset{Cl}{C}-\overset{Cl}{C}=NNHC_6H_5 \xrightarrow{-HCl} (C_6H_5\overset{-}{N}N=\overset{+}{C})_{\frac{1}{2}}$$

3-168a

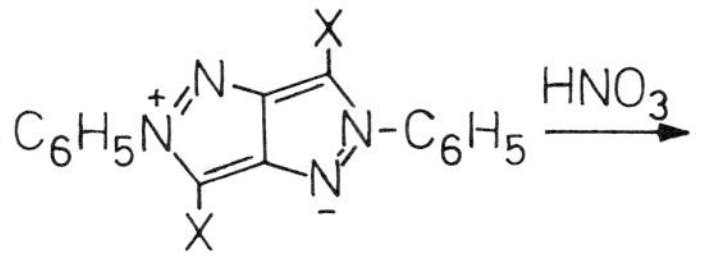

2(**3-168a**) ⟶ [C$_6$H$_5$N⁺=N–C(X)=C(X)–N=N–C$_6$H$_5$ ring] $\xrightarrow{HNO_3}$ **3-168b** X = NO$_2$

3-168b
X = C$_6$H$_5$N=N   (3-40)

REACTIONS OF NITROPYRAZOLES

N-Alkylation and N-Arylation

A survey of N-alkylation and N-arylation of pyrazoles introduced a description of the methylation of six 3-substituted and 3,6-disubstituted pyrazoles. The reactions were carried out with dimethyl sulfate in methanol, dimethyl sulfate in a basic medium, and diazomethane in acetonitrile. Product ratios were given, but yields were not.[99]

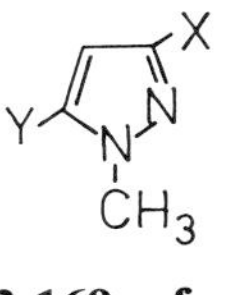

3-169a–f 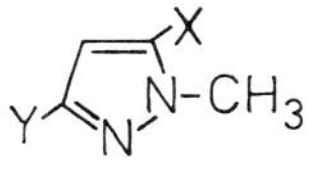   **3-170a–f**

Methylation ratios 3-169/3-170

	X	Y	(CH$_3$)$_2$SO$_4$ CH$_3$OH	(CH$_3$)$_2$SO$_4$ NaOCH$_3$	CH$_2$N$_2$
a	C$_2$H$_5$	H	1.47	0.40	1.07
b	C$_6$H$_5$	H	6.25	3.16	1.48
c	C$_6$H$_5$	CH$_3$	4.07	3.17	0.86
d	NO$_2$	H	1.25	4.10	0.56
e	NO$_2$	C$_6$H$_5$	0.01	0.45	0.15
f	CO$_2$H	CH$_3$	1	2.00	0

Orientation of N-alkylation and N-arylation was expected to be affected by reagent, solvent, pH, nucleophilicity of the pyrazole ring nitrogen atoms, steric effects, pyrazole tautomerism, and quaternary salt formation. Although an S_E2 process (**3-171**) for methylation by methyl sulfate in the neutral medium and an S_E2cB process (**3-172**) in the alkaline medium were envisaged, the events did not appear to be so simple. Electronic and steric effects and the intermediacy (undetected) of a quaternary salt presumably facilitated the formation of 1-methyl-3-nitropyrazole (**3-169d**); however, a nearly equal amount of the isomeric pyrazole (**3-170d**) was also formed in methanol. Formation of the latter was an expected S_E2' product from 3-nitro-1H-pyrazole, the probable major tautomer. A nearly exclusive formation of 1-methyl-3-phenyl-5-nitropyrazole (**3-170e**) was attributed to the combination of a steric factor from the phenyl group and the directional effect of the nitro group. Steric effects generally outweighed electronic effects.[99]

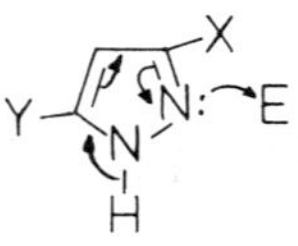

3-171 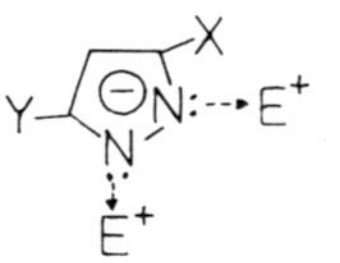   **3-172**

With the pyrazole anion — see (**3-172**) — as the likely substrate for methylation in a basic medium, electronic and steric effects (but not tautomerism and quaternary salt formation) presumably controlled product formation. In 3-nitropyrazole the electron-withdrawing substituent should afford methylation at the 1-position (the more nucleophilic ring nitrogen atom). This was supported by the formation of 1-methyl-3-nitropyrazole (**3-168d**), which accounted for 80% of the product. On the other hand, the preferential formation of 1-methyl-3-phenyl-5-nitropyrazole (**3-170e**) reflected the more powerful steric control in product formation exerted by the phenyl substituent. An electronic effect bolstered the steric effect of the phenyl substituent when electron-withdrawing para-substituents brought about an increase in the ratio (**3-173**)/(**3-174**), Eq. (3-41).[99]

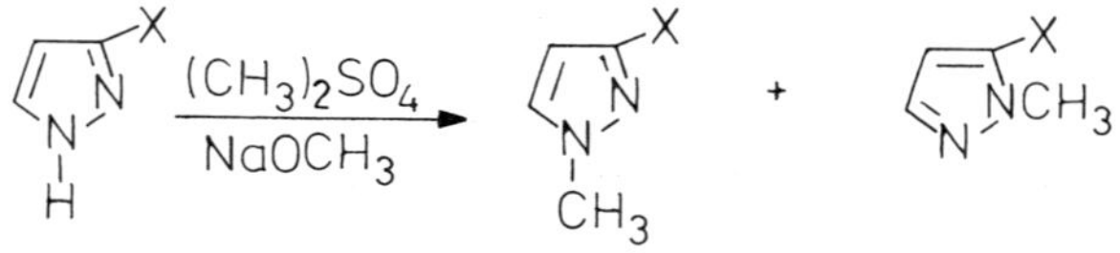

$X = C_6H_4Y\text{-}p$ **3-173**   **3-174**

Y	H	CH$_3$	OCH$_3$	Br	Cl	NO$_2$
3-173/**3-174**	3.20	2.50	3.17	3.80	3.44	5.00

(3-41)

A treatment of 3(5)-methyl-4-nitro-5(3)-cyanopyrazole with dimethyl sulfate in sodium hydroxide resulted in the formation of 1,5-dimethyl-3-cyano-4-nitropyrazole (**3-175a**) (69%) and a small amount of the isomeric 1,3-dimethyl-4-nitro-5-cyanopyrazole (**3-175b**),[186,187] the results expected from a predominant S_E2cB process.

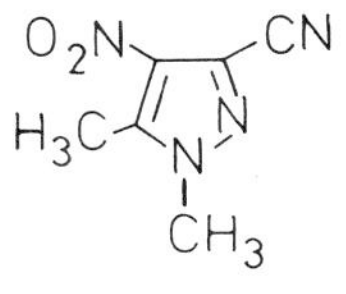

3-175a 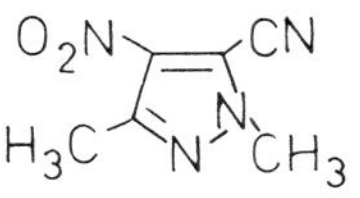   **3-175b** 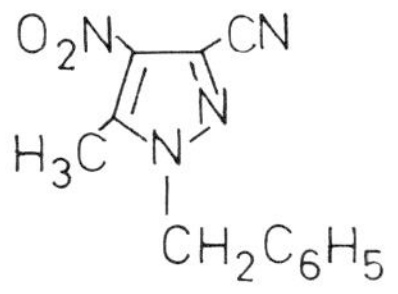   **3-176**

Benzylation of 3(5)-methyl-4-nitro-5(3)-cyanopyrazole by benzyl bromide and triethylamine in benzene at 80° for 30 min gave exclusively 1-benzyl-3-cyano-4-nitro-5-methylpyrazole (**3-176**) (75%), an expected S_E2cB product. Formation of the corresponding 1-benzyl-3-cyano-4-nitro-5-(2,3,5-O-triacetyl-β-D-ribofuranosyl)pyrazole (**3-177a**) (19%) was hindered, presumably by steric factors, when its 3-triacetylribofuranosyl-5-cyano isomer (**3-177b**) (73%) became the predominant product in a similar benzylation.[188]

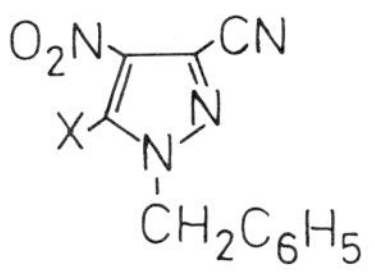

3-177a 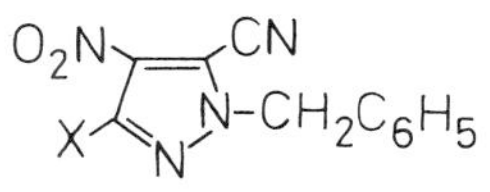   **3-177b**

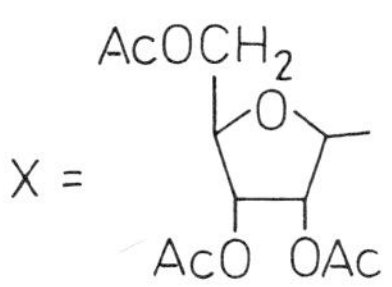

In a preparation of 1,3,5-trimethyl-4-nitropyrazole (75%), 3,5-dimethyl-4-nitropyrazole was treated with a fourfold excess of dimethyl sulfate in 2N sodium hydroxide.[10] In the presence of sodium hydride, 1,2-dimethoxyethane converted 3(5)-nitro-4-cyanopyrazole to 1-β-hydroxyethyl-3-nitro-4-cyanopyrazole.[106] An interesting conversion of 1-acetyl-3-fluoropyrazole to 1-methyl-5-fluoropyrazole (**3-178a**) (64%) was achieved in methyl fluorosulfonate (no other solvent) at 70° for 20 h, Eq. (3-42). A mixture of 3(5)-fluoropyrazole and dimethyl sulfate at 50° for 15 h gave a mixture of product (**3-178a**) (by S_E2') and the isomer 1-methyl-3-fluoropyrazole in a ratio of 19:1.[67]

Dimethyl sulfate alkylated both pyrazole nitrogen atoms in 3(5)-phenyl-4-nitro-5(3)-(4-diethylaminophenylazo)pyrazole to give the cation (**3-178b**).[188]

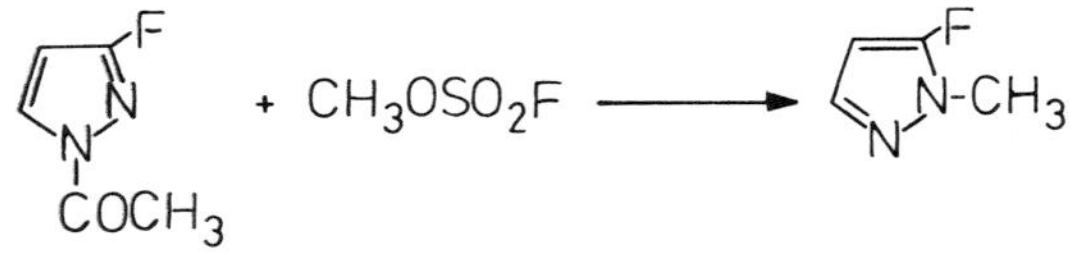

[Structure: 3-fluoro-1-acetylpyrazole + CH₃OSO₂F → 3-178a (1-methyl-3-fluoropyrazole with N-CH₃)]

3-178a (3-42)

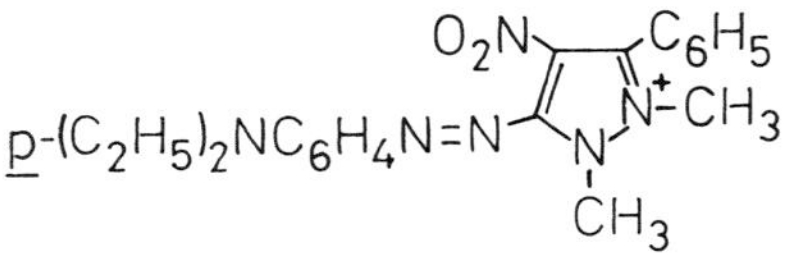

[Structure: p-(C₂H₅)₂NC₆H₄N=N— attached to pyrazole bearing O₂N, C₆H₅, N⁺-CH₃, and N-CH₃]

3-178b

Less specificity in methylation by diazomethane was found; however, there appeared to be a tendency for the methyl group to prefer attachment close to an electron-withdrawing group — eg, (**3-170d,e**).[99]

In a study of the Mannich reaction of pyrazoles, a mixture of 4-nitropyrazole and formaldehyde (30%) in water at 25° for 4 h gave the 1-hydroxymethyl derivative (**3-179a**) (77%). Similar reactions gave 1-hydroxymethyl-3(5)-methyl-4-nitropyrazole (**3-179b**) (96%) and 1-hydroxymethyl-3,5-dimethyl-4-nitropyrazole (**3-179c**) (93%). When piperidine was also present, the Mannich bases (**3-179d**) (89%), (**3-179e**) (97%), and (**3-179f**) (93%) were obtained.[189]

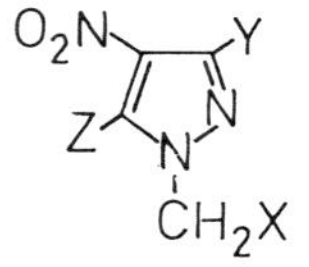

179	X	Y	Z
a	OH	H	H
b	OH	CH_3	H
c	OH	CH_3	CH_3
d	$N(CH_2)_5$	H	H
e	$N(CH_2)_5$	CH_3	H
f	$N(CH_2)_5$	CH_3	CH_3

[Structure **3-179**: pyrazole with O₂N, Y, Z substituents and N-CH₂X]

Michael adducts (**3-179g**) (60%), (**3-179h**) (63%), and (**3-179i**) (31%) were obtained from the appropriate nitropyrazole and acrylic acid or benzalacetophenone in a reaction catalyzed by Triton B.[190]

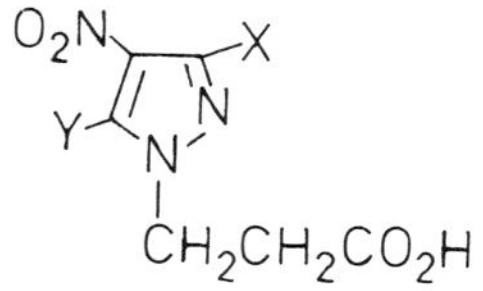

[Structures: 3-179g/h with N-CH₂CH₂CO₂H; 3-179i with N-CH(C₆H₅)CH₂COC₆H₅]

3-179g X = Y = H
3-179h X = Y = CH_3

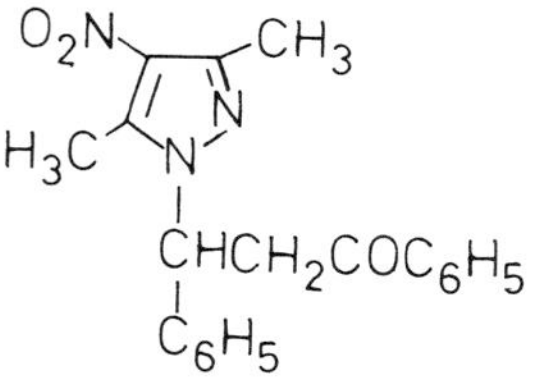

3-179i

Ullmann phenylation of 3-phenyl-, 3-nitro-, and 3-methylpyrazole by treatment with bromobenzene, potassium carbonate, and cuprous bromide in nitrobenzene gave exclusively 1,3-diphenyl- and 1-phenyl-3-nitropyrazoles (28%, 68%) and a mixture (47%) of 1-phenyl-3-methyl- and 1-phenyl-5-methylpyrazole (8:1). It appeared that the pyrazole anion — see (3-172) — displaced the bromide anion from bromobenzene, so the orientation was expected to follow that of the corresponding S_E2cB reaction. In each of the three examples the major product was the 1,3-disubstituted isomer. Again steric factors were dominant and precluded the formation of any phenylated pyrazole that contained both 3- and 5-substituents.[191] Other arylations produced 1-(2-pyridyl)- and 1-(5-nitro-2-pyridyl)-3-nitropyrazoles (3-180a,b) (48%, 47%) from 3(5)-nitropyrazole and the appropriate halopyridine.[191]

Nucleophilic displacement by pyrazole in dimethyl sulfoxide in the presence of sodium hydride converted 1-phenyl-3-methyl-4-nitro-5-chloropyrazole to the corresponding 5-pyrazol-3(5)ylpyrazole (3-180c).[192]

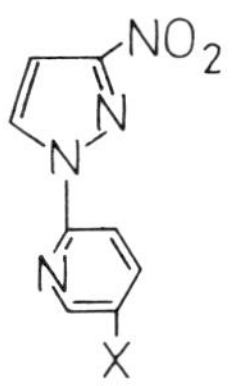

3-180a X = H
3-180b X = NO_2

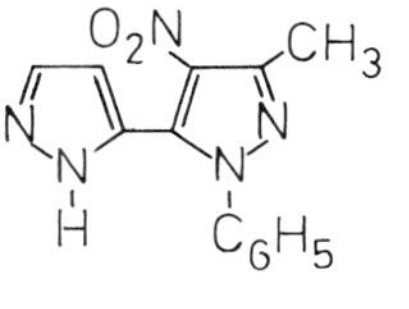

3-180c

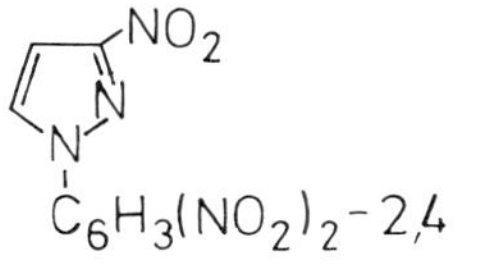

3-180d

In a related reaction, 1-(2,4-dinitrophenyl)-3-nitropyrazole (3-180d) (75%) was obtained from 3(5)-nitropyrazole, 2,4-dinitrofluorobenzene, and triethylamine in dimethyl sulfoxide at 20° for 1 h. Since nitration of 1-phenylpyrazole gave 1-(2,4-dinitrophenyl)-4-nitropyrazole, the two reactions offer complementary preparative value.[193]

A mixture of mercuric acetate, vinyl acetate, and 3,5-dimethyl-4-nitropyrazole in concentrated sulfuric acid afforded the 1-vinyl derivative (3-180e) (85%).[194]

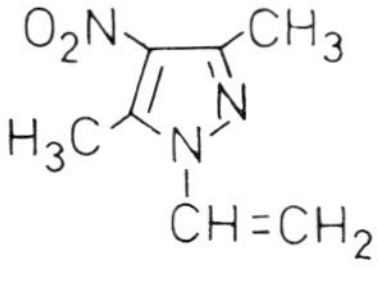

3-180e

Glycosylation of 4-nitropyrazoles by fusion of the components under vacuum gave α- and β-anomers as well as 1,3,4- and the isomeric 1,4,5-trisubstituted pyrazoles. Thus a mixture of ethyl 4-nitropyrazole-3(5)-carboxylate and 1,2,3,5-tetra-O-acetyl-β-D-ribose was heated at 150° under vacuum for 3 h to produce a mixture of the ethyl esters of 1-(2,3,5-tri-O-acetyl-β-D-ribofuranosyl)-4-nitropyrazole-3- and 5-carboxylic acids (**3-181a**) (34%) and (**3-181b**) (31%). When iodine was present during the reaction, products (**3-181a**) (39%), (**3-181b**) (19%), and the corresponding α-anomers (10% and 5%) were obtained. Glycosylation of 3(5)-cyano-4-nitropyrazole gave similar mixtures.[195] These and related ribosides have been utilized in the production of antimetabolites and antibiotics.[185-189, 196-203]

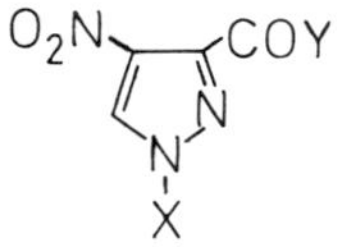

3-181a Y = OC$_2$H$_5$
3-181c Y = NH$_2$

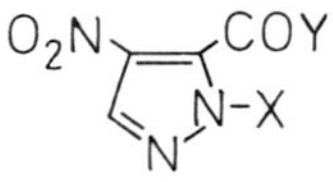

3-181b Y = OC$_2$H$_5$
3-181d Y = NH$_2$

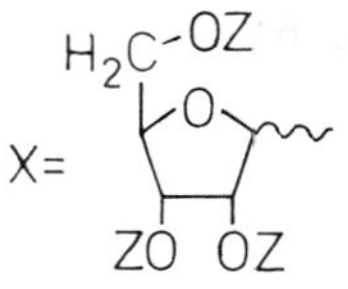

3-181a,b Z = Ac
3-181c,d Z = H

Acylation

Eleven of over 100 1-aryloxyacylpyrazoles investigated as herbicides were derivatives of 3,5-dimethyl-4-nitropyrazole. Treatment of a substituted pyrazole with an aryloxyacyl chloride with or without pyridine gave the 1-acyl derivatives (**3-182a–f**) and (**3-183a–e**).[204] The unexplained variations in yield are of interest, particularly the marked decreases in the series (**3-182d,e,f**) and (**3-183c,d,e**), in which the increasingly remote nitro group appears to be affecting the reaction. Although unsymmetrical substitution in the pyrazole ring afforded the formation of isomeric products, each reaction gave just one — eg, (**3-183a–e**).[204] No explanation for the choice is available at this time.

Acylation with dimethylaminocarbamoyl chloride gave theureas (**3-184a,b**), and other acyl chlorides gave the amides (**3-184c–g**).[63, 205]

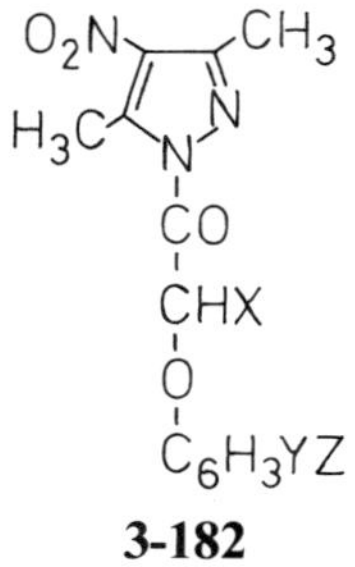

3-182	X	Y	Z	Yield%
a	H	3-Cl	H	57
b	CH$_3$	2-Cl	4-Cl	12
c	H	2-Cl	5-Cl	71
d	H	2-NO$_2$	H	80
e	H	3-NO$_2$	H	62
f	H	4-NO$_2$	H	27

3-182

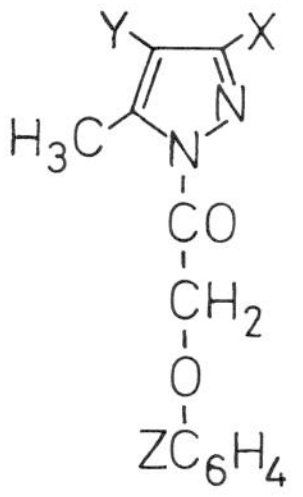

3-183	X	Y	Z	Yield%
a	OC_2H_5	H	2-Cl	47
b	OC_2H_5	H	2-NO_2	66
c	C_6H_5	H	2-NO_2	80
d	C_6H_5	H	3-NO_2	72
e	C_6H_5	H	4-NO_2	46

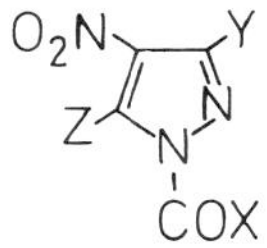

3-183

3-184	X	Y	Z
a	$N(CH_3)_2$	H	H
b	$N(CH_3)_2$	CH_3	CH_3
c	CH_3	Cl	CH_3
d	CH_2Cl	Cl	CH_3
e	C_6H_5	Cl	CH_3
f	OC_2H_5	Cl	CH_3
g	$CH_2OC_6H_5$	Cl	CH_3

3-184

Treatment with N,N-dimethylformamide and phosphoryl chloride at 100° for 4 h converted 1-phenylpyrazole to 1-phenyl-4-formylpyrazole.[206]

Acetic anhydride chose one of three available nitrogen atoms in converting 3(5)-anilino-4-nitro-5(3)-dimethylaminopyrazole (**3-121a**) to 1-acyl-3-anilino-4-nitro-5-dimethylaminopyrazole (**3-185**) (69%). Neither of the other possible isomeric products was reported.[147]

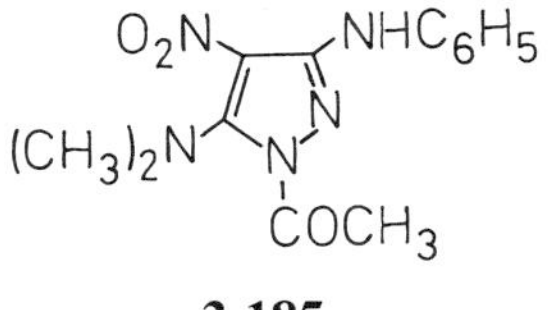

3-185

Halogenation

The treatment of 4-nitropyrazole in acetic acid with sodium hypochlorite resulted in the formation of 1-chloro-4-nitropyrazole (**3-186**).[207] Migration of chlorine to another position has not been investigated and may not occur, since no product was reported for thermal decomposition at 106°. On the other hand, iodine in an aqueous solution of potassium iodide converted 1-methylpyrazole into the 4-iodo derivative in low yield and 1,3-dimethyl-5-aminopyrazole gave the 4-iodo derivative in 80% yield.[208] A mixture of iodine and iodic acid proved to be an effective iodinating agent for many 1-alkylpyrazoles.[209] With 1-methyl-

3-nitropyrazole at 80° it gave the 4-iodo derivative (**3-187**) (79%) after 1.5 h and the 4,5-diiodo derivative (**3-188**) (52%) after 43 h. In contrast, 1-methyl-4-nitropyrazole failed to react and was recovered quantitatively after a period of 21 h.[209] A similar reactivity toward bromine accounted for the conversion of 3(5)-nitro-5(3)-(3-pyridyl)pyrazole to its 4-bromo derivative (**3-189**) and provided partial proof that the heterocycle was not the isomeric 3(5)-(3-pyridyl)-4-nitropyrazole.[51] A bromination of 1-(1-phenyl-3-methyl-4-nitropyrazol-5-yl)-3,5-dimethylpyrazole to the 4-bromo derivative (**3-190**) was reported.[210]

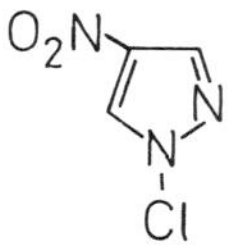

3-186

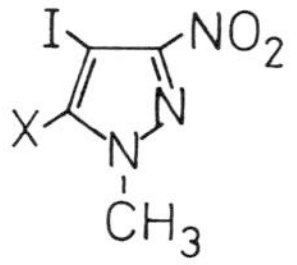

3-187 X = H
3-188 X = I

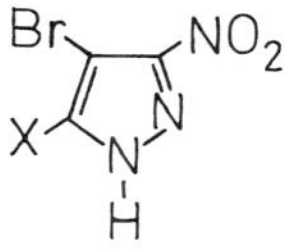

3-189

X =

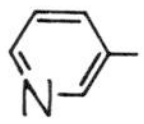

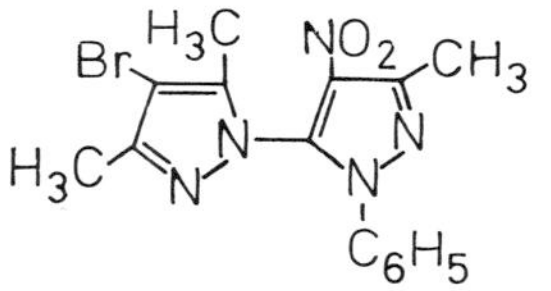

3-190

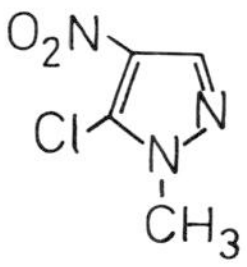

3-191

Acetyl chloride and 1-methyl-5-nitropyrazole-2-oxide (**3-91**) in an unexplained reaction gave 1-methyl-4-nitro-5-chloropyrazole (**3-191**) as the main product. Deoxygenation by phosphorus trichloride gave 1-methyl-5-nitropyrazole (**3-93**) as expected.[90]

Halogen Replacement

Without activation by an adjacent strong electron-withdrawing group, pyrazolyl halogen substituents tend to be inert toward replacement by nucleophiles.[59] The first preparation of a hydrazino derivative of a nitropyrazole by means of a nucleophilic displacement of halogen by hydrazine was reported in 1974. Anhydrous hydrazine in ethanol at 80° for 12 h converted 3(5)-methyl-4-nitro-5(3)-bromopyrazole to 3(5)-methyl-4-nitro-5(3)-hydrazinopyrazole (**3-192**) (92%).[59,62] This result established the incredibility of an earlier claim that the analogous 3(5)-methyl-4-nitro-5(3)-chloropyrazole and hydrazine gave 1,2-bis-[3(5)-methyl-4-nitropyrazol-5(3)-yl]hydrazine (**3-193**), an adduct (**3-193**) · 2HCl that failed to give a test (unspecified) for a chloride anion and a dimer (?) of compound (**3-193**).[211]

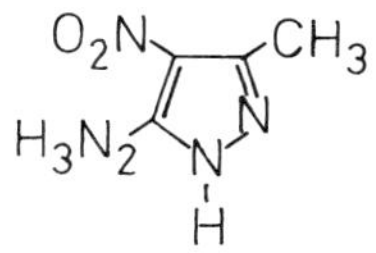

3-192

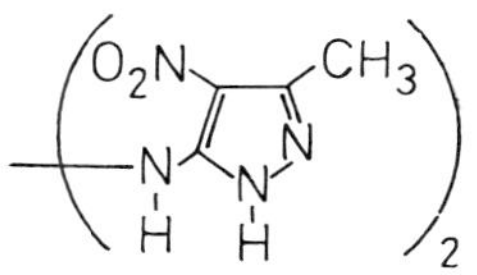

3-193

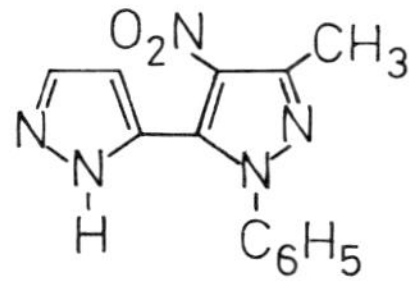

3-194

Sodium hydride in dimethyl sulfoxide at 25° for 24 h brought about the formation of 1-phenyl-3-methyl-4-nitro-5-(5-pyrazolyl)pyrazole (**3-194**) (77%) from 1-phenyl-3-methyl-4-nitro-5-chloropyrazole and pyrazole.[192] An aminolysis of 1-p-nitrophenyl-3-methyl-4-nitro-5-chloropyrazole gave the 5-benzylamino derivative (**3-195**).[212] A mixture of 1-isopropyl-3-methyl-4-nitro-5-chloropyrazole, mercaptoacetic acid, and sodium bicarbonate in ethanol at 80° for 5 h gave the pyrazol-5-ylthioacetic acid (**3-196**).[213] The activated chloro group in 1-phenyl-3-methyl-4-nitro-5-chloropyrazole was also replaced by the cyano group to give compound (**3-197**).[192]

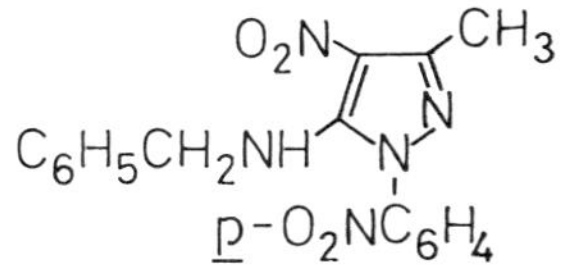

3-195

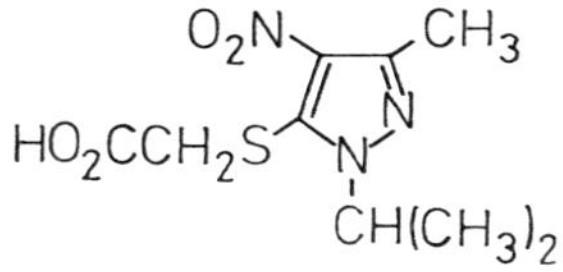

3-196

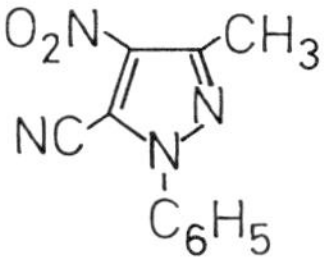

3-197

There is one example in which the halogen to be replaced is at the pyrazole 4- rather than the 3- or 5-position. An ethanol solution of 1-methyl-3,5-dinitro-4-bromopyrazole in ethanol saturated with ammonia was heated at 120° for 16 h to give the 4-amino derivative (**3-198a**) (62%). The same bromopyrazole and aniline in DMF at reflux temperature for 20 h gave the 4-anilino derivative (**3-198b**) (42%).[45]

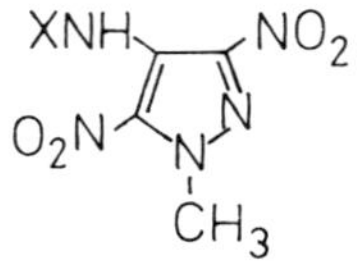

3-198a X = H
3-198b X = C$_6$H$_5$

Replacement of Nitro Groups

An investigation of the reactivity of 1-nitropyrazole toward nucleophiles revealed the cine replacement of the 1-nitro group in 1,4-dinitropyrazoles. An ethanol solution of a 1,4-dinitropyrazole and a secondary amine at 30° gave the corresponding derivative of 4-nitro-5-dialkylaminopyrazole (**3-199a–e**) (20%–85%). Under more strenuous conditions 1-nitro-4-bromo- and 1,3-dinitropyrazole (**3-199f**) reacted with cyclic secondary amines to give cyclic secondary nitramines, Eqs. (3-43), (3-44). When 1,4-dinitro-3-methylpyrazole reacted with piperidine to give 3(5)-methyl-4-nitro-5(3)-piperidylpyrazole (**3-199d**) and with morphiline to give 3(5)-methyl-4-nitro-5(3)-morpholinylpyra-

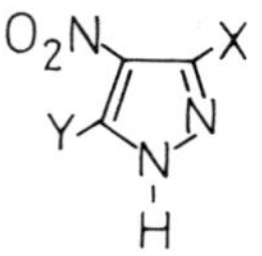

3-199a X = H   Y = piperidyl
3-199b X = H   Y = morpholino
3-199c X = H   Y = (C$_2$H$_5$)$_2$N
3-199d X = CH$_3$ Y = piperidyl
3-199e X = CH$_3$ Y = morpholino

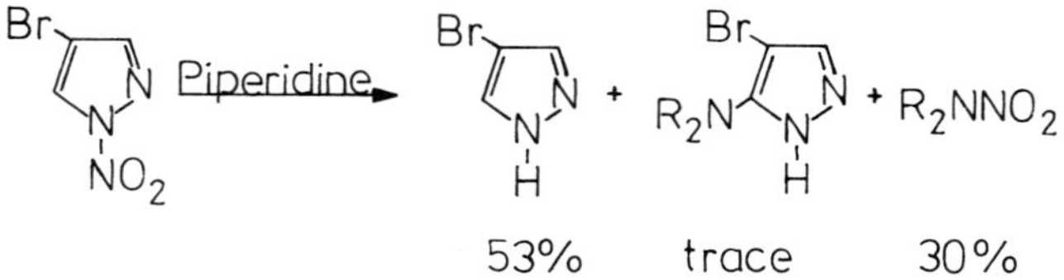

53%   trace   30%   (3-43)

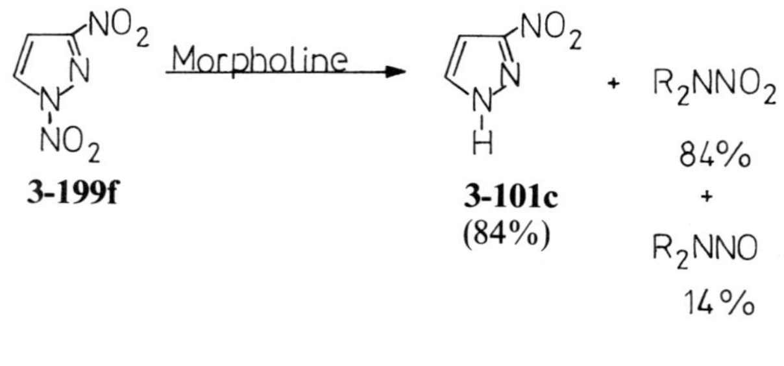

3-199f   **3-101c**
               (84%)

84%
+
R$_2$NNO
14%   (3-44)

zole (**3-199e**), it was concluded that a pyrazole 5- rather than a 3-position participated in the reaction. A mechanism was proposed, Eq. (3-45).[214]

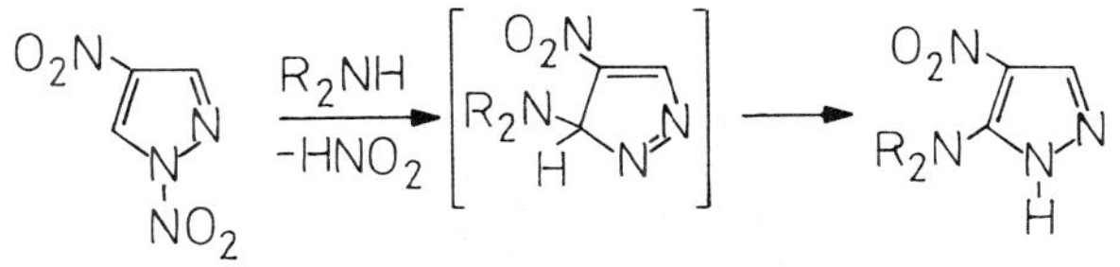

3-199g (3-45)

An extension of the reaction, Eq. (3-44), provided an attractive route to 3(5)-(pyrazol-1-yl)pyrazoles (**3-200a–k**) (80%–95%). A solution of the N-nitropyrazole and a pyrazole in ethanol was heated at 80° for 1–3 h.[215]

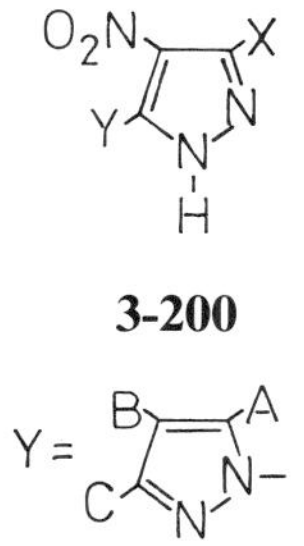

3-200	X	A	B	C
a	H	H	H	H
b	H	H	H	CH$_3$
c	H	CH$_3$	H	H
d	H	CH$_3$	H	CH$_3$
e	H	H	C$_2$H$_5$	H
f	H	H	Br	H
g	H	H	H	C$_6$H$_5$
h	CH$_3$	H	H	H
i	CH$_3$	H	H	CH$_3$
j	CH$_3$	CH$_3$	H	H
k	CH$_3$	CH$_3$	H	CH$_3$

Each of the unsymmetrically substituted 3(5)-methyl- and 3(5)-phenylpyrazoles presented the possibility of the formation of two isomeric pyrazoles. In fact both tautomers of 3(5)-methylpyrazole reacted to give a mixture of products (**3-200b,c**) (4:1) and (**3-200i,j**) (2:1); when 3(5)-phenylpyrazole gave only (**3-200g**), it was attributed to a steric control by the phenyl group.[215]

Besides secondary amines, other nucleophiles also participated; 1,4-dinitro-3-methylpyrazole reacted with ammonia, potassium ethoxide, potassium ethane thiolate, trimethylphosphite, and potassium cyanide to give the products (**3-201a–e**) (77%–98%). The last example became the model for extending cine substitution to 1,4-dinitro-3-(2,3,5-tri-*O*-acetyl-β-D-ribofuranosyl)pyrazole for the preparation of the 5(3)-cyano derivative (**3-201f**) (89%), needed for transformation in three steps to the C-nucleoside antibiotic formycin (**3-202**).[216]

Cine substitution of 2-nitroindazoles has also occurred; 2,5-dinitroindazoles reacted with primary and secondary amines to give 3-alkylamino-5-nitroindazoles (**3-203**) and 3,5-dinitroindazole (**3-141**). A similar formation of 3-alkylamino-6-nitroindazoles (**3-204**) and 3,6-dinitroindazole (**3-143**) was also realized. The amines failed to react with 3-chlorodinitroindazoles.[217]

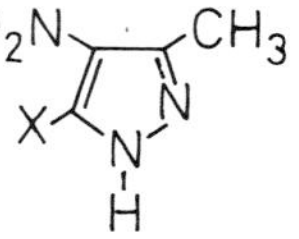

3-201a X = NH$_2$
3-201b X = OC$_2$H$_5$
3-201c X = SC$_2$H$_5$
3-201d X = P(O)(OCH$_3$)$_2$
3-201e X = CN

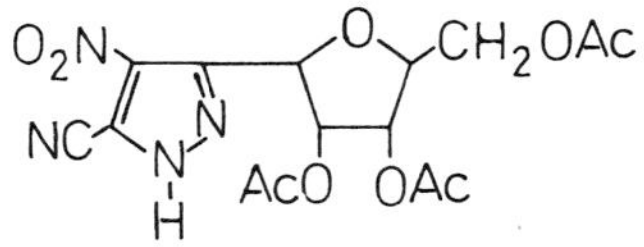

3-201f

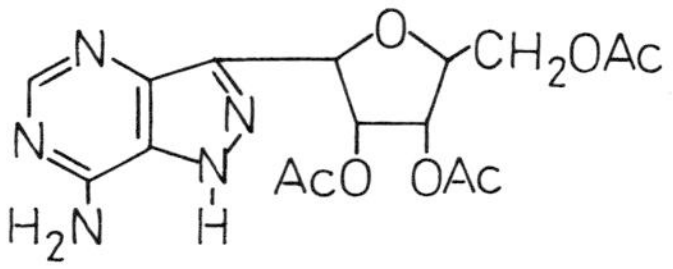

3-202

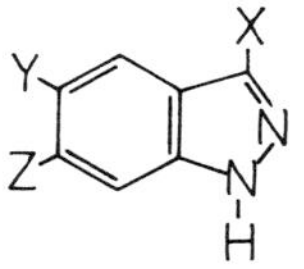

3-203 X = RNH, R$_2$N, Y = NO$_2$, Z = H
3-204 X = RNH, R$_2$N, Y = H, Z = NO$_2$

An oxidative coupling reaction between 1-phenyl-3-methyl-4-nitro-4H-pyrazolin-5-one and N,N-diethylquinone-1,4-diimine gave the same azomethine dye that was similarly obtained by the elimination of a wide variety of other 4-substituents, Eq. (3-46).[218]

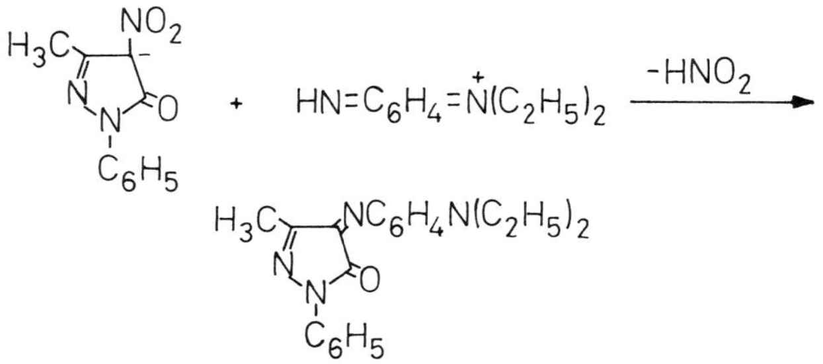

(3-46)

Treatment with bromine converted 4-nitro-5-hydroxypyrazole to 3,4,5-tribromopyrazole.[143]

Catalytic Reduction

Hydrogenation over palladium on charcoal reduced nitropyrazoles (**3-2**),[142] (**3-5**),[45,142] (**3-9**),[142] (**3-10**),[142] (**3-35**),[23] (**3-68b**),[68] (**3-68e,f**),[23,71] (**3-97a**),[93] (**3-181c,d**),[195] (**3-200a–e,h–k**),[215] and (**3-201f**)[216] to the corresponding aminopyrazoles in good to excellent yields. In a similar manner the amines (**3-205**),[219] (**3-206**),[220] (**3-207**),[67] and (**3-208a–f**)[221] were obtained from the corresponding nitropyrazoles. When acetone was also present a reductive alkylation converted the 4-nitro derivative to 2H-1,5-dimethyl-2-phenyl-4-isopropylaminopyrazol-3-one (**3-209**).[222]

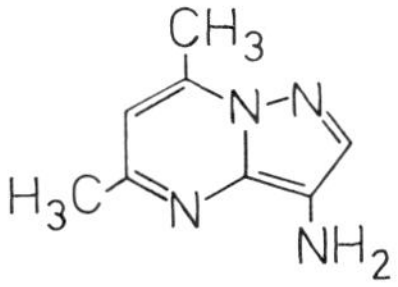

3-205

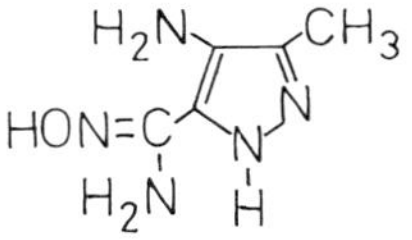

3-206

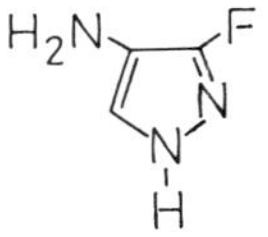

3-207

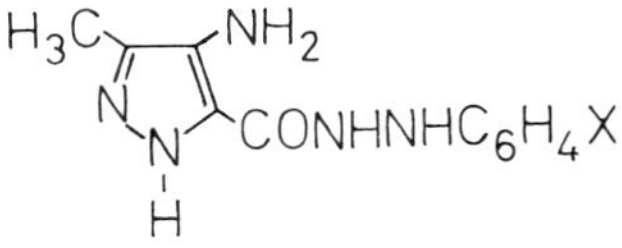

3-208a–f
X = H, 4-CH$_3$, 3-CH$_3$, 2-CH$_3$O, 3-Cl, 2-Cl

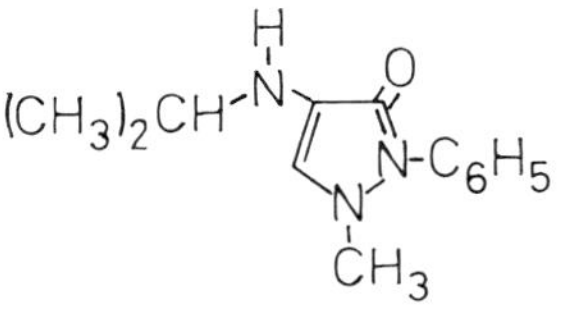

3-209

Hydrogenation over Raney nickel reduced nitropyrazoles to 1-ethyl-3-methyl-4-amino-5-benzoylpyrazole (**3-210a**),[69,223] 1-ethyl-3-methyl-4-amino-5-cyanopyrazole (**3-210b**),[69,224] and 1,3-dimethyl-4-aminopyrazole (**3-210c**),[24] also in good to excellent yields.

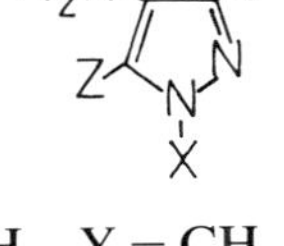

3-210a X = C$_2$H$_5$, Y = CH$_3$, Z = C$_6$H$_5$CO
3-210b X = C$_2$H$_5$, Y = CH$_3$, Z = CN
3-210c X = Y = CH$_3$, Z = H

Hydrogenation over platinum reduced 1-methyl- and 1-cyclohexyl-3-nitro-4-hydroxypyrazole[3,4-d]pyrimidines to the amines (**3-211a,b**) (66%, 70%).[176] A similar reaction over ruthenium on charcoal reduced 1-(2-chlorophenyl)-4-nitropyrazole to the amine (**3-212a**) (62%).[141] Palladium-platinum, platinum-nickel, and palladium-rhodium catalysts were also investigated for the reduction to compound (**3-209**).[222] A catalytic reduction of the nitro precursor gave 1-β-D-ribofuranosyl-3-carbamoyl-4-aminopyrazole (**3-212b**) (96%).[197]

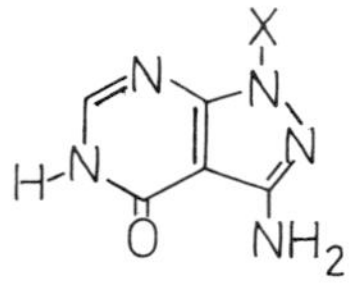

3-211a X = CH$_3$
3-211b X = cyclo-C$_6$H$_{11}$

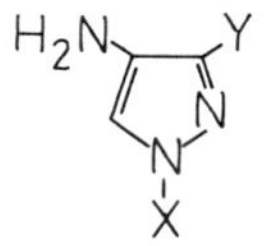

3-212a X = C$_6$H$_4$Cl-2, Y = H
3-212b X = HOCH$_2$

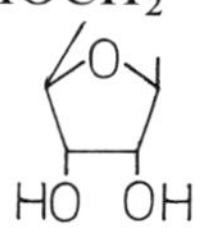

Y = CONH$_2$

Six 3(5)-methyl-4-nitro-5(3)-N-alkylcarbamoylpyrazoles gave the corresponding amines (**3-212c–h**) (65%–72%) on treatment with an aqueous mixture of sodium borohydride and 10% palladium on carbon at 0°–25° for 3 h.[225,226]

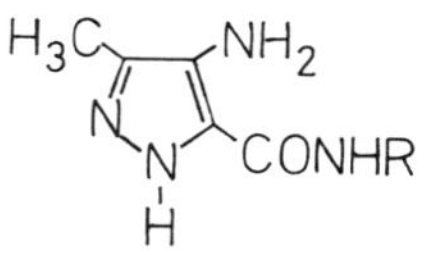

3-212c–h
R = n-C$_3$H$_7$, n-C$_4$H$_9$, i-C$_5$H$_{11}$
C$_6$H$_5$CH$_2$CH$_2$, 4-H$_3$CC$_6$H$_4$CH$_2$
C$_6$H$_5$CH$_2$

Chemical Reduction

Tin in hydrochloric acid at 100° reduced 1-phenyl-3-methyl-4-nitropyrazole (**3-42**) to the corresponding 4-aminopyrazole (**3-213a**),[36] but stannous chloride in hydrochloric acid reduced 3(5),4-dinitro-5(3)-(3-pyridyl)pyrazole (**3-33**) to the hydroxylamine (**3-213b**) (quantitative yield).[52] Air or oxygen in aqueous sodium hydroxide converted the hydroxylamine to the azoxy compound (quantitative yield), and potassium bromate converted the hydroxylamine to the corresponding 3-nitrosopyrazole (**3-213c**).[52]

Alkaline stannite solutions reduced 3,5-dimethyl-4-nitropyrazole (**3-8**) to the nitrosopyrazole (**3-213d**) (30%) and/or aminopyrazole (**3-213e**) depending on the conditions. A similar reaction observed for the 3,5-diethyl and 3,5-diphenyl

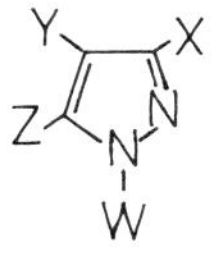

3-213	W	X	Y	Z
a	C_6H_5	CH_3	NH_2	H
b	H	NHOH	NO_2	$3-C_5H_4N$
c	H	NO	NO_2	$3-C_5H_4N$
d	H	CH_3	NO	CH_3
e	H	CH_3	NH_2	CH_3
f	H	C_2H_5	NO	C_2H_5
g	H	C_6H_5	NO	C_6H_5
h	H	CH_3	NH_2	H
i	H	NH_2	NH_2	$3-C_5H_4N$
j	H	NH_2	H	$3-C_5H_4N$

analogs of the pyrazole (**3-8**) gave the nitrosopyrazoles (**3-213f**) (30%) and (**3-213g**) (84%).[13]

In an aqueous ethanolic solution aluminum amalgam reduced 3(5)-methyl-4-nitropyrazole (**3-7**) to the amine (**3-213h**) (55%).[8] Sodium hydrosulfite converted the 3(5),4-dinitropyrazole (**3-33**) to the diaminopyrazole (**3-213i**) (quantitative yield);[52,227] alkaline dithionite reduced 3(5)-nitro-5(3)-(3-pyridyl)pyrazole (**3-135**) to the amine (**3-213j**) (quantitative yield).[51]

Zinc in hydrochloric acid reduced a dinitroantipyrine to the corresponding diamine (**3-214**).[81] Zinc in a mixture of acetic acid and acetic anhydride reduced the nitropyrazoles (**3-121a,d,e**) to acetamidopyrazoles (**3-215a–c**).[147]

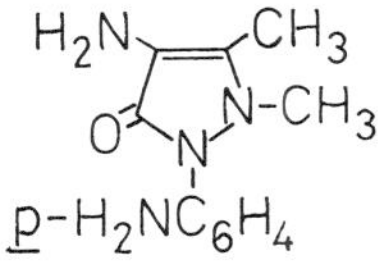

3-215	R	R'	R''
a	CH_3	CH_3	C_6H_5
b	CH_3	CH_3	$CH_2CH=CH_2$
c	$-(CH_2)_4-$		C_6H_5

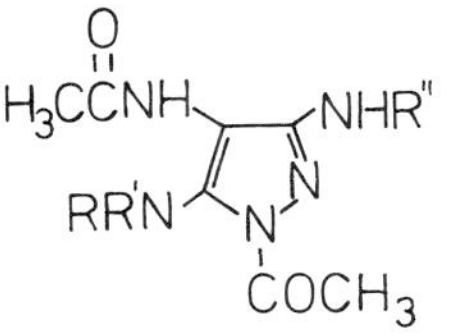

An interesting reductive cleavage of the nitro substituent in 3,5-diphenyl-4-methyl-4-nitropyrazolenine-1,2-dioxide (**3-104c**) (X = Z = C_6H_5, Y = CH_3) by sodium dithionite to give 1-hydroxy-3,5-diphenyl-4-methylpyrazole-2-oxide (**3-216**) (94%) was attributed to the loss of the nitrogen dioxide radical from an intermediate radical anion, Eq. (3-47).[111] Sodium hydrosulfite ($Na_2S_2O_4$) reduced 4-nitropyrazole-3(5)-carboxylic acid to the corresponding amino acid.[227]

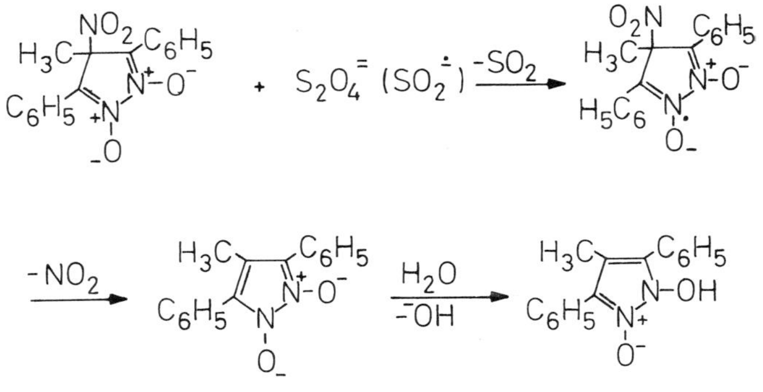

[Scheme showing reduction of nitropyrazole with $S_2O_4^{2-}$ ($SO_2^{-\cdot}$), $-SO_2$, then $-NO_2$, then $H_2O/-OH$, yielding **3-216**] (3-47)

Electrochemical reduction of 3(5)- (**3-97a**) and 4-nitropyrazole (**3-2**) occurred in two steps: the first polarographic wave corresponded to an irreversible one-electron transfer, the second to a reversible one-electron transfer. Electron paramagnetic resonance (EPR) signals were obtained at the potentials of the second wave and by their character and hyperfine structure (hfs) correlated with interaction between the unpaired electron and all of the magnetic nuclei of the molecule and were assigned to the corresponding dianion radicals. Transfer of the first electron to the heterocycle was presumably accompanied with an ejection of atomic hydrogen from the 1-position, Eq. (3-48). The half-wave potentials and EPR signal parameters of the ion radicals of 3(5)-nitropyrazole (**3-97a**) and 4-nitropyrazole (**3-2**) revealed that (1) compound (**3-97a**) reduced more readily than the isomer (**3-2**) [attributed to intramolecular hydrogen bonding as shown in the model (**3-217a**)]; (2) second-wave potentials for isomers (**3-97a**) and (**3-2**) were nearly identical and thereby revealed insignificant differences in the structures of the anions; and (3) a symmetrical distribution of the spin density corresponded to the symmetry of the anion radical obtained from 4-nitropyrazole (**3-2**).[228]

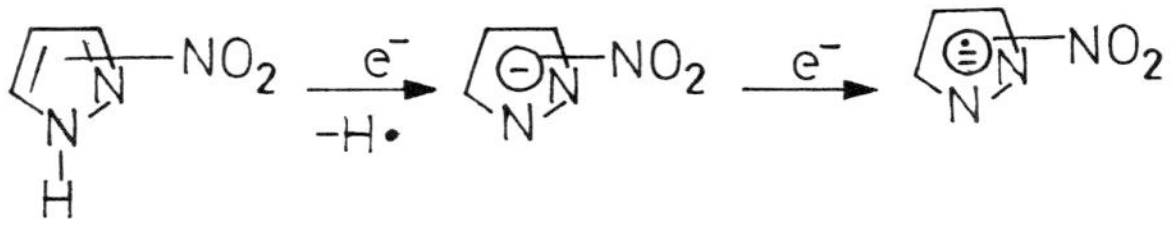

3-97a 3(5)-NO$_2$
3-2 4-NO$_2$
(3-48)

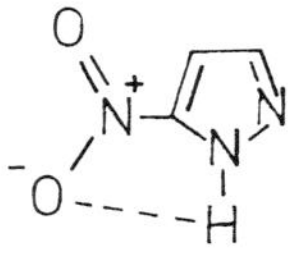

3-217a 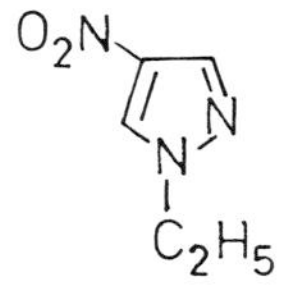 **3-217b**

The first step in the reduction of 1-methyl- and 1-ethyl-4-nitropyrazole (**3-5**) and (**3-217b**) corresponded to a reversible one-electron transfer; the second wave was similar to the second wave for nitrobenzene under identical conditions, Eq. (3-49). Signals in the EPR spectra of the alkylnitropyrazoles (**3-5**) and (**3-217b**) were observed at the potentials of the first half waves and corresponded to primary anion radicals, Eq. (3-49).[228]

$$O_2N\text{-pyrazole-}R \xrightarrow{e^-} [O_2N\text{-pyrazole-}R]^{\cdot -} \xrightarrow[4H^+]{3e^-} \text{Products}$$

3-5 R = CH$_3$
3-217b R = C$_2$H$_5$

(3-49)

The first of two waves in the polarogram of 1-nitropyrazole (**3-3**) corresponded to an irreversible transfer of one electron; however, an EPR signal was not detected at either reduction potential. A pH study revealed that the second reduction wave in neutral and alkaline media corresponded to the reduction of the nitrite anion (or nitrous acid), Eq. (3-50).[228]

$$\textbf{3-3} \xrightarrow{e^-} \left[\text{pyrazole-NO}_2 \right]^{\cdot -} \longrightarrow \text{pyrazolyl} + \text{NO}_2^-$$

(3-50)

Of the six nitropyrazoles investigated 1,4-dinitropyrazole (**3-199g**) was the most readily reduced and gave the most complicated polarogram with five waves. The first wave corresponded to an irreversible transfer of one electron and presumably led to cleavage of the molecule to the nitrite anion and a 4-nitropyrazolyl radical ($E_{\frac{1}{2}} = -1.7$ V); however, the character of the hfs of the EPR signal for this compound precluded its assignment to a neutral symmetrical 4-nitropyrazolyl radical. Since azolyl radicals have shown a high tendency to dimerize[229] and the EPR signal showed principal hfs constants similar to those for 1-alkylpyrazoles (**3-5**) and (**3-217b**), radical dimerization followed by reduction was proposed, Eq. (3-51).[228]

An advantage from a strong negative shift for $E_{\frac{1}{2}}$ for the anion of an N-unsubstituted pyrazole relative to an N-substituted pyrazole afforded a scheme for polarographic determination of one species in the presence of the other.[96,230] Polarographic reduction of 4-nitroantipyrine (see **3-214**) gave a single six-electron reduction wave independent of pH values.[231]

Reductions of 4-nitro-1,2-diethylpyrazolidine,[134] the amide of 3(5)-methyl-4-nitropyrazole-5(3)-carboxylic acid,[232] and 1,3-dimethyl-4-nitro-5-benzoyl-pyrazole[233] are reported without any indication of the method employed in the patent abstracts.

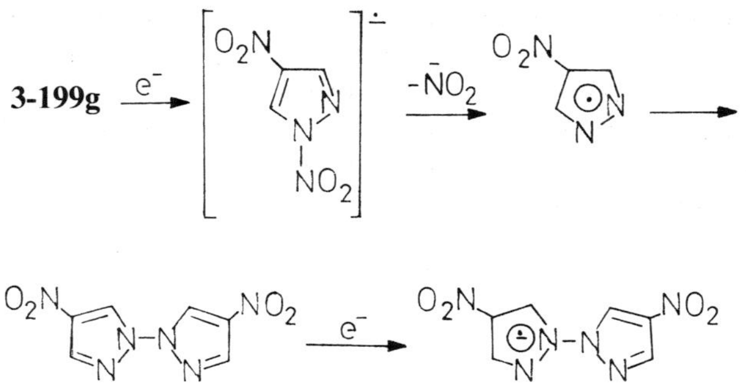

$$\text{3-199g} \xrightarrow{e^-} \left[\begin{array}{c} O_2N \\ \diagup \diagdown \\ N-N \\ | \\ NO_2 \end{array} \right]^{\cdot -} \xrightarrow{-\bar{N}O_2} \begin{array}{c} O_2N \\ \diagup \diagdown \\ \stackrel{\oplus}{N}-N \end{array} \longrightarrow$$

$$\begin{array}{c} O_2N \\ \diagup \diagdown \\ N-N \end{array} - \begin{array}{c} NO_2 \\ \diagup \diagdown \\ N-N \end{array} \xrightarrow{e^-} \begin{array}{c} O_2N \\ \diagup \diagdown \\ \stackrel{\oplus}{N}-N \end{array} - \begin{array}{c} NO_2 \\ \diagup \diagdown \\ N-N \end{array} \qquad (3\text{-}51)$$

The nitro group remained unchanged during the conversion of 1-methyl-3-nitro-4-chlorocarbonylpyrazole to 1-methyl-3-nitro-4-formylpyrazole by treatment with LiAlH(OC(CH$_3$)$_3$)$_3$ then with lead tetraacetate.[105] The nitro group was also unaffected during a Fischer reduction with sodium sulfite of a 4-nitro-5(3)-methylpyrazole-3-diazonium salt to 3(5)-hydrazino-4-nitro-5(3)-methylpyrazole.[234]

Conversions of Aminonitropyrazoles

Diazotized 3(5)-amino-4-nitropyrazole readily underwent nucleophilic substitution with sodium nitrite and sodium azide to give 3(5),4-dinitropyrazole (36%) and 3(5)-azido-4-nitropyrazole (71%). In methanol, loss of the diazonium group left 4-nitropyrazole (94%). Attempts to replace a diazo group in either of the two diazotized 3,5-diamino-4-nitropyrazole isomers with nitro and azido groups in similar reactions were unsuccessful. Tetrazotization of 3,5-diamino-4-nitropyrazole brought about removal of the nitro group and the formation of a bis-diazopyrazolone (**3-217c**) (53%),[235] an interesting compound with no hydrogen content; explosive properties precluded elemental analysis. Pyrazolopyrimidinediols were obtained from 3-aminopyrazoles and alkylmalonate esters; eg, 3(5)-amino-4-nitropyrazole afforded the nitro compound (**3-217d**).[87,83]

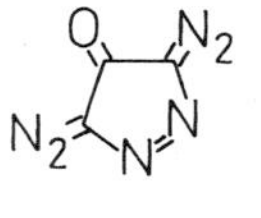

3-217c 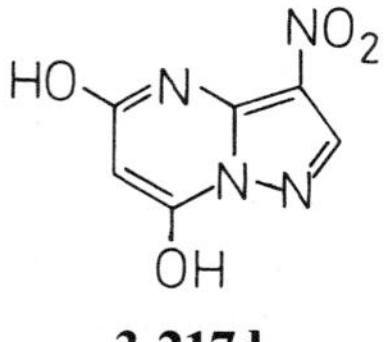   **3-217d**

Oxidation

Potassium permanganate at 100° for 8 h oxidized 3-methyl-4-nitropyrazole-5-carboxylic acid to 4-nitropyrazole-3,5-dicarboxylic acid (**3-218a**) isolated as its ethyl ester (42%).[131] A similar oxidation of 3(5)-methyl-4-nitropyrazole gave 4-nitropyrazole-3(5)-carboxylic acid (**3-218b**) (36%); 3-nitropyrazolo[1,5-*a*]-pyridine (**3-158a**) also gave the acid (**3-218b**) (49%).[114] Ozone in sulfuric acid containing a "salt of a metal of variable valence" was reported to produce pyrazole-3(5)-carboxylic acid and its 4-nitro- and 4-sulfonic acid derivatives by oxidation of the corresponding 3(5)-methylpyrazoles.[236]

3-218a X = CO_2H
3-218b X = H

Metal Salts of Nitropyrazoles

Two types of compounds $M(L)_2^+X^-$ and M^+L^- (M for metal, L for ligand) have been obtained from pyrazoles and Cu(I) or Ag(I) salts. Compounds $M(L)_2^+X^-$, in which M = Cu or Ag, L = a pyrazole (eg, 4-nitropyrazole), and X = BF_4, NO_3, Cl, or Br, are produced from MX and L in acidic solution. Analysis of the infrared spectra revealed the presence of linear cations ML_2^+ and uncoordinated anions. In less acidic media compounds M^+L^- were obtained. The silver salts were easily obtained from a solution of silver nitrate in ammonium hydroxide on addition to an alcoholic solution of the pyrazole. A recommended preparation of a Cu^+L^- consisted in the reduction of $Cu(L)_4X_2$ by copper under nitrogen. As shown by infrared and mass spectra and X-ray powder diagrams, these compounds were at least trimeric and possibly polymeric with pyrazolate anions alternating with metal ions.[237] Stoichiometric amounts of a pyrazole and a hydrated Cu(II) salt in ethanol gave compounds $Cu(L)_x^{++}X_2^-$ where x varied from 2 to 4.[238]

Pyrazolato salts were obtained from 4-nitropyrazole and a metal chloride, MCl_2, in aqueous ammonia (25%). For the cobalt salt infrared and ligand field spectra agreed with the formula $Co(L)_2(NH_3)_2 \cdot xH_2O$ where L = 4-nitropyrazole. In a similar manner salts of Cd, Zn, Mn, Ni, and Cu were prepared. The pyrazolato anion coordinated as a bidentate ligand (C_{2v} symmetry).[239]

Gold(I) pyrazolates were obtained from eight pyrazoles (azH), including 4-nitropyrazole, according to Eq. (3-52) and/or (3-53) and/or (3-54). Osmometric

$$(CH_3)_2SAuCl + azH + KOH \rightarrow$$
$$1/n[Au(az)]_n + H_2O + KCl + (CH_3)_2S \qquad (3\text{-}52)$$

$$(C_6H_5)_3PAuCl + azH + KOH \rightarrow$$
$$(C_6H_5)_3PAu(az) + KCl + H_2O \qquad (3\text{-}53)$$

$$(C_6H_5)_3PAu(az) \rightleftharpoons (C_6H_5)_3P + (1/n)[(az)Au]_n \qquad (3-54)$$

and mass spectrometric determinations showed that in many examples the value of n in Eq. (3-52) and (3-54) was 3; therefore a cyclic structure such as (3-219) was suggested. Similar formulations for Ag and Cu(I) salts were considered; however, the trimeric structures were not clearly differentiated from polymeric forms.[240]

3-219	X	Y	Z
a	H	H	H
b	CH_3	H	CH_3
c	CH_3	H	C_6H_5
d	CH_3	C_2H_5	CH_3
e	CH_3	I	CH_3
f	CH_3	NO_2	CH_3
g	CH_3	$N_2C_6H_4CH_3$	CH_3
h	CH_3	-$CH_2CH_2CH_2$-	

Pyrazolato complexes of platinum included (dppe)Pt(L)$_2$ where dppe was (($C_6H_5)_2PCH_2)_2$ and L was 3,5-dimethyl-4-nitropyrazole.[241] A photochemical reaction between Fe(CO)$_5$ and a 1-allylpyrazole (L) gave the complex Fe(CO)$_3$L and included the example where L was 1-allyl-3,5-dimethyl-4-nitropyrazole.[242]

Acids and Bases

Ultraviolet spectroscopy afforded the determination of pK_a values for eight derivatives of 4-nitropyrazole (3-2): (3-2), 9.67; 3(5)-methyl-(3-7), 10.06; 3,5-dimethyl- (3-8), 10.65; 3(5)-tert-butyl- (3-11), 10.27; 3(5)-phenyl- (3-40), 9.11; and (3-220)–(3-222). Similarly pK_a values for the conjugate bases of eight derivatives of 3(5)-nitropyrazole (3-101c) were determined: (3-101c), 9.81; and (3-223)–(3-229), and of five dinitropyrazoles (3-230)–(3-234) (Table 3-1). Nearly identical values for (3-2) and (3-107c), for (3-7) and (3-223), and for (3-40) and (3-225) showed that a 3(5)- and a 4-nitro substituent brought about closely comparable effects in acid-base equilibria. A larger difference between the pK_a values for 3,5-dinitro- (3-233) (3.14) and 3(5),4-dinitropyrazole (3-230) (5.48) was attributed to a steric inhibition of resonance in the anion of the latter isomer.[243]

Analysis of the absorption data for 4-nitropyrazole (3-2) at low acidities afforded the suggested pK_a of about −2 for the neutral molecule.[12] pK_a values for 3,5-dimethylpyrazole (4.12) and its 4-nitro derivative (3-8) (−0.95) revealed a base weakening effect of about five pK_a units brought about by the 4-nitro substituent.[18,244]

From the relationship p$K_a = mH_0^{1/2}$, m values in the range 0.37 to 0.73 for 1-arylpyrazolones and 0.64 to 1.22 for 1-arylpyrazoles were determined. The pyrazolones protonated at oxygen and followed the H_A acidity function defined for amides.[75]

C-alkylnitropyrazoles were slightly less acidic than 4- and 3(5)-nitropyrazoles (**3-2**) and (**3-101c**). Thus ΔpK_a values from +0.28 to +0.60 were introduced with a C-alkyl substituent (methyl, ethyl, or *tert*-butyl) either adjacent or nonadjacent to the nitro group.[10,243,245] With the introduction of a second C-alkyl substituent the acidity was again decreased as shown by ΔpK_a values from +0.98 to +1.62. A similar decrease in acidity of 3(5)-methyl-4,5(3)-dinitropyrazole (**3-231**) pK_a 6.35 relative to 3(5),4-dinitropyrazole (**3-230**) pK_a 5.48 (ΔpKa +0.87) was noted. The larger ΔpK_a values of +1.62 for 4-nitro-3,5-di-*tert*-butylpyrazole (**3-221**) and +0.87 for compound (**3-231**) were attributed to a buttressing effect derived from three vicinal substituents.[243] More observations are needed.

A steric hindrance was also invoked to account for ΔpK_a values of −0.70, −0.39, and −0.56 for 3(5)-nitro-4-phenylpyrazole (**3-225**), 3(5),4-dinitro-5(3)-phenylpyrazole (**3-232**), and 3(5)-phenyl-4-nitropyrazole (**3-40**) and their relation to ΔpK_a of −1.06 for 3(5)-nitro-5(3)-phenylpyrazole (**3-228**), a compound with no steric effects.[243] Again more observations are needed.

Table 3-1. Acid Strengths of Nitropyrazoles

Nitropyrazole		pK_a
O_2N—pyrazole—X, Y, N-H	**3-220** X = CH$_3$, Y = H	10.92
	3-221 X = Y = (CH$_3$)$_3$C	11.29
	3-222 X = 4-O$_2$NC$_6$H$_4$, Y = H	8.46
X—pyrazole—NO$_2$, Y, N-H	**3-223** X = CH$_3$, Y = H	10.10
	3-224 X = C$_2$H$_5$, Y = H	10.09
	3-225 X = C$_6$H$_5$, Y = H	9.11
	3-226 X = H, Y = CH$_3$	10.25
	3-227 X = H, Y = C(CH$_3$)$_3$	10.35
	3-228 X = H, Y = C$_6$H$_5$	8.75
	3-229 X = H, Y = 4-O$_2$NC$_6$H$_4$	7.59
	3-230 X = NO$_2$, Y = H	5.48
	3-231 X = NO$_2$, Y = CH$_3$	6.35
	3-232 X = NO$_2$, Y = C$_6$H$_5$	5.09
	3-233 X = H, Y = NO$_2$	3.14
	3-234 X = C$_2$H$_5$, Y = NO$_2$	3.80

An unexplained color reaction between sodium hydroxide (5%) in acetone and various di- and trinitro derivatives of 1-phenylpyrazole was noted.[32] In other studies, pK_{BH}^+ values (Table 3-2) for pyrazole, 1-alkylpyrazoles, and ten 1-substituted nitropyrazoles were determined.[244,245] Imidazole values are shown for comparison.[246]

Substituent effects, b, expressed as ratios between pK_{BH}^+ values for the

Table 3-2. pK_{BH^+} Values for Selected Pyrazoles and Imidazoles

Pyrazole	pK_{BH^+}	Pyrazole	pK_{BH^+}
—	2.52	5-NO$_2$, 1-CH$_3$	−2.38
1-CH$_3$	2.09	5-NO$_2$, 1-C$_2$H$_5$	−2.35
1-C$_2$H$_5$	2.06	1-NO$_2$	−4.21
4-NO$_2$	−1.96	3(5)-NO$_2$	−4.66
4-NO$_2$, 1-CH$_3$	−2.21	3-NO$_2$, 1-CH$_3$	−4.64
4-NO$_2$, 1-C$_2$H$_5$	−2.16	3-NO$_2$, 1-C$_2$H$_5$	−4.78
Imidazole	pK_{BH^+}	Imidazole	pK_{BH^+}
—	6.95	1-CH$_3$, 4-NO$_2$	−0.58
1-CH$_3$	7.25	2-NO$_2$	−0.81
1-CH$_3$, 5-NO$_2$	2.12	1-CH$_3$, 2-NO$_2$	−0.48
4(5)-NO$_2$	−0.16		

substituted and the unsubstituted heterocycles, $b = \log(K/K_o)_{BH^+}$, revealed a relationship between the heterocycles, $b_p = 0.85 b_i - 0.7$, where p stands for pyrazole and i for imidazole. It was noted that the effect of the nitro group depended on its position. Equivalent effects for pyrazole positions 1 and 3 were larger than the equivalent effects for positions 4 and 5. The substitution effect constants supported predominance of 3-nitro-1-unsubstituted pyrazole over its 5-nitro tautomer and 4-nitro-1-unsubstituted imidazole over its 5-nitro tautomer.[244]

An amine catalyzed elimination of nitrous acid from 1,4-dinitro-3,5-dimethylpyrazole presumably brought about the formation of an intermediate diazafulvene which then combined with alcohol (solvent) to produce 3(5)-methyl-4-nitro-5(3)-hydroxymethylpyrazole (**3-235a**) (44% to 81%) and the ether (**3-235b**) (15% to 40%). The reaction demonstrated the lability of hydrogen in the 5-methyl group, Eq. (3-55).[247]

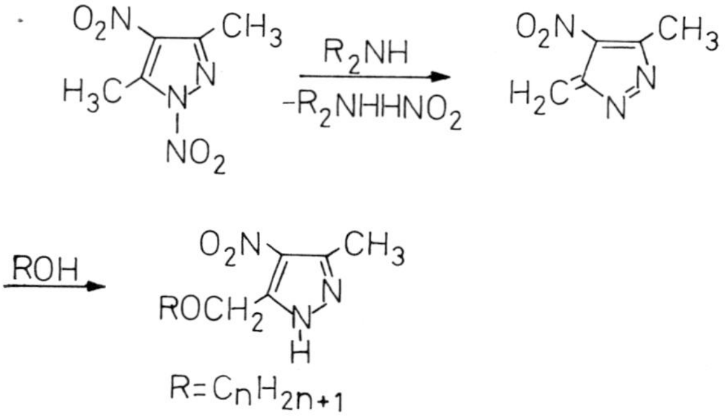

$R = C_nH_{2n+1}$

3-235a $n = 0$
3-235b $n = 1-3$

(3-55)

In an investigation of isotopic exchange of the hydrogen atom of methyl substituents in 38 five-membered aromatic heterocycles it was revealed that exchange occurred readily at 25° in the 5-methyl group and occurred on heating to 80°–110° in the 3-methyl group of 1,3,5-trimethyl-4-nitropyrazole (**3-236**).[248] A quantum chemical study of the CH acidities of methyl derivatives of five-membered heterocycles also investigated compound (**3-236**).[249]

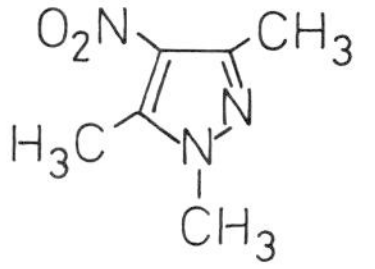

3-236 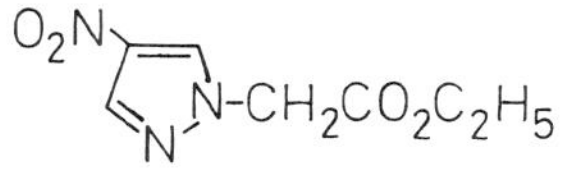   **3-237**

Thermolysis

Ethyl 4,4-dinitropyrazolidine-1-acetate (**3-115b**) at 25° gradually lost the elements of nitrous acid and hydrogen to form ethyl 4-nitropyrazole-1-acetate (**3-237**). It was suggested that nitrous acid was evolved first and facilitated the dehydrogenation of an intermediate pyrazoline.[135] This result is reminiscent of the conversion of a nitropyrazoline (**3-109a**) to a nitropyrazole (**3-109b**) by treatment with nitric acid. Thermal evolution of nitrous acid from a nitropyrazoline (**3-105**) derived from a nitro olefin and diazomethane also gave a pyrazole, but the 3-ethyl-4-nitropyrazoline derived from diphenyldiazomethane and 1-nitrobutene thermolyzed to 1-nitro-2-ethyl-3,3-diphenylpropane and nitrogen, Eq. (3-56). Nitrocyclopropanes were similarly obtained when the 3-ethyl group was replaced by either methyl or hydrogen, but replacement by phenyl led to a complicated thermolysis in which nitrogen and nitrogen dioxide were formed.[114] A nitropyrazoline intermediate was postulated for the reaction between phenyldiazomethane and 6-nitro-2,1-benzisoxazole which led to the formation of a nitrocyclopropane derivative Eq. (3-57).[114-119]

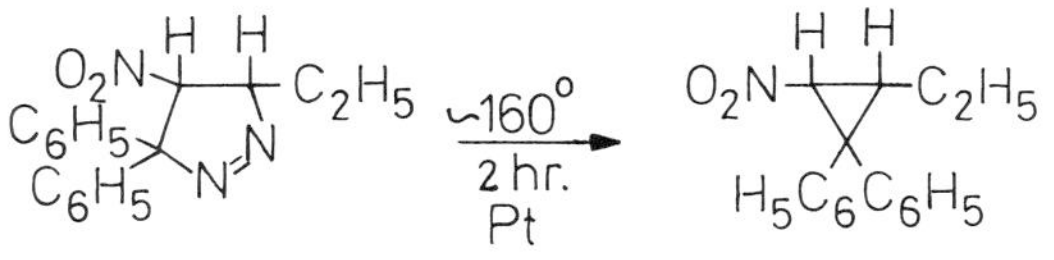

3-238   (3-56)

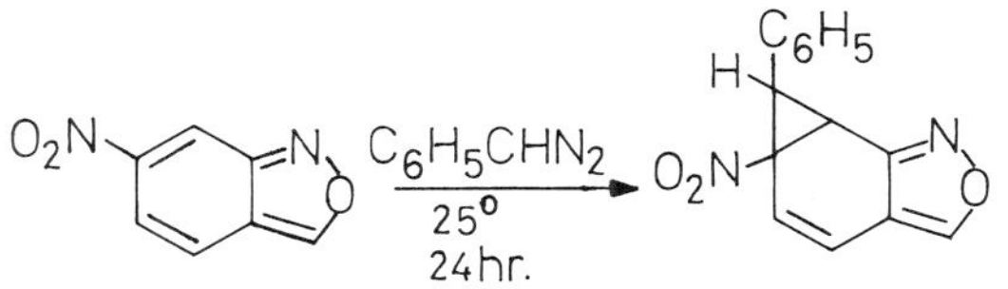

(3-57)

A stability of the potassium salts of 2-nitropyrrole, 4-nitropyrazole, and 4(5)-nitroimidazole in the solid state was shown in a study of the kinetics of thermal degradation to diminish as the acidity of the nitroheterocycle increased (corresponding pK_a values of 10.6, 9.64, and 9.35).[250]

Photolysis

Irradiation with ultraviolet light in the presence of cyanide ions converted 1-methyl-3-nitropyrazole to the 4-cyano derivative (**3-239a**) (16%), 1-methyl-3-nitrosopyrazole (**3-239b**) (14%), and 1-methyl-3-nitroso-4-cyanopyrazole (**3-239c**). This was the first example of photocyanation of an aromatic nitro compound in which nucleophilic substitution and photodeoxygenation occurred simultaneously. Since the former has been attributed to a π,π^* triplet state and the latter to an n,π^* triplet state, the question of competitive reaction by two excited states or competitive reaction from one state with mixed character was left for further investigation. Similar experiments showed the 4- and 5-nitro derivatives of 1-methylpyrazole to be photounstable in the presence of cyanide ions, but photosubstitution products were not found.[251]

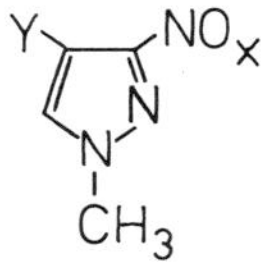

3-239a x = 2, Y = CN
3-239b x = 1, Y = H
3-239c x = 1, Y = CN

Photolysis of nitropyrazolines (**3-240**) gave the cyclopropanes (**3-241**), the pyrazoles (**3-242**), the nitrosamines (**3-243**), and the nitramine (**3-244**).[252]

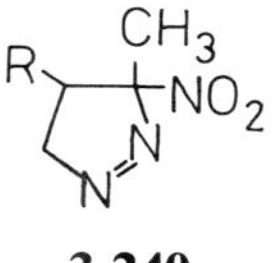

3-240

3-241

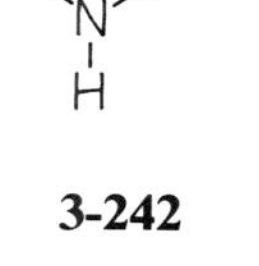

3-242

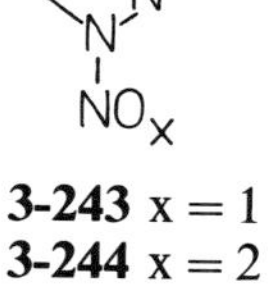

3-243 x = 1
3-244 x = 2

Picrolonic Acid

As known since 1892, 1-(4-nitrophenyl)-3-methyl-4-nitro-5-hydroxypyrazole (picrolonic acid) (**3-83**) ⇌ (**3-245a**) ⇌ (**3-245b**) has received substantial attention for its applications in analytical chemistry.[253–256] Investigations on complex formation between picrolonic acid and alkaline earth and transition metal ions have been carried out with conductometric methods. Picrolonates of amines (particularly alkaloids) are solid derivatives useful for identification and assay purposes.[257,258] An X-ray diffraction study of the complex thiamine picrolonate showed two different types of stacking interactions between the picrolonate anion and the neutral pyrimidine ring of thiamine and an absence of planar overlap with the positively charged thiazolium ring.[259]

Metal picrolonates were characterized through analytical, ir, and magnetic susceptibility data.[260] Determinations of calcium,[261,262] magnesium, and potassium were based on picrolonate formation.[262] The extraction of uranium from inorganic acid solutions[263] and the separation of americium from other transplutonium elements has been achieved in schemes utilizing picrolonic acid.[264] The acid has also played a role as a metal-complexing agent in the corrosion and protection of metals[265,266] and in the design of a liquid ion-selective electrode.[267]

The electronic spectra of picrolonic acid (2×10^{-5} M solution in 5% ethanol, 25°) at different pH values (2.08 to 10.8) showed only one absorption peak at 338 nm with an absorbance that was also pH-independent. In addition, the electronic spectra of the sodium, potassium, and rubidium salts gave absorption values identical with those obtained for the acid itself. That picrolonic acid, pK_a −2.20, behaved as a strong acid was further supported by nearly identical titration curves for 10^{-3} M solutions (5% ethanol) of picrolonic and hydrochloric acids and by a protonation degree that deviated only slightly from zero over a range pH 2.0–8.5. An assumption that the proton was released from the keto tautomer (**3-245b**) rather than from either the OH (**3-245a**) or NH tautomer (**3-83**) was correlated with an absence of absorption at v(OH) [and presumably v(NH)] frequencies and the presence of a v(CO) frequency at 1,600 cm^{-1} in the ir spectra (KBr discs) for the acid and its metal salts. The conclusion may be erroneous, since pK_a data supported the coexistence of comparable amounts of NH and OH tautomers.[75]

Picrolonic acid has been investigated for its control in the hatching rate of free eggs of larvae,[268–270] as a mutagen for modifying DNA in vivo,[271] and as a hardening agent for epoxy resins.[272]

3-245a

3-245b

Miscellaneous

Electron-withdrawing nitro and azo substituents in the pyrazoline derivative (**3-246a**) presumably activated the ester carbonyl in the addition-elimination reaction with diazoethane to produce the oxiran (**3-246b**). This was reported as the first example of the formation of an oxiran by the action of a diazoalkane upon an ester.[116]

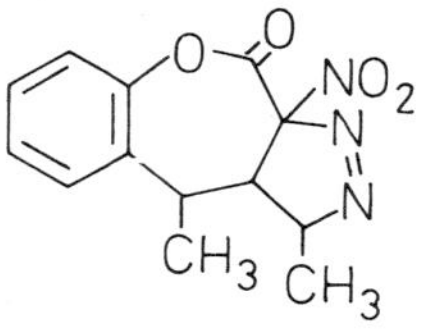

3-246a 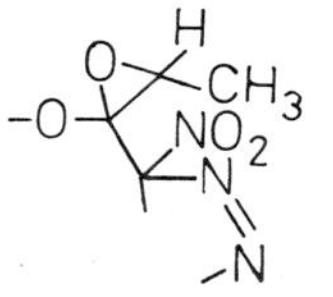   **3-246b**

Thionyl chloride converted 3-methyl-4-nitropyrazole-5-carboxylic acid to an anhydride.[226] An intermolecular anhydride (**3-246c**) rather than an intramolecular anhydride (**3-246d,e**) was assumed.

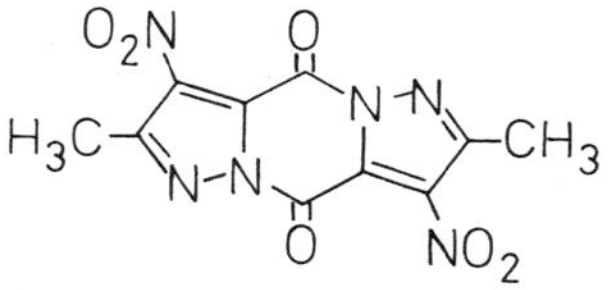

3-246c 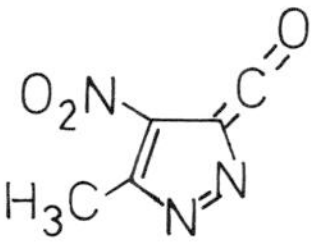   **3-246d**

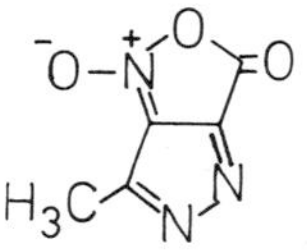

3-246e

PROPERTIES

Nuclear Magnetic Resonance Spectroscopy

Isochronous 3- and 5-positions in 4-nitropyrazole (**3-2**) were revealed in the proton magnetic resonance (pmr) signal at δ 8.54 (deuterated dimethyl sulfoxide, DMSO-d_6) or at δ 8.44 (acetone-d_6) and in the ^{13}C nuclear magnetic resonance (NMR) signal at δ 132.5 (acetone-d_6) with coupling constants ^1J 198.7 and ^3J 3.0.

A slow prototropic exchange at 30° was confirmed when the signal at δ 8.54 split into two signals at δ 8.25 and 8.84 upon dilution of the sample; a $\Delta G^{\ddagger}$ of 10 kcal/mol for the exchange was determined from the CH signal for the pyrazole (**3-2**) in acetone ($\Delta v = 33$ Hz, $T_c = -70°$). For the unsubstituted pyrazole an intermolecular exchange was assigned when E_a values of 4.8 ($-70°$) and 1.65 ($-118°$) kcal/mol were obtained. Spin-spin coupling constants $J_{13} = 1.2$ and $J_{15} = 1.4$ for the nitropyrazole (**3-2**) in acetone-d_6 at $-103°$ were reported as a rare example of coupling with an NH proton of an aromatic heterocycle and reflected the absence of tautomerization.[273]

Both pmr and ^{13}C NMR spectroscopy showed an increase in the population of an adduct between a nitropyrazole and acetone as the temperature decreased. The addition was inhibited by 3,5-disubstitution. In acetone-d_6 at $-90°$, an adduct (**3-247a**) with 3-nitropyrazole (**3-97a**) was detected by an OH signal at δ 7.30; at $-103°$ an adduct (**3-247b**) with 4-nitropyrazole gave a $J_{35} = 0.7$. For each isomer small, downfiled shifts in CH signals resulted from the reversible addition of acetone.[273]

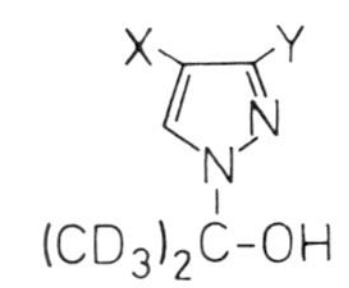

3-247a X = H, Y = NO$_2$
3-247b X = NO$_2$, Y = H

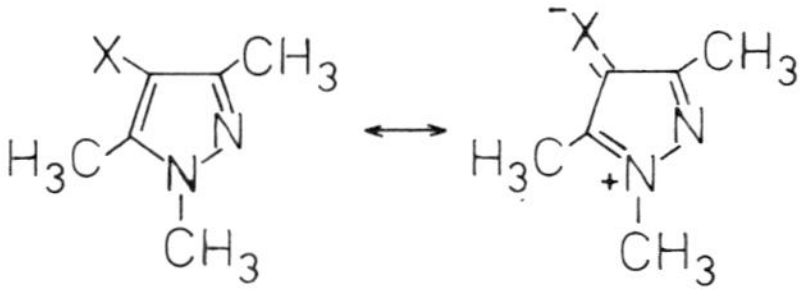

3-248a
X = NO$_2$

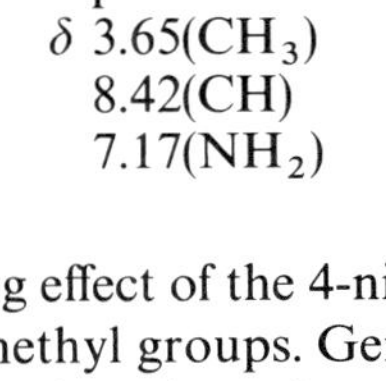

3-248b
mp 200°
δ 3.65(CH$_3$)
8.42(CH)
7.17(NH$_2$)

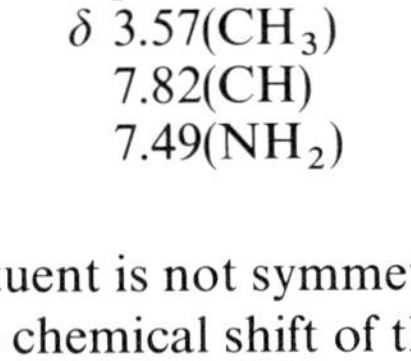

3-90a
mp 265–266°
δ 3.57(CH$_3$)
7.82(CH)
7.49(NH$_2$)

The deshielding effect of the 4-nitro substituent is not symmetrically extended to protons and methyl groups. Generally the chemical shift of the substituent at the 5-position was found at a lower magnetic field as expected by attributing to the deshielding effect a resonance interaction in which a positive charge resided mainly at N-1 (**3-248b**) (X = NO$_2$).[8] The chemical shift differences between C5 and C3 methyl protons was determined for a series of 4-substituted derivatives of 1,3,5-trimethylpyrazole (**3-248a**) (Table 3-3).[274]

The introduction of a second *peri* nitro group brought about minor reversals in the deshielding effect for 3,4-dinitropyrazolo[1,5-*a*] pyridine (**3-158b**). This

Table 3-3. Chemical Shift Differences in 4-Substituted Pyrazoles (3-248a)

X	NH$_2$	CH$_3$	H	Br	NO$_2$	NO
δ(3-CH$_3$)^a	2.010	2.017	2.083	2.100	2.408	2.252
δ(5-CH$_3$)^a	2.073	2.067	2.163	2.193	2.573	2.783
Δδ	0.063	0.050	0.080	0.093	0.165	0.531

^{a}CCl$_4$ solution.

was attributed to anisotropic changes which resulted from mutual rotations of the nitro groups out of the plane of the aromatic heterocycle.[171]

The signal of the C-5 proton in a 1,3-disubstituted 4-nitro pyrazole is more downfield and more sensitive to solvent than the signal of the C-3 proton in a 1,5-disubstituted 4-nitropyrazole.[195,275] From the established differences in their chemical shifts, ΔδH(C5) > ΔδH(C3), determined from chloroform-d and DMSO-d$_6$, assignments for the predominant product 1-methyl-3-nitropyrazole (**3-6a**) (ΔδH = 0.45) and an isomer 1-methyl-5-nitropyrazole (**3-93**) (ΔδH = 0.22) obtained in lesser amount were made. Corresponding ΔδH(C4) values of 0.13 and 0.22 ppm were observed; the δNCH$_3$ signal was independent of product.[14,41,209]

The assignment 1-methyl-3-nitro-4-phenylpyrazole (**3-56b**) was supported by an intramolecular nuclear Overhauser effect. An increase (13% ± 2%) in the ratio of the pmr signal of the heterocyclic proton compared to the signal for phenyl protons when the methyl protons were saturated indicated a nitro substituent at C3.[38]

To accommodate established correlations between charge density and chemical shift, structure assignments were determined as shown for the 3- and 5-amino derivatives (**3-248b**) and (**3-90a**)[89] of 1-methyl-4-nitropyrazole. The isomers were obtained from 4-methoxy-5-nitropyrimidine and hydrazine but were given the reverse assignment, see Eq. (3-25).[151] The revised assignments were in good agreement with the proton chemical shifts for the closely related 3- and 5-amino derivatives (**3-249a**) and (**3-250a**) of 1-methyl-4-cyanopyrazole. Compound (**3-90a**) was also obtained from 1-methyl-5-aminopyrazole by nitration. Compounds derived from (**3-248b**) and (**3-90a**) were also reassigned: 1-methyl-3,4-diaminopyrazole (not 1-methyl-4,5-

3-249a
δ 3.60 (CH$_3$)
8.00 (CH)
5.53 (NH$_2$)

3-250a
δ 3.57 (CH$_3$)
7.57 (CH)
6.57 (NH$_2$)

diaminopyrazole), 2-methylpyrazolo[3,4-*b*]pyrazine (not the 1-methyl isomer), 2,5,6-trimethylpyrazolo-[3,4-*b*]pyrazine (not 1,5,6-trimethyl), 11-methyldibenzo[f,h]pyrazolo[3,4-*b*]quinoxaline (not 10-methyl), 3-acetamido- and 3-diacetylamino-1-methyl-4-nitropyrazole (not the 5-amido isomers), and 1-methyl-3-acetamido-4-aminopyrazole (not the 5-acetamido isomer).[89]

A reversible intramolecular migration of the phenylmercury substituent from position N-1 to position N-2 with a probable transition state (**3-249c**) in which the mercury atom was bound to both pyrazole nitrogen atoms was proposed for the 4-chloro, 4-bromo, and 4-nitro (**3-249b**) derivatives of 1-phenylmercury-3,5-dimethylpyrazole. The cationotropy for the 4-nitro compound was typical for the group and permitted each methyl group to be detected by pmr (3:1 mixture of methylene chloride and chloroform) at $-100°$ at 154.9 (3-CH$_3$) and 139.0 Hz (5-CH$_3$) from the TMS signal with coalescence to 142.1 Hz at $-90°$.[276]

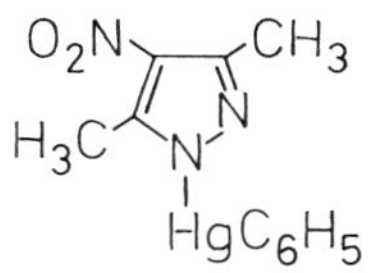

3-249b 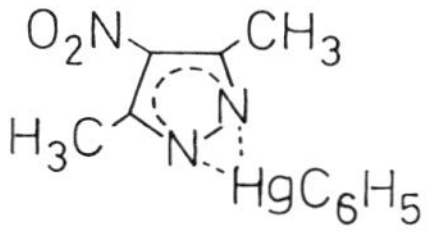   **3-249c**

Chemical shifts for the pyrazole ring carbon atoms C-3, C-4, and C-5 and the 1-phenyl carbon atoms C$_{ipso}$ and C$_{ortho}$ provided a determination of the preferred tautomers for the 4-nitro derivative (**3-250b**) of 1-phenyl-3-methyl-5-hydroxypyrazole and for 25 other 4-substituted derivatives. In polar solvents — eg, dimethylsulfoxide — the majority species was the hydroxy tautomer (**3-250b**) and the minority species was the C-H tautomer (**3-250c**). In nonpolar solvents — eg, deuteriochloroform — the CH tautomer was dominant, although a small amount of the NH tautomer (**3-250d**) was also present. The concentration of the NH tautomer (**3-250c**) was increased by cooling or by the addition of methanol. A controlling factor was attributed to the electronegativity of the substituents.[277]

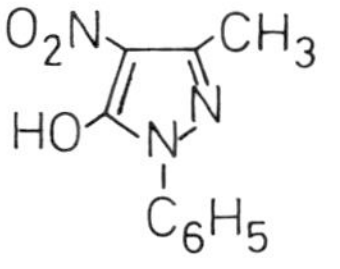

3-250b 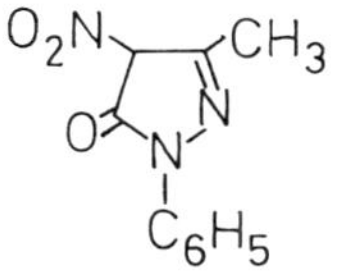   **3-250c** 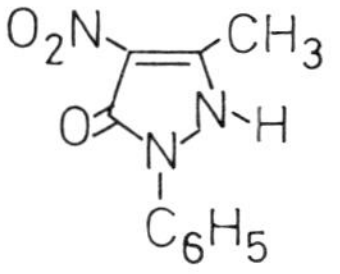   **3-250d**

Infrared Spectroscopy

The characteristic asymmetric (v_{as}) and symmetric (v_s) stretching vibration modes for the nitro group were also found for nitropyrazoles and nitropyrazolones: v_{as} 1540–1490 cm^{-1} and v_s 1355–1315 cm^{-1}.[1,75,77,126,191,213,238]

Similar absorptions were found at 1575–1535 cm^{-1} and at 1370–1360 cm^{-1} for 3-nitro-Δ^2-pyrazolines.[125] The far infrared spectra (450–200 cm^{-1}) of nitropyrazoles as ligands[237,238] and salts[239] have been recorded.

Ultraviolet Spectroscopy

Absorption in aqueous sodium hydroxide, in ethanol, and in concentrated sulfuric acid gave respectively λ_{max} 320 (log ε4.07), 275 (log ε3.91), and 238 (log ε3.89) for 4-nitropyrazole (3-2). The unsubstituted pyrazole gave λ_{max} 210 (log 3.61).[12,244] An equilibrium mixture of the conjugate base and the neutral molecule (3-2) in buffered solutions at about pH 9 showed an isosbestic point (290 nm) and a pK_a of 9.64.[12] 4-Nitropyrazole (3-2) and seven derivatives (3-7), (3-8), (3-11), (3-40), (3-220), (3-221), (3-222), 3-nitropyrazole (3-101c) and seven derivatives (3-223) through (3-229), and five dinitropyrazoles (3-230) through (3-234) in 0.05 N hydrochloric acid absorbed at λ_{max} 260–306 and in 0.05 N sodium hydroxide at λ_{max} 305–346.[10,243] Similar results were obtained for a group of 4-nitro derivatives of pyrazolecarboxylic acid derivatives.[68,221]

A polar resonance structure — eg, (3-248a) — accounted for a large solvent effect observed for 1-methyl-4-nitropyrazole (3-5) in water (λ_{max} 280) when compared with the compound in *n*-heptane (λ_{max} 264).[8] Methyl groups adjacent to a nitro substituent apparently did not introduce a steric hindrance to conjugation,[8,18] but a small hypsochromic shift for 1,3-dimethyl-4-nitro-5-bromopyrazole (3-65) in alcohol (λ_{max} 275) was attributed to a steric hindrance.[24,66] Additional examples are needed to confirm this claim.

Substitution of a second nitro group brought no appreciable change to the absorption for 1,5-dimethyl-3,4-dinitropyrazole (λ_{max} 270)[24] or for the dinitropyrazoles (3-230) through (3-234) (λ_{max} 265–280).[243] In methanol 3,5-diphenyl-4-nitropyrazole (3-108a) provided absorption at λ_{max} 266 (log 4.66[119], 3.68[120]) and 274 (log 4.14[119], 3.14[120]).

In alcohol solution 3-amino (3-126), 1-methyl-3-amino, 1-methyl-5-amino, and five 3(5)-*sec*-amino-5(3)-*tert*-amino (3-121a–e) derivatives of 4-nitropyrazole showed absorption at λ_{max}, 328–349,[147,151] characteristic of the ionic species, eg, (3-248a), expected from an electronic interaction between amino and nitro groups attached to an aromatic ring. Similar absorption was found for an alcoholic solution of the 5-amino derivative (3-251a) at λ_{max} 335 and for an alkaline alcoholic solution of the 5-hydroxy derivative (3-251b) at λ_{max} 356, of 1,3-dimethyl-4-nitropyrazole.[66] In ethanol and alkaline ethanol the hydroxy

3-251a X = NH$_2$
3-251b X = OH

compound (**3-251b**) gave absorption at λ_{max} 285 characteristic of a nitropyrazole.[66]

Absorption values were obtained for the neutral molecules: pyrazole at pH 6.1 λ_{max} 210, 1-nitropyrazole in 1 M sulfuric acid λ_{max} 267, and 4-nitropyrazole in 1 M sulfuric acid λ_{max} 274.[244] That the absorption attributed to 1-nitropyrazole should be reassigned to 4-nitropyrazole is suggested by the authentic values reported for the 4-nitro isomer and supported by the known rearrangement of 1- to 4-nitropyrazole in ice-cold concentrated sulfuric acid. Absorption assigned to the cation of 1-nitropyrazole in 16.0 M sulfuric acid λ_{max} 216[244] may be incorrectly assigned, since it was indistinguishable from the authentic absorption λ_{max} 215[214] obtained for the cation of pyrazole in 12.0 M sulfuric acid. Since the absorption λ_{max} 234[244] for the cation of 4-nitropyrazole in 17.1 M sulfuric acid required an appreciably higher wavelength, it is suggested that 1-nitropyrazole in 16.0 M sulfuric acid was converted to the cation of pyrazole but not to 4-nitropyrazole.

Comparable absorptions were reported for the 1-methyl (**3-252a**) (water, λ_{max} 290)[10] and the 1-vinyl derivative (**3-252b**) (methanol, λ_{max} 294)[194] of 3,5-dimethyl-4-nitroimidazole.

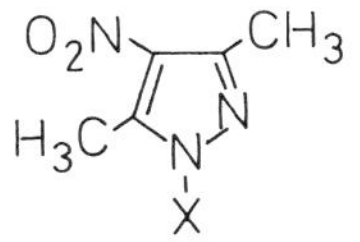

3-252a X = CH$_3$
3-252b X = CH = CH$_2$

Structure assignments of certain 3- or 5-nitropyrazoles were partially determined by uv absorption: 3(5)-nitro-5(3)-(3′-pyridyl)pyrazole (**3-135**) in 0.1 N hydrochloric acid, λ_{max} 222, 253, and 275[157]; 1-methyl-5-nitropyrazole-4-carboxylic acid at pH 1, λ_{max} 269 and at pH 11 λ_{max} 278[101]; and 1-methyl-3-nitro-4-cyanopyrazole (**3-239a**) in water λ_{max} 268.[251]

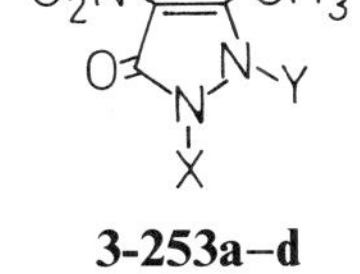

3-253a–d

3-253	X	Y	Neutral species λ_{max}nm	log ε	Cation species λ_{max}nm	log ε
a	C$_6$H$_5$	H	245	4.19	260	4.03
b	C$_6$H$_5$	CH$_3$	330	3.91	278	3.94
c	p-O$_2$NC$_6$H$_4$	H	320	4.04	315	3.96
d	p-O$_2$NC$_6$H$_4$	CH$_3$	225	4.17	270	4.24

Absorption values reported for four derivatives (**3-253a–d**) of 4-nitropyrazol-5-one (solvent not specified)[79] and four derivatives (**3-254a–d**) of 3,4-dinitro-Δ^1-pyrazoline in methanol[119,120] are shown. The absorption dissimilarities among (**3-253a–d**) (neutral species) and between (**3-253c**) and (**3-253d**) (cation species) were unexplained.

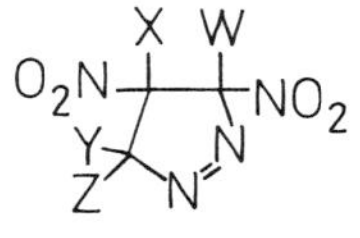

$$O_2N-\overset{X}{\underset{Y}{\bigg|}}-\overset{W}{\underset{Z}{\bigg|}}-NO_2$$

3-254a–d

3-254	W	X	Y	Z	λ_{max} nm	log ε
a	C_6H_5	H	C_6H_5	C_6H_5	272	4.01
					336	3.58
b	C_6H_5	H	biphenylene		239	4.78
c	CH_3	CH_3	C_6H_5	H	330	2.38
d	CH_3	CH_3	C_6H_5	C_6H_5	272	3.30
					422	3.63

Ultraviolet fluorescence for over 300 pyrazole (including nitropyrazoles) has been reported.[278]

Mass Spectroscopy

An apparently intact ring characterized the primary fragments $[M-O]^+$, $[M-NO]^+$, and $[M-NO_2]^+$ found in the mass spectra of nitropyrazoles.[27,195] From 4-nitropyrazole **2** two routes to $[C_2H_2N]^+$ were identified, Eq. (3-58).[279]

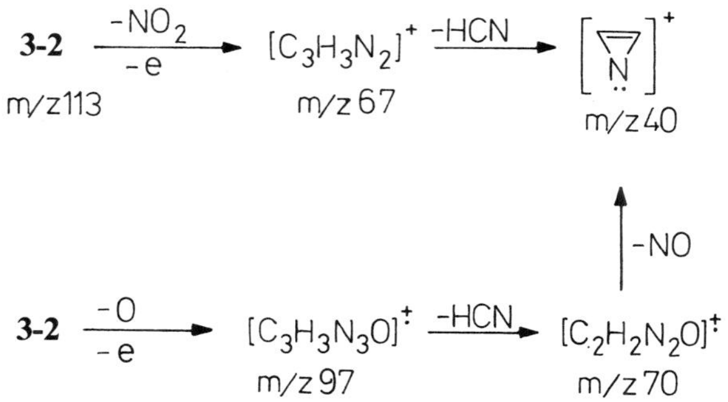

$$\underset{m/z\,113}{\textbf{3-2}} \xrightarrow[-e]{-NO_2} \underset{m/z\,67}{[C_3H_3N_2]^+} \xrightarrow{-HCN} \underset{m/z\,40}{\left[\text{N}\right]^+}$$

$$\underset{m/z\,70}{[C_2H_2N_2O]^{\cdot+}} \xrightarrow{-NO} \text{(above)}$$

$$\textbf{3-2} \xrightarrow[-e]{-O} \underset{m/z\,97}{[C_3H_3N_3O]^{\cdot+}} \xrightarrow{-HCN} \underset{m/z\,70}{[C_2H_2N_2O]^{\cdot+}} \qquad (3\text{-}58)$$

Fragmentation of methylnitropyrazoles also afforded intact pyrazole rings with barely detectable $[M-H]^+$ fragments and their ring enlargement characteristic of methylpyrazoles. Adjacent group effects also led to the loss of HO·,

Properties 255

H_2O, CHO·, and/or CH_2O, Eqs. (3-59) through (3-62). The loss of HO· and CHO· from 1-methyl-5-nitropyrazole (**3-255**) tended to obscure the presence of a nitro group. Strong $[M-OH]^+$ ions were detected from 3(5)-methyl-5(3)-nitropyrazole (**3-256**) and from 4-methyl-5(3)-nitropyrazole (**3-97b**). A difficulty in distinguishing the pyrazole (**3-256**) from its isomers the 3- and 4-nitro derivatives (**3-6b**), (**3-5**) of 1-methylpyrazole, in which the substituents are nonadjacent, illustrates the importance of adjacent group effects in structure determination. Failure to detect an adjacent group effect in 1-nitro-5-methylpyrazole was attributed to the preferred cleavage of the weak N-N nitramine bond with the formation of $[M-NO]^+$ ions, comparable in abundance to the $[M]^+$ ions.[27,28,280]

Adjacent group effects accounted for the loss of HO· and HON from 1-phenyl-3-methyl-4-nitro-5-hydroxypyrazole (**3-257**), Eqs. (3-63), (3-64).[281]

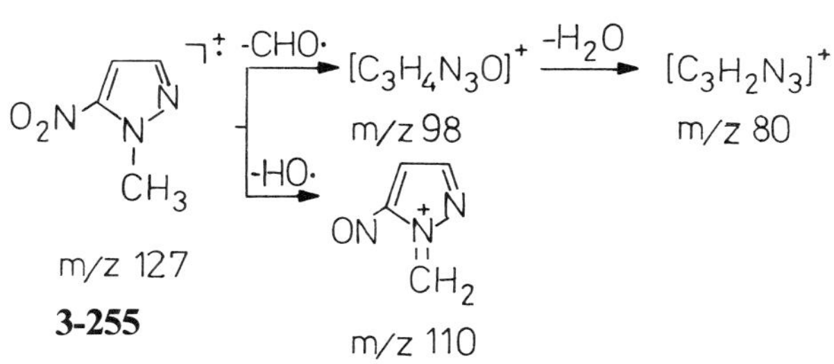

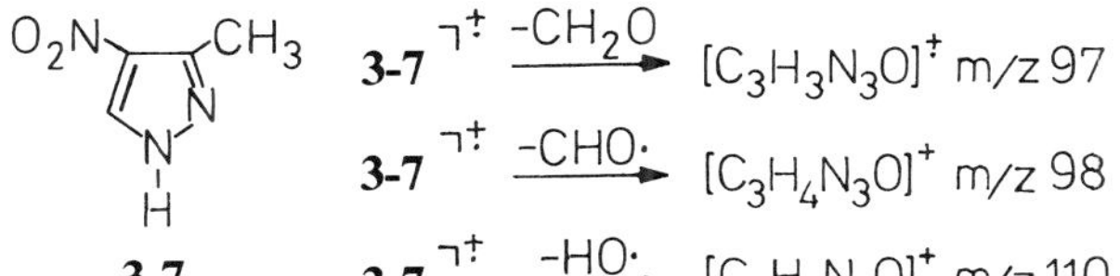

(3-59)

(3-60)

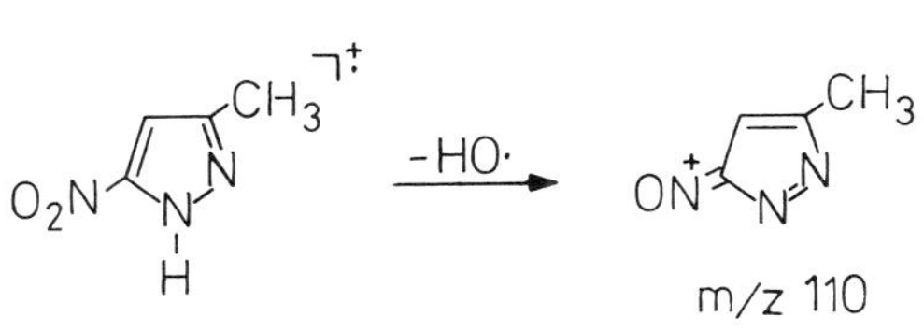

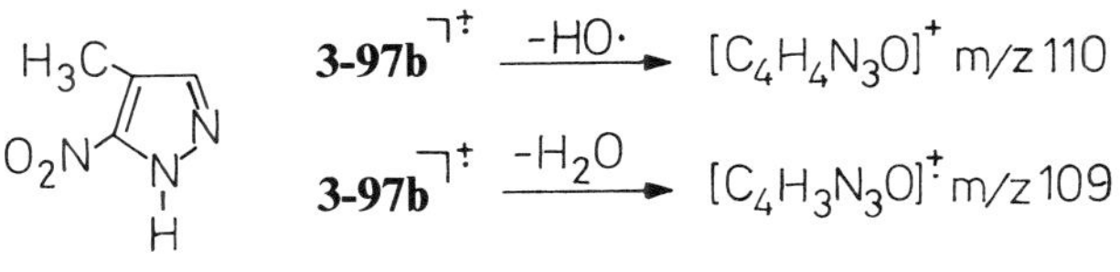

(3-61)

(3-62)

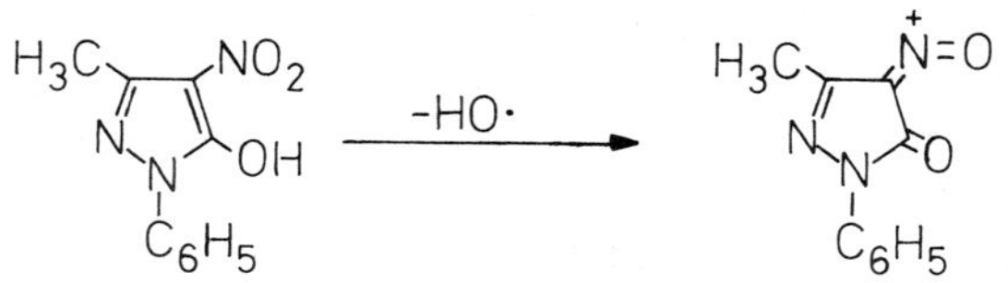

[Structures: 3-257 → (−HO·) product, equation (3-63)]

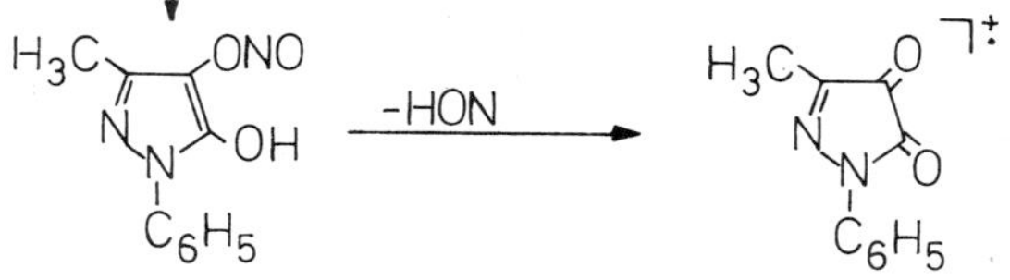

[Structures: 3-258 → (−HON) product, equation (3-64)]

X-Ray Structure Analysis

A regiospecific thermal 1,3-dipolar cycloaddition of E-β-nitrostyrene to 3-pyrazolidoneazomethinimines gave 1(Ref),3-*trans*-bisaryl-2-*cis*-nitro-5-oxoperhydropyrazolo[1,2-*a*]pyrazole (**3-259a**) (major product) and its 2-*trans*-nitro isomer (**3-259b**). The structures were confirmed by X-ray analysis. Each product formation was attributed to a [$_\pi 4_s + _\pi 2_s$] mechanism; however, it was noted that the minor "forbidden" product required a reversal of the configuration of the dipolarophile.[282,283]

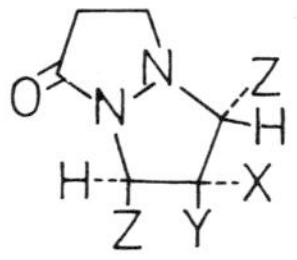

[Structure 3-259]

3-259	X	Y	Z
a	NO$_2$	H	C$_6$H$_5$, *p*-ClC$_6$H$_4$
b	H	NO$_2$	C$_6$H$_5$, *p*-ClC$_6$H$_4$

A report on N-nitropyrazole (**3-3**) described an X-ray analysis of the structure and its ^{14}N nuclear quadrupole resonance data.[284]

Other Physical Properties

A dipole moment, u(D)3.88, was determined for 3,5-dimethyl-4-nitropyrazole (**3-8**).[285]

A systematic study of the gas chromatographic separation of pyrazoles included data on 21 nitropyrazoles.[286]

Bifunctional catalysts that accelerated peptide synthesis via aminolysis of amino acid esters included 4-nitropyrazole (**3-2**).[203,287]

Nitropyrazole couplers gave magenta dyes resistant to fading and color fog useful for color photographic images.[288]

Bulk polymerization of methyl methacrylate was inhibited increasingly by 1-methyl-3-nitropyrazole, 1,2-dimethyl-5-nitroimidazole, and 1-methyl-5-nitropyrazole.[289]

An effective inhibitor of hydrogen sulfide corrosion of steel was 1-butyl-3(5)-methyl-4-nitropyrazole.[290]

Earlier assignments of hyperfine splitting constants for 3-nitropyrazole and for other nitroheterocycles were corrected, and spin-density distributions were calculated by the CNDO/2 method.[291]

Biological Properties

Over 50 references had appeared by 1981 on nitropyrazoles and other abscission chemicals for processing oranges. *Chemical Abstracts* references for 1980 and 1981 are given. Apparently the human metabolic fate of these chemicals retained in the oranges was not determined. That the pyrazoles (or derivatives) were retained in the oranges was suggested by a better flavor for the untreated oranges. Typical abscission chemicals included *Release* [5-chloro-3-methyl-4-nitro-1H-pyrazole (**3-260**) (6814-58-O)], *Pik-Off* [glyoxal dioxime (557-30-2)], Ethephon-[16672-87-0], dithiin tetraoxide[55290-64-7], *Acti-Aid* [cycloheximide (66-81-9)], *Sweep*[chlorothalonil (1897-45-6)], Triton X-100[9002-93-1], and *Chevron* X-77 [11097-66-8]. The abscission chemicals induced the production of ethylene in the orange.[292-301]

Nitropyrazoles and other pyrazole derivatives were listed as herbicides and plant growth regulators,[204,302] insecticides,[303,304] and bactericides and fungicides.[305-307] Nitropyrazoles have aided the conservation of nitrogen in soil by preventing or controlling the nitrification of soil ammonia.[308,309]

Analgesic, anti-inflammatory, and anti-pyretic carbamoylpyrazoles included several nitropyrazole derivatives.[204,310] Bacteriostatic activity of nitropyrazoles against *Escherichia coli* and *Staphylococcus aureus* was determined.[311] The amide oxime ethers (**3-261**) derived from 1,3-dimethyl-4-nitro-5-cyanopyrazole and hydroxylamine followed by O-alkylation were found useful as analgesics

3-260

3-261
R = H, CH_3, CH_2OCH_3

and psychotropics.[312-314] For treatment of infections caused by eg *Trichomonas vaginalis* a series of 4-nitropyrazole-1-alkane carboxylic acids (**3-262**) was developed.[315] A group of pyrazole anticonvulsants included 1,3,5-trimethyl-4-nitropyrazole (**3-263**) which was four times more active than meprobamate.[316] One of the antibiotic (pyrazolyacetamido)[pyridylthio)methyl]cephem-carboxylates (**3-264**) contained a 4-nitro substituent in the pyrazole ring.[317] Psychosedative 4-nitropyrazole derivatives have been reported.[60]

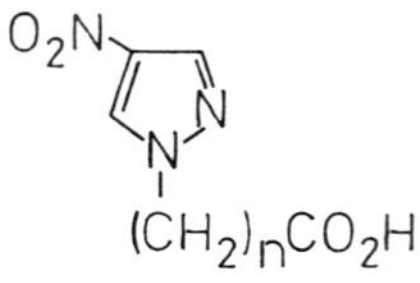

3-262
n = 1,3

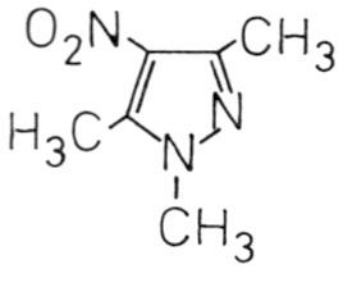

3-263

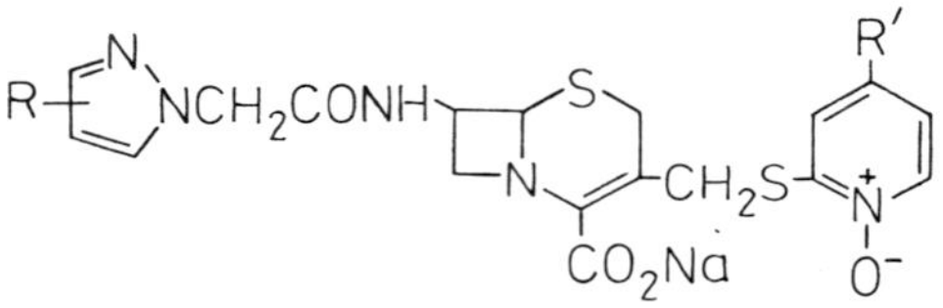

3-264

Radiosensitization of hypoxic bacteria and mammalian cells in vitro by some nitropyrazoles has been described.[318-321]

REFERENCES

1. Jacobs, T. L., In R. C. Elderfield, ed., *Heterocyclic Compounds*, Vol. 5. John Wiley, New York, **1957**, pp. 45–161.
2. Fusco, R., In A. Weissberger, ed., *The Chemistry of Heterocyclic Compounds*, Vol. 22. Interscience, New York, **1967**, pp. 3–176.
3. Jarboe, C. H., *Ibid.*, pp. 177–288.
4. Behr, L. C., *Ibid.*, pp. 289–384.
5. Kost, A. N., Grandberg, I. I., In A. R. Katritzky and A. J. Boulton, eds., *Advances in Heterocyclic Chemistry* **1966**, Vol. 6, p. 347.
6. Buchner, E., Fritsch, M., *J. Liebigs Ann. Chem.*, **1893**, *273*, 256.
7. Huttel, R., Buchele, F., Jochum, P., *Chem. Ber.*, **1955**, *88*, 1577.
8. Habraken, C. L., Munter, H. J., Westgeest, J. C. P., *Rec. Trav. Chim.*, **1967**, *86*, 56.
9. Fabra, F., Vilarrasa, J., Coll, J., *J. Heterocycl. Chem.*, **1978**, *15*, 1447.
10. Habraken, C. L., Van Woerkom, P. C. M., De Wind, H. W., Kallenberg, G. M., *Rec. Trav. Chim.*, **1966**, *85*, 1191.
11. Austin, M. W., *Chem. Ind.*, **1982**, 57.
12. Austin, M. W., Blackborow, J. R., Ridd, J. H., Smith, B. V., *J. Chem. Soc.*, **1965**, 1051.
13. Hüttel, R., Büchele, F., *Chem. Ber.*, **1955**, *88*, 1586.
14. Janssen, J. W. A. M., Habraken, C. L., *J. Org. Chem.*, **1971**, *36*, 3081.
15. Olah, G. A., Narang, S. C., Fung, A. P., *J. Org. Chem.*, **1981**, *46*, 2706.

16. Butovetskii, D. N., Sharnin, G. P., Falyakhov, I. F., Razgulyaeva, L. V., *Tr. Kazan. Khim.-Teknol. Inst.*, **1974**, *54*, 27; *Chem. Abstr.*, **1975**, *82*, 72340b.
17. Katritzky, A. R., Terem, B., Scriven, E. V., Clementi, S., Tarhan, H. O., *J. Chem. Soc. Perkin Trans. II*, **1975**, 1600.
18. Burton, A. G., Forsythe, P. P., Johnson, C. D., Katritzky, A. R., *J. Chem. Soc. (B)*, **1971**, 2365.
19. Finar, J. L., *J. Chem. Soc.*, **1968**, 725.
20. Grimmett, M. R., Lim, K. H. R., *Aust. J. Chem.*, **1978**, *31*, 689.
21. Knorr, L., *Liebigs Ann. Chem.*, **1894**, *79*, 217.
22. Morgan, J. T., Ackerman, I., *J. Chem. Soc.*, **1923**, *123*, 1308.
23. Gerster, J. F., *Ger. Offen.* 1,945,430; *Chem. Abstr.*, **1970**, *72*, 121526w.
24. Andreeva, M. A., Manaev, Y. A., Mushii, R. Y., Perevalov, V. P., Seraya, V. I., Stepanov, B. I., *Zh. Obshch. Khim.*, **1980**, *50*, 2106; Eng. 1705.
25. Manaev, Y. A., Andreeva, M. A., Perevalov, V. P., Stepanov, B. I., Dubrovskaya, V. A., Seraya, V. I., *Zh. Obshch. Khim.*, **1982**, *52*, 2592; Eng. 2291.
26. Janssen, J. W. A. M., Koeners, H. J., Kruse, C. G., Habraken, C. L., *J. Org. Chem.*, **1973**, *38*, 1777.
27. Luijten, W. C. M. M., Van Thuijl, J., *Org. Mass Spectrom.*, **1982**, *17*, 299.
28. *Ibid.*, **1979**, *14*, 577.
29. Katritzky, A. R., Tarhan, H. O., Terem, B., *J. Chem. Soc. Perkin Trans. II*, **1975**, 1632.
30. Péglion, J.-L., Pastor, R., Greiner, J., Cambon, A., *Bull. Soc. Chim. Fr. II*, **1982**, 89.
31. Claramunt, R. M., Hernandez, H., Elguero, J., Julia, S., *Bull. Soc. Chim. Fr. II*, **1983**, 5.
32. Finar, I. L., Hurlock, R. J., *J. Chem. Soc.*, **1957**, 3024.
33. Barry, W. J., Birkett, P., Finar, I. L., *Ibid.*, **1969**, 1328.
34. Parrini, V., *Ann. Chim. (Rome)*, **1957**, *47*, 929.
35. Khan, M. A., Lynch, B. M., Hung, Y.-Y., *Can. J. Chem.*, **1963**, *41*, 1540.
36. Lynch, B. M., Hung, Y.-Y., *Can. J. Chem.*, **1964**, *42*, 1605.
37. Dal Monte Casoni, D., *Gazz. Chim. Ital.*, **1959**, *89*, 1539.
38. Cohen-Fernandez, P., Habraken, C. L., *Rec. Trav. Chim.*, **1972**, *91*, 1185.
39. Burton, A. G., Katritzky, A. R., Konya, M., Tarhan, H. O., *J. Chem. Soc. Perkin Trans. II*, **1974**, 389.
40. Grimmett, M. R., Hartshorn, S. R., Schofield, K., Weston, J. B., *J. Chem. Soc. Perkin Trans. II*, **1972**, 1654.
41. Elguero, J., Jacquier, R., Duc, H. C. N. T., *Bull. Soc. Chim. Fr.*, **1966**, 3727.
42. Coburn, M. D., *J. Heterocycl. Chem.*, **1970**, *7*, 345.
43. Coburn, M. D., *J. Heterocycl. Chem.*, **1970**, *7*, 707.
44. Coburn, M. D., *J. Heterocycl. Chem.*, **1971**, *8*, 293.
45. *Ibid.*, 153.
46. Hattori, K., Horibe, S., *Japan 11,944; Chem. Abstr.*, **1968**, *68*, 68982f.
47. Granowitz, Liljefors, S., *Chem. Scr.*, **1978–1979**, *13*, 157.
48. Liljefors, S., Gronowitz, S., *Chem. Scr.*, **1980**, *15*, 102.
49. Gronowitz, S., Hallberg, A., Liljefors, S., Forsgren, U., Sjoberg, B., Westerbergh, S. E., *Acta Pharm. Sueicca*, **1968**, *5*, 163.
50. Berbee, R. P. M., Habraken, C. L., *J. Heterocycl. Chem.*, **1981**, *18*, 559.
51. Lund, H., *Ibid.*, **1935**, 418.
52. Lund, H., *J. Chem. Soc.*, **1933**, 686.
53. Perevalov, V. P., Andreeva, M. A., Baryshnenkova, L. I., Manaev, Y. A., Yamburg, G. S., Stepanov, B. I., Dubrovskaya, V. A., *Khim. Geterotsikl Soedin.*, **1983**, 1676; *Chem. Abstr.*, **1984**, *100*, 209684d.
54. Dal Monte Casoni, D., *Ann. Chim. (Rome)*, **1958**, *48*, 783.
55. Habraken, C. L., Cohen-Fernandez, P., Balian, S., Van Erk, K. C., *Tetrahedron Lett.*, **1970**, 479.
56. Buchanan, J. G., Edgar, A. R., Hutchison, R. J., Stobie, A., Wightman, R. H., *J. Chem. Soc. Chem. Commun.*, **1980**, 237.
57. Musante, C., *Gazz. Chim. Ital.*, **1948**, *78*, 178.
58. Perevalov, V. P., Baryshnenkova, L. I., Andreeva, M. A., Manaev, Y. A., Denisova, I. A., Stepanov, B. I., Seraya, V. I., *Khim. Geterotsikl. Soedin.*, **1983**, 1672; *Chem. Abstr.*, **1984**, *100*, 209683c.
59. Crovetti, A. J., Kenney, D. S., Lynch, D. M., Stein, R. G., *S. African* 73 52,15; *Chem. Abstr.*, **1975**, *82*, 98005w.
60. Bruderer, H., Richle, R., Ruegg, R., *Ger. Offen.* 2,250,316; *Chem. Abstr.*, **1973**, *79*, 18706a.
61. Alcalde, E., Garcia-Marquina, J. M., DeMendoza, J., *An. Quim.*, **1974**, *70*, 959.

62. Treuner, U. D., *U.S. 4,077,956; Chem. Abstr.*, **1978**, *89*, 109572b.
63. Crovetti, A. J., Kenney, D. S., Lynch, D. M., Stein, R. G., U.S. 3,932,453; *Chem. Abstr.*, **1976**, *84*, 164770m.
64. *Ibid.*, U.S. 1,025,530; *Chem. Abstr.*, **1977**, *87*, 102322v.
65. Chang, K.-C., Grimmett, M. R., Ward, D. D., Weavers, R. T., *Aust. J. Chem.*, **1979**, *32*, 1727.
66. Andreeva, M. A., Bolotov, M. I., Isaev, S. G., Mushii, R. Y., Perevalov, V. P., Seraya, V. I., Stepanov, B. I., *Zh. Obshch. Khim.*, **1980**, *50*, 2116; Eng. 1714.
67. Fabra, F., Fos, E., Vilarrasa, J., *Tetrahedron Lett.*, **1979**, 3179.
68. Lewis, A. F., Townsend, L. B., *Nucleic Acid Chem.*, **1978**, *1*, 121.
69. DeWald, H. A., Nordin, I. C., L'Italien, Y. J., Parcell, R. F., *J. Med. Chem.*, **1973**, *16*, 1346.
70. Musante, C., *Gazz. Chim. Ital.*, **1945**, *75*, 121.
71. Wierzchowski, J., Kuśmierek, J., Giziewicz, J., Salvi, D., Shugar, D., *Acta Biochim. Pol.*, **1980**, *27*, 35.
72. Scott, F. L., Kennedy, M. T., Reilly, J., *J. Am. Chem. Soc.*, **1953**, *75*, 1294.
73. Scott, F. L., *Angew. Chem.*, **1957**, *69*, 506.
74. Dereli, M., Katritzky, A. R., Tarhan, H. O., *J. Chem. Soc. Perkin Trans. II*, **1975**, 1609.
75. Konya, M., Tarhan, H. O., *Chim. Acta Turc.*, **1979**, *7*, 225.
76. Arriau, J., Deschamps, J., Teyseyrre, J., Maquestiau, A., Van Haverbeke, Y., *Tetrahedron*, **1974**, *30*, 1225.
77. Maquestiau, A., Van Haverbeke, Y., Jacquerye, R., *Bull. Soc. Chim. Belg.*, **1973**, *82*, 233.
78. Ajello, T., *Gazz. Chim. Ital.*, **1940**, *70*, 401.
79. Burton, A. G., Dereli, M., Katritzky, A. R., Tarhan, H. O., *J. Chem. Soc. Perkin Trans. II*, **1974**, 382.
80. Kovar, K. A., Rohlfes, W., Auterhoff, H., *Arch. Pharm. (Weinheim, Ger.)*, **1981**, *314*, 532.
81. Batz, H. G., *Eur. Pat. Appl.* EP 33,359; *Chem. Abstr.*, **1982**, *96*, 20096w.
82. Cardillo, G., Merlini, L., Boeri, E., *Gazz. Chim. Ital.*, **1966**, *96*, 973.
83. Auwers, K., Niemeyer, F., *J. Prakt. Chem.*, **1925**, *110*, 153.
84. Hirota, K., Yamada, Y., Assav, T., Senda, S., *J. Chem. Soc., Perkin Trans. I*, **1982**, 277.
85. Dorn, H., Dilcher, H., *J. Liebigs Ann. Chem.*, **1967**, *707*, 141.
86. Mühmel, G., Hanke, R., Breitmaier, E., *Synthesis*, **1982**, 673.
87. Springer, R. H., Dimmitt, M. K., Novinson, T., O'Brien, D. E., Robins, R. K., Simon, L. N., Miller, J. P., *J. Med. Chem.*, **1976**, *19*, 291.
88. Springer, R. H., Scholten, M. B., O'Brien, D. E., Novinson, T., Miller, J. P., *J. Med. Chem.*, **1982**, *25*, 235.
89. Khan, M. A., Lynch, B. M., *Can. J. Chem.*, **1971**, *49*, 3566.
90. Ferguson, I. J., Schofield, K., Barnett, J. W., Grimmett, M. R., *J. Chem. Soc. Perkin Trans. I*, **1977**, 672.
91. Bozzi, E. G., Clapp, L. B., *J. Heterocycl. Chem.*, **1978**, *15*, 1525.
92. Birkofer, L., Franz, M., *Chem. Ber.*, **1972**, *105*, 1759.
93. Klebe, K. J., Habraken, C. L., *Synthesis*, **1973**, 294.
94. Janssen, J. W. A. M., Habraken, C. L., Louw, R., *J. Org. Chem.*, **1976**, *41*, 1758.
95. Janssen, J. W. A. M., Cohen-Fernandez, P., Louw, R., *J. Org. Chem.*, **1975**, *40*, 915.
96. Dumanović, D., Maksimović, R., Cirić, J., Jeremić, D., *Talanta*, **1974**, *21*, 455.
97. Dumanović, D., Cirić, J., Miek, A., Nikolić, V., *Talanta*, **1975**, *22*, 819.
98. Grimmett, M. R., Lim, K. H. R., Weavers, R. T., *Aust. J. Chem.*, **1979**, *32*, 2203.
99. Rufer, C., Schwarz, K., *J. Liebigs Ann. Chem.*, **1975**, 1465.
100. Ketari, R., Foucaud, A., *Synthesis*, **1982**, 844.
101. Cheng, C. C., *J. Heterocycl. Chem.*, **1968**, *5*, 195.
102. Bagal, L. I., Pevzner, M. S., Frolov, A. N., Sheludyakova, N. I., *Khim. Geterotsikl. Soedin.*, **1970**, 259; Eng. 240.
103. Ward, E. R., Johnson, C. D., Hawkins, J. G., *J. Chem. Soc.*, **1960**, 864.
104. Jones, J. W., Robins, R. K., *J. Am. Chem. Soc.*, **1960**, *82*, 3773.
105. Berger, H., Gall, R., Stach, K., Thiel, M., Voemel, W., *Ger. Offen.* 2,558,117; *Chem. Abstr.*, **1977**, *87*, 135320w.
106. Jones, R. G., Terando, N. H., *Ger. Offen.* 2,212,080; *Chem. Abstr.*, **1973**, *78*, 4247u.
107. *Ibid.*, U.S. 4,066,776; *Chem. Abstr.*, **1978**, *88*, 152614h.
108. Coburn, M. D., *J. Heterocycl. Chem.*, **1970**, *7*, 455.
109. Birkofer, L., Franz, M., *Chem. Ber.*, **1971**, *104*, 3062.
110. Freeman, J. P., Gannon, J. J., *J. Org. Chem.*, **1969**, *34*, 194.
111. Freeman, J. P., Gannon, J. J., Surbey, D. L., *J. Org. Chem.*, **1969**, *34*, 187.

112. Torf, S. F., Kudryashova, N. I., Khromov-Borisov, N. V., Mikhailova, T. A., *Zh. Obshch. Khim.*, **1962**, *32*, 1740; *Chem. Abstr.*, **1963**, *58*, 4540e.
113. Ref. 2, p. 97.
114. Parham, W. E., Braxton, H. G., Serres, C., *J. Org. Chem.*, **1961**, *26*, 1831.
115. Gabitov, F. A., Kremleva, O. B., Fridman, A. L., *Khim. Geterotsikl Soedin.*, **1976**, 997; Eng. 826.
116. Dean, F. M., Park, B. K., *J. Chem. Soc. Chem. Commun.*, **1974**, 162.
117. Bourah, R. C., Sandhu, J. S., *Indian J. Chem.*, **1982**, *21B*, 374.
118. *Ibid. Synthesis*, **1982**, 677.
119. Gabitov, F. A., Kremleva, O. B., Fridman, A. L., *Zh. Org. Khim.*, **1977**, *13*, 1117; Eng. 1026.
120. Gabitov, F. A., Fridman, A. L., Kremleva, O. B., *Khim. Geterotsikl. Soedin.*, **1975**, 1577; Eng. 1342.
121. Shebaldova, A. D., Ryzhenko, L. M., Khidekel, M. L., *Tezisy Vses. Soveshch. Khim. Nitrosoedinenii, 5th*, **1974**, 69; *Chem. Abstr.*, **1977**, *87*, 53144c.
122. Bouchet, P., Elguero, J., Jacquier, R., *Bull. Soc. Chim. Fr.*, **1967**, 4716.
123. Kremleva, O. B., Gabitov, F. A., Fridman, A. L., *Khim. Geterotsikl. Soedin.*, **1977**, 703; Eng. 575.
124. Ranganathan, D., Rao, C. B., Ranganathan, S., Mehotra, A. K., Iyengar, R., *J. Org. Chem.*, **1980**, *45*, 1185.
125. Franck-Neumann, M., Miesch, M., *Tetrahedron*, **1983**, *39*, 1247.
126. Verbruggen, R., Viehe, H. G., *Chimia*, **1975**, *29*, 350.
127. Fridman, A. L., Gabitov, F. A., *USSR 367,097*; *Chem. Abstr.*, **1973**, *79*, 5334c.
128. Devi, P., Sandhu, J. S., *J. Chem. Soc. Chem. Commun.*, **1983**, 990.
129. Reddy, G. D., *Chem. Ind. London*, **1984**, 144.
130. Snider, B. B., Conn, R. S. E., Sealfon, S., *J. Org. Chem.*, **1979**, *44*, 218.
131. Dorn, H., Ozegowski, R., Gründemann, E., *J. Prakt. Chem.*, **1979**, *321*, 555.
132. Dorn, H., Ozegowski, R., *Ger. (East) 143,617*; *Chem. Abstr.*, **1981**, *95*, 43100m.
133. Kutschabsky, L., Dorn, H., *Krist. Tech.*, **1979**, *14*, 1107.
134. Lo, Y. S., Munson, H. R., *U.S. 4,309,552*; *Chem. Abstr.*, **1982**, *96*, 122790k.
135. Simonenko, L. S., Korsakova, I. S., Novikov, S. S., *Izv. Akad. Nauk SSSR, Ser. Khim.*, **1972**, 2333; Eng. 2268.
136. Korsakova, I. S., Simonenko, L. S., Novikov, S. S., *USSR 349,690*; *Chem. Abstr.*, **1973**, *78*, 16177j.
137. *Ibid. USSR 382,626*; *Chem. Abstr.*, **1973**, *79*, 66358k.
138. Huisgen, R., Granhey, R., Krischke, R., *Tetrahedron Lett.*, **1962**, 381.
139. Pocar, D., Maiorana, S., Dalla Croce, P., *Gazz. Chim. Ital.*, **1968**, *98*, 949.
140. Dychenko, A. I., Pupko, L. S., Pel'kis, P. S., *Khim. Geterotsikl. Soedin.*, **1974**, 425; Eng. 372.
141. Alberti, C., Thironi, C., *Farmaco (Pavia) Ed. Sci.*, **1966**, *21*, 883; **1971**, *26*, 66.
142. Foster, H. E., Hurst, J., *J. Chem. Soc. Perkin Trans. 1*, **1976**, 507.
143. Hill, H. B., Black, A. F., *Am. Chem. J.*, **1905**, *33*, 292.
144. Thankarajan, N., Sen, D. N., *Indian J. Chem.*, **1964**, *2*, 64.
145. Rajappa, S., Advani, B. G., *Indian J. Chem.*, **1977**, *15B*, 890.
146. Rajappa, S., *Heterocycles*, **1977**, *7*, 507.
147. Schäfer, H., Gewald, K., *J. Prakt. Chem.*, **1981**, *323*, 332.
148. Pilgram, K., *J. Heterocycl. Chem.*, **1980**, *17*, 1413.
149. Wolfbeis, O. S., *Synthesis*, **1977**, 136.
150. Matsumura, E., Ariga, M., Tohda, Y., *Bull. Chem. Soc. Jpn.*, **1980**, *53*, 2891.
151. Biffin, M. E. C., Brown, D. J., Porter, O. N., *J. Chem. Soc. (C)*, **1968**, 2159.
152. Ivashchenko, A. V., Garicheva, O. N., *Khim. Geterotsikl. Soedin.*, **1982**, 579; Eng. 429.
153. Temple, Jr., C., Kussner, C. L., Montgomery, J. A., *J. Org. Chem.*, **1969**, *34*, 2102.
154. Van der Plas, H. C., Jongejan, H., Koudijs, A., *J. Heterocyclic Chem.*, **1978**, *15*, 485.
155. Senda, S., Hirota, K., Asao, T., Yamada, Y., *Heterocycles*, **1976**, *4*, 1765.
156. Clark, J., Curphey, M., *J. Chem. Soc. Chem. Comm.*, **1974**, 184.
157. Leete, E., Isaacson, H. V., *J. Indian Chem. Soc.*, **1978**, *55*, 1125.
158. Bell, F., *J. Chem. Soc.*, **1941**, 285.
159. Van der Plas, H. C., *Ring Transformations of Heterocycles*, Vol. 1. Academic Press, London, **1973**, pp. 293–296.
160. Musante, C., *Gazz. Chim. Ital.*, **1942**, *72*, 537.
161. *Ibid.*, **1943**, *73*, 355.
162. Musante, C., Stener, A., *Ibid.*, **1959**, *89*, 1579.
163. Fusco, R., Zumin, S., *Gazz. Chim. Ital.*, **1946**, *76*, 223.

164. Berenyi, E., Szirt, E., Gorog, P., Petocz, L., Kosoczky, I., Kovacs, A., Urmos, G., *Ger. Offen. 3,019,019; Chem. Abstr.*, **1981**, *94*, 156963t.
165. Reicheneder, F., Dury, K., *Belg. 660,636; Chem. Abstr.*, **1965**, *63*, 18096e.
166. Ach, F., *Liebigs Ann. Chem.*, **1889**, *253*, 44.
167. Cohen-Fernandes, P., Habraken, C. L., *J. Org. Chem.*, **1971**, *36*, 3084.
168. Pevzner, M. S., Gladkova, N. V., Lopukhova, G. A., Bedin, M. P., Dolmatov, V. Y., *Zh. Org. Khim.*, **1977**, *13*, 1300; Eng. 1192.
169. Wrzeciono, U., Linkowska, E., *Pharmazie*, **1980**, *35*, 593.
170. Fries, K., Fabel, K., Eckhardt, H., *J. Liebigs Ann.*, **1941**, *550*, 31.
171. Lynch, B. M., Lam, B. P.-L., *J. Heterocycl. Chem.*, **1974**, *11*, 223.
172. Grandberg, L. I., Nikitina, S. B., Moskalenko, V. A., Minkin, V. I., *Khim. Geterotsikl. Soedin.*, **1967**, *3*, 1076; *Chem. Abstr.*, **1968**, *69*, 52067w.
173. Ochi, H., Miyasaka, T., Kanada, K., Arakawa, K., *Bull. Chem. Soc. Jpn.*, **1976**, *49*, 1980.
174. Okamoto, T., Hirobe, M., Minamoto, Y., Irikura, T., *Japan 75 11,399; Chem. Abstr.* **1975**, *83*, 193305y.
175. Foster, H. E., Hurst, J., *J. Chem. Soc. Perkin Trans. I*, **1973**, 2901.
176. Chu, I., Lynch, B. M., *J. Med. Chem.*, **1975**, *18*, 161.
177. Lynch, B. M., Khan, M. A., Sharma, S. C., Teo, H. C., *Can. J. Chem.*, **1975**, *53*, 119.
178. Novinson, T., Miller, J. P., Scholten, M., Robins, R. K., Simon, L. N., O'Brien, D. E., Meyer, R. B., *J. Med. Chem.*, **1975**, *18*, 460.
179. Novinson, T., Hanson, R., Dimmitt, M. K., Simon, L. N., Robins, R. K., O'Brien, D. E., *Ibid.*, **1974**, *17*, 645.
180. Senga, K., Novinson, T., Springer, R. H., Rao, R. P., O'Brien, D. E., Robins, R. K., Wilson, H. R., *J. Med. Chem.*, **1975**, *18*, 312.
181. Khan, M. A., Ribeiro, V. L. T., *Monatsh. Chem.*, **1983**, *114*, 425.
182. Cartwright, D., Collins, D. J., Urlwin-Smith, P. L., *S. African 79 01,138; Chem. Abstr.*, **1981**, *94*, 175173b.
183. Cartwright, D., Urlwin-Smith, P., Collins, D. J., *Eur. Pat. Appl. 4,171; Chem. Abstr.* **1980**, *92*, 94442v.
184. Vogel, A., Troxler, F., *Ger. Offen. 2,424,334; Chem. Abstr.*, **1975**, *83*, 10160e.
185. Grundmann, C., Datta, S. K., Sprecher, R. F., *J. Liebigs Ann. Chem.*, **1971**, *744*, 88.
186. Townsend, L. B., Long, R. A., McGraw, J. P., Miles, D. W., Robins, R. K., Eyring, H., *J. Org. Chem.*, **1974**, *39*, 2023.
187. Buchanan, J. G., Stobie, A., Wightman, R. W., *J. Chem. Soc. Perkin Trans. I*, **1981**, 2374.
188. Coispeau, G. E. E., *Ger. Offen. 2,435,921; Chem. Abstr.*, **1975**, *83*, 12184g.
189. Hüttel, R., Jochum, P., *Chem. Ber.*, **1952**, *85*, 820.
190. Wiley, R. W., Smith, N. R., Johnson, D. M., Moffat, J., *J. Am. Chem. Soc.*, **1955**, *77*, 2572.
191. Khan, M. A., Pinto, A. A. A., *Monatsh. Chem.*, **1980**, *111*, 883.
192. Khan, M. A., Freitas, A. C. C., *Monatsh. Chem.*, **1981**, *112*, 675.
193. Khan, M. A., Lynch, B. M., *J. Heterocycl. Chem.*, **1970**, *7*, 1237.
194. Grandberg, I. I., Sharova, G. I., *Khim. Geterotsikl. Soedin.*, **1968**, 1097; *Chem. Abstr.*, **1969**, *70*, 77856m.
195. Korbukh, I. A., Budanova, O. V., Yakunina, N. G., Seraya, V. I., Preobrazhenskaya, M. N., *Zh. Org. Khim.*, **1976**, *12*, 1560; Eng. 1538.
196. Korbukh, I. A., Yakunina, N. G., Preobrazhenskoya, M. N., *Zh. Org. Khim.* **1975**, *11*, 463; *Chem. Abstr.*, **1975**, *83*, 28500a.
197. Korbukh, I. A., Budanova, O. V., Preobrazhenskoya, M. N., *Nucl. Acid Chem.*, **1978**, *1*, 469.
198. Korbukh, I. A., Kitaev, S. V., Preobrazhenskoya, M. N., *Zh. Org. Khim.*, **1976**, *12*, 682; Eng. 673.
199. Korbukh, I. A., Abramova, L. N., Preobrazhenskoya, M. N., *Zh. Org. Khim.*, **1977**, *13*, 731; Eng. 668.
200. Korbukh, I. A., Abramova, L. N., Stepanenko, B. N., Preobrazhenskoya, M. N., *Zh. Org. Khim.*, **1978**, *14*, 2169; Eng. 2007.
201. Chavis, C., Grodenic, F., Imbach, J.-L., *Eur. J. Med. Chem.-Chim. Ther.*, **1979**, *14*, 123.
202. Makabe, O., Nakamura, M., Umezawa, S., *Bull. Chem. Soc. Jpn.*, **1975**, *48*, 3210.
203. Makabe, O., Yajima, J., Umezawa, S., *Bull. Chem. Soc. Jpn.*, **1976**, *49*, 3552.
204. Ochiai, M., Kamikado, T., *Chem. Pharm. Bull. (Tokyo)*, **1966**, *14*, 628.
205. Bousquet, E. W., *U.S. 3,308,130; Chem. Abstr.*, **1968**, *68*, 95811f.
206. Finar, I. L., Lord, G. H., *J. Chem. Soc.*, **1957**, 3314.
207. Hüttel, R., Schäfer, O., Welzel, G., *J. Liebigs Ann. Chem.*, **1956**, *598*, 186.
208. Hüttel, R., Schäfer, O., Jochum, P., *Ibid.*, **1955**, *593*, 200.

209. Vasilevskii, S. F., Shvartsberg, M. S., *Izv. Akad. Nauk SSSR, Ser. Khim.*, **1980**, 1071; Eng. 778.
210. Khan, M. A., Freitas, A. C. C., *J. Heterocycl. Chem.*, **1983**, *20*, 277.
211. Zauhar, J., Ladouceur, B. F., *Can. J. Chem.*, **1968**, *46*, 1079.
212. Lange, M., Quell, R., Lettau, H., Schubert, H., *Z. Chem.*, **1977**, *17*, 94.
213. Stein, R. G., Shoepke, H. G., *U.S. 3,282,954; Chem. Abstr.*, **1967**, *66*, 37921q.
214. Habraken, C. L., Poels, E. K., *J. Org. Chem.*, **1977**, *42*, 2893.
215. Cohen-Fernandes, P., Erkelens, C., Van Eendenburg, C. G. M., Verhoeven, J. J., Habraken, C. L., *J. Org. Chem.*, **1979**, *44*, 4156.
216. Buchanan, J. G., Stobie, A., Wightman, R. H., *Can. J. Chem.*, **1980**, *58*, 2624.
217. Wrzeciono, U., Linkowska, E., Jankowiak, D., *Pharmazie*, **1981**, *36*, 673.
218. Wilde, H., Mann, G., Burkharat, U., Weber, G., *J. Prakt. Chem.*, **1979**, *321*, 495.
219. Kirkpatrick, W. E., Okabe, T., Hillyard, I. W., Robins, R. K., Dren, A. T., Novinson, T., *J. Med. Chem.*, **1977**, *20*, 386.
220. Long, R. A., Gerster, J. F., Townsend, L. B., *J. Heterocycl. Chem.*, **1970**, *7*, 863.
221. Baraldi, P. G., Guarneri, M., Moroder, F., Simoni, D., Vicentini, C. B., *Synthesis*, **1982**, 70.
222. Paryjczak, T., Falak, B., Grzywna, R., Kotlicki, S., Pieta, S., *Pol. 104,578; Chem. Abstr.*, **1980**, *93*, 95268a.
223. Nordin, I. C., *U.S. 3,553,207; Chem. Abstr.*, **1971**, *75*, 5972b.
224. L'Italien, Y. J., Nordin, I. C., *U.S. 3,553,209; Chem. Abstr.*, **1971**, *75*, 5967d.
225. Baraldi, P. G., Guarneri, M., Moroder, F., Simoni, D., *Synthesis*, **1981**, 727.
226. Baraldi, P. G., Vincentini, C. B., Simoni, D., Guarneri, M., *Farm. Ed. Sci.*, **1983**, *38*, 369.
227. Robins, R. K., Furcht, F. W., Grauer, A. D., Jones, J. W., *J. Am. Chem. Soc.*, **1956**, *78*, 2418.
228. Vakul'skaya, T. I., Larina, L. I., Nefedova, O. B., Lopyrev, V. A., *Khim. Geterotsikl. Soedin.*, **1982**, 523; Eng. 400.
229. De Mendoza, J., Millan, C., Bull, P., *J. Chem. Soc., Perkin Trans. II*, **1966**, 403.
230. Dumanović, D., Cirić, J., *Talanta*, **1973**, *20*, 525.
231. Hamel, D. M., Oelschlager, H., *J. Electroanal. Chem. Interfacial Electrochem.*, **1970**, *28*, 197.
232. Hecht, S. M., Jordis, U., *U.S. 4,282,361; Chem. Abstr.*, **1981**, *95*, 187295b.
233. Swett, L. B., *U.S. 3,657,271; Chem. Abstr.*, **1972**, *77*, 19691n.
234. De Mendoza, J., Garcia-Marquina, R. J. M., *An. Quim.*, **1970**, *66*, 911; *Chem. Abstr.*, **1971**, *74*, 125655b.
235. Latypov, N. V., Silevich, V. A., Ivanov, P. A., Pevzner, M. S., *Khim. Geterotsikl. Soedin.*, **1976**, 1649; Eng. 1355.
236. Tyupalo, N. F., Yakobi, V. A., Bernashevskii, N. V., Stepanyan, A. A., *USSR 453,404; Chem. Abstr.*, **1975**, *82*, 140122j.
237. Okkersen, H., Groeneveld, W. L., Reedijk, J., *Rec. Trav. Chim.*, **1973**, *92*, 945.
238. Reedijk, J., Windhorst, J. C. A., Van Ham, N. H. M., Groeneveld, W. L., *Rec. Trav. Chim.*, **1971**, *90*, 234.
239. Nieuwpoort, G., Vos, J. G., Groeneveld, W. L., *Inorg. Chim. Acta*, **1978**, *29*, 117.
240. Minghetti, G., Banditelli, G., Bonati, F., *Inorg. Chem.*, **1979**, *18*, 658.
241. Banditelli, G., Bandini, A. L., Bonati, F., Minghetti, G., *Inorg. Chim. Acta*, **1982**, *60*, 93.
242. Fukushima, K., Miyamoto, T., Sasaki, Y., *Bull. Chem. Soc. Jpn.*, **1978**, *51*, 499.
243. Janssen, J. W. A. M., Kruse, C. G., Koeners, H. J., Habraken, C. L., *J. Heterocycl. Chem.*, **1973**, *10*, 1055.
244. Dumanović, D., Cirić, J., Muk, A., Nikolić, V., *Talanta*, **1975**, *22*, 819.
245. Habraken, C. L., Beenakker, C. I. M., Brussee, J., *J. Heterocycl. Chem.*, **1972**, *9*, 939.
246. Gallo, G. G., Pasqualucci, C. P., Radaelli, P., Lancini, G., *J. Org. Chem.*, **1964**, *29*, 862.
247. Habraken, C. L., Bonser, S. M., *Heterocycles*, **1977**, *7*, 259.
248. Zatsepina, N. N., Tupitsyn, I. F., Belyashova, A. I., Kirova, A. V., Konyakhina, E. Y., *Khim. Geterotsikl. Soedin.*, **1977**, 1196; Eng. 963.
249. Zatsepina, N. N., Tupitsyn, I. F., Kane, A. A., Sudakova, G. N., *Khim. Geterotsikl. Soedin.*, **1977**, 1192; Eng. 959.
250. Sharnin, G. P., Khabirov, R. A., Nurgatin, V. V., Khmel'nitskii, L. I., Lebedev, O. V., Prikhod'ko, A. S., *Izv. Akad. Nauk SSSR, Ser. Khim.*, **1977**, 2711; Eng. 2506.
251. Oldenhof, C., Cornelisse, J., *J. Neth. Chem. Soc.*, **1978**, *97*, 35.
252. Valades, L., Jiminez, M., Rodriquez-Hahn, L., *Rev. Latinoam. Quim.*, **1975**, *6*, 152; *Chem. Abstr.*, **1976**, *84*, 30090f.
253. Altschul, J., *Chem. Ber.*, **1892**, *25*, 1842.
254. Vogel, A. I., *A Textbook of Macro and Semimicro Quantitative Inorganic Analysis*, 5th ed. by G. Svehla. Longman, London, 1979, p. 284.

255. Joshi, D. P., Jain, D. V., *J. Indian Chem. Soc.*, **1964**, *41*, 711; 1965, *42*, 871.
256. Osman, M. M., Hafez, A. M., Makhyoun, M. A., Tadros, A. B., *Acta Chim. Acad. Sci. Hung.*, **1982**, *109*, 83; *Chem. Abstr.*, **1982**, *97*, 144214y.
257. Pedersen, E., *Phytochemistry*, **1975**, *14*, 2086.
258. Taha, A. M., Gomaa, C. S., *J. Assoc. Off. Anal. Chem.*, **1976**, *59*, 683; *Chem. Abstr.*, **1976**, *85*, 37305j.
259. Shin, W., Pletcher, J., Blank, G., Sax, M., *J. Am. Chem. Soc.*, **1977**, *99*, 3491.
260. Lorenzutti, A., Cingolani, A., Leonesi, D., Bonati, F., *Syn. Reactiv. Inorg. Metal-Org. Compounds*, **1983**, *13*, 263.
261. Komarek, J., Jambor, J., Sommer, L., *Scr. Fac. Sci. Natur. Univ. Purkynianae Brun.*, **1972**, *2*, 33; *Chem. Abstr.*, **1973**, *79*, 38217v.
262. Barbero, L., Cavalli, G., *Tec. Molitoria*, **1981**, *32*, 607; *Chem. Abstr.*, **1982**, *96*, 84171z.
263. Nikitin, Y. E., Kuvatov, Y. G., Murinov, Y. I., Egutkin, N. L., *USSR 602,476*; *Chem. Abstr.*, **1979**, *91*, 42576b.
264. Myasoedov, B. F., Molochnikova, N. P., Kuvatov, Y. G., Nikitin, Y. E., *Radiokhimiya*, **1981**, *23*, 43; *Chem. Abstr.*, **1981**, *94*, 163421t.
265. Klyuchnikov, N. G., Ushenina, V. F., *Uch. Zap. Mosk. Gos. Pedagog. Inst.*, **1969**, 214; *Chem. Abstr.*, **1972**, *77*, 23913d.
266. *Ibid.*, **1969**, 229; *Chem. Abstr.*, **1972**, *77*, 24721b.
267. Gur'ev, I. A., Kalugin, A. A., Gushchina, E. A., *Zavod. Lab.*, **1980**, *46*, 497; *Chem. Abstr.*, **1980**, *93*, 88136c.
268. Greet, D. N., *Nematologica*, **1974**, *20*, 363.
269. Okada, T., *Noyaku Kensasho Hokoku*, **1972**, 93; *Chem. Abstr.*, **1974**, *80*, 128799m.
270. Whitehead, A. G., *Ann. Appl. Biol.*, **1977**, *87*, 225.
271. Rosenkranz, H. S., Stein, A. B., *Mutat. Res.*, **1975**, *28*, 127.
272. Fujiki, S., Tsukuni, H., Yamazaki, Y., Nakajima, K., *Japan Kokai 75 07,899*; *Chem. Abstr.*, **1975**, *83*, 29157f.
273. Claramunt, R. M., Elguero, J., Marzin, C., Seita, J., *Ann. Quim.*, **1979**, *75*, 701.
274. Carlsson, L.-O., *Acta Chem. Scand.*, **1975**, *B29*, 325.
275. Batterham, T. J., *NMR Spectra of Simple Heterocycles*. Wiley (Interscience), New York, **1973**, p. 169.
276. Kravtsov, D. N., Fedorov, L. A., Peregudov, A. S., Nesmeyanov, A. N., *Dokl. Akad. Nauk SSSR*, **1971**, *196*, 110; Eng. 18.
277. Zeigan, D., *J. Prakt. Chem.*, **1981**, *323*, 188.
278. Grandberg, I. I., Tabak, S. V., Kost, A. N., *Zh. Obshch. Khim.*, **1963**, *33*, 506; *Chem. Abstr.*, **1963**, *59*, 1616c.
279. Van Thuijl, J., Klebe, K. J., Van Houte, J. J., *Org. Mass Spec.*, **1970**, *3*, 1549.
280. Luijten, W. C. M. M., Van Thuijl, J., *Org. Mass Spec.*, **1982**, *17*, 304.
281. Wilde, H., Herzschuh, R., Lepom, P., Mann, G., *J. Prakt. Chem.*, **1981**, *323*, 14.
282. Kutscharsky, L., Dorn, H., *Krist. Tech.*, **1979**, *14*, 1107.
283. Kutscharsky, L., Kretschmer, R. G., Dorn, H., *Krist. Tech.*, **1979**, *14*, 1429.
284. Taramici, C., Schempp, E., *Acta Cryst.*, **1977**, *B33*, 240.
285. Hillers, S., Mazheika, I. B., Grandberg, I. I., Gorbacheva, L. I., *Khim. Geterotsikl. Soedin.*, **1967**, 130; *Chem. Abstr.*, **1967**, *67*, 63621g.
286. Ward, D. D., Grimmett, M. R., *J. Chromatogr.*, **1979**, *174*, 221.
287. Beyerman, H. C., Van der Brink, W. M., Weygand, F., Prox, A., Konig, W., Schmidhammer, L., Nintz, E., *Rec. Trav. Chim.*, **1965**, *84*, 213.
288. Ichijima, S., Furutachi, N., *Ger. Offen. 3,033,499*; *Chem. Abstr.*, **1981**, *95*, 88924z.
289. Stankovic, R., Jovanovic, S., Kosanovic, D., *Glas. Hem. Drus. Beograd.*, **1979**, *44*, 581; *Chem. Abstr.*, **1980**, *92*, 94766d.
290. Bocharnikov, A. I., Sorochenko, A. A., Mudrakova, V. Y., Seraya, V. I., Dubrovskaya, V. A., *Zashch. Met.*, **1979**, *15*, 705; *Chem. Abstr.*, **1980**, *92*, 151076e.
291. Larina, L. I., Dubnikov, V. M., Lur'e, F. S., Vakul'skaya, T. I., Vitkovskaya, N. M., Lopyrev, V. A., Voronkov, M. G., *Zh. Strukt. Khim.*, **1980**, *21*, 203.
292. El-Zeftawi, B. M., *Proc. Int. Soc. Citric.*, **1978**, 255; *Chem. Abstr.*, **1980**, *93*, 90112k.
293. Hutton, R. J., *Proc. Int. Soc. Citric.*, **1978**, 257; *Chem. Abstr.*, **1980**, *93*, 90113m.
294. Moshonas, M. G., Shaw, P. E., Sims, D. A., *Proc. Int. Soc. Citric.*, **1977**, 802; *Chem. Abstr.*, **1980**, *92*, 53285c.
295. Wilson, W. C., Donhaiser, J. R., Coppock, G. E., *Proc. Fla. State Hortic. Soc.*, **1979**, *92*, 56; *Chem. Abstr.*, **1980**, *93*, 90101f.

296. Rasmussen, G. K., *Proc. Fla. State Hortic. Soc.*, **1979**, *92*, 51; *Chem. Abstr.*, **1980**, *93*, 63513k.
297. Rasmussen, G. K., *J. Am. Soc. Hortic. Sci.*, **1980**, *105*, 145; *Chem. Abstr.*, **1980**, *93*, 2107j.
298. Evensen, K. B., Bausher, M. G., Biggs, R. H., *HortScience*, **1981**, *16*, 43; *Chem. Abstr.*, **1981**, *94*, 205664g.
299. Evensen, K. B., Bausher, M. G., Biggs, H. R., *J. Am. Soc. Hortic. Sci.*, **1981**, *106*, 57; *Chem. Abstr.*, **1981**, *94*, 151793c.
300. Rasmussen, G. K., *Proc. Fla. State Hortic. Soc.*, **1981**, *93*, 27; *Chem. Abstr.*, **1981**, *95*, 92212h.
301. Wilson, W. C., Coppock, G. E., *HortScience*, **1981**, *16*, 299; *Chem. Abstr.*, **1981**, *95*, 75329j.
302. Toyosato, T., Ochiai, M., Hagimoti, H., Tamura, H., Kamikado, T., *U.S. 3,326,662*; *Chem. Abstr.*, **1968**, *69*, 19148a.
303. *Japan Kokai Tokkyo Koho* 80 47,995; *Chem. Abstr.*, **1981**, *94*, 47318r.
304. Ruefenacht, K., *Helv. Chim. Acta*, **1973**, *56*, 2186.
305. Fenyes, J. G. E., *U.S. 3,577,545*; *Chem. Abstr.*, **1971**, *75*, 4607z.
306. Berger, H., Gall, R., Thiel, M., Schaumann, W., Voemel, W., *U.S. 4,069,330*; *Chem. Abstr.*, **1978**, *88*, 136617f.
307. Berger, H., Gall, R., Stach, K., Volmel, W., Veser, J., *Ger. Offen. 2,522,082*; *Chem. Abstr.*, **1977**, *86*, 72633g.
308. Griffith, J. D., *U.S. 3,635,690*; *Chem. Abstr.*, **1972**, *72*, 112099g.
309. Kaestner, G., Lang, S., Gross, M., Hartbrich, H. J., Klepel, M., Geilhufe, A., Jumar, A., Walter, R., Held, P., Ackermann, H., *Ger. (East) 133,088*; *Chem. Abstr.*, **1979**, *91*, 107164n.
310. Bianchi, M., Bonacina, F., Osvaldo, A., Pirola, C., *Farmaco, Ed. Sci.*, **1970**, *25*, 592; *Chem. Abstr.*, **1970**, *73*, 66498g.
311. Alberti, C., Tironi, C., Bainotti, F., Deleide, G., *Farmaco, Ed. Sci.*, **1977**, *32*, 92; *Chem. Abstr.*, **1977**, *86*, 171321k.
312. Bruderer, H., Richle, R., Ruegg, R., *Swiss 385,768*; *Chem. Abstr.*, **1975**, *83*, 206258n.
313. *Ibid. Swiss 566,993*; *Chem. Abstr.*, **1970**, *84*, 59480u.
314. Bruderer, H., Richle, R., Ruegg, R., *U.S. 3,822,283*; *Chem. Abstr.*, **1974**, *81*, 105495r.
315. Domaschke, L., *Ger. Offen. 1,950,329*; *Chem. Abstr.*, **1971**, *75*, 20395j.
316. Batulin, Y. M., *Farmakol. Toksikol.*, **1968**, *31*, 533; *Chem. Abstr.*, **1969**, *70*, 2236a.
317. Ochiai, M., Aki, O., Morimoto, A., Okada, T., Masuda, K., *Ger. Offen. 2,233,499*; *Chem. Abstr.*, **1973**, *78*, 124609k.
318. Asquith, J. C., Watts, M. E., Patel, K., Smithen, C. E., Adams, G. E., *Radiat. Res.*, **1974**, *60*, 108.
319. Ruddock, G. W., Greenstock, C. L., *Biochim. Biophys. Acta*, **1977**, *496*, 197.
320. Kimler, B. F., McDonald, T., Cheng, C. C., Podrebarac, E. G., Mansfield, C. M., *Radiology*, **1979**, *133*, 515; *Chem. Abstr.*, **1980**, *92*, 51763h.
321. Greenstock, C. L., Ruddock, G. W., Neta, P., *Radiat. Res.*, **1976**, *66*, 472.

BIBLIOGRAPHY

Baraldi, P. G., Vicentini, C. B., Simoni, D., Menziani, E., SYNTHESIS OF PYRAZOLE AND PYRAZOLO(4,3-d)PYRIMIDINONE DERIVATIVES. PART II., *Farmaco, Ed. Sci.*, **1983**, *38*, 508; *Chem. Abstr.*, **1983**, *99*, 82058f.

Baraldi, P. G., Vicentini, C. B., Simoni, D., Guarneri, M., SYNTHESIS OF PYRAZOLE AND PYRAZOLE(4,3-d)PYRIMIDONE DERIVATIVES, *Farmaco, Ed. Sci.*, **1983**, *38*, 369; *Chem. Abstr.* **1983**, *99*, 139889j.

Buchanan, J. G., Saxena, N. K., Wightman, R. H., C-NUCLEOSIDE STUDIES. 17. THE SYNTHESIS OF 3(5)-CARBAMOYL-5(3)-β-D-RIBOFURANOSYLPYRAZOLE(4-DEOXYPYRAZOFURIN) AND 4-AMINO-3(5)-AMINO-3(5)-CARBAMOYL 5(3)-β-D-RIBOFURANOSYL-PYRAZOLE, *J. Chem. Soc. Perkin Trans. I*, **1984**, 2367.

Cabildo, P., Claramunt, R. M., Elguero, J., SYNTHESIS AND REACTIVITY

OF NEW 1-(1-ADAMANTYL)PYRAZOLES, *J. Heterocycl. Chem.*, **1984**, *21*, 249.
Chatgilialoglue, C., Ingold, K. U., Tse-Sheepy, I., Warkentin, J., A HOMOALLYL RADICAL REARRANGEMENT. KINETICS OF THE ISOMERIZATION OF THE 2,2-DIMETHYL-3-BUTEN-1-YL RADICAL TO THE 1,1-DIMETHYL-3-BUTEN-1-YL RADICAL, *Can. J. Chem.*, **1983**, *61*, 1077.
Claramunt, R. M., Hernandez, H., Elguero, J., Julia, S., (N-POLYAZOLYL)METHANES. II. SYNTHESIS AND REACTIVITY OF 1,1-METHYLENEDIPYRAZOLES, *Bull. Soc. Chim. Fr.*, **1983**, *1–2, Pt. 2*), 5.
Cornell, N. W., Hansch, C., Kim, K. H., Henegar, K., THE INHIBITION OF ALCOHOL DEHYDROGENASE IN VITRO AND IN ISOLATED HEPATOCYTES BY 4-SUBSTITUTED PYRAZOLES. *Arch. Biochem. Biophys.*, **1983**, *227*, 81.
Franck-Neumann, M., Miesch, M., SYNTHESIS OF 5,5-DIMETHYLPYRAZOLENINES, PRECURSORS OF ISOPRENIC VINYLCARBENES, BY DIRECT OXIDATION OF Δ^2-PYRAZOLINES, *Tetrahedron Lett.*, **1983**, *39*, 1247.
Franck-Neumann, M., Miesch, M., NITROVINYLCARBENES: PHOTOLYSIS OF THE 3H-PYRAZOLE PRECURSORS, INTRAMOLECULAR EVOLUTION AND INTERMOLECULAR REACTIVITY, *Tetrahedron Lett.*, **1984**, *25*, 2909.
Khan, M. A., Freitas, A. C. C., PYRAZOLE DERIVATIVES. I. SYNTHESIS OF SOME CYANOPYRAZOLES, *Rev. Latinoam. Quim.*, **1982**, *13*, 100, *Chem. Abstr.*, **1983**, *99*, 70618a.
Lorenzotti, A., Cingolani, A., Leonesi, D., Bonati, F., ADDUCTS OF SOME NITROGEN BASES WITH NICKEL(II) PICROLONATES. *Congr. Naz. Chim. Inorg. (Atti) 16th*, **1983**, 160; *Chem. Abstr.*, **1984**, *100*, 28859u.
Lorenzotti, A., Cingolani, A., Leonesi, D., Bonati, F., METAL PICROLONATES, *Congr. Naz. Chim. Inorg. (Atti) 15th*, **1982**, 148; *Chem. Abstr.*, **1984**, *100*, 131423w.
Newton, C. G., Ollis, W. D., Podmore, M. L., Wright, D. E., CYCLIC MESOIONIC COMPOUNDS. PART 21. THE EXAMINATION OF NITRO DERIVATIVES OF MESO-IONIC HETEROCYCLES AS POTENTIAL PHARMACEUTICALS, *J. Chem. Soc., Perkin Trans. I*, **1984**, 63.
Perevalov, V. P., Baryshnenkova, L. I., Andreeva, M. A., Manaev, Y. A., Denisova, I. A., Stepanov, B. I., Seraya, V. I., EXAMINATION OF ISOMERIC BROMO-1-METHYLNITROPYRAZOLES, *Khim. Geterotsikl. Soedin.* **1983**, 1672; *Chem. Abstr.*, **1984**, *100*, 209683c.
Perevalov, V. P., Andreeva, M. A., Baryshnenkova, L. I., Manaev, Y. A., Yamburg, G. S., Stepanov, B. I., Dubrovskaya, V. A., SYNTHESIS OF 3-AMINO-4-NITROPYRAZOLES, *Khim. Geterotsikl. Soedin.*, **1983**, 1676; *Chem. Abstr.*, **1984**, *100*, 209684d.
Perevalov, V. P., Baryshnenkova, L. I., Denisova, E. A., Andreeva, M. A., Stepanov, B. I., ABNORMAL DIAZOTIZATION OF 5,4- AND 4,5-

AMINONITROPYRAZOLES AND REDUCTION OF 5-ARYLAZO-1-METHYL-4-NITROPYRAZOLE, *Khim. Geterotsikl. Soedin.*, **1984**, 1691; *Curr. Abstr. Chem.*, **1984**, *97*, 371451.

Wrzeciono, U., Linkowska, E., Rogowska, A., Lakoma, M., Kubiak, D., AZOLES. PART 11. BEHAVIOR of 2,7-DINITROINDAZOLE IN RELATION TO CYCLIC AND ALIPHATIC AMINES, *Pharmazie*, **1984**, *39*, 389.

Wrzeciono, U., Linkowska, E., Majewska, K., AZOLES. 15. STRUCTURAL CLARIFICATION OF SOME N-NITROINDAZOLE DERIVATIVES THROUGH C-13 NMR SPECTROSCOPY, *Pharmazie* **1984**, *39*, 498.

Zibuck, R., Stahl, M. A., Barchiesi, B., Waalwijk, P. S., Cogenfernandes, P., Habraken, C. L., INDAZOLE STUDIES. 4. THERMOLYSIS OF 3-BROMO-1-NITRO-1H-INDAZOLES IN BENZENE AND TOLUENE. FORMATION OF 1-PHENYL-1H-INDAZOLES, *J. Org. Chem.*, **1984**, *49*, 3310.

Chapter 4

Nitrotriazoles and Nitrotetrazoles

INTRODUCTION

Mono- and polycyclic nitrotriazole, triazoline, and tetrazole derivatives were briefly discussed in earlier reviews of the heterocycles.[1-5] This chapter covers the literature through *Chem. Abstr.*, **1983**, *98*; there is a bibliographic addendum for Vols. *99*, and *100*.

The literature assignments to 3- and/or 5-positions in these mono- and disubstituted 1,2,4-triazoles and their derivatives are reported here. Many of these assignments have not been rigorously established and may be better described as 3(5)- and or 5(3).

NITROTRIAZOLE PREPARATIONS

Nitration

Direct C-nitration of 1,2,3-triazole (**4-1**) (*v*-triazole) and 1,2,4-triazole (**4-2**) (*s*-triazole) and their C- and N-alkyl derivatives is unknown.[1-6]

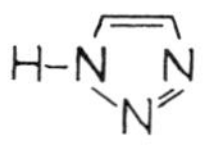

4-1 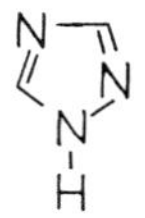   **4-2**

A modest success in C-nitration of the heterocyclic ring in certain aryltriazoles has accompanied the more facile nitration in the other aromatic ring, usually phenyl. Thus a mixture of concentrated nitric and sulfuric acids at 20° for 1 h converted 2-phenyl-1,2,3-2H-triazole to both the 2-(4-nitrophenyl)- (**4-3**) (80%) and the 2-(4-nitrophenyl)-4-nitro derivatives (**4-4**) (10%).[7,8] In nearly quantitative yield the 2-phenyltriazole was converted to the mononitro derivative (**4-3**) without a trace of the formation of the dinitro compound (**4-4**) by treatment with nitric acid in acetic anhydride.[8]

Mixed acid at 25° efficiently nitrated the mononitro compound (**4-3**) to the dinitro compound (**4-4**), and further nitration at 80° gave 2-(2,4-dinitrophenyl)-4-nitro-1,2,3-triazole (**4-5**).[8] In a similar reaction nitration by the mixed acid at 100° for 2 h converted 4-methyl-2-phenyl-1,2,3-triazole to the trinitro derivative (**4-6**) (90%).[9] The tetranitro compound (**4-7**) was obtained in 36% yield from the trinitro compound (**4-5**) and sodium nitrate in concentrated sulfuric acid at 145° for 16 h; it was also obtained by the nitration of 2-picryl-1,2,3-triazole in mixed acid.[10]

In sharp contrast 1-picryl- (**4-8**), 1-(4-nitrophenyl)- (**4-9**), and 1-(4-nitrophenyl)-4-nitro-1H-1,2,3-triazole (**4-10**) resisted nitration even in the presence of sodium nitrate and concentrated sulfuric acid at 140°, at which temperature degradation to water-soluble products occurred.[10]

4-3 W = Y = Z = H, X = NO_2
4-4 W = Y = H, X = Z = NO_2
4-5 W = H, X = Y = Z = NO_2
4-7 W = X = Y = Z = NO_2

4-6

4-8 X = H, Y = Picryl
4-9 X = H, Y = 4-$O_2NC_6H_4$
4-10 X = NO_2, Y = 4-$O_2NC_6H_4$

Monitration of 4-phenyl-1H-1,2,3-triazole in mixed acid at 0° gave the p-nitrophenyl derivative (50%), and dinitration at 90° for 20 h gave the op-dinitrophenyl derivative (35%). Attempts to nitrate further by treatment with sodium nitrate in concentrated sulfuric acid brought about degradation to water-soluble products.[10] There was no indication of nitration at the 5-position.

Mixed acid mononitration of phenyl-1,2,4-triazoles gave exclusively *p*-nitrophenyl derivatives. The two N-phenyltriazoles gave unspecified amounts, and an 84% yield was obtained from the C-phenyltriazole.[11] Apparently phenyl ring nitration was activated by the triazolyl substituents. Nitric acid also preferentially attacked N-alkenyl substituents in 1,2,4-triazoles.[12-17]

In a 1905 report[18], fuming nitric acid converted a hydroxytriazole to 3-hydroxy-5-nitro-1,2,4-triazole later formulated as the tautomeric 3-nitro-1,2,4-triazolin-5-one (**4-11**).[19-23] Other 3-nitro derivatives (**4-12a-g**) of 1,2,4-triazolin-5-ones were also obtained (56%-95%) from 1,4-disubstituted-Δ^2-1,2,4-triazolin-5-ones and fuming nitric acid. Nitration of 1-phenyl- and 1-benzyl-1,2,4-triazolin-3-ones exclusively gave *p*-nitrophenyl derivatives (**4-13a**) (50%) (and **4-13b**) (78%).[23] A free base mechanism was postulated for the nitration of 1,2,4-triazolin-5-one and its 1,4-dimethyl derivative in sulfuric acid (73-95%) under pseudo first-order conditions at 70° and 25°, respectively. The higher reactivity of the dimethyl derivative (about 9,000 times more reactive) was attributed to activation by electron release from the methyl substituent.[24]

4-11 X = Y = H
4-12a X = CH$_3$, Y = H
4.12b X = H, Y = CH$_3$
4-12c X = Y = CH$_3$
4-12d X = *p*-O$_2$NC$_6$H$_4$CH$_2$, Y = H
4-12e X = H, Y = *p*-O$_2$NC$_6$H$_4$CH$_2$
4-12f X = *p*-O$_2$NC$_6$H$_4$, Y = H
4-12g X = H, Y = *p*-O$_2$NC$_6$H$_4$

4-13a X = *p*-O$_2$NC$_6$H$_4$CH$_2$
4-13b X = *p*-O$_2$NC$_6$H$_4$

C-aminotriazoles are also activated to electrophilic substitution. Nitration of 4-picrylamino-1,2,3-1*H*-triazole by a mixture of sulfuric (96%) and nitric acid (100%) at 25° for 1 h gave the 5-nitro derivative (**4-14**) (30%).[25] On the other hand, nitration was not competitive with the oxidation by nitric acid (71%) of 1,2,4-triazole-3-thiol to 1,2,4-triazole (**4-2**) (55%).[26]

In a recent Russian patent the conversion of 1-trimethylsilyl-3-substituted 1,2,4-triazole by treatment with either dinitrogen pentoxide or a nitronium salt in an inert aprotic solvent to the 3(5)-nitro derivative (**4-15**) was described.[27] Excellent yields (75%-80%) were reported.[28]

4-14

4-15 X = H, Cl, CH$_3$

A suggested, enhanced nitration at the 5-position in 2-phenyl-1,2,3-triazole-1-oxide (**4-16a**), anticipated by (**4-16b**), was apparently diminished in an acid environment by protonation on the N-oxide oxygen atom. The product mixture obtained at 25° contained 2-*p*-nitrophenyltriazole-1-oxide, 5-nitro-2-phenyltriazole-1-oxide, and 5-nitro-2-*p*-nitrophenyltriazole-1-oxide after 1 min. After 2 h the final product, 2-(2,4-dinitrophenyl)-5-nitrotriazole-1-oxide (**4-17**) was present, and the starting material was consumed after 13 days. In a preparative run with an excess of nitric acid at 100° for 2 h the trinitro compound (**4-17**) was obtained in 50% yield. The substitution patterns for nitrating 2-phenyltriazole (above) and its 1-oxide (**4-16a**) were remarkably comparable. Nonacidic nitration of the 1-oxide (**4-16a**) was not investigated.[29]

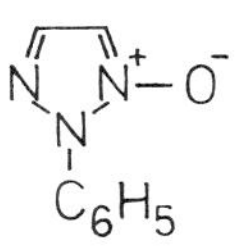

4-16a

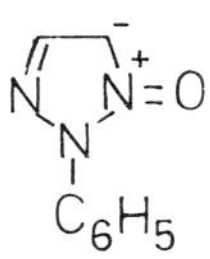

4-16b

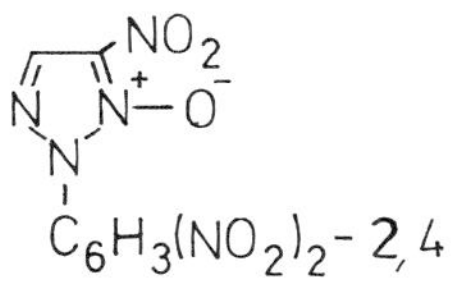

4-17

Rearrangement of Nitramines

Nonacidic nitration of 1,2,4-triazoles by nitronium salts[28,30] or by acetyl nitrate[30,31] produced nitramines. Nitronium tetrafluoroborate in acetonitrile at 0° converted 1,2,4-triazole (**4-2**) to the 1-nitro derivative (**4-18a**) (58%). A similar reaction afforded 1-nitro-3-methyl-1,2,4-triazole (**4-18b**).[28] Nitric acid in acetic anhydride at 0° converted 5-amino-1,2,4-triazole to its 1-nitro derivative (**4-19a**) (54%), and nitronium tetrafluoroborate afforded the isomeric nitramine (**4-19b**) (42%). In an interesting variation nitronium tetrafluoroborate converted 5-acetamido-1,2,4-triazole to the azo compound (**4-20**), mp > 400°.[30]

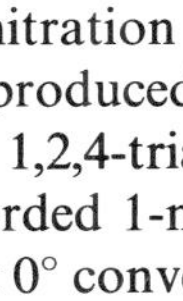

4-18a X = H
4-18b X = CH_3

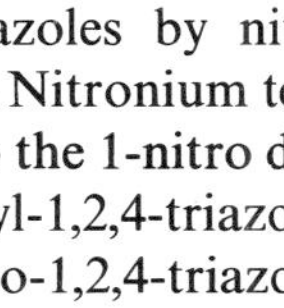

4-19a X = NO_2, Y = H
4-19b X = H, Y = NO_2

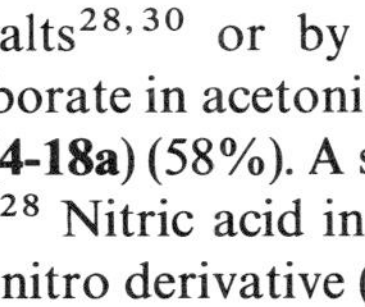

4-20

Thermal isomerization of 1-nitro-1,2,4-triazoles gave 3-nitro-1,2,4-triazoles. For this reaction Habraken suggested a [1,5] sigmatropic shift.[28,31,32] Related rearrangements of N-nitropyrazoles, also discovered by Habraken and co-workers, are discussed in Chapter 3. High yields were reported for the rearrangements to 3-nitro- (**4-21a**), 5-phenyl-3-nitro- (**4-21b**), and 3-*p*-nitrophenyl-5-nitro-1,2,4-triazole (**4-21c**). Insofar as there had been some previous success in the nitration of hydroxy and amino-1,2,4-triazoles (see above), the claim[22] that these rearrangements provided the first examples of direct nitration of the 1,2,4-triazole ring was incorrect.

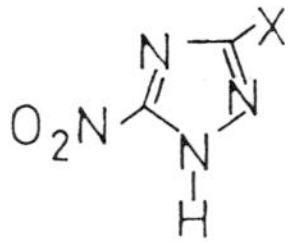

4-21a X = H
4-21b X = C_6H_5
4-21c X = *p*-$O_2NC_6H_4$

Thermolysis of 1-nitro-5-amino-1,2,4-triazole followed a different path. In benzonitrile at 100° for 36 h, an aminotriazolinone (**4-22**) (50%) was obtained and the nitramino-1,2,4-triazole (**4-19b**) was detected.[30]

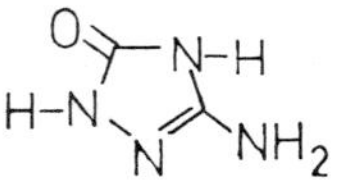

4-22

Nitro group migration accounted for the formation of 5-nitro-3-halo-1,2,4-triazoles (**4-23**) when 1-nitro-3-halo-1,2,4-triazoles were treated with bases and reducing agents.[33]

Conversion of Diazonium Salts

A patented preparation of 3,5-dinitro-1,2,4-triazole (**4-24**) apparently depended on a Sandmeyer reaction when the 3,5-diaminotriazole was treated with a hot aqueous solution of a mixture of sodium nitrite and cupric nitrate. In a similar reaction the dinitro compound (**4-24**) was also obtained from 3,4,5-triamino-1,2,4-triazine (guanazine).[34-36] The reaction was extended to a preparation of 3-nitro-1,2,4-triazole-5-carboxylic acid **4-25** (40%).[37] Decarboxylation at 115° gave 3-nitro-1,2,4-triazole (**4-26**).

Improved efficiency was claimed for an interesting modification of the reaction.[38] When 3-amino-1,2,4-triazole was diazotized by sodium nitrite (excess) in aqueous nitric acid below 10° for 30 min, 3-nitro-1,2,4-triazole (**4-26**) was produced in yields up to 70%.[39] Extended reactions gave (**4-25**) (25%) from 3-diazo-1,2,4-triazole-5-carboxylic acid and 5-methyl-3-nitro-1,2,4-triazole

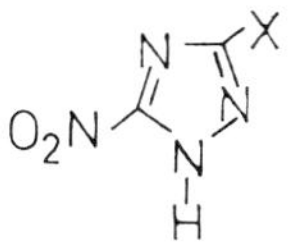

4-23 X = Cl, Br

(4-27) (61%).[38] In several instances the conversion was brought about by simply combining an aqueous solution of the amine and a mineral acid with a solution of an excess of sodium nitrite.[40] Not only was the presence of copper salts not needed, but it interfered with product isolation.[38]

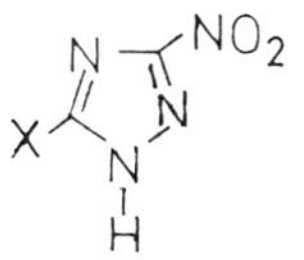

4-24 X = NO$_2$
4-25 X = CO$_2$H
4-26 X = H
4-27 X = CH$_3$

Microquantities of 3-nitro-1,2,4-triazole (**4-26**) were obtained from the pesticide Weedazol (3-amino-1,2,4-triazole) and nitrous acid.[41] The same method was chosen to prepare the nitro compound (**4-26**) with a ^{14}C label.[42]

Ring Formations

Addition of phenyl azide to nitroethenes to produce triazolines and, by an elimination, triazoles was simultaneously reported from Italy, Germany, and England.[43-46] In a sealed tube at 80° for several days 1-morpholino-2-nitroethylene and phenyl azide gave 1-phenyl-4-nitro-1,2,3-triazole (**4-28**) (60%); a similar reaction with *p*-nitrophenyl azide in ethanol at 80° for 100 h gave 1-(4-nitrophenyl)-4-nitro-1,2,3-triazole (**4-29**) (75%).[43] β-Nitrostyrene and phenyl azide in toluene at 110° for 17 h gave 4-nitro-1,5-diphenyl-1,2,3-triazole (**4-30**) via an intermediate Δ^2-1,2,3-triazoline.[45,46] The addition was sluggish and nonspecific, giving both (**4-30**) and 1,4-diphenyl-1,2,3-triazole.[46]

A 50% yield of 4-nitro-1,2,3-triazole was obtained from 1-morpholino-2-nitroethene and tosyl azide.[44]

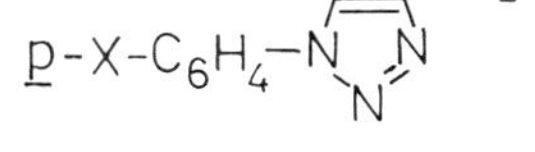

4-28 X = H
4-29 X = NO$_2$

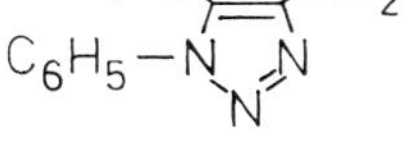

4-30

gem-Diamino-β-nitroenamines (**4-31a–e**) were similarly converted by treatment with *p*-chlorobenzenesulfonyl azide to 4-nitro-1,2,3-triazoles (**4-32a–e**) (15%–75%).[47]

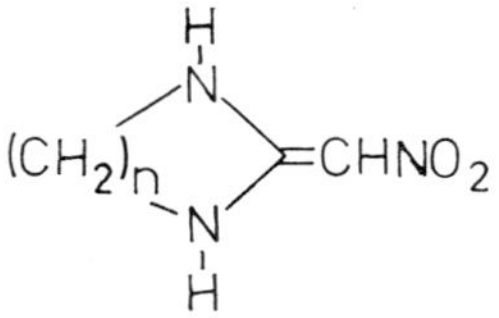

4-31a n = 2
4-31b n = 3

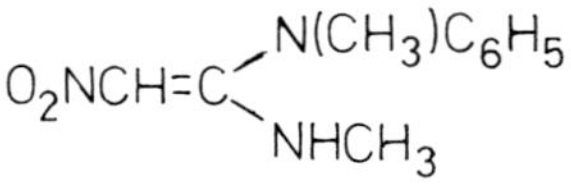

4-31c

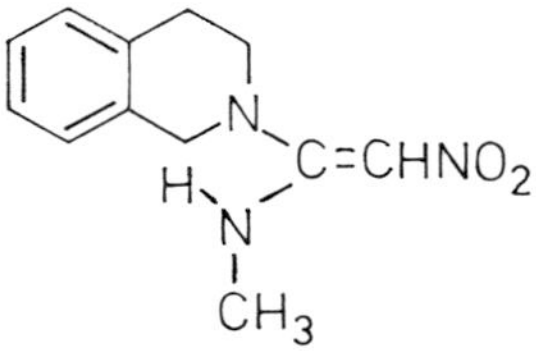

4-31d

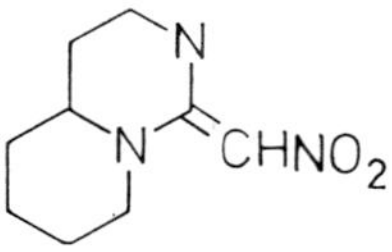

4-31e

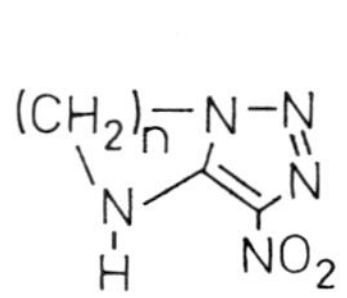

4-32a n = 2
4-32b n = 3

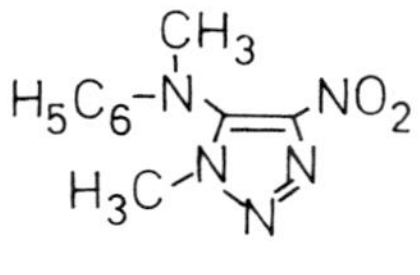

4-32c

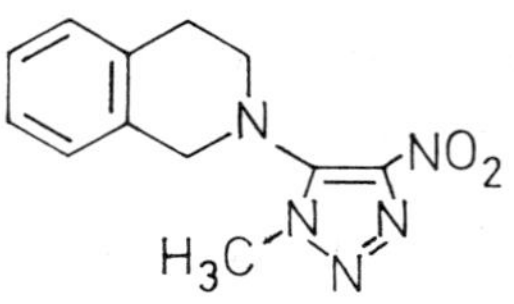

4-32d

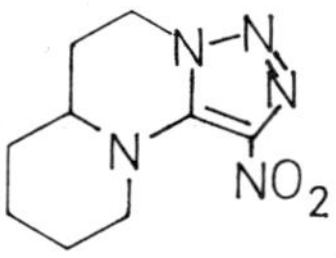

4-32e

Nitroethenes and sodium azide or hydrazoic acid produced triazoles without nitro substituents;[48–50] however, intermediate nitrotriazolines were assumed.[50] gem-Bromonitroethenes and sodium azide combined to form 4-bromo-4-nitrotriazoline salts which converted to 4-nitro-1,2,3-triazoles with ejection of the bromide anion.[51,52]

Acetic anhydride and alkali brought about ring closure of the coupling product (**4-33**) from diazotized 3-amino-1,2,4-triazole and methazonic acid to

4-nitro-2-(1,2,4-triazol-3-yl)-1,2,3-triazole (**4-34**) (63%).[38] A related reaction afforded 2-phenyl-4-nitro-1,2,3-triazole (**4-35**).[53,54]

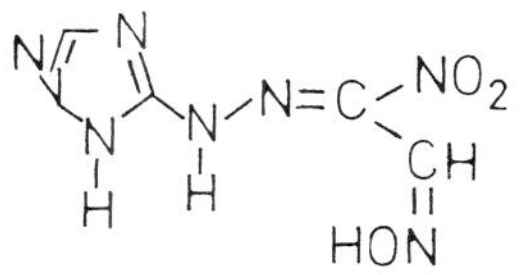

4-33

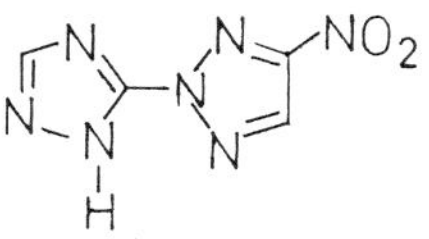

4-34

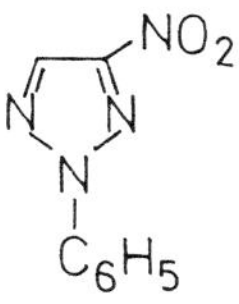

4-35

Nucleophilic Substitution

In facile attacks at the C-5 position in the 1-methoxy-2-phenyl-1,2,3-triazolium cation (**4-36**), a wide variety of nucleophiles brought about the formation of 5-substituted triazoles. Potassium and silver nitrite afforded 2-phenyl-4-nitro-1,2,3-triazole (**4-37a**) (18%, 23%) along with 2-phenyl-4-hydroxy-1,2,3-triazole (**4-37b**) (presumably by hydrolysis of the nitrite ester) (1%, 9%). Nitrosation of the latter compound and nitrite oxidation gave 2-phenyl-4-hydroxy-5-nitro-1,2,3-triazole (**4-38**) (32%, 46%).[29,55]

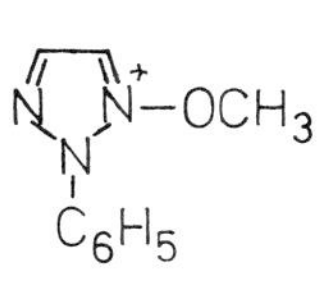

4-36

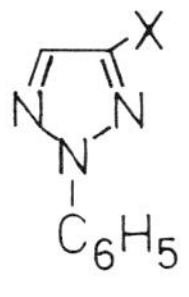

4-37a X = NO_2
4-37b X = OH

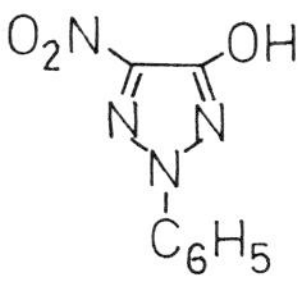

4-38

Oxidation of Aminotriazoles

Peroxytrifluoroacetic acid oxidized 3-amino- to 3-nitro-1,2,4-triazole (**4-15**), X = H (45%)[56] and the triazolylamine (**4-39a**) to the nitrotriazole (**4-39b**).[57]

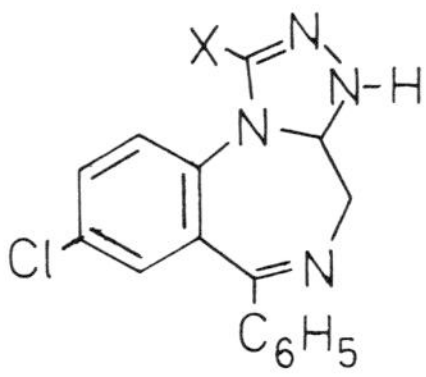

4-39a X = NH$_2$
4-39b X = NO$_2$

Fused Ring Triazoles

Electrophilic attack by chlorine, bromine, or mercuric acetate brought about degradation of the triazole ring in 1,2,3-triazolo[1,5-a]pyridine, but substitution in the Vilsmeier reaction and in nitration by a mixture of acetic anhydride and nitric acid gave the 3-formyl- (**4-40a**) and the 3-nitro derivative (**4-40b**) (25%).[58] In contrast, nitric acid in acetic anhydride converted 6,7-diphenyl-1,2,4-triazolo[4,3-b][1,2,4]triazine to the triazolotriazinone (**4-41**) (50%).[59]

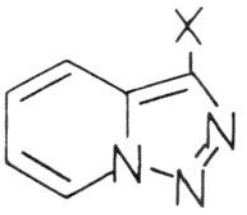

4-40a X = CHO
4-40b X = NO$_2$

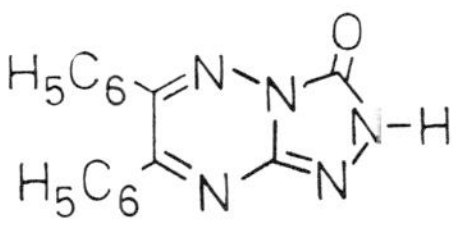

4-41

A mixture of concentrated nitric and sulfuric acids at about 0° for 45 min converted 2,6-dimethyl-3,7-dicarboxy-1,5-dehydro-1,2,3-triazolo[2.1-a][1,2,3]-triazole to the corresponding 3,7-dinitro compound (**4-42**) (50%).[60] Modifications of the reaction produced the related 3-nitro- and 3-nitro-7-carboxy compounds.

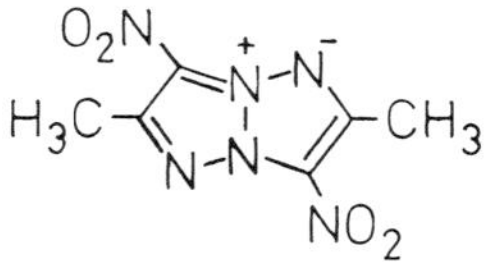

4-42

NITROTETRAZOLE PREPARATIONS

Nitration of the tetrazole ring has not been reported; however C-nitrotetrazoles have been prepared by the replacement of a diazonium with a nitro group. The copper salt (90% yield) of nitrotetrazole (**4-43**), from diazotized aminotetrazole and cupric nitrate, was described in 1931, 1937, and 1978 patents.[61-63] Apparently the formation of a nitrotetrazole from the diazonium salt does not require the presence of a cupric salt. Diazotization of 2-methyl-5-aminotetrazole in aqueous acetic acid containing an excess of nitrous acid gave 2-methyl-5-nitrotetrazole (**4-44**) (76%).[40,64,65] A similar reaction gave the 1-methyl isomer (**4-45**) (57%).[40] Nitration of 1-phenyltetrazole with fuming nitric acid at 25° gave 1-*p*-nitrophenyltetrazole (**4-46**) (92%). A similar reaction gave 2-*p*-nitrophenyltetrazole (95%).[66]

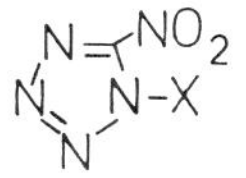

4-43 X = H
4-45 X = CH$_3$

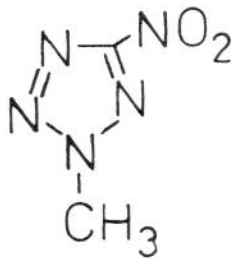

4-44

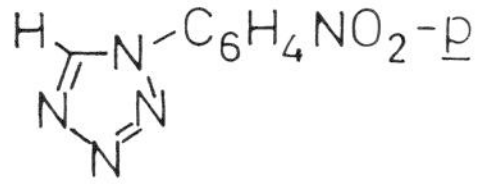

4-46

REACTIONS

Alkylation Reactions

Alkylation of 3-nitro-1,2,4-triazoles with dimethylsulfate and alkali in acetone for 8 h gave 1-alkyl derivatives: 1-methyl- (**4-47a**) (66%), 1,5-dimethyl- (**4-47b**) (55%), and 1-methyl-5-carbomethoxy- (**4-47c**) (51%). Dimethyl sulfate and the sodium salt of 3,5-dinitro-1,2,4-triazole in acetone gave the 1-methyl derivative (**4-47d**) (64%); the same product was obtained exclusively from the corresponding silver salt and methyl iodide in the dark in acetonitrile. When treated with an excess of diazomethane in ether, a derivative of 3-nitro-1,2,4-triazole in dioxane gave mixtures of 1-methyl-3-nitro- and 1-methyl-5-nitro- compounds (**4-47a**) (76%) and the isomer (**4-48a**) (24%), and (**4-47b**) (64%) and the isomer (**4-48b**) (36%). Substitution at N-4 was not detected.[67] Similar conversions gave triazoleacetates (**4-49**) from 3-nitro-1,2,4-triazoles[68] and allyltriazoles (**4-50**) from the sodium salt of dinitrotriazoles.[69,70] Ferrocenylmethyltriazoles and tetrazoles were obtained from the anions of 3-substituted and 3,5-disubstituted triazoles and 5-substituted tetrazoles.[71,72] Second-order rate constants and activation parameters were obtained for the methylation of 3-nitro-1,2,4-triazoles by dimethyl sulfate.[73]

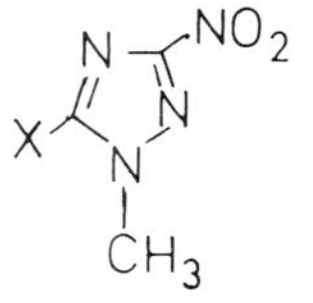

4-47a X = H
4-47b X = CH$_3$
4-47c X = CO$_2$CH$_3$
4-47d X = NO$_2$

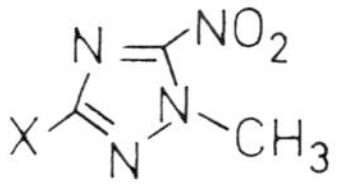

4-48a X = H
4-48b X = CH$_3$

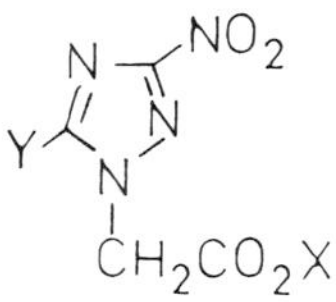

4-49 X = CH$_3$, C$_2$H$_5$,
n-C$_4$H$_9$
Y = H, CH$_3$, C$_2$H$_5$,
Br, NO$_2$

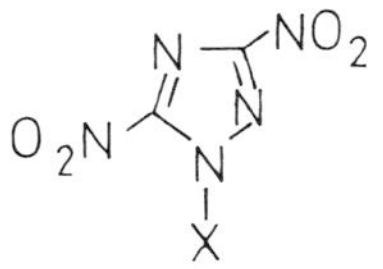

4-50
X = CH$_2$CH = CH$_2$
CH$_2$C(CH$_3$) = CH$_2$
CH$_2$CH = C(CH$_3$)$_2$

Sodium ethoxide catalyzed alkylations of 3-nitrotriazole to the 1-cyanomethyl and 1-acetomethyl derivatives.[74]

Oxiranes alkylated 3-nitro-,[75] 3,5-dinitro-,[75-79] and 3-nitro-5-bromo-1,2,4-triazoles.[75,80] Propylene oxide in aqueous sodium hydroxide was added to a solution of 3-nitrotriazole in ethanol. The mixture was stored in a sealed container until the pH reached 7.5–8.0. Dilution and extraction afforded the secondary alcohol which, without isolation, was oxidized by chromium trioxide to the triazolylpropanone (**4-51a**) (70%). Similar reactions gave the ketones (**4-51b**) (70%) and (**4-51c**) (82%). Triazolyl butanones (**4-52a–d**) (65%–90%) were obtained by Michael additions of an appropriate triazole to an unsaturated ketone.[75]

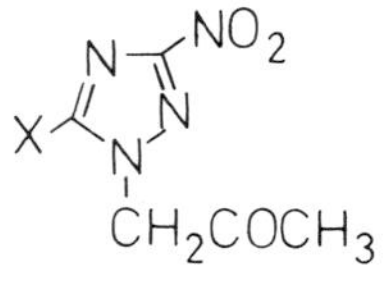

4-51a X = H
4-51b X = NO$_2$
4-51c X = Br

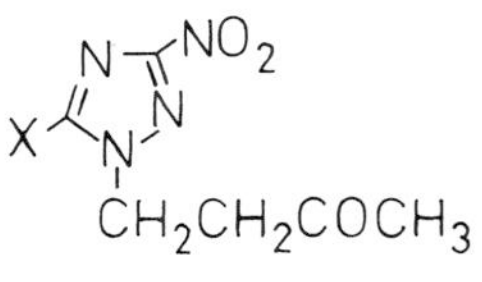

4-52a X = H
4-52b X = CH$_3$
4-52c X = Br
4-52d X = NO$_2$

Base-catalyzed cyclization of 5-bromo- and 5-nitrotriazolylpropanols gave 5,6-dihydrooxazolo[3,2-b]-1,2,4-triazole derivatives (**4-53**) (78%–89%).[76-80]

The product mixture obtained from a reaction between picryl chloride and potassium or lithium salts of 3,5-dinitrotriazole in acetonitrile at reflux temper-

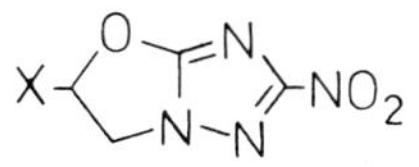

4-53 X = H, CH$_3$, CH$_2$Cl, CH$_2$OH, CH$_2$ONO$_2$, CH$_2$OCH$_3$

ature for 46 h contained 1-picryl-3-nitro-5-chloro-1,2,4-triazole (**4-55**) (20%), 1-picryl-3-nitro-1,2,4-triazol-5-one (**4-58**) (10%), and picric acid. A rationale for product formation was given, Eq. (4.1).[81]

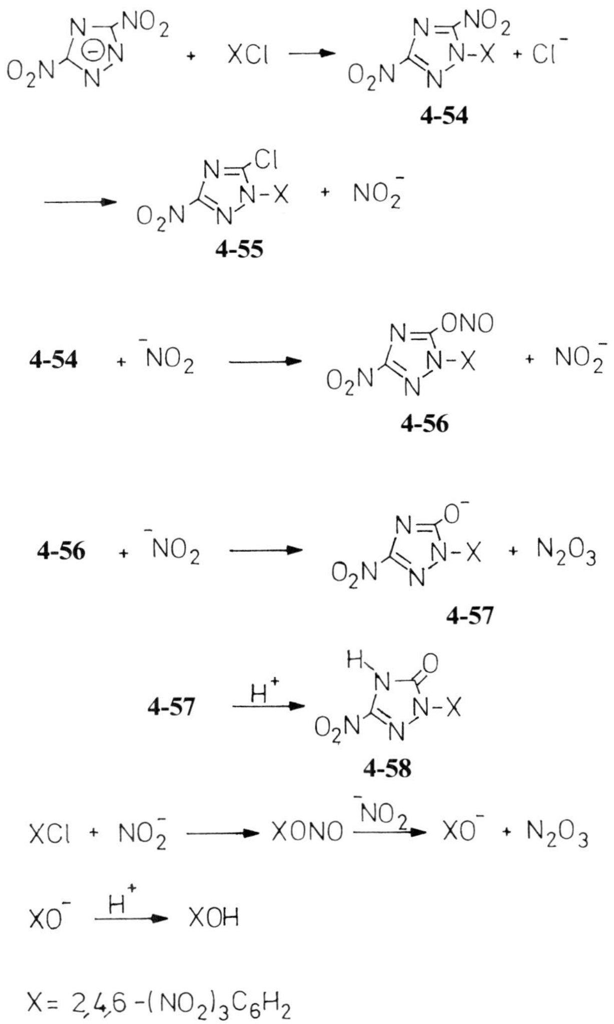

$$X = 2,4,6-(NO_2)_3C_6H_2 \qquad (4\text{-}1)$$

The first two reports of the alkylation of 3-nitro-1,2,4-triazolin-5-one appeared in 1969.[23,82] Dimethyl sulfate or methyl iodide in an alkaline

(1 equivalent of alkali) solution afforded 4-methyl-3-nitro-1,2,4-triazolin-5-one (**4-59**) (50%); in the presence of an excess of alkali (3–5 equivalents) 1-methyl-3-nitro-1,2,4-triazolin-5-one (**4-60**) (93%) was produced. In a similar reaction diethyl sulfate gave 4-ethyl-3-nitro-1,2,4-triazolin-5-one (yield not stated); however, all higher homologs of methyl iodide failed to alkylate the nitrotriazolinone. The nitrotriazolinone was regenerated from an unstable, unidentified material derived from the treatment of its silver salt with methyl iodide. Attempts to dialkylate the nitrotriazolinone were unsuccessful. The new alkyl nitrotriazolinones were identified as the same products obtained from nitrations of corresponding alkyltriazolinones.[82]

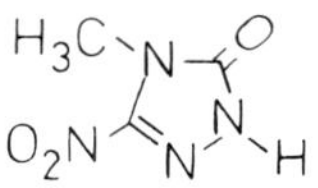

4-59

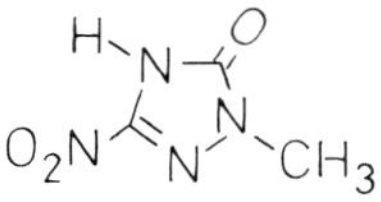

4-60

A controversy over dialkylation[14,62] was partially resolved in the efficient conversion of 1-methyl-3-nitro-1,2,4-triazolin-5-one to the 4-(2-hydroxyalkyl)-derivatives (**4-61a–e**) (30%–57%) or their nitrate esters by treatment with epoxides in ethanolic sodium hydroxide stored in a closed vessel at 25° until the pH reached 6.5–6.8.[83]

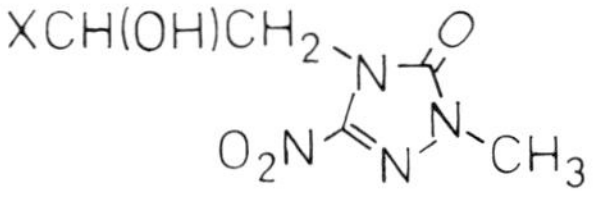

4-61a–e
X = H, CH_3, CH_2Cl
CH_2ONO_2, CH_2OH

In 1980 a reexamination of the methylation of 3-nitro-1,2,4-triazolin-5-one by dimethyl sulfate in alkali revealed monomethylation (97%) to the N-methyl compounds (**4-59**) and (**4-60**) and also the O-methyl isomer (**4-62**) in the ratio of 21:70:9. Further methylation gave the dimethyl compounds (**4-63**) and (**4-64**).[84]

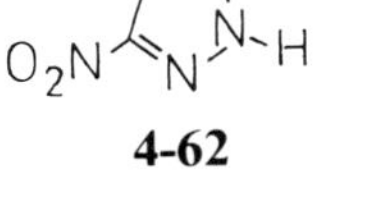

4-62

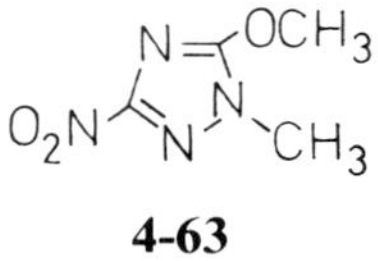

4-63

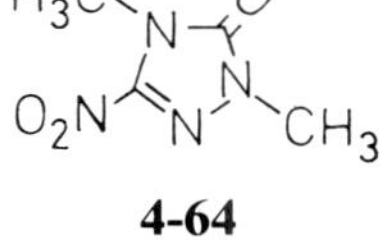

4-64

Michael reaction adducts from 3-nitro-1,2,4-triazolin-5-one and its monomethyl derivatives with methyl vinyl ketone in the presence of bases included 1,4-di-(3-oxobutyl)-3-nitro-1,2,4-triazolin-5-one (**4-65a**), 1-methyl-3-nitro-4-(3-oxobutyl)-1,2,4-triazolin-5-one (**4-65b**), and 1-(3-oxobutyl)-3-nitro-4-methyl-1,2,4-triazolin-5-one (**4-65c**).[85] In a Schmidt reaction (sulfuric acid and sodium azide), the diketone (**4-65a**) was converted to the diamide (**4-65d**) (40%). Attempts to obtain similar adducts with acrylonitrile and methyl acrylate were unsuccessful. In the Michael reaction 3-nitro-5-alkyl-1,2,4-triazoles were more active than 3-nitro-1,2,4-triazolin-5-ones.[75,85] Replacement of the 3-oxobutyl substituent by hydrogen was brought about by treatment with sodium hydroxide.[86]

Formaldehyde and N-methylolacetamide did not need a catalyst to convert 3-nitro-1,2,4-triazolin-5-one in acetone to the 1-hydroxymethyl (**4-65e**) (85%) and the 1-acetamidomethyl derivative (**4-65f**) (53%). In a similar reaction 3-nitro-4-methyl-1,2,4-triazolin-5-one (**4-59**) in acetone gave the 1-hydroxymethyl derivative (**4-65g**) (60%), but 1-methyl-3-nitro-1,2,4-triazolin-5-one (**4-60**) failed to react. The occurrence of substitution at N-1 but not at N-4 in hydroxymethylation reactions was proposed as a means to identify N-alkyl derivatives of 3-nitrotriazolinones.[85,86] In contrast, 2-nitropropene in ether converted 3,5-dinitro-1,2,4-triazole to the 4-(2-nitropropyl) derivative (**4-66a**).[87] Similar reactions gave the 4-(2-nitroethyl)- (57%) and the 4-(2-nitro-2-phenylethyl)-derivatives (**4-66b,c**). Formaldehyde also converted 3-nitrotriazoles to their 1-hydroxymethyl derivatives.[88]

4-65a X = Y = CH$_2$CH$_2$COCH$_3$
4-65b X = CH$_3$, Y = CH$_2$CH$_2$COCH$_3$
4-65c X = CH$_2$CH$_2$COCH$_3$, Y = CH$_3$
4-65d X = Y = CH$_2$CH$_2$NHCOCH$_3$
4-65e X = CH$_2$OH, Y = H
4-65f X = CH$_2$NHCOCH$_3$, Y = H
4-65g X = CH$_2$OH, Y = CH$_3$

4-66a X = CH$_3$
4-66b X = H
4-66c X = C$_6$H$_5$

Methyl iodide in acetone converted sodium 5-nitrotetrazole trihydrate to a mixture (79%) of 1- and 2-methyl-5-nitrotetrazole. Recrystallization from a mixture of benzene and petroleum ether gave the predominant 2-methyl isomer.[89] Anhydrous sodium 5-nitrotetrazole and picryl chloride gave 2-picryl-5-nitrotetrazole.[90]

Replacement of Nitro and Halo Groups

Soon after 3-nitro-1,2,4-triazoles were converted by treatment with halogen acids to the 3-chloro (bromo) derivative,[22,37,91] they were transformed to the 3-fluoro derivative (**4-67a**) (80%) by treatment with liquid hydrogen fluoride at 150° in a Monel or Teflon-lined bomb for 48 h. Similar reactions gave the 5-bromo (**4-67b**) (98%), 5-acetyl (**4-67c**) (70%), and 5-hydroxy (**4-67d**) (21%) derivatives of 3-fluoro-1,2,4-triazole.[92]

4-67a X = H
4-67b X = Br
4-67c X = $COCH_3$
4-67d X = OH

Hydrochloric acid at reflux temperature for 8 h converted 3-nitro-1,2,4-triazole to 3-chloro-1,2,4-triazole (**4-68a**) (60%). Comparable treatment with the appropriate acid afforded the 3-bromo derivative (**4-68b**) (50%) and related compounds (**4-68c–f**) (40%–68%) from 3- and 5-nitrotriazoles and 3,5-dinitrotriazoles. Heating briefly with sodium hydroxide converted the halides (**4-68e,f**) to 1-methyl-3-nitro-5-hydroxy-1,2,4-triazole (**4-68g**) (or its triazolinone tautomer) (72%, 69%). As expected, each halide (**4-68e,f**) gave 1-methyl-3-nitro-5-methoxy-1,2,4-triazole (**4-68h**) (40%) by treatment with sodium methoxide in methanol at 70° for 3 h. Sodium ethoxide afforded the ethyl ether (44%), and sodium phenoxide in methanol gave the phenyl ether (74%).[93] Hydrohalic acids reacted with 1-alkeynl-3,5-dinitro-1,2,4-triazoles by replacing the 5-nitro group with chlorine and by adding hydrogen halide to the unsaturated side chain.[94]

Aqueous methylamine at 90° for 3 h afforded 1-methyl-3-nitro-5-methylamino-1,2,4-triazole (**4-68i**) (62%) from either 1-methyl-3-nitro-5-bromo-1,2,4-triazole or the corresponding chloride; dimethylamine gave the corresponding 5-dimethylamino derivative (**4-68j**) (60%), and methylhydrazine gave 1-methyl-3-nitro-5-(1-methylhydrazino)-1,2,4-triazole (**4-68k**) (56%). In addition to the expected formation of 1-methyl-3-chloro-5-amino-1,2,4-triazole (**4-68m**), the formation of 1-methyl-3-nitro-5-chloro-1,2,4-triazole (**4-68e**) was reported for the reaction between 1-methyl-3-nitro-5-amino-1,2,4-triazole and urea in concentrated hydrochloric acid at reflux temperature for 6 h. The formation of compound (**4-68e**) was attributed to the diazotization of the starting amine by nitrous acid made available during the substitution of a nitro group with subsequent replacement of a diazonium group by chlorine in a reaction with hydrochloric acid.[93]

Aqueous ammonia and the dinitrotriazole (**4-47d**) in an autoclave at 80° for 4 h gave 1-methyl-3-nitro-5-amino-1,2,4-triazole, confirmed by reduction of the diazotized amine with hypophosphorus acid to 1-methyl-3-nitro-1,2,4-triazole. The dinitrotriazole (**4-47d**) was converted by methylamine and dimethylamine to the 5-methylamino and 5-dimethylamino derivatives of 1-methyl-3-nitro-1,2,4-triazole.[95]

Hydrazine hydrate in dioxane at 15° converted 1-methyl-3,5-dinitro-1,2,4-triazole (**4-47d**) to a mixture of the 5-hydrazino derivative (up to 70%) and, by reduction, to the 5-amino derivative (up to 86%) of 1-methyl-3-nitro-1,2,4-triazole (**4-47a**).[96]

Ammonia, methylamine, and ethylamine in alcohol failed to convert the dinitrotriazole (**4-47d**) to an aminotriazole; instead, 1-methyl-3-nitro-5-ethoxy-1,2,4-triazole (**4-68n**) was produced. In preparative procedures of 5-alkoxy (phenoxy) triazoles (**4-68n**) through (**4-68s**) (10%–72%), the presence of a tertiary amine was chosen to avoid the possible formation of an aminotriazole as an impurity. Hydrolysis of the ethoxy and the propoxy derivatives (**4-68n**) and (**4-68p**) with hydrobromic acid replaced the 3-nitro substituent with bromine to give 1-methyl-3-bromo-1,2,4-triazolin-5-one (**4-69a**) (45%, 25%), a compound also obtained from 1-methyl-3-nitro-1,2,4-triazolin-5-one (**4-69b**) and hydrobromic acid.[91]

4-68a   X = Z = H, Y = Cl
4-68b   X = Z = H, Y = Br
4-68c   X = Br, Y = H, Z = CH$_3$
4-68d   X = Y = Cl, Z = H
4-68e   X = Cl, Y = NO$_2$, Z = CH$_3$
4-68f   X = Br, Y = NO$_2$, Z = CH$_3$
4-68g   X = OH, Y = NO$_2$, Z = CH$_3$
4-68h   X = OCH$_3$, Y = NO$_2$, Z = CH$_3$
4-68i   X = NHCH$_3$, Y = NO$_2$, Z = CH$_3$
4-68j   X = N(CH$_3$)$_2$, Y = NO$_2$, Z = CH$_3$
4-68k   X = N(CH$_3$)NH$_2$, Y = NO$_2$, Z = CH$_3$
4-68m   X = NH$_2$, Y = Cl, Z = CH$_3$
4-68n   X = OC$_2$H$_5$, Y = NO$_2$, Z = CH$_3$
4-68p   X = OC$_3$H$_7$-n, Y = NO$_2$, Z = CH$_3$
4-68q   X = OC$_6$H$_5$, Y = NO$_2$, Z = CH$_3$
4-68r   X = OC$_6$H$_4$NO$_2$-m, Y = NO$_2$, Z = CH$_3$
4-68s   X = OC$_6$H$_4$NO$_2$-p, Y = NO$_2$, Z = CH$_3$

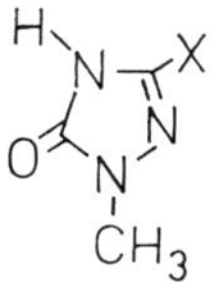

4-69a X = Br
4-69b X = NO$_2$

Both aqueous alkali hydroxides and triethylamine in acetonitrile promoted the conversion of 1-nitro-3-X-1,2,4-triazoles (X = Cl, Br, NO$_2$) (**4-70a**) to 1,5'-bis-1,2,4-triazolyls (**4-70b**) (20%–70%), the corresponding 3(5)-X-1,2,4-triazole (**4-70c**), and, as previously noted, 3-X-5-nitro-1,2,4-triazole (**4-70d**) by rearrangement. The conversions were also effected by reducing agents (KI, Fe^{+2}, H$_2$PO$_2^-$). In one explanation the formation of the bistriazolyl was attributed to a nucleophilic attack at C-5 of the starting nitramine by the anion of the triazole (**4-70c**) followed by loss of the nitrite anion, Eq. (4-2). In a second proposed pathway the loss of the nitrite anion from the radical anion of the starting nitramine left an azole radical that could convert to a bis-triazolyl product by attacking the starting nitramine or by dimerization, Eq. (4-3).[33]

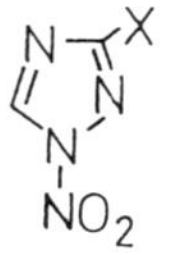

4-70a 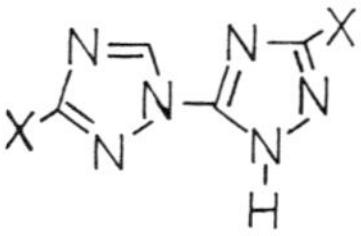   **4-70b** 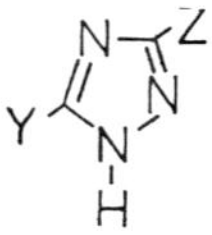   **4-70c** Y = H, Z = X
                                   4-70d Y = X, Z = NO$_2$

$$\textbf{4-70a} \xrightarrow[-\text{BONO}]{B^-} \textbf{4-70a} \xrightarrow[-\text{NO}_2^-]{} \textbf{4-70b} \qquad (4\text{-}2)$$

X = Cl, Br, NO$_2$

$$[\textbf{4-70a}]^{\bar{\cdot}} \xrightarrow{-\text{NO}_2^-} \left[\begin{array}{c} \text{triazolyl radical} \end{array} \right]^{\cdot} \xrightarrow{i/ii} \textbf{4-70b}$$

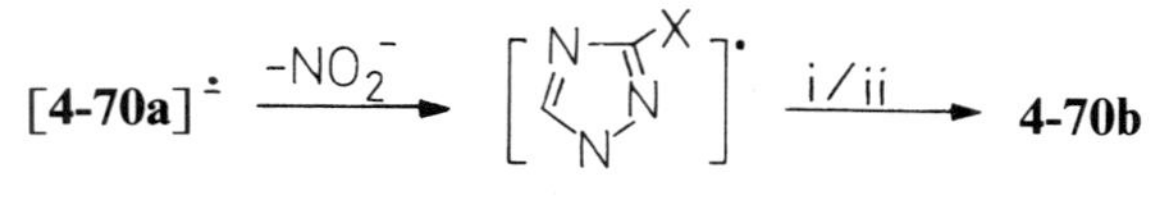

i = **4-70a**, -NO$_2$; ii = dimerization   (4-3)

Azole N-anions converted 1-substituted-3,5-dinitro-1,2,4-triazoles to a mixture of 1-substituted-3-nitro-5-(N-azolyl)-1,2,4-triazoles (predominant) and 1-substituted-3-nitro-1,2,4-triazolin-5-ones. Thus the dinitrotriazole (**4-47d**) and 1,2,4-triazole in sodium hydroxide gave 1-methyl-3-nitro-5-(1,2,4-triazol-1-yl)-1,2,4-triazole (**4-71**) and 1-methyl-3-nitro-1,2,4-triazolin-5-one (**4-69b**). Other bis-triazoles were similarly obtained when the 1-substituent was carbomethoxy-methyl, acetyl, vinyl, and carbomethoxy and the anion of 1,2,4-triazole contained 3-nitro, 3-chloro, 3-acetamido, 3-nitro-5-methyl, or 3,5-dichloro substituents. Similar reactions were afforded by anions of derivatives of 1,2,3-triazoles, pyrazoles, and imidazoles. But the base-promoted reaction between the dinitrotriazole (**4-47d**) and tetrazole gave 1-methyl-3-nitro-5-amino-1,2,4-triazole (60%) instead of the expected 5-tetrazolyltriazole. When the reaction was extended to 1-methyltetrazole, the expected product, 1-methyl-3-nitro-5-(5-methyltetrazolyl)-1,2,4-triazole, was obtained.[97]

The analogous replacement of a nitro group in 1-(2-nitroethyl)-3,5-dinitro-1,2,4-triazole did not occur on treatment with an alkaline solution of 3-nitro-1,2,4-triazole. Instead trans alkylation gave 1-(2-nitroethyl)-3-nitro-1,2,4-triazole (**4-72**) and the sodium salt of 3,5-dinitro-1,2,4-triazole. The transfer was attributed to a reverse Michael reaction followed by a Michael reaction. The ketone (**4-52**) (X = NO$_2$) proved to be more resistant to alkyl transfer and gave the expected base-catalyzed nucleophilic displacement of its 5-nitro group with methoxide, phenoxide, and both substituted and unsubstituted triazol-1-yl anions. The by-product formation of 1-(3-oxobutyl)-3-nitro-1,2,4-triazolin-5-one (**4-73**) by alkaline hydrolysis of the dinitrotriazole (**4-52**) (X = NO$_2$) was not unexpected. Each example of a 1-(3-oxobutyl)-nitrotriazole or triazolinone on storage in aqueous sodium hydroxide until the mixture became homogeneous lost methyl vinyl ketone from the 1-position by a reverse Michael reaction.[98]

4-71

4-72

4-73

Competitive second-order replacement of halogen and nitro groups in 1-methyl-3-nitro-5-halo-1,2,4-triazole (**4-68e,f**) by hydroxide ions,[99] alkoxide ions,[100] and amines[101,102] has been examined. Rates of substitution by the hydroxide ion were 30 times faster for replacement of the 5-chloro (bromo) than for the 3-nitro substituent. Similar substitution of the 5-nitro substituent in the dinitrotriazole (**4-47d**) occurred 300,000 times faster than replacement of a 5-chloro (bromo) group in (**4-68e,f**), a result in qualitative agreement with related nucleophilic substitutions in other substrates where the rate ratios were about 1000:1.[99]

An interesting quantitative comparison of chlorine and nitro groups in different positions was made. Substitution of chlorine in (**4-68e**) proceeded 30 times faster than substitution of the nitro group, but in 1-methyl-3-chloro-5-nitro-1,2,4-triazole (**4-74a**) there was practically no substitution of chlorine. The ratio of substitution rate constants for the nitro groups in compounds (**4-74a**) and (**4-74b**) was 117:1 (compare 50:1 for (**4-68e**) and (**4-47d**)). Apparently the replacement of a nitro group was enhanced more by a 3-chloro than by a 5-chloro substituent.[99]

4-74a X = Cl
4-74b X = H

Lithium, sodium, and potassium methoxide and ethoxide in the corresponding alcohol replaced a 5-halogen and 3-nitro substituent in the triazoles (**4-68e,f**) in the variable ratio 300:1–700:1. The second-order reaction rate increased as the metal cation changed from lithium to sodium to potassium.[100]

Piperidine also replaced the 5-chloro group in the triazole (**4-68e**) in benzene, dioxane, dimethylformamide (DMF), and mixtures of these solvents with methanol. A catalysis by piperidine in benzene, dioxane, and their mixtures with small amounts of methanol was detected but was absent in DMF, methanol, and water. Rate constants and activation energies for the formation of the intermediate complex between the substrate and the amine were determined.[101] In similar displacements by methylamine, dimethylamine, and diethylamine, only 1-methyl-3-nitro-5-dialkylamino-1,2,4-triazoles were produced.[102,103]

Although the cyanide anion did not displace a 5-nitro substituent, it did convert the dinitrotriazolyl ketones (**4-51b**) and (**4-52**) (X = NO$_2$) to 5,6-dihydrooxazolo and 5,6-dihydrooxazino[3,2-*b*]-1,2,4-triazole (**4-75a**) (84%) and (**4-75b**) (66%), respectively. Ring closure with displacement of a nitro group was attributed to a base catalyzed reaction of an intermediate cyanohydrin.[104]

4-75a n = 1
4-75b n = 2

Among other 1,2,3- and 1,2,4-triazole derivatives, 4-nitro-1,2,3-triazole underwent an acid catalyzed fusion with an acylated ribofuranose to give the corresponding 2-β-**D**-ribofuranosyl-4-nitro-1,2,3-triazole **4-76** and the isomeric 1-β-**D**-ribofuranosyl derivative **4.77**.[105–107] The fusion procedure with methyl

3-nitro-1,2,4-triazole-5-carboxylate and 1,2,3,5-tetra-*O*-acetyl-*β*-**D**-ribofuranose gave methyl 3-nitro-1-(2,3,5-tri-*O*-acetyl-*β*-**D**-ribofuranosyl)-1,2,4-triazole-5-carboxylate (**4-78**).[108] Glucosyl derivatives of 5-nitro-1,2,4-triazoles were similarly prepared.[109,110] In related reactions, protected guanosine and uridine derivatives were converted to compounds (**4-79**) and (**4-80**) by treatment with 1-(mesitylene-2-sulfonyl)-3-nitro-1,2,4-triazole.[111,112] Utilization of the reaction in related systems is becoming widespread.[113–118]

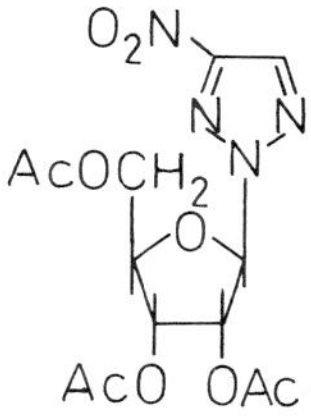

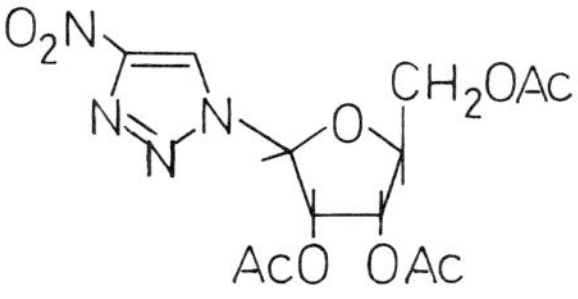

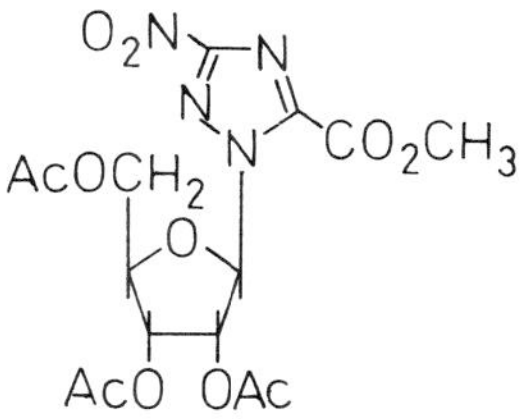

Recently 1-(8-quinolinesulfonyl)-3-nitro-1,2,4-triazole was found to be an effective condensing agent for oligonucleotide syntheses.[119,120]

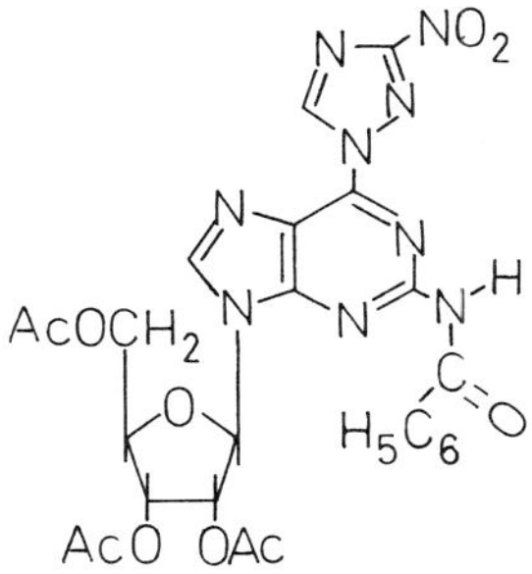

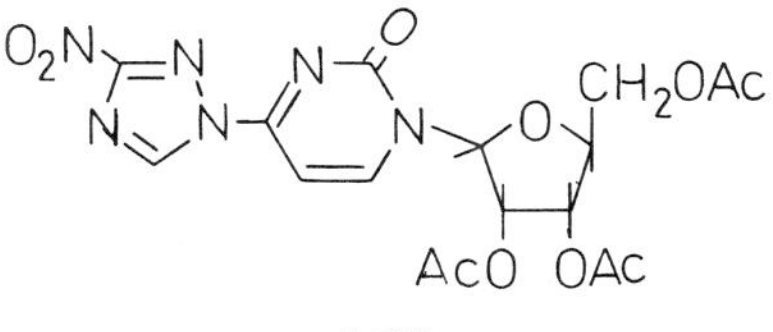

Reduction

Catalytic reduction of nitro-1,2,3-triazoles,[25,43,44,46,58,60,105] 1,2,4-triazoles,[95] tetrazoles,[64] Δ^2-1,2,3-triazolines,[45] and Δ^5-1,2,4-triazolin-3-ones[21,82] by hydrogen over palladium[25,44,58,60,95,105] or platinum[21,45,64,82] gave the corresponding aminoazoles/aminoazolones in good yields. Transfer hydrogenation by heating a solution of the 3-nitrotriazolopyridine (**4-40b**) over palladium on charcoal gave 3-(2-pyridyl)imidazo-[1,5-*a*]pyridine (**4-81**) (20%); hydrogenation over palladium gave a mixture that contained the 3-aminotetrahydrotriazolopyridine (**4-82**). A rationale for the deep-seated change (**4-40b**) → (**4-81**) via carbene intermediates and loss of nitrogen was presented.[58]

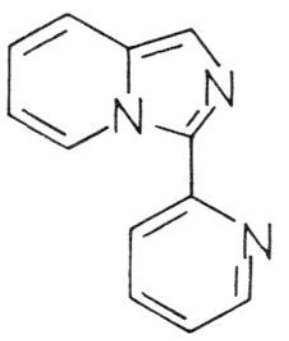

4-81

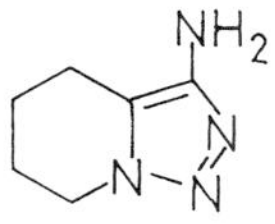

4-82

It was noted above that hydrazine converted 1-methyl-3,5-dinitro-1,2,4-triazole (**4-47d**) to the 5-hydrazino derivative and the 5-amino derivative of 1-methyl-3-nitro-1,2,4-triazole. Reduction to the amine (27%) was also brought about by 1,1-dimethylhydrazine; phenylhydrazine brought about reduction to 1-methyl-3-nitro-5-hydroxylamino-1,2,4-triazole (**4-83a**) (23%), which was oxidized by nitric acid to the corresponding azoxy compound (10%) and to the dinitrotriazole (**4-47d**) (52%). Conversion of the latter to the 5-azidotriazole (**4-83b**) (33%) was brought about by treatment with acetylhydrazine; the azide was also obtained from 1-methyl-3-nitro-5-hydrazino-1,2,4-triazole and nitrous acid.[96]

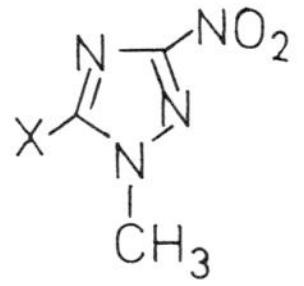

4-83a X = NHOH
4-83b X = N$_3$

Hydrazine also reduced a nitroribofuranosyl-1,2,4-triazole[56] and 3-nitro-Δ^2-1,2,4-triazolin-5-ones[23] to corresponding amines. In its conversion of 1-(oxoalkyl)-3,5-dinitro-1,2,4-triazole to triazolo[3,2-*c*]-1,2,4-triazines (**4-84a**) and triazepines (**4-84b**), a differentiation between an initial attack by hydrazine at the C-NO$_2$ function or at the carbonyl function was not made. Hydroxylamine reduced the dinitrotriazole to an amine (**4-85**) (70%).[121]

An expected exchange of an azido group for a methylamino group did not occur when 1-alkyl-3-nitro-5-azido-1,2,4-triazole was treated with methyl amine. Instead a reduction gave the corresponding 5-aminotriazole. Similar reductions were also effected by dimethyl and triethyl amines but not by ammonia or ethylene imine, which afforded recovery of the azide. Various 3-azido-1,2,4-triazoles were neither reduced nor replaced when subjected to the action of methyl amine.[122]

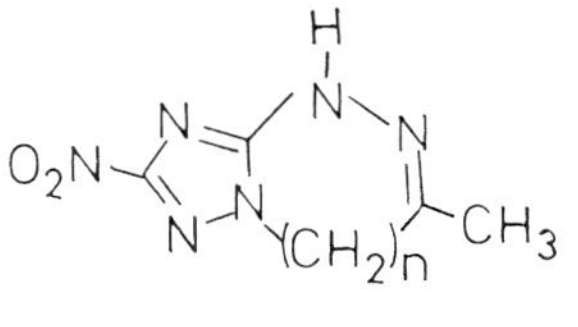

4-84a n = 1
4-84b n = 2

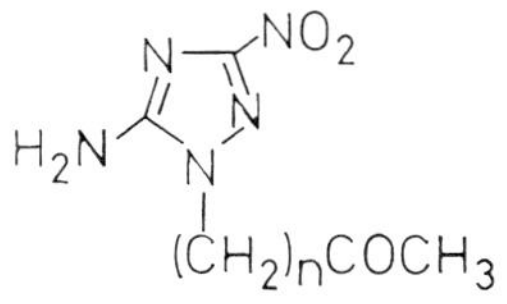

4-85 n = 1,2

An attempt to repeat a conversion of 3-nitro-1,2,4-triazolin-5-one to the corresponding aminotriazolone[18] by zinc and hydrochloric acid was unsuccessful and apparently brought about ring cleavage. The desired reduction was achieved in 74% yield by catalytic hydrogenation over platinum.[21]

The electrochemical reduction of nitroazoles, including 3-nitro-1,2,4-triazole, showed two one-electron irreversible changes at -0.8 to 1.5 V and at -1.9 to -2.4 V. It was contended that the latter confirmed the existence of a dianion radical.[123]

Miscellaneous

Bromine converted 3-nitro-1,2,4-triazole in an aqueous alkaline solution to the 1-bromo derivative (**4-86**) (90%).[124]

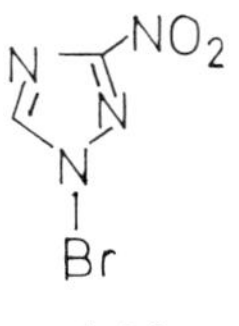

4-86

An expected nucleophilic substitution of the nitro group was brought about by treating 1-alkyl-3-nitro-1,2,4-triazoles (**4-87a–d**) with hydroxide, methoxide, and ethoxide anions. Demethylation by *t*-butoxide ions to 3(5)-nitro-1,2,4-triazole (**4-15**) (X = H) was reminiscent of the dealkylation of alkylheterocycles by the action of sodium naphthalide, solutions of metals in liquid ammonia, and other reducing agents. It was suggested that dealkylation was initiated by an electron transfer from the *t*-butoxide anion to convert the alkyltriazole to its radical anion, which dissociated to the anion of the product (**4-15**) (X = H) isolated after acidification, Eq. (4-4). Abstraction of hydrogen from the solvent presumably accounted for the conversion of 1-(3-nitro-1,2,4-triazol-1-ylmethyl)-

3-nitro-1,2,4-triazole (**4-87d**) to both (**4-15**) (X = H) and its 1-methyl derivative
4-47a (X = H) (20%).[125]

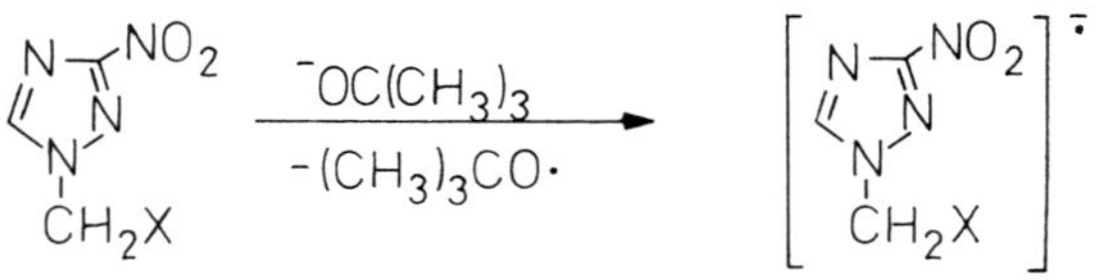

4-87a X = H
4-87b X = CO$_2$C$_2$H$_5$
4-87c X = CN

4-87d X =

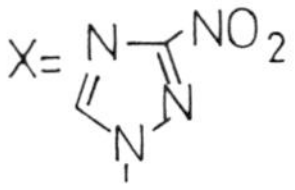

$$\mathbf{4\text{-}88} \xrightarrow{-\dot{C}H_2X} \quad \xrightarrow{H^+} \quad \mathbf{4\text{-}15}\ (X = H) \qquad (4\text{-}4)$$

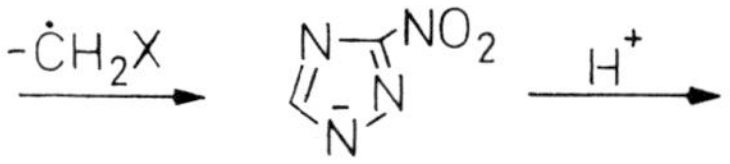

In contrast with the nucleophilic replacement of a nitro group or a halogen at C-5 in a 1,2,4-triazole (above)—eg, by treatment with an aliphatic amine—an attempt to replace a C-5 azido group by treatment with methylamine unexpectedly afforded instead a reduction to the C-5 amine (**4-89a–d**). A yield of 82% was reported for 1-methyl-3-nitro-5-amino-1,2,4-triazole (**4-89a**). Similar conversions were effected by treatment with dimethyl- and triethylamines. Attempts to reduce 3-azido-1,2,3-triazoles in this manner were unsuccessful.[126] A comparable reduction of an azidopyridine by amines was recently reported.[127]

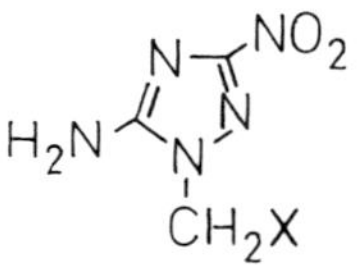

4-89a X = H
4-89b X = CO$_2$H
4-89c X = CO$_2$CH$_3$
4-89d X = CONHCH$_3$

A Schmidt reaction with hydrazoic acid in concentrated sulfuric acid converted 1-oxoalkyl-3-nitro-5-substituted-1,2,4-triazoles to 1-acetamidoalkyltriazoles (**4-90a–e**) (38%–75%), Eq. (4-5). Acid hydrolysis followed by neutralization with alkali converted the amides (**4-90a,e**) to corresponding amines; a spontaneous ring closure of the latter when a 5-nitro substituent was present gave 2-nitro-5,6-dihydro-4H-imidazo[3,2-b]1,2,4-triazole **4-91** (51%),[128] Eq. (4-6) a structure supported by ir spectroscopy, elemental analysis, an ebullioscopic mw 153 (calculated 153), and analogous reactions.[77,95]

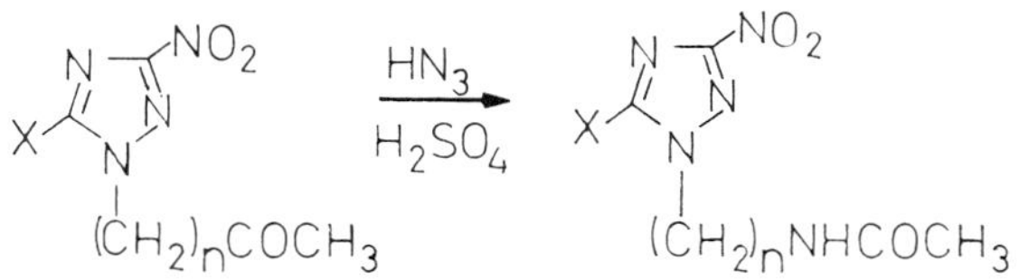

4-90a X = H, n = 1
4-90b X = Br, n = 1
4-90c X = H, n = 2
4-90d X = Br, n = 2
4-90e X = NO_2, n = 2

(4-5)

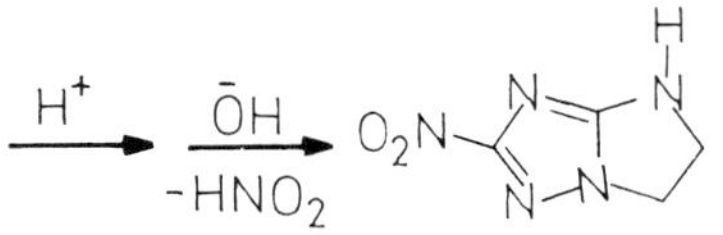

4-91

(4-6)

PROPERTIES

Acid-Base Properties

Acidities in pK_a values are known for at least 30 nitro derivatives of 1,2,4-triazoles[98,129-131] and 1,2,4-triazol-5-ones[19,24] and for 5-nitrotetrazole.[132]

Good correlations with Hammett's σ_I, σ_m, and σ_p constants were found for the acidities of seventeen nitro-1,2,4-triazoles. High ρ values revealed a great sensitivity of the triazole system to substituent effects. Alkyl substitution at C-5 lowered the acidity of 3-nitro-1,2,4-triazole (pK_a 6.05) by 0.55–0.70 pK_a units, and a 5-phenyl substituent raised the acidity to pK_a 5.65. The higher acidity of 3,5-dinitro-1,2,4-triazole, pK_a −0.66, was attributed to the expected electron-withdrawing effect of the nitro groups. From a spectroscopic analysis of related structures, a dissociation of carboxylic hydrogen accounting for $pK_{a1} < 2$, and a dissociation of imine hydrogen accounting for pK_{a2} 6.10 was established for 3-nitro-1,2,4-triazole-5-carboxylic acid (**4-25**).[129,130]

It was not possible to obtain a constant pK_{BH^+} by using the H_o scale; however, the amide scale of acidities H_A permitted the calculation of pK_{BH^+} values between −2.86 and −3.14 for 3-nitro-5-methyl-1,2,4-triazole.[129,130] A spectroscopic determination gave pK_{BH^+} −7.14 for 3-azido-5-nitro-1,2,4-triazole (**4-92a**) and pK_{BH^+} −6.51 for the 1-methyl derivative (**4-92b**). Although organic azides generally decompose rapidly after protonation by mineral acids, the azides (**4-92a,b**) remained unchanged at 90° in concentrated sulfuric acid. Apparently protonation of the azido group was not competitive with protonation of a ring nitrogen atom.[131]

Information on acidity determination for nitrotriazololones is limited to

4-92a X = H
4-92b X = CH$_3$

3-nitro-1,2,4-triazol-5-one (**4-11**) pK$_a$ 3.63, pK$_{BH^+}$ −3.41, and the 1,4-dimethyl derivative (**4-12c**), pK$_{BH^+}$ −3.68.[19,24] The pK$_{BH^+}$ differences were too small to conclude that the dimethyl derivative (**4-12c**) was less basic than the unmethylated compound (**4-11**).

Acidities for a series of 5-substituted tetrazoles were determined spectroscopically and were found to follow the linear relationship pK$_a$ = −0.78 pK$_{BH^+}$ + 6.37. Values for 5-nitro-tetrazole were pK$_a$ −0.83, pK$_{BH^+}$ 9.2.[132]

From 3-nitro-1,2,4-triazole (HL) complexes ML$_2$ (M = Zn, Co, Cu), NiL$_2$(H$_2$O)$_2$, ZnL$_2$(NH$_3$)(H$_2$O), ML$_2$(NH$_3$)$_4$ (M = Cd, Ni), and CuL$_2$(NH$_3$)$_2$(H$_2$O)$_2$ were prepared. In ML$_2$ and NiL$_2$(H$_2$O)$_2$ a bidentate triazole ligand with coordination via one ring nitrogen atom and one oxygen atom of the nitro group was proposed.[133] An iron complex FeL$_2$(H$_2$O)$_2$ was obtained from 5-nitrotetrazole (HL).[134]

There has been appreciable interest in the ammonium salt of 1,2,4-triazole[135-137] and in the ammonium,[137] mercuric,[138-142] and silver salts[139] of 5-nitrotetrazoles for their high energy content.

Spectroscopy

Generally the replacement of a ring CH unit by N has little effect on the ultraviolet (uv) absorption of the heterocyclic system. This proved advantageous in providing support for the assigned structure for 4-nitro-2-(4-nitrophenyl)-1,2,3-triazole (**4-4**), λ_{max} 304 nm (log ε 4.36), based on comparison with the spectrum for 4-nitro-1-(4-nitrophenyl)pyrazole, λ_{max} 304 nm (log ε 4.37).[8] On the other hand an analysis of uv absorption may offer little assistance in differentiating triazole substituent positions. For example 1- and 2-alkyl derivatives of 4-nitro-1,2,3-triazole gave nearly identical uv absorption at λ_{max} 257 (log ε 3.87) from solutions at pH 1 or pH 11 or in ethanol.[105] In five examples of 1-alkyl-4-nitro-5-amino-1,2,3-triazoles (**4-32a–e**), "quinonoid" structures from amino and nitro group interaction explained the absorption in ethanol at higher wavelengths for λ_{max} 332–356 nm (log ε 3.21–4.00).[47]

An interesting photochromic property of the bistriazolyl ether (**4-93**) was noted. When irradiated with light below 540 nm the compound on a tlc plate became intensively emerald green.[55]

Although there is a claim (probably erroneous) to the contrary,[38] the addition of a proton to a neutral 3-nitro-1,2,4-triazole brought about a hypsochromic shift of the uv absorption maximum by 25–35 nm.[128,129] For example, λ_{max} 270

$$\text{4-93}$$

and 245 were respectively obtained for 1-methyl-5-nitro-1,2,4-triazole (**4-48a**) at pH 7 and in sulfuric acid (83.4%); under similar conditions λ_{max} 255 and 230 were obtained for the isomer 1-methyl-3-nitro-1,2,4-triazole (**4-47a**). A change in the solution of 3-nitro-1,2,4-triazole (**4-26**) (X = H) and its 5-methyl derivative (**4-27**) (X = CH$_3$) from pH 7 to pH 13 brought about a bathochromic shift in the absorption maxima by 40–60 nm.[38,128,129] In water the anion of 3,5-dinitro-1,2,4-triazole (**4-24**) absorbed at λ_{max} 285 nm (log ε 3.90) and in sulfuric acid (40%) absorbed at λ_{max} 255 nm (log ε 3.79).[128,129]

Ultraviolet absorption for 3-nitro-1,2,4-triazol-5-one (**4-11**) and its 1,4-dimethyl derivative (**4-12c**) in buffer solutions of pH 11 gave λ_{max} 350 and 328 nm (log ε 3.46 and 3.70) and in sulfuric acid (98%) gave λ_{max} 240 and 260 nm (log ε 3.68 and 3.66).[24]

In aqueous sulfuric acid 5-nitrotetrazole showed uv absorption at λ_{max} 235 nm.[132]

Six infrared absorption assignments are widely used for the characterization and identification of nitro-1,2,4-triazoles. Asymmetric and symmetric stretching vibrations for the nitro group occur in the ranges 1,580–1,550 cm^{-1} (v_{as}) and 1,360–1,300 cm^{-1} (v_s). Four assignments to ring stretching vibrations include 1,525–1,490, 1,450–1,340, 1,370–1,260, and 1,190–1,120 cm^{-1}. References cited present tabulated information.[40,71,75,78,83,91,98,128]

Potential energy and force constants were calculated for 1,2,4-triazoles including the 3-nitro-5-azido derivative and for tetrazoles including the 1-methyl-5-nitro derivative.[143,144] Infrared and Raman spectra of anions of nitro derivatives of 1,2,4-triazoles have been obtained.[145]

C-substituted 1,2,4-triazoles were shown by ir spectroscopy to prefer one tautomeric form in the solid state, whereas two tautomeric forms were detected in the molten solid, in the trapped gas at 22 K, and in solution in various polar solvents.[146]

The proton at C-5 in a variety of 3-nitro-1,2,4-triazoles was detected in ^{1}H nmr by a signal at δ8.75–9.10. For 3-nitro-1,2,4-triazole (**4-26**), the signal was solvent-dependent and ranged from δ8.52 in D$_2$O-NaOD to δ9.04 in D$_2$O.[10,41,85,128]

The signals at δ9.00 and 8.84 were not differentiated between the C-5' and C-5 protons in 4-nitro-2-(3-triazolyl)-1,2,3-triazole (**4-34**).[38] In 1-(2,4-dinitrophenyl)-1,2,3-triazole, δ8.10 was assigned to the proton at C-4 and δ8.88 to the proton at C-5. After substitution of a nitro group at C-4 converted the dinitro compound to a trinitro compound (**4-5**), the signal for the proton at C-5 was found at δ10.10. These results compared with δ8.36 and 9.35 for the C-3 and C-5 protons in 1-(2,4-dinitrophenyl)-1,2,4-triazole and δ9.55 for the C-5 proton in 1-(2,4-dinitrophenyl)-3-nitro-1,2,4-triazole.[6]

A shift in the ^{1}H NMR signal from the N-methyl group toward the weaker field for 1-methyl-3-nitro-5-substituted-1,2,4-triazoles correlated with σ constants for the substituents.[147]

Miscellaneous

The dipole moments of seven 3(5)-nitro-1,2,4-triazoles were determined from solutions in chloroform (lower values) and in dioxane (higher values) (Table 4-1).[148] Tautomeric forms for compounds (**4-26**), (**4-27**), (**4-47a**), (**4-48a**), and (**4-94**) are shown. The results were compatible with acid-base properties of nitro-1,2,4-triazoles[129] and with the Charton assumption that the predominating tautomer in aromatic heterocycles that have several nitrogen atoms in the ring and are capable of tautomerism will have a proton bound to the ring nitrogen atom farthest removed from an electron-withdrawing substituent.[149]

Calculated π-electron densities, bond orders, and the energies of π-electron interactions of ^{1}H-1,2,4-triazole and 10 derivatives including four nitro derivatives, and 3-nitro-4-methyl-1,2,4-triazole (**4-94**) were obtained by the Huckel MO LCAO method. From the data π-electron, total dipole moments and bond lengths were calculated.[150]

Transmission of substituent effects were described in correlation equations for properties and reactivity in 14 series of 1,2,4-triazoles. Several nitro derivatives were included in the 69 compounds examined. A decreased contribution from

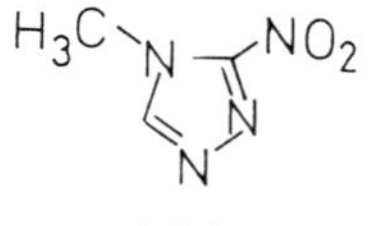

4-94

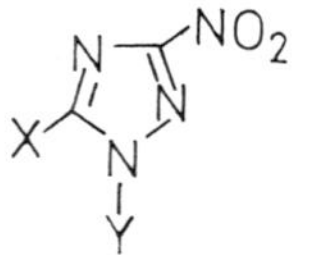

  α   β   γ

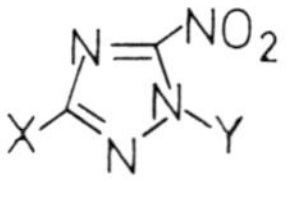

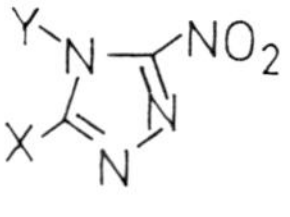

4-26,	α,β,γ,X = Y = H	
4-27,	α,β,γ,X = CH$_3$, Y = H	
4-47a,	α,X = H, Y = CH$_3$	
4-48a,	β,X = H, Y = CH$_3$	
4-94,	γ,X = H, Y = CH$_3$	
4-47d,	α,X = NO$_2$, Y = CH$_3$	
4-68e,	α,X = Cl, Y = CH$_3$	

Table 4-1. Experimental and Calculated Dipole Moments (μ, D) of Nitro-1,2,4-Triazoles

Compound	Tautomeric form	Dipole moment (dioxane)	
		Found	Calculated
4-26	α		7.02
	β	6.74	3.52
	γ		5.61
4-27	α		7.28
	β	7.19	3.57
	γ		5.93
4-47a	α	6.78	6.95
4-48a	β	3.30	3.67
4-94	γ	5.96	5.80
4-47d	α	4.96	5.10
4-68e	α	6.05	5.98

the conjugation effect correlated with destruction of aromaticity on going from a triazole to a triazolone structure.[151]

Spin-density distribution calculated by CNDO/2 for anion radicals and dianion radicals of nitro derivatives of pyrrole, pyrazole, imidazole, 1,2,3-triazole, and 1,2,4-triazole brought about revisions in earlier assignments of hyperfine splitting constants for 3-nitropyrazole, 4-nitroimidazole, and 3-nitro-1,2,4-triazole dianion radicals.[152]

A semiempirical variant of the SCF LCAO MO method in the CNDO approximation provided quantum chemical calculations on tetrazole and its derivatives including 1-methyl-5-nitrotetrazole and 2-methyl-5-nitrotetrazole. Charge distributions for neutral molecules, their anions, and protonated forms were given. A maximum electron density located on N-1 and N-4 correlated linearly with σ substitution values.[153]

An X-ray analysis established the structure of 1-(mesitylsulfonyl)-3-nitro-1,2,4-triazole (**4-95**).[154] The crystal structure for the mercuric salt of 5-nitrotetrazole has been described.[155]

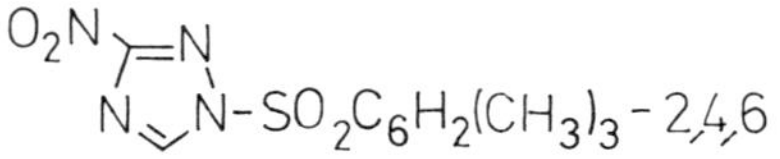

$$O_2N-C=N\diagdown N=N-SO_2C_6H_2(CH_3)_3-2,4,6$$

4-95

In a broad examination of inhibitors of hypoxanthine metabolism in Ehrlich ascites tumor cells in vitro, 1-β-**D**-ribofuranosyl-4-nitro-1,2,3-triazole (see **4-77**) reduced the GTP:GDP + GMP ratio to 50% of the control value (GTP, GDP, and GMP are guanosine tri-, di-, and monophosphates).[156] A nitrotriazolobenzodiazepine was one of several derivatives of the bicyclic heterocycles that improved feed utilization for lambs and increased the gain in weight.[157]

To facilitate the correlation between the reduction potential of heterocyclic compounds and their effectiveness as radiosensitizing agents in cancer chemotherapy, the polarographic analysis of various nitroimidazoles and nitrotriazoles was examined.[158]

Considerable interest has been shown in 3,5-dinitro-1,2,4-triazole[135–137] and in nitrotetrazoles[137–142,158–166] as high-energy material. Nitrotriazoles have been investigated as light stabilizers for cellulose acetate.[167]

REFERENCES

1. Boyer, J. H., In R. C. Elderfield, *Heterocyclic Compounds*, Vol. 7. Wiley, New York, 1961, pp. 384–461.
2. Finley, K. T., In J. A. Montgomery, Vol. ed., *The Chemistry of Heterocyclic Compounds*. Wiley, New York, 1980, pp. 177–184.
3. Temple, Jr., C., In J. A. Montgomery, Vol. ed., *The Chemistry of Hetrocyclic Compounds*. Wiley, New York, 1981, pp. 225–238.
4. Gilchrist, T. L., Gymer, G. E., *Adv. Heterocycl. Chem.*, **1974**, *16*, 33–86.
5. Schofield, K., Grimmett, M. R., Keene, B. R. T., *The Azoles*. Cambridge University Press, Cambridge, 1976, p. 350.
6. Khan, M. A., Lynch, B. M., *J. Heterocycl. Chem.*, **1970**, *7*, 1237.
7. Riebsomer, J. L., *J. Org. Chem.*, **1948**, *13*, 815.
8. Lynch, B. M., Chan, T.-L., *Can. J. Chem.*, **1963**, *41*, 274.
9. Eagles, T. E., Khan, M. A., Lynch, B. M., *Org. Prep. Proc.*, **1970**, *2*, 117.
10. Neuman, P. N., *J. Heterocycl. Chem.*, **1971**, *8*, 51.
11. Kroeger, C. F., Miethchen, R., *Z. Chem.*, **1969**, *9*, 378.
12. Lebedeva, N. P., Stotskii, A. A., Fomina, V. V., *Zh. Org. Khim.*, **1981**, *17*, 1786; Eng. 1599.
13. Stotskii, A. A., Tkacheva, N. P., *Zh. Org. Khim.*, **1976**, *12*, 235; Eng. 231.
14. *Ibid.*, p. 655; Eng. 646.
15. Stotskii, A. A., Tkackeva, N. P., Shostakovskii, M. F., *Dokl. Akad. Nauk SSSR*, **1976**, *231*, 1370; Eng. 768.
16. *Ibid*, **1977**, *233*, 386; Eng. 184.
17. Stotskii, A. A., *Deposited Doc.*, **1980** SPSTL 583 khp D80, 8 pp.; *Chem. Abstr.*, **1981**, *95*, 186154t.
18. Manchot, W., Noll, R., *J. Liebigs Ann. Chem.*, **1905**, *343*, 1.
19. Schmidt, J., Gehlen, H., *Z. Chem.*, **1965**, *5*, 304.
20. Gehlen, H., Schmidt, J., *Ann. Chem.*, **1965**, *682*, 123.
21. Chipens, G. I., Bokalders, R. K., Grinsteins, V., *Khim. Geterotsikl. Soedin.*, **1966**, 110; Eng. 79.
22. Kroeger, C. F., Miethchen, R., Frank, H., Siemer, M., Pilz, S., *Chem. Ber.*, **1969**, *102*, 755.
23. Kroeger, C. F., Miethchen, R., Frank, H., Siemer, M., Pilz, S., *Chem. Ber.*, **1969**, *102*, 755.
24. Katritzky, A. R., Ogretir, C., *Chim. Acta Turc.*, **1982**, *10*, 137; *Chem. Abstr.*, **1983**, *98*, 159930y.
25. Neuman, P. N., *J. Heterocycl. Chem.*, **1970**, *7*, 1159.
26. Ainsworth, C., *Org. Syn.*, **1973**, *V*, 1070.
27. Gidaspov, B. V., Tartakovskii, V. A., Pevzner, M. S., Ioffe, S. L., Kulibabina, T. N., Maslina, I. A., *USSR 662,551; Chem. Abstr.*, **1979**, *91*, 107978f.
28. Pevzner, M. S., Kulibabina, T. N., Ioffe, S. L., Maslina, I. A., Gidaspov, B. V., Tartakovskii, V. A., *Khim. Geterotsikl. Soedin.*, **1979**, 550; Eng. 451.
29. Begtrup, M., Holm, J., *J. Chem. Soc. Perkin Trans I*, **1981**, 503.
30. Pevzner, M. S., Kulababina, T. N., Povarova, N. A., Kilina, L. V., *Khim. Geterotsikl. Soedin.*, **1979**, 1132; Eng. 929.
31. Habraken, C. L., Cohen-Fernandez, P., *J. Chem. Soc. Chem. Commun.*, **1972**, 37.
32. Janssen, J. W. A. M., Habraken, C. L., *J. Org. Chem.*, **1971**, *36*, 3081.
33. Pevzner, M. S., Kulibabina, T. N., Malinina, L. A., *Khim. Geterotsikl. Soedin.*, **1979**, 555; Eng. 456.

34. Burchfield, H. P., Gullstrom, D. K., *U.S. 3,054,800; Chem. Abstr.*, **1963**, *60*, 10220g.
35. Wiley, R. W., Smith, N. R., *U.S. 3,111,524; Chem. Abstr.*, **1963**, *60*, 2951c.
36. Smith, N. R., Wiley, R. W., *U.S. 3,165,753; Chem. Abstr.*, **1965**, *62*, 7770d.
37. Bagal, L. I., Pevzner, M. S., Lopyrev, V. A., *Khim. Geterosikl. Soedin., Sb.1: Azotsoderzhashchie Geterotsikly*, **1967**, 180; *Chem. Abstr.*, **1969**, *70*, 77876t.
38. Browne, E. J., *Aust. J. Chem.*, **1969**, *22*, 2251.
39. Jones, S. S., Rayner, B., Reese, C. B., Ubasawa, A., Ubasawa, M., *Tetrahedron*, **1980**, *36*, 3075.
40. Bagal, L. I., Pevzner, M. S., Frolov, A. N., Sheludyakova, N. I., *Khim. Geterotsikl. Soedin.*, **1970**, 259; Eng. 240.
41. Closset, J.-L., Copin, A., Drèze, P., Alderweireldt, F., Durant, F., Evrard, G., Michel, A., *Bull. Soc. Chim. Belg.*, **1975**, *84*, 1023.
42. Copin, A., Drèze, P., *Ann. Gembloux*, **1976**, *82*, 61.
43. Maiorana, S., Pocar, D., Dalla Croce, P., *Tetrahedron Lett.*, **1966**, 6043.
44. Pocar, D., Maiorana, S., Dalla Croce, P., *Gazz. Chim. Ital.*, **1968**, *98*, 949.
45. Rembarz, G., Kirchhoff, B., Dongowski, G., *J. Prakt. Chem.*, **1966**, *33*, 199.
46. Callaghan, P. D., Gibson, M. S., *J. Chem. Soc. Chem. Commun.*, **1967**, 918.
47. Viswanathan, N. I., Balakrishnan, V., *J. Chem. Soc. Perkin I*, **1979**, 2361.
48. Zefirov, N. S., Chapovskaya, N. N., Kolesnikov, V. V., *J. Chem. Soc. Chem. Commun.*, **1971**, 1001.
49. Blank, H. U., Fox, J. J., *J. Am. Chem. Soc.*, **1968**, *90*, 7175.
50. Shin, C., Yonezawa, Y., Yoshimura, J., *Tetrahedron Lett.*, **1974**, 7.
51. Khisamutdinov, G. K., Bondarenko, O. A., Kupriyanova, L. A., Klimenko, V. G., Demina, L. A., *Zh. Org. Khim.*, **1979**, *15*, 1307; Eng. 1168.
52. Khisamutdinov, G. K., Bondarenko, O. A., Kupriyanova, L. A., *Zh. Org. Khim.*, **1975**, *11*, 2445; Eng. 2506.
53. Mohr, R., Zimmermann, M., *Ger. 1,168,437; Chem. Abstr.*, **1964**, *61*, 1873h.
54. *Fr. 1,577,760; Chem. Abstr.*, **1970**, *72*, 100716t.
55. Begtrup, M., Knudsen, N. O., *Acta Chem. Scand.*, **1983**, *37B*, 97.
56. Witkowski, J. T., Robins, R. K., *J. Org. Chem.*, **1970**, *35*, 2635.
57. Hester, Jr., J. B., *Ger. Offen. 2,220,694; Chem. Abstr.*, **1973**, *78*, 84418p.
58. Jones, G., Sliskovic, D. R., Foster, B., Rogers, J., Smith, A. K., Wong, M. Y., Yarham, A. C., *J. Chem. Soc. Perkin Trans. I*, **1981**, 78.
59. Gray, E. J., Stevens, M. F. G., *J. Chem. Soc. Perkin Trans. I*, **1976**, 1492.
60. Pfleger, R., Garthe, E., Rauer, K., *Chem. Ber.*, **1963**, *96*, 1827.
61. Von Herz, E., *Ger. 562,511; Chem. Abstr.*, **1933**, *27*, 1013.
62. *Ibid., U.S. 2,066,945; Chem. Abstr.*, **1937**, *31*, 1212[4].
63. Gilligan, W. H., Kamlet, M. J., *U.S. 4,093,623; Chem. Abstr.*, **1978**, *89*, 199863z.
64. Hattori, K., Lieber, E., Horwitz, J. P., *J. Am. Chem. Soc.*, **1956**, *78*, 411.
65. Spear, R. J., *Rep.-Mater. Res. Lab. (Aust.)*, **1980**, MRL-R-780, 21 pp.; *Chem. Abstr.*, **1981**, *95*, 153126g.
66. Lippmann, E., Könnecke, A., *Z. Chem.*, **1975**, *15*, 226.
67. Bagal, L. I., Pevzner, M. S., Sheludyakova, N. I., Kerusov, V. M., *Khim. Geterotsikl. Soedin.*, **1970**, 265; Eng. 245.
68. Ostapkovich, A. M., Kofman, T. P., Lisitsyna, L. V., Pevzner, M. V., *Izv. Vyssh. Uchebn. Zaved., Khim. Khim. Teknol.*, **1979**, *22*, 402; *Chem. Abstr.*, **1979**, *91*, 157666y.
69. Stotskii, A. A., Tkacheva, N. P., Kirov, S. M., *USSR 446,507; Chem. Abstr.*, **1975**, *82*, 43428f.
70. Stotskii, A. A., Tkacheva, N. P., *Zh. Org. Khim.*, **1974**, *10*, 2232; Eng. 2250.
71. Tverdokhlebov, V. P., Tselinskii, I. V., Vasil'eva, N. Y., *Zh. Org. Khim.*, **1978**, *14*, 1056; Eng. 985.
72. Tverdokhlebov, V. B., Tselinskii, I. V., Vasil'eva, N. Y., Polyakov, B. V., Frolova, G. M., *Zh. Org. Khim.*, **1080**, *16*, 218; Eng. 207.
73. Kerusov, V. M., Pevzner, M. S., *Khim. Geterotsikl. Soedin.*, **1974**, 1564; Eng. 1375.
74. Tirpigarev, A. N., Shcherbinin, M. B., Bazanov, A. G., Tselinski, I. V., *Zh. Org. Khim.*, **1982**, *18*, 463; *Chem. Abstr.*, **1982**, *96*, 181213x.
75. Kofman, T. P., Uspenskaya, T. L., Medvedeva, N. Y., Pevzner, M. S., *Khim. Geterotsikl. Soedin.*, **1976**, 991; Eng. 821.
76. Kofman, T. P., Pevzner, M. S., Manuilova, V. I., *USSR 432,147; Chem. Abstr.*, **1974**, *81*, 77930h.
77. *Ibid. USSR 425,911; Chem. Abstr.*, **1974**, *81*, 4967r.
78. Kofman, T. P., Manuilova, V. I., Pevzner, M. S., Timofeeva, T. N., *Khim. Geterotsikl. Soedin.*, **1975**, 705; Eng. 612.
79. Kofman, T. P., Pevzner, M. S., Manuilova, V. I., *USSR 425,911; Chem. Abstr.*, **1974**, *81*, 49687r.

80. Kofman, T. P., Zykova, G. A., Manuilova, V. I., Timofeeva, T. N., Pevzner, M. S., *Khim. Geterotsikl. Soedin.*, **1974**, 997; Eng. 871.
81. Sitzmann, M. E., *J. Org. Chem.*, **1978**, *43*, 3389.
82. Chipen, G. I., Bokaldere, R. P., *Khim. Geterotsikl. Soedin.*, **1969**, 159; Eng. 123.
83. Kofman, T. P., Vasil'eva, I. V., Pevzner, M. S., *Khim. Geterotsikl. Soedin.*, **1977**, 1407; Eng. 1129.
84. Kofman, T. P., Pevzner, M. S., Zhukova, L. N., Kravchenko, T. A., Frolova, G. M., *Zh. Org. Khim.*, **1980**, *16*, 420; Eng. 375.
85. Kofman, T. P., Zhukova, L. N., Pevzner, M. S., *Khim. Geterotsikl. Soedin.*, **1981**, 552; Eng. 406.
86. Kofman, T. P., Pevzner, M. V., Uspenskaya, T. L., Sushchenko, L. F., *USSR 753,849; Chem. Abstr.*, **1981**, *94*, 47336v.
87. Sickman, D. V., *U.S. 2,987,520; Chem. Abstr.*, **1962**, *56*, 4776e.
88. Pevzner, M. S., Ivanov, P. A., Gladkova, N. V., Sushchenko, O. N., Tverdokhlebov, V. P., Myasnikova, Z. S., *Khim. Geterotsikl. Soedin.*, **1980**, 251; Eng. 189.
89. Henry, R. A., Finnegan, W. G., *J. Am. Chem. Soc.*, **1954**, *76*, 923.
90. Spear, R. J., Elischer, P. P., *Rep.-Mater. Res. Lab. (Aust.)*, **1982**, MRLR 859, 21 pp.; *Chem. Abstr.*, **1983**, *98*, 91969k.
91. Bagal, L. I., Pevzner, M. S., Samarenko, V. Y., Ergorov, A. P., *Khim. Geterotsikl. Soedin.*, **1970**, 702; Eng. 650.
92. Naik, S. R., Witkowsik, J. T., Robins, R. K., *J. Org. Chem.*, **1973**, *38*, 4353.
93. Bagal, L. I., Pevzner, M. S., Samarenko, V. Y., Ergorov, A. P., *Khim. Geterotsikl. Soedin.*, **1970**, 1701; Eng. 1588.
94. Stotskii, A. A., Lebedeva, N. P., *Deposited Doc.*, **1981**, SPSTL 851 Khp-D81, 6 pp.; *Chem. Abstr.*, **1983**, *98*, 107221y.
95. Bagal, L. I., Pevzner, M. S., Samarenko, V. Y., *Ibid*, **1970**, 269; Eng. 249.
96. Bagal, L. I., Pevzner, M. S., Egorov, A. P., Samarenko, V. Y., *Khim. Geterotsikl. Soedin.*, **1970**, 997; Eng. 928.
97. Pevzner, M. S., Kofman, T. P., Kibasova, E. N., Sushchenko, L. F., Uspenskaya, T. L., *Khim. Geterotsikl. Soedin.*, **1980**, 257; Eng. 194.
98. Kofman, T. P., Sushchenko, L. F., Pevzner, M. S., *Ibid*, **1980**, 1553; Eng. 1186.
99. Pevzner, M. S., Samarenko, V. Y., Bagal, L. I., *Khim. Geterotsikl. Soedin.*, **1972**, 117, 568, 848; Eng. 108, 518, 770.
100. Mel'nikova, N. N., Pevzner, M. S., Bagal, L. I., *Zh. Org. Khim.*, **1973**, *9*, 799; Eng. 823.
101. Mel'nikova, N. N., Pevzner, M. S., Malysheva, N. M., Bagal, L. I., *Zh. Org. Khim.*, **1973**, *9*, 2535; Eng. 2555.
102. Mel'nikova, N. N., Pevzner, M. S., Bagal, L. A., *Reakts. Sposobnost. Org. Soedin.*, **1972**, *9*, 553; *Chem. Abstr.*, **1973**, *78*, 158591.
103. *Ibid*, 563; *Chem. Abstr.*, **1973**, *78*, 135254s.
104. Kofman, T. P., Pevzner, M. S., *Khim. Geterotsikl. Soedin.*, **1981**, 1403; Eng. 1055.
105. Lehmkuhl, F. A., Witkowski, J. T., Robins, R. K., *J. Heterocycl. Chem.*, **1972**, *9*, 1195.
106. Barascut, J. L., Tamby, C., Imbach, J. L., *J. Carbohydr. Nucleosides Nucleotides*, **1974**, *1*, 77; *Chem. Abstr.*, **1975**, *82*, 156620f.
107. Robins, R. K., Witkowski, J. T., *U.S. 3,968,103; Chem. Abstr.*, **1976**, *85*, 94660v.
108. Naik, S. R., Witkowski, J. T., Robins, R. K., *J. Heterocycl. Chem.*, **1974**, *11*, 57.
109. Camarasa, M. J., De las Heras, F. G., *J. Heterocycl. Chem.*, **1983**, *20*, 1307.
110. De las Heras, F. G., Camarasa, M. J., Martinezfernandez, A. R., Escario, J. A., *Eur. J. Med. Chem.-Chim. Ther.*, **1984**, *19*, 89.
111. Reese, C. B., Ubasawa, A., *Tetrahedron Lett.*, **1980**, *21*, 2265.
112. Reese, C. B., Ubasawa, A., *Nucleic Acids Symp. Ser.*, **1980**, *7*, 5; *Chem. Abstr.*, **1981**, *94*, 140090v.
113. Engels, J., *Tetrahedron Lett.*, **1980**, *21*, 4339.
114. Gait, M. J., Popov, S. G., *Tetrahedron Lett.*, **1980**, *21*, 2841.
115. Reese, C. B., Titmas, R. C., Valente, L., *J. Chem. Soc. Perkin Trans I*, **1981**, 2451.
116. De Rooij, J. F. M., Wille-Hazeleger, G., Vink, A. B. J., Van Boom, J. H., *Tetrahedron*, **1979**, *35*, 2913.
117. Chattopadhyaya, J. B., Reese, C. B., *Tetrahedron Lett.*, **1979**, 5059.
118. De Rooij, J. F. M., Wille-Hazeleger, G., van Deursen, P. H., Serdijn, J., Van Boom, J. H., *Recl. Trav. Chim. Pays-Bas*, **1979**, *98*, 537.
119. Engels, J., Krahmer, U., Zsolnai, L., Huttner, G., *J. Liebigs Ann. Chem.*, **1982**, 745.
120. Takaku, H., Kamaike, K., Kasuga, K., *Chem. Lett.*, **1982**, 197.
121. Kofman, T. P., Kirpenko, Z. V., Pevzner, M. S., *Khim. Geterotsikl. Soedin.*, **1982**, 1113; Eng. 854.
122. Kofman, T. P., Pakhomov, K. E., Pevzner, M. S., *Khim. Geterotsikl. Soedin.*, **1982**, 848; Eng. 647.

123. Lopyrev, V. A., Larina, L. I., Rakhmatulina, T. N., Shibanova, E. F., Vakul'skaya, T. L., Voronkov, M. G., *Dokl. Akad. Nauk SSSR*, **1978**, *242*, 142; *Chem. Abstr.*, **1979**, *90*, 22900d.
124. Miethchen, R., Seipt, H. U., Kroeger, C. F., *Z. Chem.*, *1969*, *9*, 300.
125. Terpigorev, A. N., Shcherbinin, M. B., Bazanov, A. G., Tselinskii, I. V., *Zh. Org. Khim.*, **1982**, *18*, 676; Eng. 587.
126. Kofman, T. F., Pakhomov, K. E., Pevzner, M. S., *Khim. Geterotsikl. Soedin.*, **1982**, 848; Eng. 647.
127. Mokrushina, G. A., Kotovskaya, S. K., Postovskii, I. Y., *Khim. Geterotsikl. Soedin.*, **1979**, 131; Eng. 118.
128. Kofman, T. P., Medvedeva, N. Y., Uspenskaya, T. L., Pevzner, M. S., *Khim. Geterotsikl. Soedin.*, **1977**, 1271; Eng. 1026.
129. Bagal, L. I., Pevzner, M. S., *Khim. Geterotsikl. Soedin.*, **1970**, 558; Eng. 517.
130. *Ibid.* **1971**, 272; Eng. 249.
131. Pevzner, M. S., Martynova, M. N., Timofeeva, T. N., *Khim. Geterotsikl. Soedin.*, **1974**, 1288; Eng. 1121.
132. Ostrovskii, V. A., Koldobskii, G. I., Shirokova, N. P., Poplavskii, V. S., *Khim. Geterotsikl. Soedin.*, **1981**, 559; Eng. 412.
133. Vos, J. G., Driessen, W. L., Van der Waal, J., Groeneveld, W. L., *Inorg. Nucl. Chem. Lett.*, **1978**, *14*, 479.
134. Harris, A. D., Herber, R. H., Jonassen, H. B., Wertheim, G. K., *J. Am. Chem. Soc.*, **1963**, *85*, 2927.
135. Ott, D. G., *U.S. 4,236,014; Chem. Abstr.*, **1981**, *94*, 159219r.
136. Lee, K. Y., Ott, D. S., Stinecipher, M. M., *Ind. Eng. Chem. Process Res. Dev.*, **1981**, *20*, 358.
137. Selig, W., *Mikrochim. Acta*, **1981**, *2*, 251.
138. Scott, C. L., Leopold, H. S., *U.S. 3,965,951; Chem. Abstr.*, **1981**, *85*, 110638b.
139. Bates, L. R., Jenkins, J. M., *U.S. 4,094,879; Chem. Abstr.*, **1979**, *89*, 163577v.
140. Gilligan, W. H., Kamlet, M. J., *U.S. NTIS, AD Rep.*, **1976**, AD-AO36086, 15 pp.; *Chem. Abstr.*, **1977**, *87*, 154288h.
141. Farncomb, R. E., Chang, M., Pisacane, F. J., *U.S. NTIS, AD Rep.*, **1976**, AD-AO42058, 23 pp.; *Chem. Abstr.*, **1977**, *87*, 203762u.
142. Glover, D. J., *U.S. NTIS, AD Rep.*, **1977**, AD-AO44844, 16 pp.; *Chem. Abstr.*, **1978**, *88*, 123386e.
143. Mel'nikov, V. V., Baeva, L. F., Stolpakova, V. V., Pevzner, M. S., Gidaspov, E. V., *Khim. Geterotsikl. Soedin.*, **1977**, 839, Eng. 686.
144. Sokolova, M. M., Mel'nikov, V. V., Mel'nikov, A. A., Gidaspov, B. V., *Khim. Geterotsikl. Soedin.*, **1977**, 843; Eng. 689. (This is Part VII in a continuing study of the vibrational spectra and structures of 1,2,4-triazoles.)
145. Mel'nikov, V. V., Stolpakova, V. V., Khor'kova, L. F., Pevzner, M. S., Mel'nikova, M. N., *Khim. Geterotsikl. Soedin.*, **1972**, 120; Eng. 111. (Earlier papers in the series are cited.)
146. Saidi Idrissi, M., Lalane, P., Garrigou-Lagrange, C., *J. Chim. Phys. Phys.-Chim. Biol.*, **1981**, *78*, 511.
147. Pevzner, M. S., Timofeeva, T. N., Gorbunova, N. N., Frolova, G. M., *Zh. Org. Khim.*, **1976**, 1361; Eng. 1352.
148. Pevzner, M. S., Fedorova, E. Y., Shokhor, I. N., Bagal, L. N., *Khim. Geterotsikl. Soedin.*, **1971**, 275; Eng. 252.
149. Charton, M., *J. Org. Chem.*, **1965**, *30*, 3346.
150. Mel'nikov, V. V., Pevzner, M. S., Stolpakova, V. V., Khor'kova, L. F., *Khim. Geterotsikl. Soedin.*, **1971**, 409; Eng. 377.
151. Pevzner, M. S., Kofman, T. P., Gorbunova, N. N., *Zh. Org. Khim.*, **1978**, *14*, 2024; Eng. 1877.
152. Larina, L. I., Dubnikov, V. M., Lur'e, F. S., Vakuls'kaya, T. I., Vitkovskaya, N. M., Lopyrev, V. A., Voronkov, M. G., *Zh. Strukt. Khim.*, **1980**, *21*, 203; *Chem. Abstr.*, **1981**, *94*, 46331c.
153. Ostrokovskii, V. A., Panina, N. S., Koldobskii, G. I., Gidaspov, B. V., Shirobokov, I. Y., *Zh. Org. Khim.*, **1979**, 844; Eng. 755.
154. Kuroda, R., Sanderson, M. R., Neidle, S., Reese, C. B., *J. Chem. Soc., Perkin Trans. II*, **1982**, 617.
155. Jin, X., Shao, M., Huang, H., Wang, J., Zhu, Y., *Huaxue Tongbao*, **1982**, *336–7*, 368; *Chem. Abstr.*, **1982**, *97*, 136922n.
156. Smith, C. M., Zombor, G., Henderson, J. F., *Cancer Treat. Rep.*, **1976**, *60*, 1567.
157. Collins, R. J., *U.S. 3,803,315; Chem. Abstr.*, **1974**, *81*, 62428b.
158. Tarver, C. M., Goodale, T. C., Shaw, R., Cowperthwaite, M., *Off. Nav. Res. [Tech. Rep.] ACR (U.S.)*, **1976**, ACR-221, Proc. Symp. (Int.) Detonation, 6th, 231–249; *Chem. Abstr.*, **1980**, *92*, 8480b.

159. Leach, S. C., Weaver, R. D., Kinoshita, K., Lee, W. W., *J. Electroanal. Chem.*, **1981**, *129*, 213.
160. Scott, C. L., Leopold, H. S., *U.S. 4,024,818; Chem. Abstr.*, **1977**, *87*, 87254d.
161. Spear, R. J., Elischer, P. P., Bird, R., *Rep.-Mater. Res. Lab. (Aust.)*, **1980**, MRL-R-771, 30 pp.; *Chem. Abstr.*, **1981**, *95,* 22350q.
162. Spear, R. J., Bentley, J. R., Wolfson, M. G., *Rep.-Mater. Res. Lab. (Aust.)*, **1981**, MRL-R-815, 20 pp.; *Chem. Abstr.*, **1982**, *96*, 183694d.
163. *Ibid., Combust. Flame*, **1983**, *50*, 249; *Chem. Abstr.*, **1983**, *98*, 200782s.
164. Spear, R. J., Elischer, P. P., *Aust. J. Chem.*, **1982**, *35*, 1.
165. Brown, M. E., Swallowe, G. M., *Thermochim. Acta*, **1981**, *49*, 383.
166. Kalontarov, I. Y., Stotskii, A. A., Makhkamov, K. M., Tkacheva, N. P., Sobolev, V. I., *Izv. Vyssh. Uchebn. Zaved., Khim. Khim. Teknol.*, **1976**, *19*, 1863; *Chem. Abstr.*, **1977**, *86*, 140852e.
167. Geigy, J. R., *Fr. 1,559,131; Chem. Abstr.*, **1969**, *71*, 12445j.

BIBLIOGRAPHY

Begtrup, M., Knudsen, N. O., ADDITION OF NITRITE IONS TO 1-METHOXY-1,2,3-TRIAZOLIUM SALTS. FORMATION OF NITRO AND HYDROXY SUBSTITUTED TRIAZOLES, *Acta Chem. Scand., Ser. B*, **1983**, *B37*, 97.

Camarasa, M. J., De las Heras, F. G., REACTIONS OF 1-GLUCOSYL-5-NITRO-1,2,4-TRIAZOLE WITH NUCLEOPHILES, *J. Heterocycl. Chem.*, **1983**, *20*, 1307.

Habraken, C. L., Erkelens, C., Mallena, J. R., Cohenfernandes, P., 1-NITROBENZOTRIAZOLE-2-(NITROIMINO)DIAZOBENZENE ISOMERIZATION: FORMATION OF TRIAZENES BY AZO COUPLING WITH CYCLIC AMINES, *J. Org. Chem.*, **1984**, *49*, 2197.

Chapter 5

Nitro Derivatives of Isoxazoles, Oxazoles, and Oxadiazoles

INTRODUCTION

Brief discussions of the nitro derivatives have appeared in reviews of mono- and polycyclic isoxazoles, isoxazolines, isoxazolidines, and oxadiazoles.[1-7] Although only a few nitro derivatives of oxazoles and oxadiazoles are known, it appears that an interest in them is on the increase. This review covers the literature through *Chem. Abstr.*, **1983**, *98*; there is a bibliographic addendum for Vols. *99* and *100*.

ISOXAZOLES

Preparations

Nitration and Nitrolysis

Isoxazole and its 3-methyl and 3,5-dimethyl derivatives reveal the resistance of the heterocyclic ring to nitration and the activation provided by alkyl substitution. A mixture of concentrated nitric and sulfuric acids at 40° for 1.5 h gave only a trace (3.5%) of 4-nitroisoxazole (**5-1**)[8] but at −5° afforded 3-methyl-4-nitroisoxazole (**5-2**) (65%)[9] and gave 3,5-dimethyl-4-nitroisoxazole (**5-3**) (90%).[10,11] Although the latter was quantitatively produced by nitration with nitric acid in acetic anhydride in the presence of a catalytic amount of sulfuric acid at 25° for 130 h, the conversion was 78%.[12] A recent patent disclosed the nitration of isoxazole by nitronium tetrafluoroborate in sulfolane to the 4-nitro derivative, but further detail was unavailable.[13]

Mixed acid (nitric and sulfuric acids) converted both 3-methyl-5-dichloromethyl- and 3-dichloromethyl-5-methyl-isoxazole to the corresponding 4-nitro derivatives (**5-4**) and (**5-5**) after treatment at 60° for 20 h.[14] In a similar way the 4-nitro derivatives of 3- and 5-*tert*-butylisoxazole were obtained.[15]

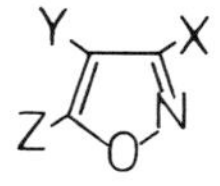

5-1 X = Z = H, Y = NO$_2$
5-2 X = CH$_3$, Y = NO$_2$, Z = H
5-3 X = Z = CH$_3$, Y = NO$_2$
5-4 X = CH$_3$, Y = NO$_2$, Z = CHCl$_2$
5-5 X = CHCl$_2$, Y = NO$_2$, Z = CH$_3$

Phenylisoxazoles in nitric and sulfuric acids nitrated first in a benzene ring. Mixed-acid reactions gave the nitrophenylisoxazoles (**5-6,7**),[16] (**5-8**),[17-19] and (**5-9**).[20-23] Further nitration in the same medium converted the latter to a mixture of dinitro derivatives from which 3-*m*-nitrophenyl-5-*p*-nitrophenylisoxazole (**5-10**) was isolated.[21] This result suggested that the 5-phenyl group was activated and the 3-phenyl group deactivated by the 5- and 3-isoxazolyl substituent toward electrophilic substitution. Mixed-acid nitration of 3-phenyl-5-methylisoxazole brought about substitution at the *meta* and *para* positions.[22]

Apparently a claim[9] that the formation of 4-nitro-5-phenylisoxazole accompanied the formation of (**5-8**) was in error.[18] Further nitration converted (**5-8**) to 4-nitro-5-(*p*-nitrophenyl)isoxazole (**5-11**) (89%), whereas nitric acid alone converted (**5-7**) to 4-nitro-3-(4-nitrophenyl)isoxazole (**5-12**) (40%) and 4-nitro-3-(2,4-dinitrophenyl)isoxazole (**5-13**) (10%). A direct conversion of 3-phenylisoxazole to the trinitro derivative (**5-13**) (68%) was brought about by treatment with mixed acid.[25]

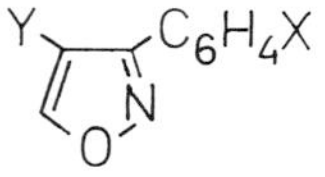

5-6 X = *m*-NO$_2$, Y = H
5-7 X = *p*-NO$_2$, Y = H
5-12 X = *p*-NO$_2$, Y = NO$_2$

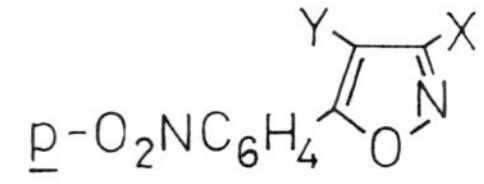

5-8 X = Y = H
5-9 X = C$_6$H$_5$, Y = H
5-10 X = C$_6$H$_4$NO$_2$-*m*, Y = H
5-11 X = H, Y = NO$_2$

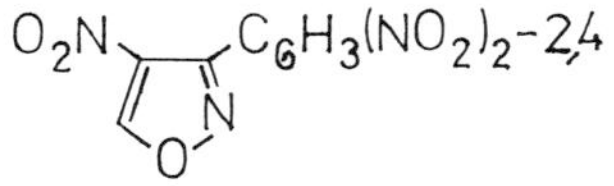

5-13

Mixed-acid nitrations at 20° of 3- and 5-(*p*-chlorophenyl)isoxazole may have revealed a slightly greater activity at the 4-position of the former. A trace of the 4-nitro derivative (**5-14**) was detected in the product mixture which also contained 3-(4-chloro-3-nitrophenyl)- (**5-15**) (63%) and 3-(4-chloro-2-nitrophenyl)isoxazoles. The other isomer gave 5-(4-chloro-3-nitrophenyl)-isoxazole (**5-16**) (58%) as the only product identified.[25]

Both the 3,3'- and 5,5'-bisisoxazoles were nitrated in mixed acids; the latter gave the mono- and dinitro derivatives (**5-17a,b**).[26,27]

5-14 X = 4-ClC$_6$H$_4$, Y = NO$_2$
5-15 X = 4-Cl-3-NO$_2$C$_6$H$_3$, Y = H

5-16 4-Cl-3-NO$_2$C$_6$H$_3$

5-17a X = (isoxazolyl)

5-17b X = (nitroisoxazolyl)

Nitric acid in acetic acid converted 3-phenylisoxazole to the 4-nitro derivative (**5-18a**) (15%) and did not nitrate the phenyl group [compare (**5-6**) and (**5-7**)].[16] More recently the same product was obtained in 57% yield from nitric acid in acetic anhydride at −5° for 70 h.[25] In contrast 3-*p*-chlorophenylisoxazole was unreactive to a similar treatment of nitric acid in acetic anhydride. Slowly 3,5-diphenylisoxazole reacted with this reagent (−20° for 8 days). The only product identified in the mixture was the 4-nitro derivative (**5-18b**) (trace); in a mixture of nitric and acetic acids 3,5-diphenylisoxazole was unreactive at 20–120°.[21] For the conversion of 1,2-diphenylcyclopropane to 3,5-diphenylisoxazole (52%) and 3,5-diphenyl-4-nitroisoxazole (**5-18b**) (15%) on treatment with cupric nitrate in acetic anhydride at 0–10°, it was believed that a conversion of 1-nitro-1,2-diphenylcyclopropane to the diphenylisoxazole was involved.[23,24] Apparently nitration to form (**5-18b**) occurred after isoxazole ring formation, but this was not experimentally verified.

Nitrolysis of 3,5-dimethyl-4-iodoisoxazole in a mixture of nitric and sulfuric acids gave the 4-nitro derivative (**5-3**) (85%). Similar nitrolyses of the 4-bromo- and the 4-chloro-3,5-dimethylisoxazoles were less efficient in producing the nitro derivative (**5-3**) (0%–26%). The reaction in a mixture of nitric acid and acetic anhydride in the presence of a catalytic amount of sulfuric acid gave the product (**5-3**) in nearly quantitative yield from the 4-iodoheterocycle (35%

recovered after 48 h at 20°), in 60% yield from the 4-bromoheterocycle (72 h at 20°), and in 49% yield from the 4-chloroheterocycle (2 weeks at 20°); other products were also isolated.[13]

A convenient preparation of a 4-nitroisoxazole was discovered in the nitration of a 5-aminoisoxazole. Nitric acid in acetic anhydride ($-10°$ to 20°, 12 h) converted 3-phenyl-5-acetamidoisoxazole to the 4-nitro derivative **(5-19)** (68%) without nitrating the benzene ring.[28] A similar reaction afforded 3-methyl-4-nitro-5-acetamidoisoxazole **(5-20)** (46%).[29]

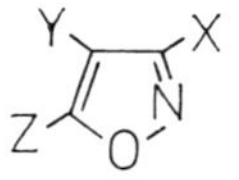

5-18a X = C_6H_5, Y = NO_2, Z = H
5-18b X = Z = C_6H_5, Y = NO_2

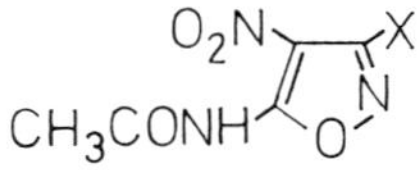

5-19 X = C_6H_5
5-20 X = CH_3

Fuming nitric acid in oleum nitrated 3-succinimido-5-methylisoxazole to the 4-nitro derivative (58%); acid hydrolysis gave the amine **(5-21)** (72%).[30] This result tends to support the claim that nitration (unspecified) products previously identified as 3-(2,4-dinitroanilino)-5-phenyl- and 3-(2,4,6-trinitroanilino)-5-phenylisoxazole were actually the isomeric 4-nitro isoxazoles **(5-22)** and **(5-23)**.[31-33]

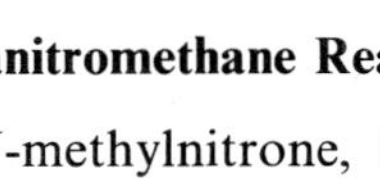

5-21 X = H, Y = CH_3
5-22 X = $C_6H_4NO_2$-p, Y = C_6H_5
5-23 X = $C_6H_3(NO_2)_2$-2,4, Y = C_6H_5

Nitrones, Nitronates, and Tetranitromethane Reactions

Nitroethylene and α-phenyl-N-methylnitrone, $C_6H_5CH=N(O)CH_3$, at 60° gave a mixture of *cis*- and *trans*-2-methyl-3-phenyl-4-nitroisoxazolidine **(5-24a)** in a ratio of 2:1. Silica gel isomerized the *cis*-product (major) to the *trans*-isomer (more stable). Since olefins with less powerful electron-withdrawing substituents gave 5-substituted isoxazolidines, the reversal in regioselectivity to form **(5-24a)** was attributed to lower orbital energies, a conclusion compatible with comparable dipolar addition reactions between electron-deficient acetylenes and nitrones. Such a reversal in regiospecificity with the predominant formation of a 4-nitroisoxazolidine was less pronounced in the addition of nitroethylene to the α-phenyl-N-*tert*-butyl nitrone. Two pairs of diastereoisomers were obtained: *cis*- and *trans*-2-*tert*-butyl-3-phenyl-4-nitroisoxazolidine (**5-24b**) (50%) were isolated in their pure stable forms, and a labile 5-nitro isomer (**5-24c**) (24%) was converted by silica gel to cinnamaldehyde and 3-phenylisoxazole.[34,35]

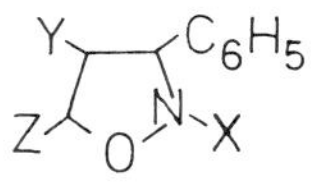

5-24	X	Y	Z
a	CH$_3$	NO$_2$	H
b	C(CH$_3$)$_3$	NO$_2$	H
c	C(CH$_3$)$_3$	H	NO$_2$
d	CH$_3$	CN	NO$_2$
e	CH$_3$	NO$_2$	CN
f	C(CH$_3$)$_3$	CN	NO$_2$
g	C(CH$_3$)$_3$	NO$_2$	CN

Similar results were obtained when cycloadditions from α-phenyl-N-alkylnitrones with *trans*-1-nitro-2-cyanoethylene afforded *cis*- (60%) and *trans*-2-methyl-3-phenyl-4-cyano-5-nitroisoxazolidine (**5-24d**) (16%) and the 4-nitro regioisomer (**5-24e**) (29%) in one example and a mixture of the regioisomers (**5-25f**) (57%) and (**5-25g**) (43%) in another. Two regioisomeric adducts (1:1) were also obtained from *trans*-methyl 3-nitroprop-2-enoate and an α-phenyl-N-alkylnitrone(alkyl = methyl or *tert*-butyl).[35]

A mixture of two bis-adducts (**5-25a**) (15%) and (**5-25b**) (20%) and a monoadduct (**5-25c**) (41%) was obtained from *trans*-1-chloro-2-nitroethylene and α-phenyl-N-methylnitrone in benzene at 25° for 18 h. From two equivalents of the olefin in the presence of strontium carbonate only the monoadduct (**5-25c**) (85%) was obtained. The bis-adducts were also produced in good yield on heating the monoadduct in the presence of the nitrone. It was accounted for by the assumption that a dipolar addition occurred after initial loss of hydrogen chloride provided 2-methyl-3-phenyl-4-nitro-4-isoxazoline. Instead of a comparable dipolar addition reaction with benzonitrile oxide, the monoadduct (**5-25c**) and benzohydroximoyl chloride in ether in the presence of triethylamine afforded 4-methyl-3,5-diphenyl-1,2,4-oxadiazoline (**5-25d**).[35]

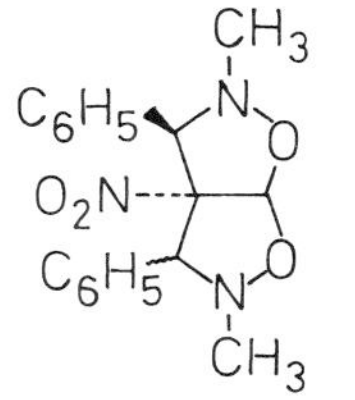

5-25a *cis*-diphenyl
5-25b *trans*-diphenyl

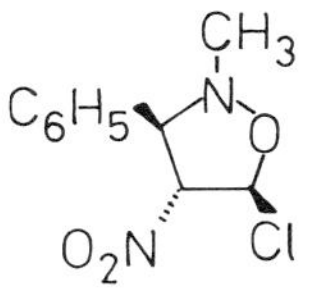

5-25c

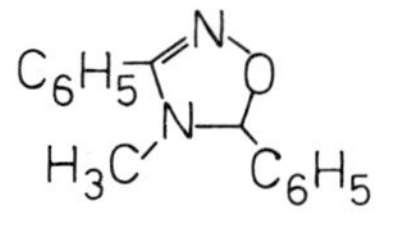

5-25d

The addition of E-β-nitrostyrene to Z-α,N-diphenylnitrone in benzene at 80° was complete after 48 h and gave a mixture of isoxazolidines (**5-26**) and (**5-27**) in a ratio of 85:15. A similar addition with Z-α-benzoyl-N-phenylnitrone was complete after 6 h and gave only the isoxazolidine (**5-28**).[36]

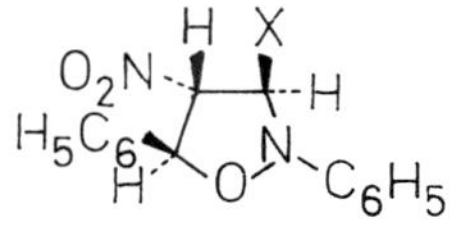

5-26 X = C_6H_5
5-28 X = C_6H_5CO

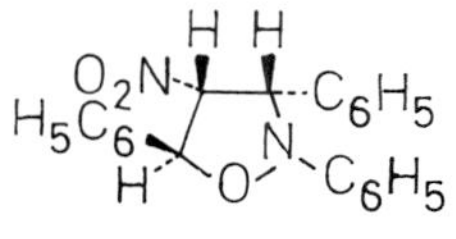

5-27

Esters (nitronates, RR'C = N(O)OR'') of nitronic acids are also N-alkoxy-nitrones. The linear nitronates will be referred to here by naming the substituents at C(α) and O atoms in the nitronate skeleton $>C = N(O)O^-$.

There has been considerable exploration of the dipolar additions of nitronates to unsaturated compounds. Addition of styrene to α-nitro-O-trimethylsilylnitronate (**5-29**) regiospecifically gave 2-trimethylsilyloxy-3-nitro-5-phenylisoxazolidine (**5-30**). Similar reactions with ethylene and methyl acrylate gave isoxazolidines (**5-31**) and (**5-32**), Eq. (5-1). On storage or by treatment with dry hydrogen chloride, elimination of trimethylhydroxysilane produced the 2-isoxazolines (**5-33**) through (**5-35**).[37] Addition of silylalkenes to α-nitro-N-methoxynitronate gave directly 5-silylalkyl-3-nitro-2-isoxazoline (**5-36**) (70%).[38] A similar reaction with styrene gave the 2-isoxazoline (**5-33**), but methyl acrylate gave 2-methoxy-3-nitro-5-carbomethoxyisoxazolidine, which was converted to the corresponding 2-isoxazoline (**5-35**) by treatment with anhydrous hydrogen chloride.[39]

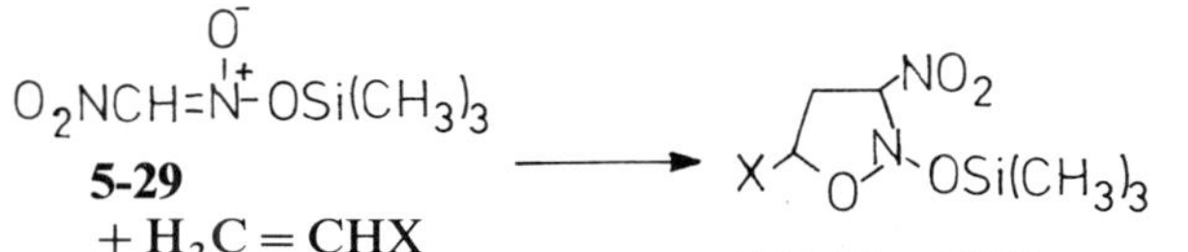

$O_2NCH=\overset{\overset{O^-}{|+}}{N}OSi(CH_3)_3$
5-29
$+ H_2C = CHX$

⟶

5-30 X = C_6H_5
5-31 X = H
5-32 X = CO_2CH_3

(5-1)

A wide variety of olefins combined with α,α-dinitro-O-methylnitronate (from nitroform and diazomethane) to give derivatives (**5-37**) of 3,3-dinitroisoxazolidine.[40] Similar adducts (**5-38**) through (**5-46**) were obtained from dienes.[41]

The results showed that (1) 1,3- and 1,4-dienes reacted as diolefins with isolated double bonds toward the α,α-dinitro-O-methylnitronate, and (2) bis-adducts with the nitronate occurred only with a diolefin of the types $CH_2 = CHCX = CH_2$ (X = H,R) or $(CH_2 = CH)_2X$ (X = CH_2, O) but not RCH = CHCH = CH_2 or $(CH_2 = CH)_2CO$. Cyclopentadiene gave two adducts, (**5-45**) and (**5-46**), in the ratio of 1:1.5. All efforts to obtain a bis-adduct from divinyl ketone failed; after monoadduct formation polymerization gave the product (**5-44**).[41]

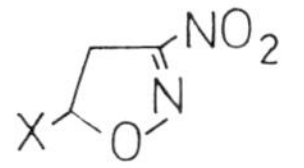

5-33 X = C_6H_5
5-34 X = H
5-35 X = CO_2CH_3
5-36 X = $(CH_2)_n Si(CH_3)_3$
  n = 2,3,4

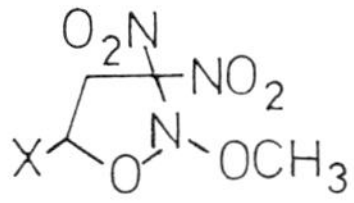

5-37 X = H, CH_3, C_6H_5, $ClCH_2$, $HOCH_2$, C_6H_5CONH, CH_3CO_2, CH_3CO, CH_3O_2C, HO_2C

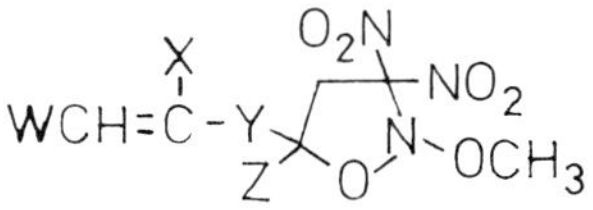

5-38 W = CH_3, CO_2H; X = Z = H; Y = bond
5-39 W = X = H; Z = H, CH_3; Y = bond
5-40 W = Z = H; X = CH_3; Y = bond
5-41 W = X = Z = H; Y = CH_2, O, CO

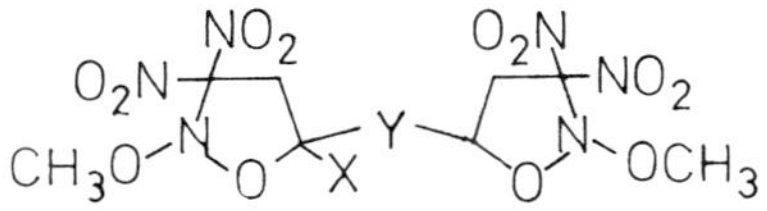

5-42 X = H, CH_3; Y = bond
5-43 X = H; Y = CH_2, O

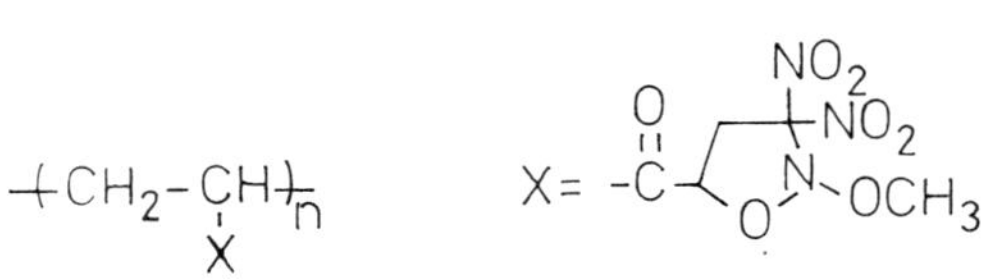

5-44

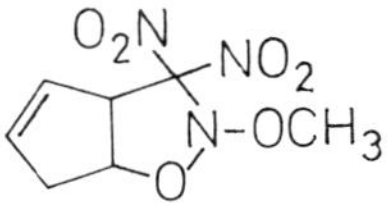

X = $-\overset{O}{\underset{\|}{C}}-$

5-45 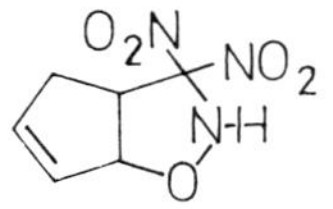   **5-46**

Styrene and the α,α-dinitro-O-trimethylsilylnitronate gave initially the unstable adduct (**5-47**) (94%) and then a small amount of 3-methoxy-5-phenyl-2-isoxazoline (**5-48**) via the 3-nitro compound (**5-33**) on treatment with potassium methoxide. The adduct (**5-47**) on storage at 20° for 6 days also gave the nitro compound (**5-33**) (30%). A search for the N-oxide of the isoxazoline (**5-48**) was unsuccessful.[42,43]

A variety of α,α-dinitro-O-alkylnitronates combined with olefins to form 2-alkoxy-3,3-dinitro-5-substituted isoxazolidines (**5-49**). In certain instances the nitronates and olefins were intermediates from the reactions between the silver salt of nitroform and alkyl iodides, Eq. (5-2).[44-46]

$$\underset{\textbf{5-47}}{\underset{H_5C_6}{\bigvee}\underset{O}{\overset{O_2N\;NO_2}{\bigvee}}N-OSi(CH_3)_3} \qquad \underset{\textbf{5-48 } X = OCH_3}{\underset{H_5C_6}{\bigvee}\underset{O}{\overset{X}{\bigvee}}N}$$

$$\begin{array}{c}(O_2N)_3C^-\;Ag^+\\+\\(CH_3)_2CXI\end{array} \longrightarrow \begin{array}{c}(O_2N)_2C=\overset{O^-}{\overset{|+}{N}}OCX(CH_3)_2\\+\\CH_3\underset{X}{C}=CH_2\end{array} \longrightarrow$$

$$\underset{\textbf{5-49 } X = H,\,CH_3}{\underset{H_3C}{\overset{X}{\bigvee}}\underset{O}{\overset{O_2N\;NO_2}{\bigvee}}N-OCX(CH_3)_2} \qquad\qquad (5\text{-}2)$$

To account for the formation of 2-isoxazoline-2-oxides (**5-51**) and (**5-52**) from C-metal derivatives of nitroform [MC(NO$_2$)$_3$, M = K, Cs, (CH$_3$)$_3$Si, (C$_4$H$_9$)$_3$Sn], the intermediacy of dinitro carbene (**5-50**) was proposed, Eq. (5-3a–d).[47,48] The product (**5-52**), also a nitrone, added a second molecule of styrene to produce 3a-nitro-2,5-diphenyl-3,3a,4,5-tetrahydroisoxazolo-2H[2,3-b]isoxazole (**5-53**). The latter was considered to be the first example of a dialkoxyamine.[49,50]

The dipolar additions to 3-nitro-2-isoxazoline-2-oxides was further investigated. The 4- and 5-methyl and the 4- and 5-phenyl derivatives of 3-nitro-2-isoxazoline-2-oxide gave the expected adducts (**5-54a,b**) and (**5-55a,b**) with ethylene. This permitted identification of the regiospecific adducts from (**5-52**) (X = H) with propylene and with styrene as the 2-substituted-3a-nitro-3,3a,4,5-tetrahydro-2H-isoxazolo[2,3-b]isoxazoles (**5-54b**) and (**5-55b**). It was established that the cycloadditions of the isoxazoline-N-oxides (**5-52**) (X = H,C$_6$H$_5$) to styrene led predominantly to the formation of adducts with the cisoid arrangement of the free pair of electrons at the ring nitrogen atom and the 2-phenyl substituent. Apparently the regiospecificity was independent of the olefin substituent and not affected by the more remote 5-substituent of the isoxazoline.[51,52] A kinetic investigation of the reaction between the heterocycle (**5-52**) (X = H) and styrene was carried out.[53]

$$MC(NO_2)_3 \xrightarrow{-MNO_2} (O_2N)_2C: \longleftrightarrow O_2N\overset{+}{C}=\overset{-}{NO_2}$$
5-50 (5-3a)

5-50 + ⬡ ⟶ [bicyclic structure] **5-51** (5-3b)

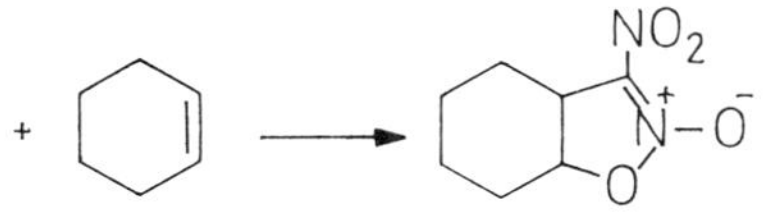

5-50 $\xrightarrow{C_6H_5CH=CH_2}$ [structure] **5-52** X = C$_6$H$_5$ (5-3c)

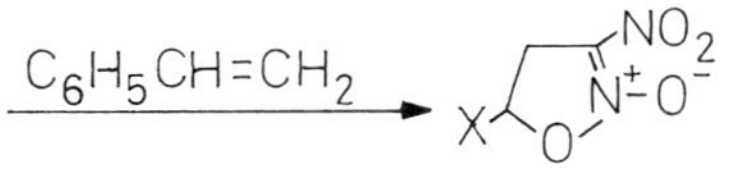

5-52 $\xrightarrow{C_6H_5CH=CH_2}$ [structure] **5-53** (5-3d)

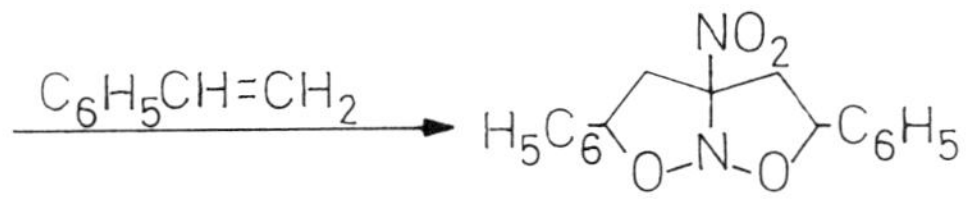

As expected the adduct (**5-56**) from methyl acrylate and 3-nitro-2-isoxazoline-2-oxide (**5-52**) (X = H) was identical with the adduct from ethylene and 3-nitro-5-carbomethoxy-2-isoxazoline-2-oxide.[51]

A further investigation of steric control from substituents led to the conclusion that 3-nitro-2-isoxazoline-2-oxides react with olefins of the following types: RCH = CHAr, CH$_2$ = CRAr, cycloolefins, bridged bicyclic systems with a strained double bond, tetraalkyl ethylenes, and dienic hydrocarbons.[54] Trimethylsilylolefins (CH$_2$ = CH(CH$_2$)$_n$Si(CH$_3$)$_3$, n = 0, 1, 2) converted 3-nitro-2-isoxazoline-2-oxide and its 4-methyl derivative in benzene at 25° for several days to 2-substituted tetrahydroisoxazoloisoxazole derivatives (**5-57**).[54] Failure to obtain addition with stilbene or with dimethyl fumarate was attributed to a steric inhibition in the transition state.[55]

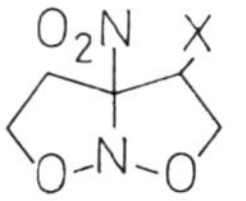

5-54a X = CH$_3$
5-55a X = C$_6$H$_5$

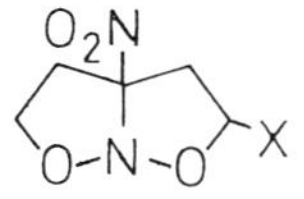

5-54b X = CH$_3$
5-55b X = C$_6$H$_5$

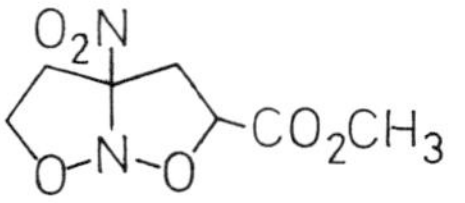

5-56

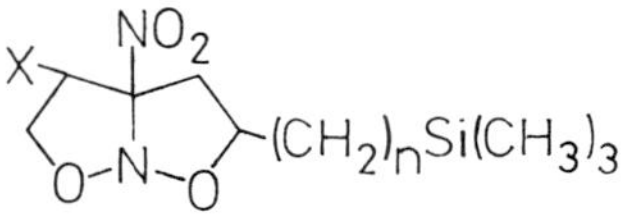

5-57 X = H, CH$_3$
n = 0,1,2

An *endo* approach occurred exclusively in the addition of methyl acrylate to 3-nitro-4,5-dihydro-6H-1,2-oxazine-2-oxide (**5-58a**) to give the *cis*-adduct (**5-58b**). When the 3-nitro group was replaced with other substituents (CH$_3$, CH$_3$CO, CN, CO$_2$CH$_3$, C$_6$H$_5$), similar additions with methyl acrylate, acrylonitrile, and styrene occurred preferably via an *exo* approach.[56] The unique control of the 3-nitro substituent was not accounted for.

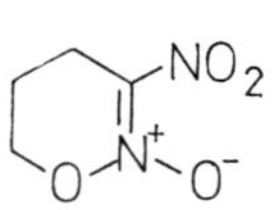

5-58a

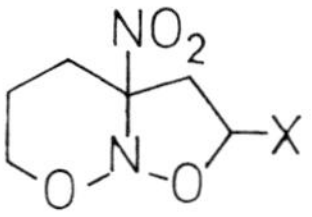

5-58b
X = CO$_2$CH$_3$, CN, C$_6$H$_5$

Table 5-1. Compounds (**5-58c**)

W	X	Y	Z	Yield (%)
H	H	H	H	25
H	H	H	CH$_3$	68
H	H	H	C$_2$H$_5$	42
H	H	CH$_3$	CH$_3$	70
H	H	H	C$_3$H$_7$	69
H	H	CH$_3$	C$_2$H$_5$	90
H	H	H	i-C$_3$H$_7$	54
H	H	H	C$_4$H$_9$	68
CH$_3$	CH$_3$	CH$_3$	CH$_3$	15
H	H	H	C$_5$H$_{11}$	49
H	H	H	C$_6$H$_5$	60

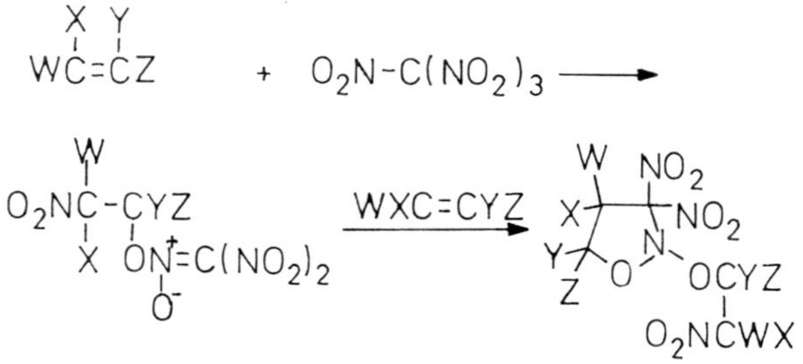

WC=CZ (X, Y) + O$_2$N-C(NO$_2$)$_3$ ⟶

O$_2$NC-CYZ (W, X, ON=C(NO$_2$)$_2$, O$^-$) $\xrightarrow{\text{WXC=CYZ}}$ **5-58c**

(5-4)

Nitroisoxazolidines have also been obtained from olefins and tetranitromethane (TNM). The reactions were carried out in the liquid phase (with or without an added solvent, 15–17°, 2–4 days, with 2 mol of an alkene per mole of tetranitromethane) and gave a 90%–98% conversion with a trace amount of the formation of nitroform. In a heterogeneous medium the reaction was slower and less efficient. The presence of ionic and radical initiators (stannic tetrachloride, benzoyl peroxide, azobisisobutyronitrile) and of γ-irradiation, changes in solvent polarity, and the relative amounts of reactants did not change product yields. Product formation was attributed to the dipolar addition of an olefin to an intermediate α,α-dinitro-O-β-nitroalkylnitronate, Eq. (5-4). Eleven 3,3-dinitro-2-(2-nitroethoxy)isoxazolidines (**5-58c**) were obtained; yields are shown in Table 5-1. Other products included 1,2-dinitroethane (7%), 1,2-dinitropropane (3%), and 1,2-dinitrohexane (3%).[57–59] In carbon tetrachloride at 25° for 2 days cyclohexene afforded the trinitro adduct (**5-59a**) (18%);[60,61] the olefin was oxidized by tetranitromethane to adipic acid (30%–40%) in the absence of a solvent. A neat mixture of stilbene and tetranitromethane gave 1,2-dinitro-1,2-diphenylethane (22%) and α-phenyl-α-nitroacetophenone (61%) but an isoxazolidine was not detected, presumably a result of a steric interference with the dipolar addition.[57–59]

5-59a X = H, n = 4
5-59b X = CH$_3$, n = 4
5-59c X = CH$_3$, n = 3

5-60a n = 4
5-60b n = 3

A reaction in ether at 25° for 5 days between 1-methylcyclohexene and TNM gave the homologous product (**5-59b**) (76%), but 1-phenylcyclohexene and TNM in ether for 1 h gave 2-nitro-1-phenyl-1-trinitromethylcyclohexane (**5-60a**) (48%); 1-phenylcyclopentene gave the related product (**5-60b**) (54%), but 1-methylcyclopentene afforded the isoxazolidine (**5-59c**) (63%). A neat mixture of 1-phenylcyclopentene, 1-hexene, and TNM gave the isoxazolidine (**5-61**) (38%);[61] apparently the cyclic olefin and TNM gave a nitronate ester which then combined preferentially with the acyclic olefin. Further investigations with sterically hindered olefins revealed that TNM did not convert 1,1-diphenylethylene and 1,1,2-triphenylethylene to isoxazolidines; instead 2-nitroethylenes were produced, Eq. (5-5); β-methylstyrene gave α-nitropropiophenone after chromatography, and β,β-dimethylstyrene gave α-nitroisobutyrophenone, Eq. (5-6); α,β-dimethylstyrene and α-methylstilbene gave nitronate esters (**5-62**), Eq. (5-7). On the other hand, α-methylstyrene, 2,4-dimethylpentene-2, and cis-pentene-2 gave the isoxazolidines (**5-63a–c**).[62]

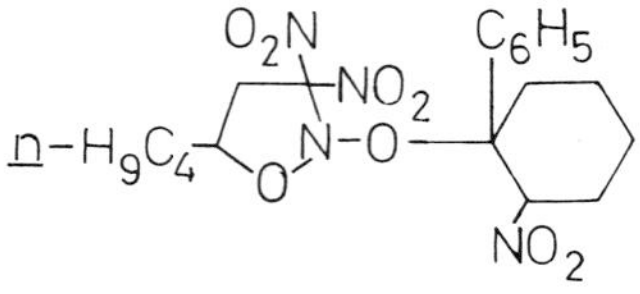

5-61

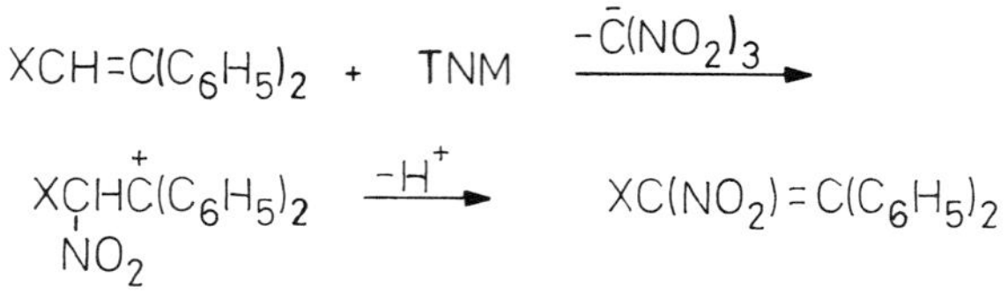

$$XCH=C(C_6H_5)_2 + TNM \xrightarrow{-\bar{C}(NO_2)_3}$$

$$X\overset{+}{C}HC(C_6H_5)_2 \xrightarrow{-H^+} XC(NO_2)=C(C_6H_5)_2$$
$$NO_2$$

$X = H, C_6H_5$ (5-5)

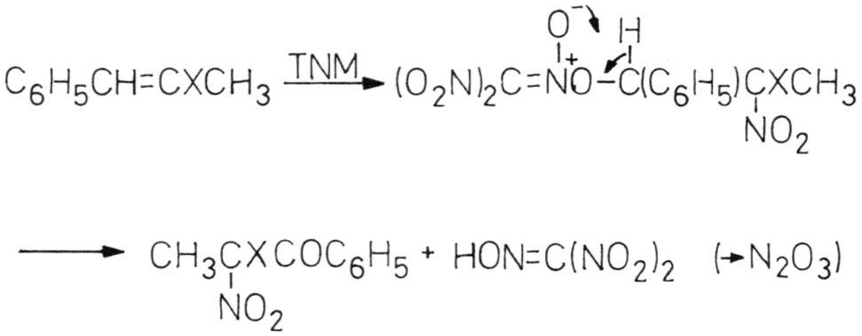

$$C_6H_5CH=CXCH_3 \xrightarrow{TNM} (O_2N)_2C=\overset{O^-}{\underset{1+}{N}O}\overset{H}{-}C(C_6H_5)\underset{NO_2}{C}XCH_3$$

$$\longrightarrow CH_3\underset{NO_2}{C}XCOC_6H_5 + HON=C(NO_2)_2 \ (\to N_2O_3)$$

$X = H, CH_3$ (5-6)

$$C_6H_5C(CH_3)=CHX \xrightarrow{TNM} X\underset{NO_2}{C}HC(CH_3)O\overset{O^-}{\underset{1+}{N}}=C(NO_2)_2$$

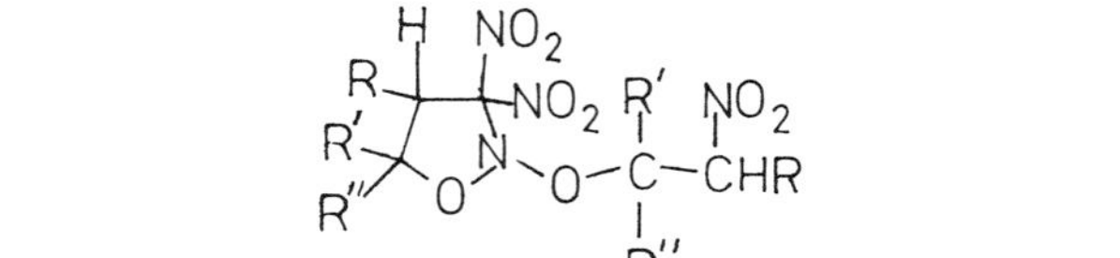

$X = CH_3, C_6H_5$ **5-62** (5-7)

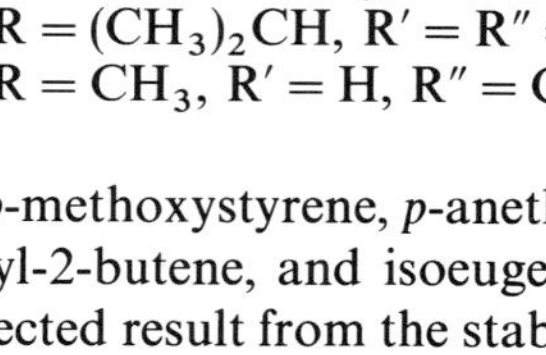

5-63a $R = H$, $R' = CH_3$, $R'' = C_6H_5$
5-63b $R = (CH_3)_2CH$, $R' = R'' = CH_3$
5-63c $R = CH_3$, $R' = H$, $R'' = C_2H_5$

Reactions of TNM with p-methoxystyrene, p-anethole, 2-p-methoxyphenyl-1-propene, 2-p-methoxyphenyl-2-butene, and isoeugenol gave 1,1,1,3-tetranitropropane derivatives, an expected result from the stabilization of an intermediate carbonium ion and its subsequent combination with the trinitromethane anion,

Eq. (5-8); *p*-methoxystyrene also gave an isoxazolidine (**5-64**), but in trace amounts.[63] Alkenes, such as vinyl chloride, acrylonitrile, β-nitrostyrene, and *p*-methoxycinnamic acid, with electron-withdrawing substituents at the olefinic carbon atoms did not react with TNM.[64] Vinyl and allyl ethers and alkenyl silanes ($(CH_3)_3Si(CH_2)_nCH=CH_2$, n = 1, 2) were converted by treatment with TNM to the expected isoxazolidines.[65,66] In one instance iodonitroform ($IC(NO_2)_3$) replaced TNM.[67,68] Dienes and TNM gave monoisoxazolidines.[69,70]

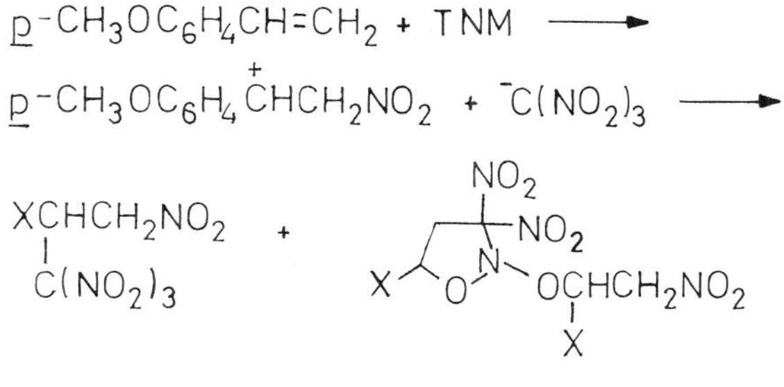

$$p\text{-}CH_3OC_6H_4CH=CH_2 + TNM \longrightarrow$$

$$p\text{-}CH_3OC_6H_4\overset{+}{C}HCH_2NO_2 + {}^-C(NO_2)_3 \longrightarrow$$

$$\underset{\underset{C(NO_2)_3}{|}}{XCHCH_2NO_2} + \underset{X}{\underset{}{\text{[isoxazolidine ring]}}}$$

$X = p\text{-}CH_3OC_6H_4$ **5-64**   (5-8)

Nitrile Oxide Reactions

Styrene derivatives ($ArCX=CH_2$) combined with benzonitrile oxide to give isoxazolines and then isoxazoles by chromic acid oxidation (X = H) or directly (X = Cl).[71,72] β-Nitrostyrene and benzonitrile oxide in ether at 37° for 1.5 h gave 3,5-diphenyl-4-nitro-2-isoxazoline. α-Nitrostilbene gave 3,4,5-triphenylisoxazole directly;[71] each of the three x,β-dinitrostyrenes (x = *o*-, *m*-, and *p*-) gave a 3-phenyl-5-x-nitrophenyl-4-nitro-2-isoxazoline.[73] Similar reactions with nitroethylene afforded isoxazolines (**5-65**).[74]

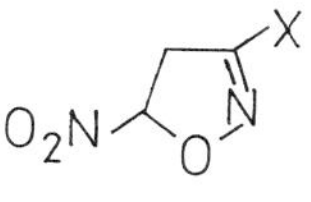

5-65
X = C_6H_5, *m*-, *p*-$O_2NC_6H_4$, *p*-$CH_3OC_6H_4$, 5-nitro-2-furyl

Other investigations on the addition of nitroethylenes to a benzonitrile oxide were reported in 1976.[75] Apparently a steric factor controlled formation of the products (**5-66**) (90%) and (**5-67**) (68%) from nitroethylene and 1-nitro-1-propene. For the formation of 5-methyl-5-nitromethyl-3-phenyl-2-isoxazoline (**5-68**) (10%), 2-methyl-1-nitropropene (a trisubstituted olefin) reacted as the tautomeric 2-methyl-3-nitropropene (a disubstituted olefin); the anticipated product (**5-69**) was not detected. The intermediate isoxazolines (**5-70**) and (**5-71**) from 2-nitropropene and 2-nitrobutene lost nitrous acid on formation to give isoxazoles. From ring-substituted benzonitrile oxides, comparable isoxazoline adducts were obtained with various nitroethylenes.[75-77] Unconjugated nitroalkenylesters combined with benzonitrile oxide to give 3-phenyl-5-nitro-5-carboethoxyalkyl-2-isoxazolines (**5-72**) ($\sim 50\%$).[78]

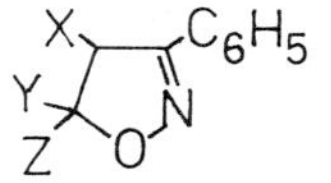

5-66 X = Y = H, Z = NO_2
5-67 X = NO_2, Y = H, Z = CH_3
5-68 X = H, Y = CH_3, Z = CH_2NO_2
5-69 X = NO_2, Y = Z = CH_3
5-70 X = H, Y = NO_2, Z = CH_3
5-71 X = Y = CH_3, Z = NO_2
5-72 X = H, Y = NO_2, Z = $(CH_2)_nCO_2C_2H_5$
n = 1,2

A variation called for an in situ dehydration of a primary nitroalkane by an isocyanate to provide a nitrile oxide. In the presence of 1-morpholino-2-nitroethene, addition and elimination in benzene gave a 3-alkyl-4-nitro-isoxazole, Eq. (5-9a).[79]

Dipolar addition of an acetylene to nitronitrile oxide apparently accounted for the formation of 3-nitro-5-substituted isoxazoles when chloronitroformoxime was treated with an acetylenic Grignard reagent, Eq. (5-9b). Two products (**5-73**) (30%) and (**5-74**) (22%) and a small amount of a by-product (**5-75**) were reported.[80] Other hydroxamic acid chlorides also served as precursors to nitrile oxides. Combinations with nitroacetonitrile and nitromethyl-ketones afforded derivatives (**5-76**) (38%) and (**5-77**) (53%–82%) of isoxazoles, Eq. (5-10).[81]

An apparent dipolar addition of an acetylene to a nitrile oxide accounted for the formation of 4-nitro- and 5-nitroisoxazoles obtained from acetylene precursors. Nitroketene aminals afforded 4-nitro- (**5-78**) (10%–25%), Eq. (5-11),[82] and β-chloronitroethylene afforded 5-nitroisoxazoles (**5-79**) (55%), Eq. (5-12a).[83] The latter reaction afforded an example of a rarely encountered 5-nitroisoxazole.

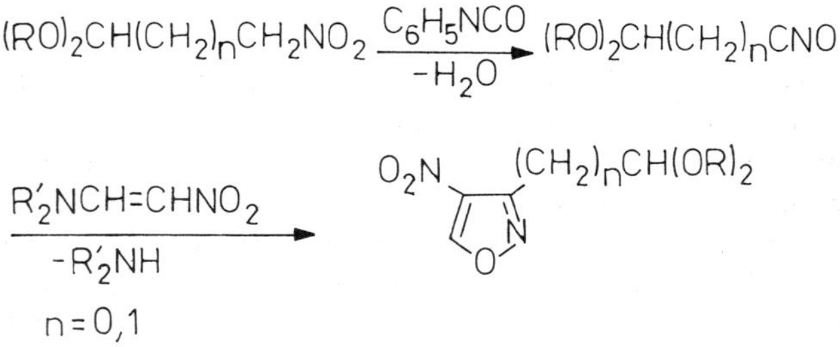

$$(RO)_2CH(CH_2)_nCH_2NO_2 \xrightarrow[-H_2O]{C_6H_5NCO} (RO)_2CH(CH_2)_nCNO$$

$$\xrightarrow[-R'_2NH]{R'_2NCH=CHNO_2}$$

n = 0,1

(5-9a)

Isoxazoles

O$_2$NC(Cl)=NOH
+
BrMgC≡CX

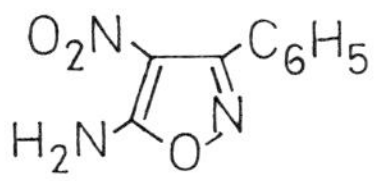

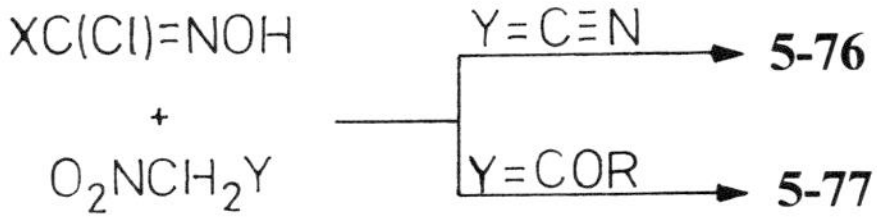

→

(structure 5-73 with NO$_2$ and X on isoxazole ring)

5-73 X = CH$_2$O-(tetrahydropyran)

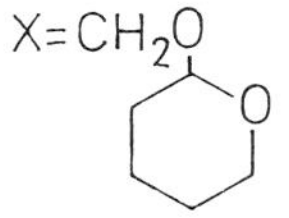

5-74 X = C$_6$H$_5$ (5-9b)

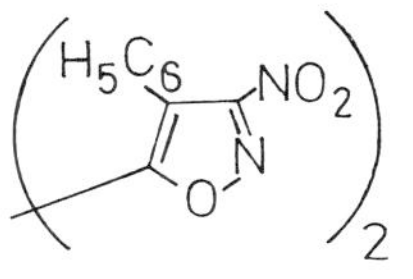

5-75 (bis-isoxazole with C$_6$H$_5$ and NO$_2$ substituents), ( )$_2$

5-76 (isoxazole with O$_2$N, C$_6$H$_5$, and H$_2$N substituents)

XC(Cl)=NOH
+
O$_2$NCH$_2$Y

$\xrightarrow{Y=C\equiv N}$ **5-76**

$\xrightarrow{Y=COR}$ **5-77**   (5-10)

(isoxazole structure with O$_2$N, X, and R substituents)

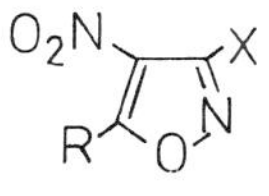

5-77a X = C$_6$H$_5$, R = CH$_3$
5-77b X = R = C$_6$H$_5$
5-77c X = p-ClC$_6$H$_4$, R = C$_6$H$_5$
5-77d X = C$_6$H$_5$CO, R = C$_6$H$_5$
5-77e X = C$_2$H$_5$O$_2$C, R = C$_6$H$_5$

Although the intermediate nitroacetylenes in Eqs. (5-11) and (5-12) have not been detected, a nitroacetylene was also considered to be an intermediate in certain formations of pyrazoles (Chapter 2) and of a pyrrole (Chapter 1).

Cycloaddition between an acylnitrile oxide and a nitroenamine in refluxing toluene for 18 h gave a 3-acyl-4-nitroisoxazole (50%). In support of the assumption that a 5-dimethylamino cycloadduct from an N,N-dimethylnitroenamine accounted for product formation by loss of dimethylamine rather than cycloaddition from nitroacetylene, it was shown that the enamine, a synthon for nitroacetylene, was recovered unchanged when subjected to comparable conditions but in the absence of the nitrile oxide, Eq. (5-12b).[84] A reported condensation between an acylhydroxamic chloride (synthon for an acylnitrile oxide) and an acylnitromethane in the presence of triethylamine to

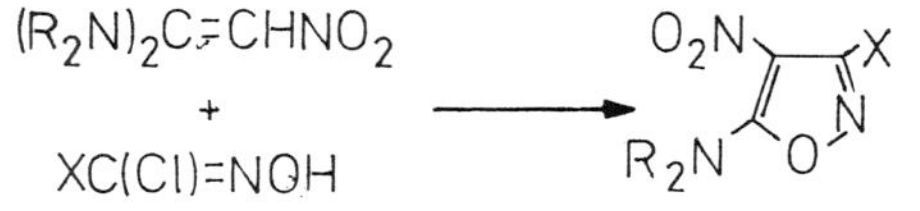

$(R_2N)_2C=CHNO_2$
+
$XC(Cl)=NOH$

⟶

[isoxazole with O_2N, R_2N, X substituents]

5-78a $X = p\text{-}ClC_6H_4$
$R_2N = (CH_2)_4N$
5-78b $X = p\text{-}ClC_6H_4$
$R_2N = CH_3NH$
5-78c $X = C_6H_5$
$R_2N = (CH_2)_4N$
5-78d $X = tert\text{-}C_4H_9$
$R_2N = (CH_2)_4N$ (5-11)

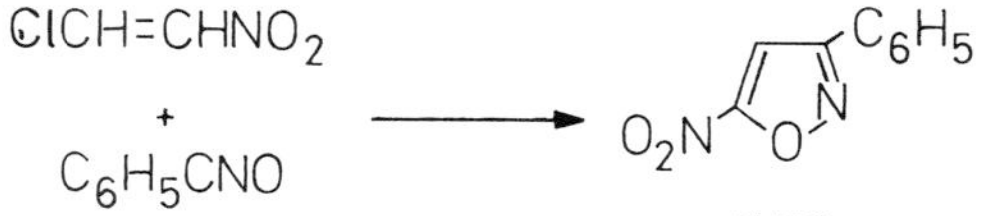

$ClCH=CHNO_2$
+
C_6H_5CNO

⟶

[isoxazole O_2N, C_6H_5] **5-79**   (5-12a)

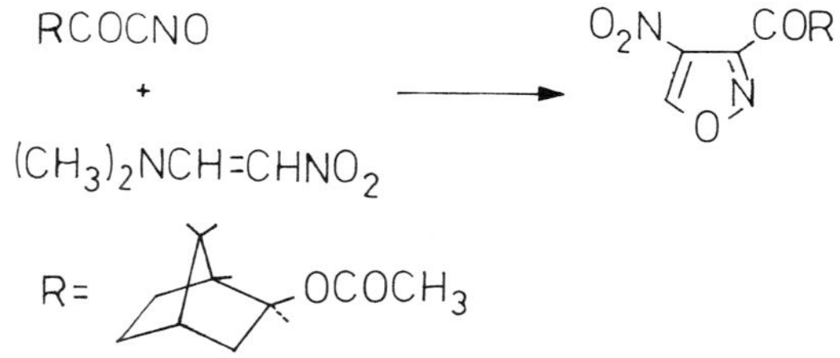

$RCOCNO$
+
$(CH_3)_2NCH=CHNO_2$

⟶

[isoxazole O_2N, COR]

$R =$ [norbornyl]$-OCOCH_3$ (5-12b)

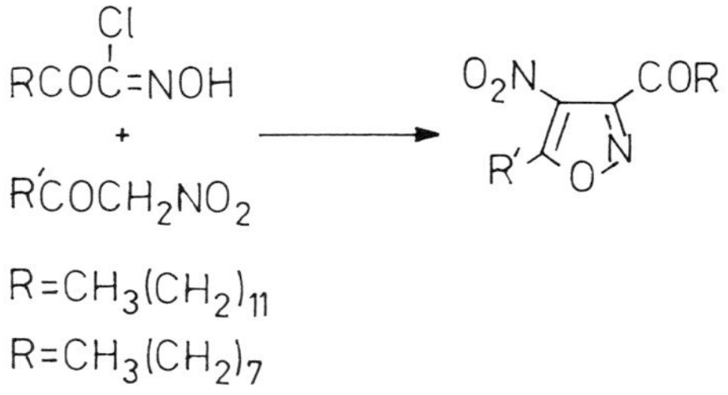

$\underset{\underset{}{}}{RCOC}\!=\!NOH$ with Cl
+
$R'COCH_2NO_2$

⟶

[isoxazole O_2N, COR, R']

$R = CH_3(CH_2)_{11}$
$R = CH_3(CH_2)_7$ (5-12c)

give a 3-acyl-4-nitro-5-alkylisoxazole, Eq. (5-12c),[85] can be envisaged as an example of 1,3-cycloaddition of a nitrile oxide to a nitroenol followed by an elimination of water.

Other Acyclic Nitro Compound Reactions

The known conversion of α,β-dinitroalkanes to isoxazoles by treatment with aqueous acids was adapted to prepare nitroisoxazoles. In water at 90° for 3 h 1,1,1,3-tetranitropropane gave 3,5-dinitroisoxazole (**5-80**) (70%); 2-methyl-1,1,1,3-tetranitropropane in nitric acid (10%) at 70° for 7 h gave the 4-methyl derivative (**5-81**) (50%). Although mechanism studies have not been carried out, it was suggested that the conversion involved dehydration of an isoxazoline-N-oxide intermediate (**5-82**).[86]

In a related reaction, 1,1,1-trinitro-2-alkyl-3-bromopropanes in an alkaline medium gave 3-nitro-2-isoxazoline-2-oxides (**5-83**).[87,88] On formation from 1,1-dinitro-1,3-dibromopropane in ethanolic potassium hydroxide, the potassium salt of 1,1-dinitro-2-ethoxy-3-bromopropane gave 3-nitro-4-ethoxy-2-isoxazoline-2-oxide (**5-84a**) (50%).[89–91] In a related reaction the adduct from the potassium salt of 4,4-dinitrobutenoic acid and N-bromosuccinimide gave the isoxazoline-N-oxide (**5-84b**).[91']

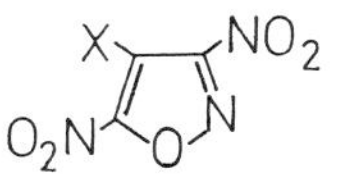

5-80 X = H
5-81 X = CH$_3$

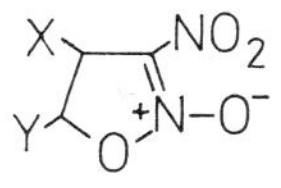

5-82 X = H, CH$_3$; Y = NO$_2$
5-83 X = R, Y = H
5-84a X = OC$_2$H$_5$, Y = H
5-84b

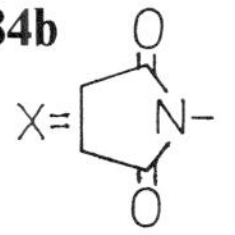

$$X = \text{(succinimidyl)}$$

$$Y = CO_2CH_3$$

There are several related intramolecular cyclizations to nitroisoxazole derivatives. Sodium methoxide (1 mol) in methanol converted 2-alkyl- and 2-aryl-1-bromo-1-nitroethenes and nitroacetonitrile or nitroacetic ester at 0–25° to 5-nitro-2-isoxazoline-2-oxide. It was assumed that cyclization proceeded from the nitronate salt of the *aci* form of a 1,3-dinitro-1-bromopropane, Eq. (5-13). Yields ran from 38% to 63%.[92]

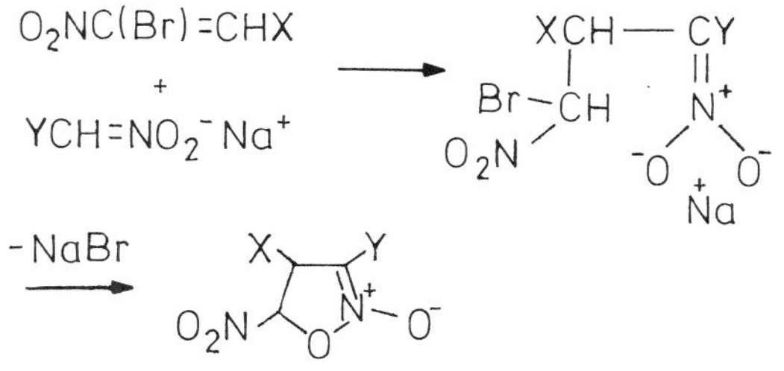

X	C$_2$H$_5$	C$_6$H$_5$	C$_6$H$_4$Z	C$_6$H$_4$Z	C$_6$H$_4$Z
Z			m-NO$_2$	p-NO$_2$	p-OCH$_3$
Y	CN	CN	CN	CN	CN

X	C$_6$H$_4$Z	C$_6$H$_4$Z	C$_6$H$_4$Z	C$_6$H$_4$Z
Z	p-OCH$_3$	m-NO$_2$	p-NO$_2$	p-OCH$_3$
Y	CO$_2$C$_2$H$_5$	CO$_2$C$_2$H$_5$	CO$_2$C$_2$H$_5$	CO$_2$C$_2$H$_5$

(5-13)

When heated in refluxing ethanol for 1 h the dimers of pseudonitrosites, obtained from α,β-unsaturated ketones of the chalcone type and dinitrogen trioxide, gave 3,5-diaryl-4-nitroisoxazoles, Eq. (5-14) (no yield data).[93-95] There was no apparent reason that the pseudonitrosite $C_6H_5COCH(NO_2)CH(NO)C_6H_4\text{-}OCH_3\text{-}o$ failed to give an isoxazole.

$$\left(\begin{array}{c} XC_6H_4COCHNO_2 \\ | \\ ONCHC_6H_4Y \end{array}\right)_2 \xrightarrow{80°} \begin{array}{c} O_2N \diagdown \diagup C_6H_4X \\ YC_6H_4 \diagup O \diagdown N \end{array}$$

X	H·	H	H	H	p-CH₃	p-OCH₃	p-Cl
Y	H	p-OR	p-CH₃	p-Cl	H	H	H

$$R = OCH_3, OC_2H_5 \qquad (5\text{-}14)$$

Nitric acid converted crotonic acid to a mixture of eulite (**5-85a**) and dislite (**5-85b**).[1]

$$\begin{array}{c} O_2N \diagdown \diagup C(NO_2)_2CH_3 \\ H_3C \diagup O \diagdown N \end{array} \qquad \left(\begin{array}{c} O_2N \diagdown \diagup \\ H_3C \diagup O \diagdown N \end{array}\right)_2$$

5-85a   **5-85b**

On treatment with a combination of *n*-propyl and sodium nitrites in dimethylsulfoxide at 35° for 18 h, 1-chloro-3-nitropropane gave 3-nitro-2-isoxazoline (**5-34**) (79%), Eq. (5-15).[96] In a comparable scheme, a propargyl bromide and sodium nitrite in N,N-dimethylformamide (DMF) at 20° for several hours gave a 5-substituted-3-nitroisoxazole (20%–63%), Eq. (5-16).[97,101,102] The intermediacy of a nitro compound, a nitrite ester, and a nitrolic acid was assumed. Ring closure to an isoxazole from an α,β-acetylenic nitrolic acid was previously known.[98-100]

The intermediacy of an oxime of an α-acyl-α-nitroketone can be envisaged for Eq. (5-14). In related ring closures to 36 4-nitro-2-isoxazolines (19%–93%), intermediate oximes of β-hydroxy-α-nitroketones were apparently produced from oximes of α-nitroketones and an aromatic or aliphatic aldehyde, acetone,

$$Cl(CH_2)_2CH_2NO_2 \xrightarrow{\overset{+}{N}O} \left[\begin{array}{c} CH_2\text{-}C^{\diagup NO_2} \\ | \quad \| \\ ClCH_2 \quad N \\ \diagdown OH \end{array}\right]$$

$$\xrightarrow{-HCl} \textbf{5-34} \qquad (5\text{-}15)$$

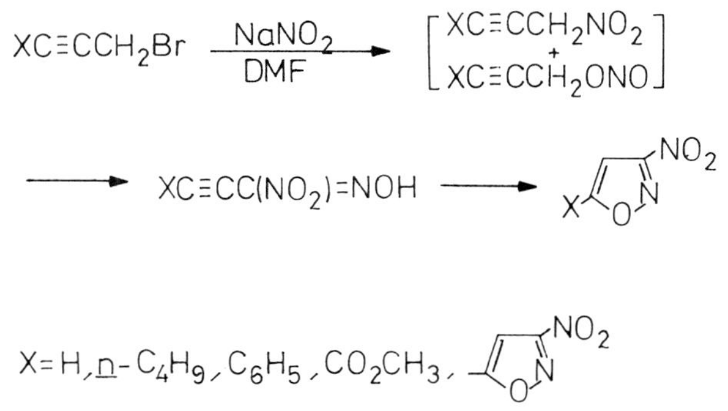

$$XC\equiv CCH_2Br \xrightarrow[DMF]{NaNO_2} \begin{bmatrix} XC\equiv CCH_2NO_2 \\ + \\ XC\equiv CCH_2ONO \end{bmatrix}$$

$$\longrightarrow XC\equiv CC(NO_2)=NOH \longrightarrow \underset{X}{\underset{O}{\underset{\|}{\overset{NO_2}{\diagup}}}}_N$$

$$X = H, \underline{n}\text{-}C_4H_9, C_6H_5, CO_2CH_3, \underset{O}{\overset{NO_2}{\diagup}}_N \qquad (5\text{-}16)$$

or a cyclic ketone, Eq. (5-17a).[103-105] The lower yields of isoxazolines described reactions from ketones and sterically hindered alkehydes. The preparation was extended to 17 3- and/or 5-fluorinated aryl derivatives of 4-nitro-2-isoxazoline.[106,107] In another example the oxime of nitromethylcyclohexyl ketone and ethyl chlorooxoacetate in ether at 25° gave ethyl 3-cyclohexyl-4-nitroisoxazole-5-carboxylate (70%), Eq. (5-17b). When the cyclohexyl group was replaced by a ribofuranosyl group, cyclization depended on the presence of two equivalents of pyridine.[107]

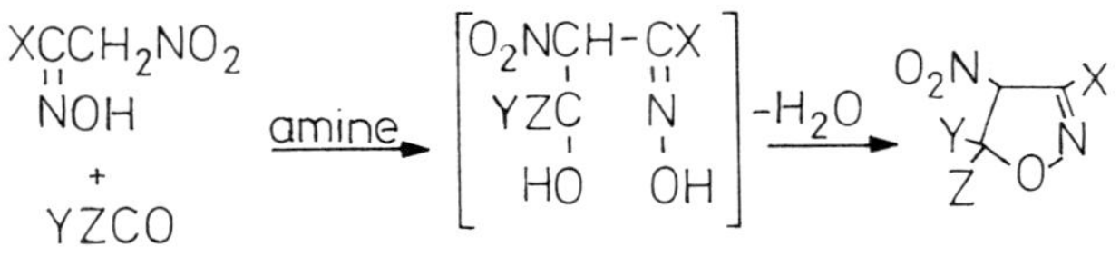

$$\begin{array}{c} XCCH_2NO_2 \\ \| \\ NOH \\ + \\ YZCO \end{array} \xrightarrow{amine} \begin{bmatrix} O_2NCH\text{-}CX \\ | \quad \| \\ YZC \quad N \\ | \quad | \\ HO \quad OH \end{bmatrix} \xrightarrow{-H_2O} \begin{array}{c} O_2N \\ Y \\ Z \end{array}\!\!\diagup\!\!\begin{array}{c} X \\ N \\ O \end{array}$$

X	C_6H_5	$p\text{-}CH_3OC_6H_5$	C_6H_5	C_6H_5	C_6H_5
Y	C_6H_5	C_6H_5	iso-C_4H_9	CH_3	$(CH_2)_{4,5}$
Z	H	H	H	CH_3	

(5-17a)

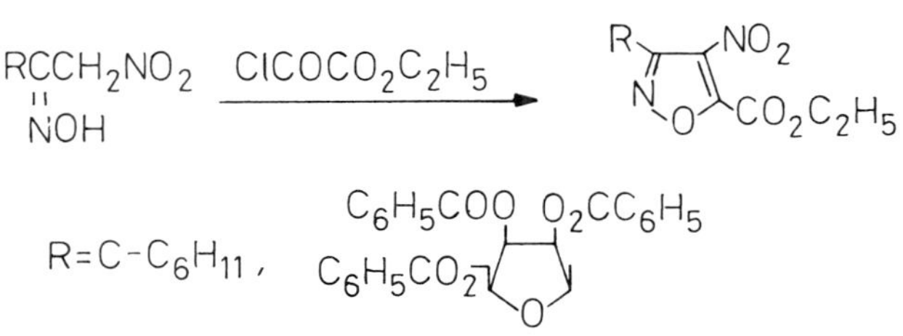

$$\begin{array}{c} RCCH_2NO_2 \\ \| \\ NOH \end{array} \xrightarrow{ClCOCO_2C_2H_5} \underset{N}{\underset{O}{\overset{R}{\diagup}}}\!\!\diagup\!\!\underset{CO_2C_2H_5}{\overset{NO_2}{}}$$

$$R = C\text{-}C_6H_{11}, \quad \underset{C_6H_5CO_2}{\overset{C_6H_5COO \quad O_2CC_6H_5}{\diagup\!\!\!\!\diagdown}}\!\!\diagup\!\!\diagdown_O$$

(5-17b)

On acidification the salt of the dioxime of nitromalonic dialdehyde gave 4-nitroisoxazole (**5-1**) quantitatively.[108]

Ring Conversion

Sodium hydroxide (1 N, 25°, 5 min) converted the oxime of formylfuroxan to 4-nitro-2-isoxazolin-5-imine, Eq. (5-18a) (83%).[109] Apparently tautomerization to 4-nitro-5-aminoisoxazole was not detected; the imine isomer was hydrolyzed to fulminuric acid, $NCCH(NO_2)CONH_2$.

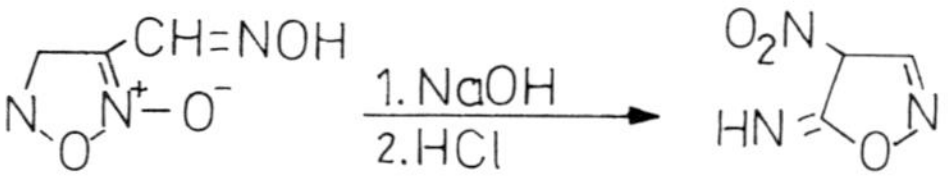

$$\text{(structure with CH=NOH)} \xrightarrow[\text{2. HCl}]{\text{1. NaOH}} \text{(4-nitro-5-aminoisoxazole structure)} \tag{5-18a}$$

An assumed 4-nitro-3-hydroxyisoxazole was obtained from either 3,5-dinitro-1H-4-pyridone or its 1-p-nitrophenyl derivative by treatment with an excess of hydroxylamine. The product was acetylated with acetyl chloride, Eq. (5-18b).[110]

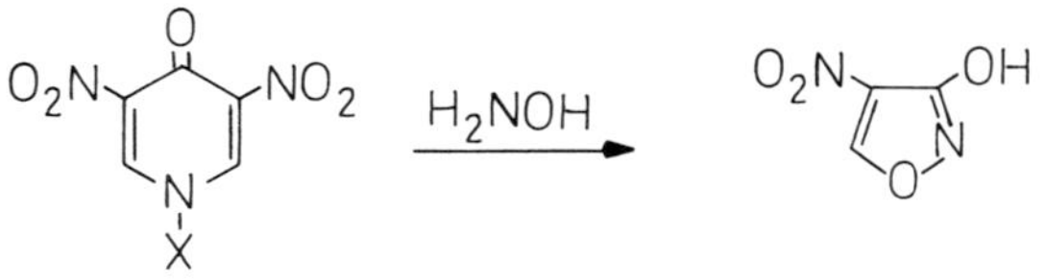

$$\text{(3,5-dinitropyridone)} \xrightarrow{H_2NOH} \text{(4-nitro-3-hydroxyisoxazole)}$$

X = H,
p-X-C_6H_4 (5-18b)

Diazoalkane Reactions

A diazoalkane added to *trans*-1-phenyl-1,2-dinitroethylene in ether at 25° for 24 h to give substituted derivatives of 4-nitro-2-isoxazoline-2-oxides, Eq. (5-19) (48%–83%).[111] From 1-(p-nitrophenyl)diazoethane both the isoxazoline (48%) (Eq. (5-19), X = CH_3, Y = p-$O_2NC_6H_4$) and the 3,4-dinitropyrazoline (**5-85c**) (20%) were obtained.

From *gem*-dinitroethylenes 3-nitro-2-isoxazoline-2-oxides (**5-86a–c**) were obtained (76%–84%).[112,113] Apparently a steric interference did not allow an addition to occur between 1,1-dinitro-2,2-diphenylethylene and diazomethane. It was suggested that a diazomethane and a *gem*-dinitroethylene initially gave a pyrazoline which then lost nitrogen and recyclized to the isoxazoline-2-oxide, Eq. (5-20a,b).[114]

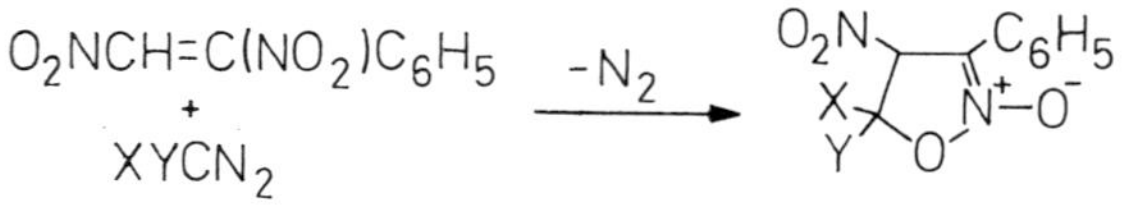

$$O_2NCH=C(NO_2)C_6H_5 + XYCN_2 \xrightarrow{-N_2} \text{(isoxazoline-2-oxide)}$$

X	C_6H_5	CH_3	CH_3	C_6H_5
Y	p-$CH_3OC_6H_4$	p-BrC_6H_4	p-$O_2NC_6H_4$	o-BrC_6H_4

(5-19)

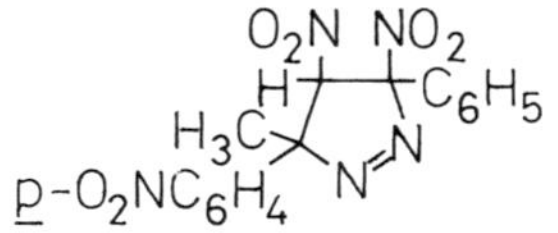

$$\underset{p\text{-}O_2NC_6H_4}{\text{(3,4-dinitropyrazoline structure with }H_3C, C_6H_5)}$$

5-85c

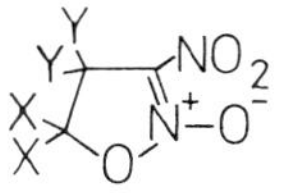

5-86a X=H, Y₂=. (fluorenylidene)

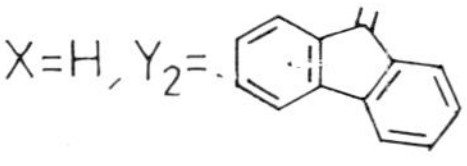

5-86b X₂= (fluorenylidene), Y=H

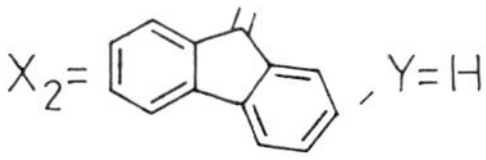

5-86c $X = C_6H_5$, $Y = H$
5-86d $X = Y = H$

On discovering the interesting formation of 3-nitro-2-isoxazoline-2-oxide (**5-86d**) from the treatment of nitroform with an excess of diazomethane, it was shown that coproduct α,α-dinitro-O-methylnitronate (**5-87a**), but not coproduct 1,1,1-trinitromethane (**5-87b**), was also converted with diazomethane to the heterocycle (**5-86d**) (25%).[114] That both methylene units in the latter apparently originated from diazomethane is reminiscent of a similar combination of benzonitrile oxide and diazomethane to give 3-phenylisoxazole (**5-87c**).[115] Each of the two reports suggested a pyrazoline precursor to its product after initial formation of an adduct with methylene, Eq. (5-20a,b). It was proposed[114] that dinitroethylene, a formal adduct from methylene and dinitrocarbene (**5-50**), arose from a replacement of the elements of methyl nitrite in the nitronate ester (**5-87a**) in a reaction with diazomethane. In one reaction 1-nitroso-3-phenyl-2-pyrazoline, an isomer of (**5-87d**), was also obtained.[115] The formation of 3-cyano-2-isoxazoline-2-oxide by the reaction between diazomethane and chloro- or bromodinitroacetonitrile was recently reported.[116] A similar sequence of reactions was proposed, Eq. (5-21).

$(O_2N)_2C = N(O)OCH_3$
5-87a

$CH_3C(NO_2)_3$
5-87b

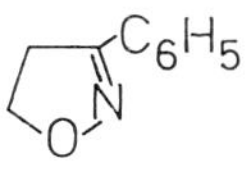

5-87c

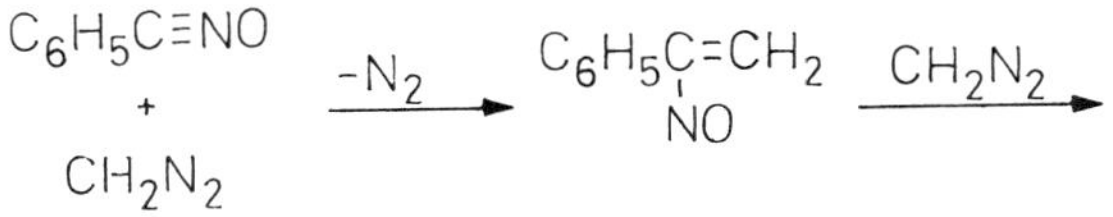

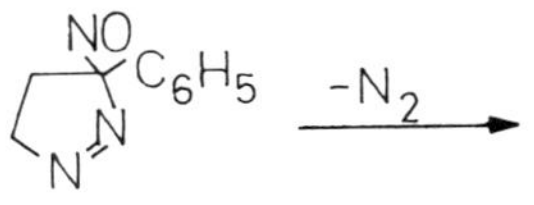

$C_6H_5C\equiv NO + CH_2N_2 \xrightarrow{-N_2} \underset{NO}{C_6H_5C=CH_2} \xrightarrow{CH_2N_2}$

(1-nitroso-3-phenyl-pyrazoline) $\xrightarrow{-N_2}$ **5-87c**

5-87d

(5-20a)

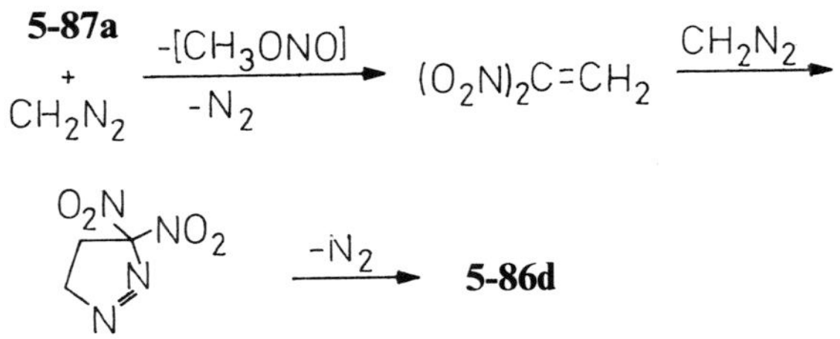

$$\text{5-87a} + CH_2N_2 \xrightarrow[-N_2]{-[CH_3ONO]} (O_2N)_2C=CH_2 \xrightarrow{CH_2N_2}$$

$$\xrightarrow{-N_2} \text{5-86d} \qquad (5\text{-}20b)$$

$$CH_2N_2 + XC(NO_2)_2CN \longrightarrow XCH_2O\overset{+}{N}=C(NO_2)CN$$
$$\underset{O^-}{|}$$

$$\xrightarrow{CH_2N_2} CH_2=C(NO_2)CN \xrightarrow{CH_2N_2} \text{[isoxazoline N-oxide with CN]} \qquad (5\text{-}21)$$

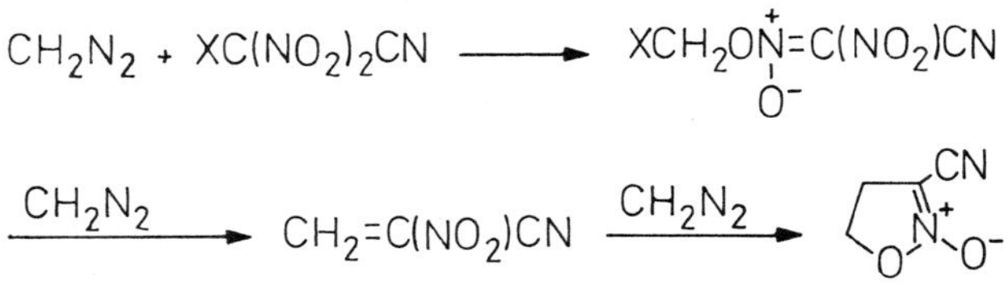

Even less understood is the formation of 3-nitro-2-isoxazoline-2-oxide (**5-86d**) (20%–25%) from halotrinitromethane and diazomethane.[117,118] There was no reaction between diazomethane and fluorotrinitromethane.[117] An investigation showed that an intermediate α,α-dinitro-O-chloromethylnitronate (**5-88**) (from chlorotrinitromethane and diazomethane) was trapped as its adduct (**5-89**) (15%) with the dimethyl ester of 7-oxabicyclo[2,2,1]heptenedicarboxylic acid, Eq. (5-22). A similar formation of the adduct (**5-89**) (Y = OCH$_3$) from bromo- or iodotrinitromethane revealed a preferential reaction with diazomethane to form the intermediate nitronate (**5-88**). It was assumed that bromo- and iododiazomethane decomposed on formation.[114–116]

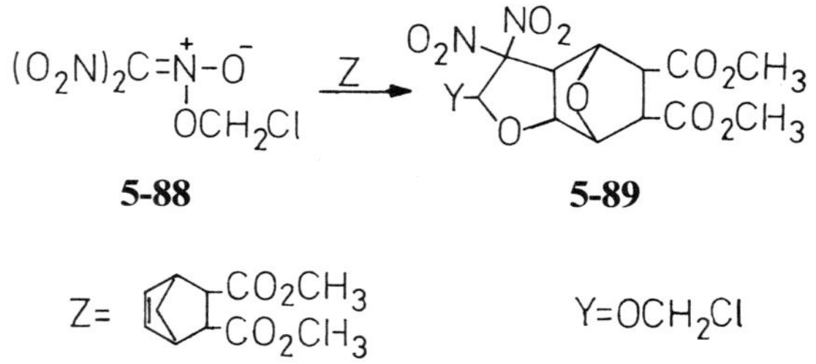

$$(O_2N)_2C=\overset{+}{N}-O^- \qquad \xrightarrow{Z}$$
$$\underset{OCH_2Cl}{|}$$

5-88 → **5-89**

$$Z = \text{[norbornene-CO}_2CH_3\text{ diester]} \qquad Y=OCH_2Cl \qquad (5\text{-}22)$$

Reactions

Isoxazolines to Isoxazoles

In general isoxazolines are dehydrogenated to isoxazoles by treatment with an oxidizing agent such as potassium permanganate[119,120] and active manganese dioxide.[121] Such a conversion afforded 2,5-diaryl-4-nitroisoxazoles in 65%[119,120] and 99% yields.[121]

Thermal dehydration of a 2-isoxazoline-2-oxide to an isoxazole can occur easily[122] and has been assumed to be spontaneous—eg, the conversion (**5-82**) to

(**5-80**), (**5-81**).[86] By treatment with dry hydrogen chloride, the elimination of methanol from a 2-methoxyisoxazolidine afforded the expected 2-isoxazoline (**5-35**).[39]

A facile conversion to an isoxazole by an elimination of nitrous acid from a 5-nitro-2-isoxazoline has often been noted—eg, (**5-66**),[75] (**5-70**),[77] (**5-71**),[75] and (**5-72**).[78,79] Conversion to an isoxazole by the elimination of nitrous acid from a 4-nitroisoxazoline can be more difficult. At 150° 3,5-diphenyl-4-nitro-2-isoxazoline, see Eq. (5-17), gave a nearly quantitative conversion to 3,5-diphenylisoxazole and nitrous fumes;[71] however, a recent report described conversion to 3,5-diphenyl-4-nitroisoxazole (60%) on heating the isoxazoline in refluxing dimethylformamide.[72] The same isoxazoline in alkali coupled with an aryldiazonium salt to give 3,5-diphenyl-4-arylazo-4-nitro-2-isoxazoline (unisolated) and a 3,5-diphenyl-4-arylazoisoxazole (75%) by an elimination of nitrous acid, Eq. (5-23).[26,123] Nitrous acid was not eliminated when 3,5-diperfluorophenyl-4-nitro-2-isoxazoline was treated with sodium methoxide; instead a fluorine was replaced to give 3-p-methoxytetrafluorophenyl-5-pentafluorophenyl-4-nitro-2-isoxazoline (**5-90a**); 2 mole of sodium methoxide replaced both p-fluoro groups with methoxyl groups.[106,107]

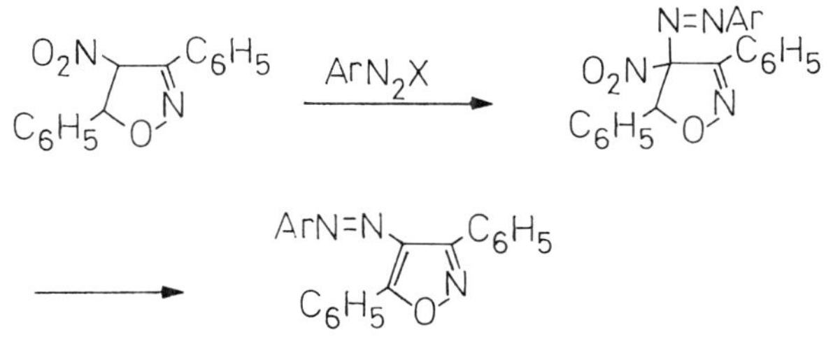

(5-23)

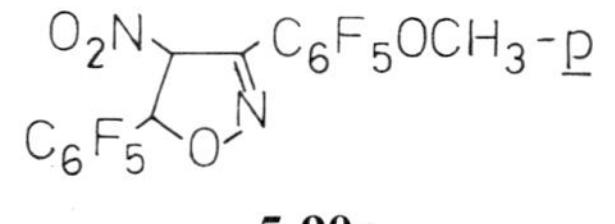

5-90a 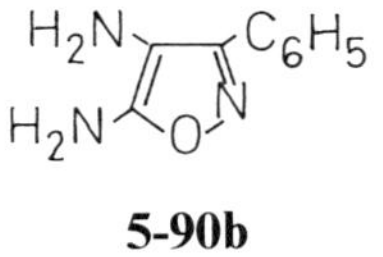   **5-90b**

Reduction

Reduction of 3-phenyl-4-nitro-5-acetamidoisoxazole (**5-19**) with sodium borohydride in the presence of palladium on charcoal gave 3-phenyl-4,5-diaminoisoxazole (**5-90b**), an unstable compound isolated as its hydrochloride salt (30%).[26,27] Stannous chloride[12] or aluminum amalgam[10] converted 3-methyl-4-nitroisoxazole (**5-2**) to the corresponding amine. Stannous chloride afforded 3-phenyl-4-amino-5-methylisoxazole in nearly quantitative yield from the nitroisoxazole.[124] Stannous chloride in hydrochloric acid reduced 3-nitro-5-[3-nitro-5-isoxazoyl]-isoxazole to the 3,3'-diamino-5,5'-bisisoxazole (**5-91**).[125] Phenylhydrazine and 3-nitroisoxazole gave 3-hydroxylaminoisoxazole (**5-92**).[101]

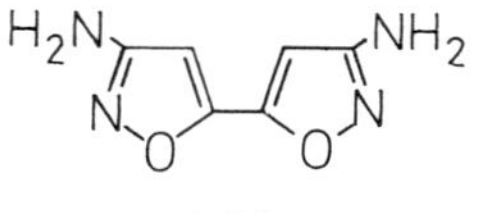

5-91 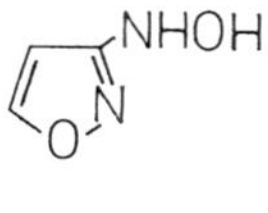 **5-92**

Tri-*n*-butyltin hydride replaced a 5-chloro substituent with hydrogen to give 2-methyl-3-phenyl-4-nitroisoxazolidine.[35]

Displacement of the Nitro Group

Various nucleophiles have replaced the nitro group in *cis*-8-substituted-6-nitro-2,9-dioxa-1-azabicyclo[4,3,0]nonane (**5-93a**).[126,127] Treatment with methanol brought about nearly quantitative replacement by methoxyl to give compounds (**5-93b–d**);[127,128] ethanol gave the ethoxy derivative (**5-93e**) (78%);[127] sodium cyanide in anhydrous dimethylformamide gave the 6-cyano derivatives (**5-93f**) (82%), (**g**) (95%), **h** (66%), (**i**) (69%), and (**j**) (81%);[129] sodium azide in aqueous acetone gave the 6-azido compounds (**5-93k**) (93%), (**l**) (92%), (**m**) (94%), and (**n**) (90%);[130] and the potassium salt of nitroethane in aqueous dimethylformamide gave a 6-hydroxy derivative (**5-93o**) presumably via the nitrite ester (**5-93p**).[131] Each of the 14 reactions gave a mixture of two diastereomers: *trans* > *cis* for (**5-93b,e,k,l,m,n**) was based on nmr analyses.[127,130] It was also shown that the ratios of diastereoisomers for products (**5-93f**) and (**5-93g**) were identical to the ratios of diastereoisomers for these products when obtained by the addition of styrene or *p*-bromostyrene to 3-nitro- and 3-cyano-4,5-dihydro-6*H*-1,2-oxazine-2-oxide, Eq. (5-24). In contrast with the formation of a mixture of stereoisomeric

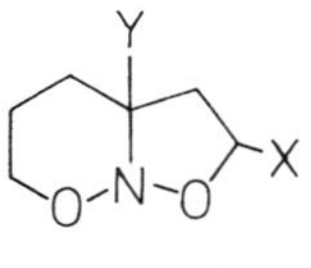

5-93a X = group, Y = NO_2
5-93b X = C_6H_5, Y = OCH_3
5-93c X = CO_2CH_3, Y = OCH_3
5-93d X = CH_2Cl, Y = OCH_3
5-93e X = C_6H_5, Y = OC_2H_5
5-93f X = C_6H_5, Y = CN
5-93g X = *p*-BrC_6H_4, Y = CN
5-93h X = CO_2CH_3, Y = CN
5-93i X = CH_2Cl, Y = CN
5-93j X = *n*-C_4H_9, Y = CN
5-93k X = C_6H_5, Y = N_3
5-93l X = CO_2CH_3, Y = N_3
5-93m X = $COCH_3$, Y = N_3
5-93n X = CN, Y = N_3
5-93o X = C_6H_5, Y = OH
5-93p X = C_6H_5, Y = ONO

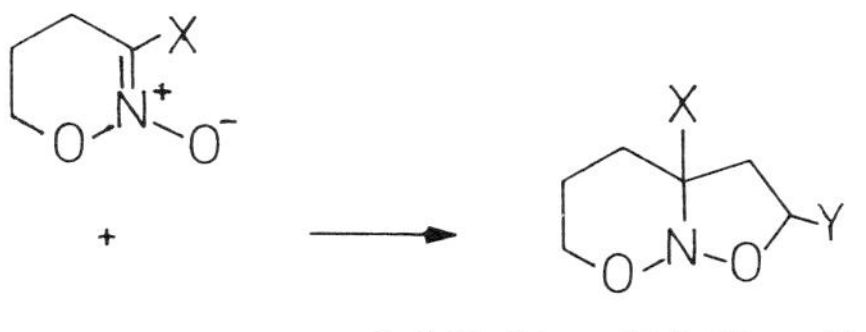

Isoxazoles

CH$_2$=CHY   5-93f X = CN, Y = C$_6$H$_5$
              5-93g X = CN, Y = C$_6$H$_4$Br-p (5-24)

ethers from the nitro compound (5-93a) and alcohols, the corresponding reaction with alcoholates gave *trans* products (5-93b) (96%) and (5-93e) (95%).[127]

An anion-radical replacement of the nitro group at a tertiary carbon atom[132,133] was abandoned in favor of an S$_N$1 displacement of the nitro substituents in the examples shown above except for the reactions with alcoholate anions, which were assigned S$_N$2.[131] Evidently the ring nitrogen atom facilitated the S$_N$ reactions. To accommodate this assistance it was assumed that the stable conformation[56] (5-94) converted to an antiperiplanar conformation (5-95) prior to the initiation of an S$_N$ process. Inversion at nitrogen, which ordinarily proceeds with great difficulty even at elevated temperatures in an —O—N—O— system,[127,134,135] may have been promoted by the presence of the α-nitro substituent; such an assumed effect was observed for the 2-alkoxy-3,3-disubstituted-isoxazolidine system (5-96a,b) where inversion barriers of 27.5 kcal/mol for (5-96a) and 16 kcal/mol for (5-96b) were observed.[131,136] To test the validity of this "α-nitro" effect, displacement reactions on the analogous compounds (5-97), (5-98), (5-99), and (5-100) were examined.[131] The nitrite (5-97) was unreactive to sodium cyanide, and the nitrile (5-98) was unreactive to sodium methoxide in methanol; apparently neither an α-ONO nor an α-CN substituent was comparable to the nitro group in promoting a displacement reaction. In compound (5-99), where the barrier to inversion at nitrogen should be low and a conformation similar to (5-95) should be easily attainable, nucleophilic displacement of the cyano group was observed.[137] In compound (5-100) the nitro group did not undergo nucleophilic displacement. This interesting development was attributed to resistance to a conversion from the stable *cis*-bicyclic arrangement[138] to an unfavorable *trans*-arrangement necessary for an antiperiplanar conformation of the type seen in (5-95).[131]

The proposed intermediacy of the nitrite (5-97) in the S$_N$1 conversion of the nitro compound (5-93a), X = C$_6$H$_5$, when treated with the potassium salt of

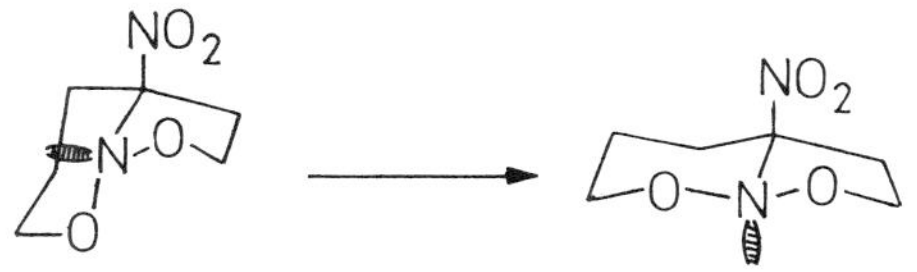

5-94   5-95

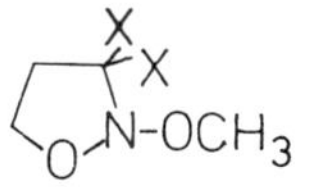

5-96a X = CO₂CH₃
5-96b X = NO₂

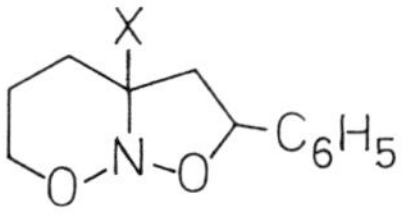

5-97 X = ONO
5-98 X = CN
5-101 X = OH

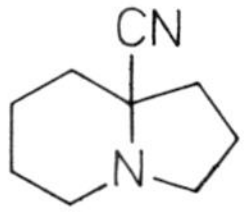

5-99

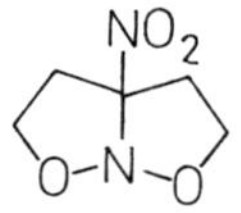

5-100

nitroethane in aqueous dimethylformamide to the alcohol (**5-101**), Eq. (5-25), was supported by an experiment in which $H_2^{18}O$ was introduced to the system. The product (**5-101**) did not contain ^{18}O and thus offered confirmation of the above assignments of activated S_N1 displacements of the nitro group.[131]

5-93a
X = C₆H₅
+
CH₃CH=NO₂K

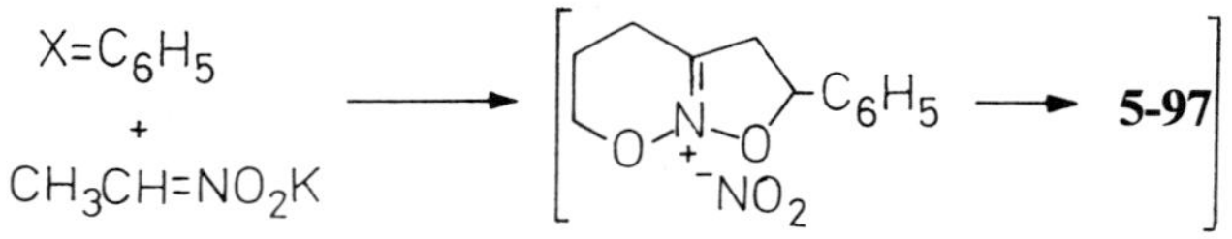

5-97 → 5-101

→ **5-97**

(5-25)

The dioxazabicyclononanes (**5-93a**) (X = CO₂CH₃, CN, COCH₃) rearranged thermally in an inert solvent at 60°–90° to spiro[isoxazoline-5],2'-tetrahydrofurans (**5-102a**) (X = CO₂CH₃, CN, COCH₃) (55%–60%); (**5-93a**) (X = C₆H₅) gave 3-phenyl-5-γ-hydroxypropylisoxazole (**5-102b**) (79%), sometimes with a small amount of (**5-102a**) (X = C₆H₅) and the isoxazoline (**5-102c**). The latter, in equilibrium with an oxime (**5-102d**), was found to be an early thermolysis product. Although it was stable in inert solvents at 70°–80° for 8 h, the isoxazoline (**5-102c**) was converted to the isoxazole (**5-102b**) by nitric acid in acetic acid and converted to a mixture of (**5-102a**) (X = C₆H₅) and (**5-102b**) by treatment with diethyloxonium tetrafluoroborate.[139,140] The formation of the isoxazoline (**5-102c**) and/or its tautomer (**5-102d**) from the nitro compound (**5-93a**) (X = C₆H₅) was not explained.

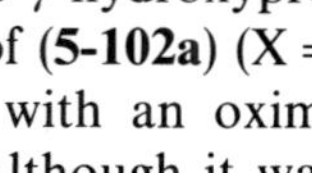

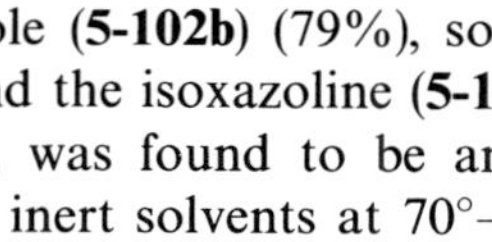

5-102a
X = CO₂CH₃, CN
COCH₃, C₆H₅

5-102b

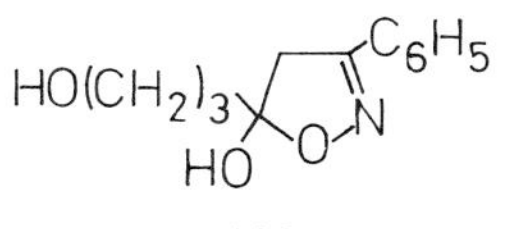

Isoxazoles

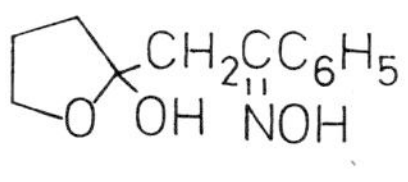

5-102c: HO(CH$_2$)$_3$, HO, C$_6$H$_5$ (isoxazoline structure)

5-102d: CH$_2$CC$_6$H$_5$, OH, NOH (tetrahydrofuran structure)

It was noted that the nitro substituent in 3a-nitrotetrahydroisoxazolo-isoxazole (**5-100**) was not subject to nucleophilic displacement. Mild treatment with base did, however, degrade 4-hydroxy-2-substituted-3a-nitrotetrahydroisoxazolo[2,3-b]isoxazole (**5-103a–f**) (from chloroacetaldehyde and the potassium salt of dinitromethane followed by a dipolar addition of an olefin to 3-nitro-4-hydroxy-2-isoxazoline-2-oxide, Eq. (5-26a)) to a 3-nitro-5-substituted-isoxazoline (**5-104a–f**).[87,142] Each of the two diastereoisomers of the 3-ethoxy compound (**5-105**) gave 3-(α-ethoxy-β-hydroxy)-2-isoxazoline (**5-106**) on treatment with anhydrous hydrogen chloride in benzene.[89]

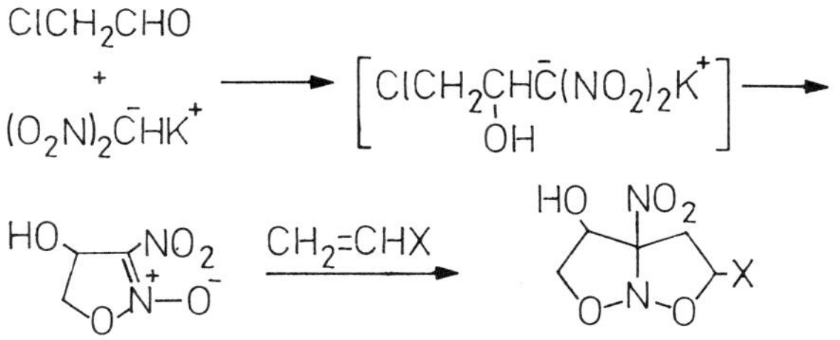

$$\text{ClCH}_2\text{CHO} + (\text{O}_2\text{N})_2\bar{\text{C}}\text{HK}^+ \longrightarrow \left[\text{ClCH}_2\underset{\text{OH}}{\text{CH}}\bar{\text{C}}(\text{NO}_2)_2\text{K}^+ \right] \longrightarrow$$

5-103a X = H, **b** X = CH$_3$
c, X = C$_6$H$_5$, **d**, X = OCOCH$_3$
e, X = CO$_2$CH$_3$,
f, X = CN   (5-26a)

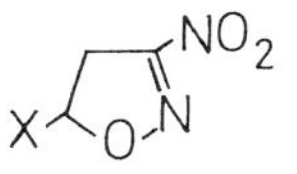

5-104a–f (as in **5-103**)

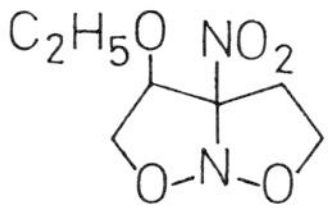

5-105

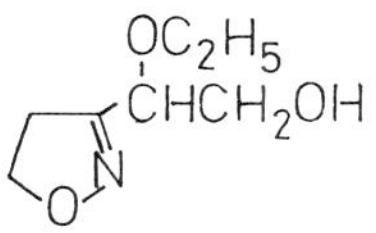

5-106

Displacement of the nitro substituent in 3-nitro-2-isoxazoline (**5-34**) by the appropriate anion in a protic solvent at 25° gave the 3-thiophenoxy (**5-107a**) (91%), 3-cyano (**5-107b**) (85%), 3-n-butyl (**5-107c**) (51%), and 3-azido (**5-107d**) (69%) derivatives.[96] The azide (**5-107d**) showed typical ir absorption. Sodium benzenesulfinate in water at 65° for 13 h afforded 3-phenylsulfonyl-2-isoxazolines

(5-107e) (28%).⁹⁶ A series of 5-trimethylsilylalkyl-3-substituted-2-isoxazolines (5-108) were similarly obtained by displacement of a 3-nitro group with ethoxy or amino groups.³⁸

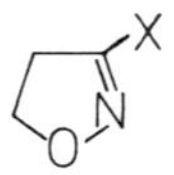

5-107a X = SC$_6$H$_5$
5-107b X = CN
5-107c X = n-C$_4$H$_9$
5-107d X = N$_3$
5-107e X = SO$_2$C$_6$H$_5$

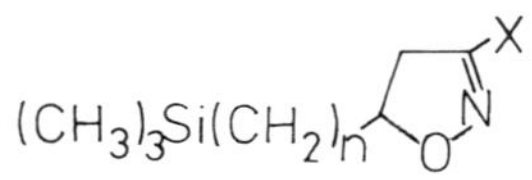

5-108 X = OC$_2$H$_5$, NH$_2$
n = 0,1,2

Hydrogen peroxide in acetic acid converted 4-nitro-2-isoxazolines to isoxazol-2-in-4-ones (5-109) (46%) and (5-110) (30%). A radical process was assumed.¹⁴³

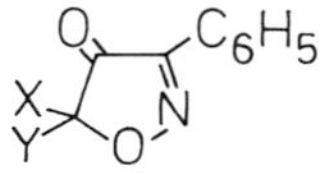

5-109 X = Y = CH$_3$
5-110 XY = (CH$_2$)$_5$

The latter product, (5-110) (69%), was also obtained when tetranitromethane reacted with the potassium salt of 3-phenyl-5-spirocyclohexyl-4-nitroisoxazoline in methanol.¹⁴⁴

Nucleophilic displacement converted 3-nitroisoxazoles to 3-hydroxy (5-111a), 3-methoxy (5-111b), and 3-ethoxy derivatives (5-111c).⁸⁰ A similar reaction afforded 3-nitro-5-(3-hydroxy-5-isoxazolyl)isoxazole (5-111d) from an alkaline treatment of the corresponding 3,3'-dinitro-5,5'-bisisoxazole.¹²⁵ Alkaline degradation of 3-hydroxyisoxazole (5-111a) to cyanoacetic acid¹²⁵ has not been confirmed.

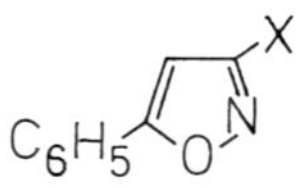

5-111a X = OH
5-111b X = OCH$_3$
5-111c X = OC$_2$H$_5$

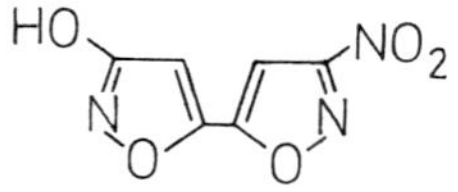

5-111d

Side Chain Reactions

Piperidine catalyzed a condensation between 3,5-dimethyl-4-nitroisoxazole (5-3) and benzaldehyde in ethanol at 80° for 2 min to give 3-methyl-4-nitro-5-styrylisoxazole (5-112a) (96%).¹⁴⁵⁻¹⁴⁷ A lack of reactivity at the 3-methyl group was substantiated by an isotopic exchange that was restricted to the 5-methyl group of compound (5-3) when treated with diethylamine enriched with deuterium in the NH group.[148] Related products (5-112b–g) were obtained in a kinetic study of the bimolecular condensation reaction catalyzed by sodium methoxide[149,150] in which the conjugate base of (5-3) added to the aldehyde in the rate-determining step, the isokinetic temperature was 350°K and ρ was 0.8.[149,150] Polar and steric factors for the reactions producing compounds (5-112h–l) were examined with the aid of Taft's equation (log $k/k_o = \sigma^*\rho^* + \delta E_s$) for ortho substituents.[151,152] The nitro compounds (5-112a,e,m,n) were reduced to derivatives of 3-aminoisoxazole.[153]

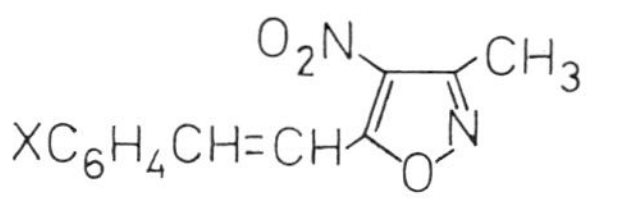

5-112 a X = H, b X = o-OH
c X = m-OH, d X = p-OH
e X = p-Cl, f X = m-NO$_2$
g X = p-NO$_2$, h X = o-OCH$_3$
i X = o-CH$_3$, j X = o-Cl
k X = o-Br, l X = o-NO$_2$
m X = p-OCH$_3$, n X = p-(CH$_3$)$_2$N
o X = m-Cl, p X = p-CH$_3$

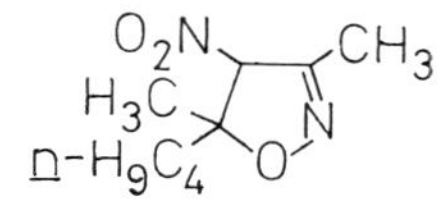

5-113a

In another attempt to generate an anion in the activated 3-methyl substituent of 3,5-dimethyl-4-nitroisoxazole (5-3) by treatment with butyl lithium, an addition occurred instead to give, after workup, 5-n-butyl-3,5-dimethyl-4-nitro-2-isoxazoline (5-113a).[146,147]

A Michael addition of acetylacetone to styrylisoxazoles (5-112a,e,j,m,p) in the presence of triethylamine gave the expected products (5-113b). Reversion to the (5-112) compounds was brought about by treatment with ethylenediamine.[154–156] Another Michael addition gave 1,3-diphenyl-4-(3-methyl-4-nitro-5-isoxazoyl)-1-butanone (5-114) (32%) when the reaction between benzalacetophenone and 3,5-dimethyl-4-nitroisoxazole (5-3) was catalyzed by piperidine at 150° for 1 h.[145]

Chalcone adducts (5-115) were reduced by stannous chloride and cyclized to 7,8-dihydro-6H-3-methyl-5,7-diarylisoxazole[4,5-b]azepines (5-116).[157]

In a Mannich reaction formaldehyde and a secondary amine converted 3,5-

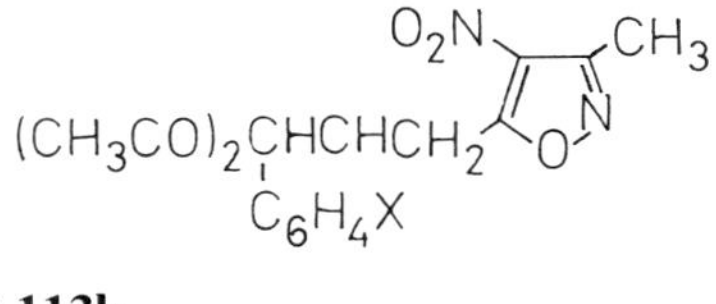

5-113b
X = H, p-Cl, o-Cl, p-OCH$_3$, p-CH$_3$

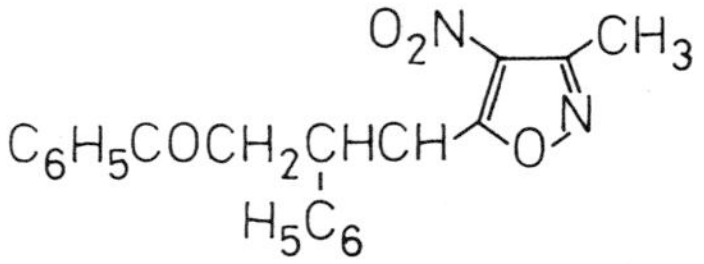

$$\underset{\textbf{5-114}}{\text{C}_6\text{H}_5\text{COCH}_2\underset{\underset{\text{H}_5\text{C}_6}{|}}{\text{CHCH}}\text{—[O}_2\text{N, CH}_3\text{-isoxazole]}}$$

5-114

$$\underset{\textbf{5-115}}{\text{O=CHCH}_2\underset{\underset{\text{X}}{|}}{\text{CHCH}_2}\text{—[O}_2\text{N, CH}_3\text{-isoxazole]}}$$
         Y

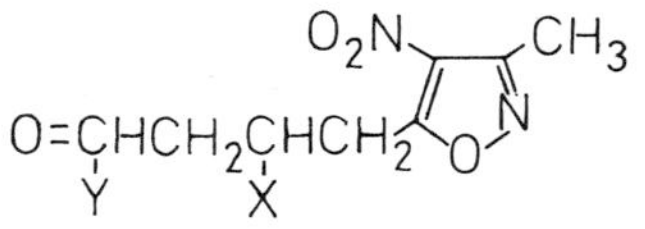

5-115
X = Y = substituted phenyl

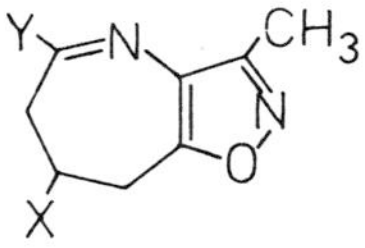

5-116

dimethyl-4-nitroisoxazole (**5-3**) to Mannich bases (**5-117**) (90%–100%).[158] The reaction kinetics were found to be third order.[159]

N,N′-diphenylformamide, p-dimethylaminonitrosobenzene, and acridine also combined with 3,5-dimethyl-4-nitroisoxazole (**5-3**) in base-catalyzed reactions to give the enamine (**5-118**), the imine (**5-119**) (which hydrolyzed to an aldehyde), and 9-acridyl(3-methyl-4-nitro-5-isoxazolyl)-methane (**5-120a**).[145] A preparation of 3-methyl-4-nitroisoxazole-5-carboxaldehyde (**5-120b**) was also brought about by an acid hydrolysis of the 5-dimorpholinomethylisoxazole, in turn obtained from the dichloromethyl derivative and morpholine.[14]

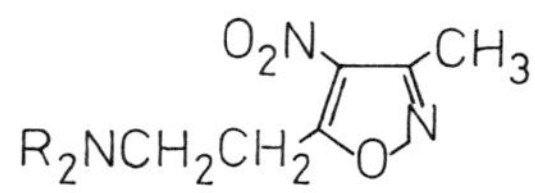

5-117
R$_2$N = (CH$_2$)$_5$N, (CH$_2$)$_4$N
(CH$_3$)$_2$N, (C$_2$H$_5$)$_2$N
O(CH$_2$CH$_2$)$_2$N

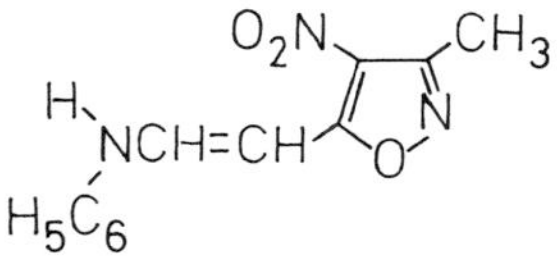

5-118

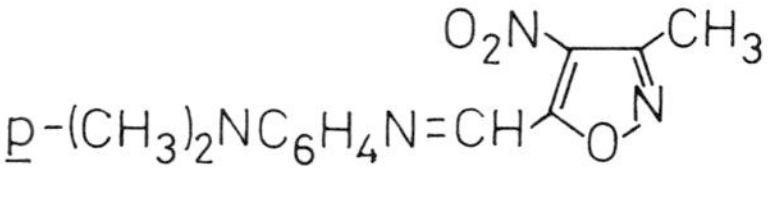

p-(CH$_3$)$_2$NC$_6$H$_4$N=CH—[isoxazole]

5-119

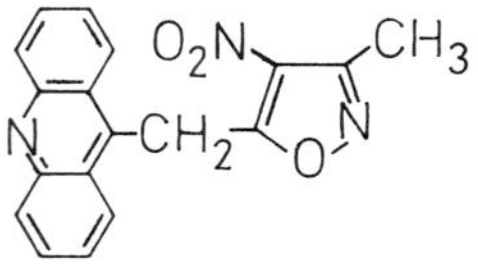

5-120a

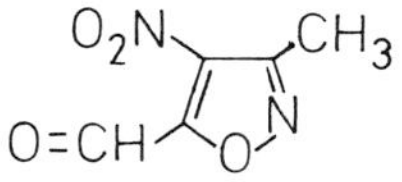

4-120b

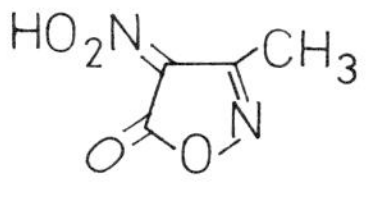

5-121

Oxidation of side chains in nitroisoxazoles has not been developed. The compound (**5-112a**) with potassium permanganate in acetone gave benzoic acid and 3-methyl-4-nitroisoxazole-5-carboxylic acid (67%). The reaction was reported to occur with a flash.[28,29] Recently it was shown that instead of an isoxazolecarboxylic acid an oxidation gave 3-methyl-4-nitro-5-hydroxyisoxazole or its nitronic acid tautomer (**5-121**). A rationale for the result was given, Eq. (5-26b),[160] but the conversion of the intermediate 3-methyl-4-nitroisoxazole-5-carboxylate anion was not experimentally verified.

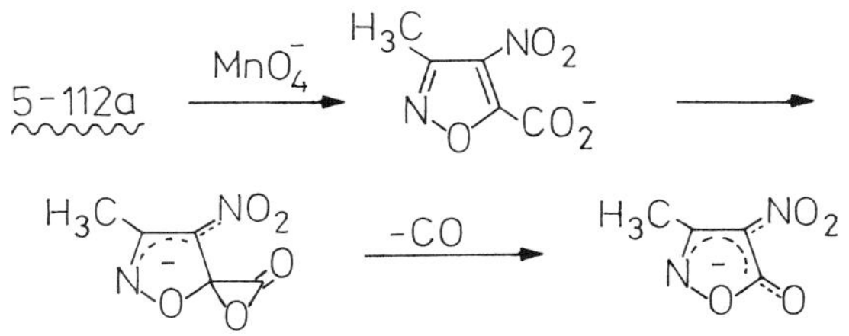

(5-26b)

Benzonitrile oxide and styrylisoxazoles (**5-112a,e,f,j,m,p**) added in benzene to give the isoxazoylisoxazolines (**5-122**) (15%–20%) in a regiospecific reaction.[161,162]

Photodimerization of 3-methyl-4-nitro-5-styrylisoxazole (**5-112a**) in the solid state gave the dimer (**5-123**), whereas in benzene solution both (**5-123**) and an isomer (**5-124**) were obtained.[163,164]

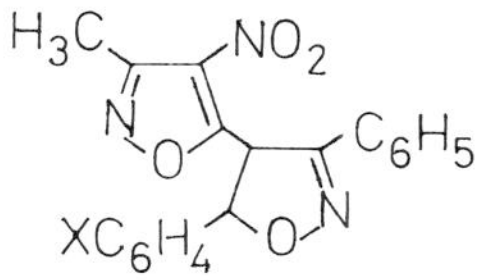

5-122
X = H, p-Cl, m-NO$_2$
o-Cl, p-OCH$_3$

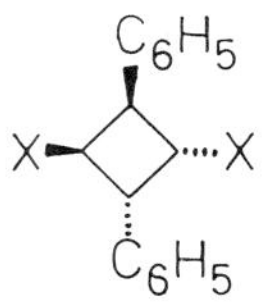

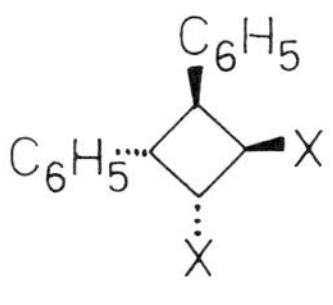

5-123   **5-124**

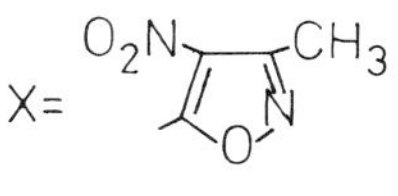

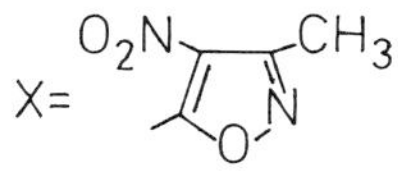

Ring Opening

Stable enolate salts of α-cyano-α-nitroacetone were obtained by the treatment of 4-nitro-5-methylisoxazole with piperidine or pyridine. The reaction was presumably initiated by base abstraction of the 3-proton, Eq. (5-27a). In accordance with the polar effects for the nitro (σ, 0.65), chloro (σ, 0.46), and acetyl groups (σ, 0.28), a similar ring-opening reaction for 4-chloro-5-methylisoxazole was also catalyzed by piperidine or pyridine, but the less reactive 4-acetyl-5-methyl required dilute sodium hydroxide or piperidine for conversion to the enolate salt of cyanoacetylacetone.[165] A methanol solution of 4-nitroisoxazole gave the potassium or sodium enolate of 2-cyano-2-nitroethanol (**5-125**) on treatment with a stoichiometric amount of potassium hydroxide or sodium methoxide in methanol. Acidification converted (**5-125**) to nitroacetonitrile and formic acid.[108] A similar ring opening in the polarographic reduction of 4-nitroisoxazole at pH > 6 gave the enolate anion (**5-126**) of the monoxime of nitromalonic dialdehyde.[166] An alcoholic solution of aniline converted 4-nitroisoxazole to nitrocyanoacetaldehyde (**5-127**).[8]

$$\underset{H_3C}{\overset{O_2N}{\diagup}}\!\!\diagdown\!\!\underset{O}{\diagup}\!\!\diagdown\!\!\overset{}{N} \xrightarrow{\text{base}} CH_3\overset{O^-}{\underset{|}{C}}=C(NO_2)CN \qquad (5\text{-}27a)$$

XC$_6$H$_4$CH=CHCO$_2$H   HC≡CCN
                                  $\overset{|}{O}$ $\overset{|}{NO_2}$

5-128   **5-125**
X = H, *p*-Cl, *o*-Cl, *p*-OCH$_3$,
    m-Cl, *p*-CH$_3$

HC=C—CH=NOH   NCCHCHO
$\overset{|}{O}$ $\overset{|}{NO_2}$ $\overset{|}{NO_2}$

5-126   **5-127**

Substitution may change the site of reaction. Alkaline hydrolysis of 3-methyl-4-nitro-5-styrylisoxazoles (**5-112a,e,j,m,o,p**) gave the cinnamic acids (**5-128**) (65%–96%) with degradation of the heterocyclic ring.[167,168]

Warm water converted 4-nitro-5-aminoisoxazole (or its imine tautomer) to fulminuric acid (**5-129**).[109,110] Hot sodium hydroxide (5%) reacted similarly with 3-phenyl-4-nitro-5-aminoisoxazole to give the oxime (**5-130**) of α-nitrobenzoylacetic acid (**5-131**).[26,27] The ketoacid (**5-131**) was otherwise obtained by the decarboxylation of 4-nitro-5-phenylisoxazole-3-carboxylic acid (**5-132**) (74%) at 55° in vacuo (0.02 torr).[81]

A hydrazine converted isoxazoles unsubstituted at the 3-position to pyrazoles; 4-nitro-5-methylisoxazole (**5-133**) gave a 1-substituted-3-methyl-4-nitro-5-aminopyrazole (**5-134**).[169] The presence of potassium hydroxide improved the

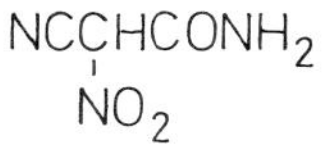

NCCHCONH₂ $C_6H_5\underset{\underset{NOH}{\|}}{C}CH(NO_2)CO_2H$
|
NO₂

5-129 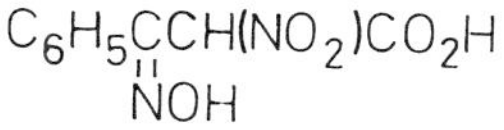   **5-130**

$C_6H_5COCH(NO_2)CO_2H$

5-131

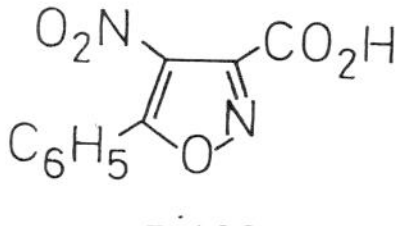

[Structure: 5-isoxazole with O_2N at 4-position, CO_2H at 3-position, C_6H_5 at 5-position]   [Structure: 5-isoxazole with O_2N at 4-position, H_3C at 5-position]

5-132 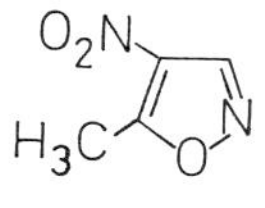   **5-133**

efficiency.[170,171] A base-catalyzed proton abstraction from the 3-position was assumed to bring about ring opening to the anion of a β-ketonitrile followed by hydrazone formation and ring closure, Eq. (5-27b). For the formation of 1-substituted-3,5-dimethyl-4-nitropyrazole (**5-135**) from 3,5-dimethyl-4-nitroisoxazole (**5-3**), a different explanation was proposed, Eq. (5-28).[171-176] A hydrazine also converted 5-methyl-4-nitro-3-(5-methyl-4-nitro-3-isoxazolyl)-isoxazole (**5-136**) (dislite) to the 4,4′-dinitro-3,3′-bispyrazole (**5-137**).[126]

Both acids and bases have degraded nitroisoxazolidines to open-chain products. Sulfuric acid (20%) at 75° for 8 h converted 2-isopropoxy-3,3-dinitro-5-methylisoxazolidine (**5-49**) (X = H) to crotonic acid.[44] When heated

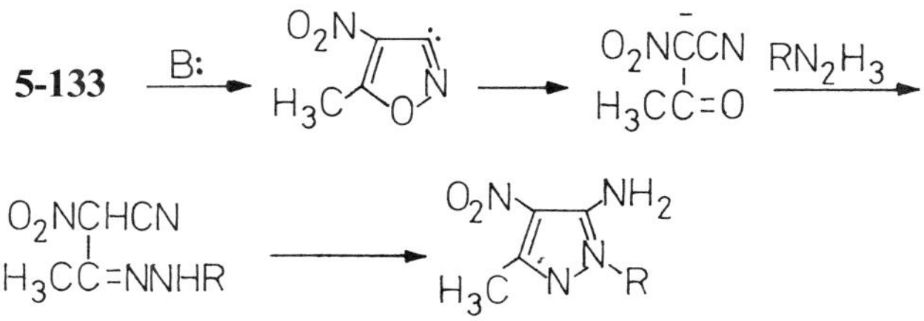

5-133 →(B:) [isoxazole intermediate] → $O_2N\bar{C}CN$ / $H_3C\overset{\|}{C}=O$ →(RN₂H₃)

O_2NCHCN / $H_3C\overset{|}{C}=NNHR$ → [pyrazole **5-134** with O_2N, NH_2, H_3C, R substituents]   (5-27b)

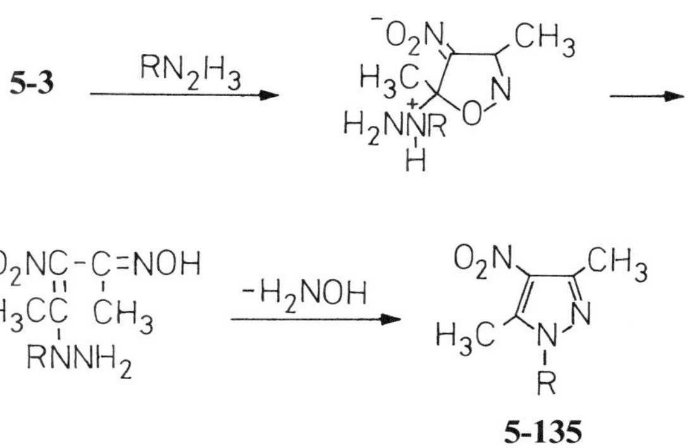

5-3 →(RN₂H₃) [isoxazolidine intermediate with $\bar{O}_2N$, CH_3, H_3C, $H_2\overset{+}{N}NR$, H] →

$O_2NC-C=NOH$ / $H_3C\overset{\|}{C}$ $\overset{|}{C}H_3$ / RNNH₂ →(−H₂NOH) [pyrazole **5-135** with O_2N, CH_3, H_3C, R]   (5-28)

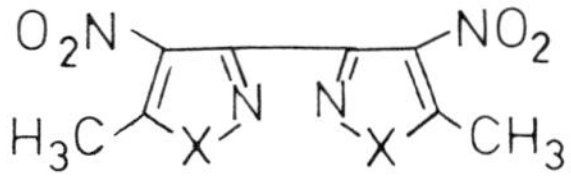

$$\text{O}_2\text{N}\underset{\underset{\text{H}_3\text{C}}{}}{\diagdown}\overset{\text{N}}{\underset{\text{X}}{\diagup}}\text{---}\overset{\text{N}}{\underset{\text{X}}{\diagdown}}\underset{\underset{\text{CH}_3}{}}{\diagup}\text{NO}_2$$

5-136 X = O
5-137 X = NR

for 50 h in concentrated hydrochloric acid, 1-(α-phenyl-β-nitroethoxy)-3,3-dinitro-5-phenylisoxazolidine (**5-58c**) (W = X = Y = H, Z = C_6H_5) gave cinnamic and benzoic acids, Eq. (5-29), but in alcoholic potassium hydroxide at 0° for 2.5 h it afforded the potassium salt of 1,1-dinitro-3-phenylpropan-3-ol (91%), Eq. (5-30).[59] Structurally related *gem*-dinitroisoxazolidines gave comparable results on treatment with acids and bases.[43,57,58,66,67] In some instances base degradation afforded both a dinitropropanol and a nitroethylene derivative, Eqs. (5-30, 5-31).[57] In one example a double-bond migration to form 6-nitro-1-phenylcyclohexene was needed.[61]

After alkaline hydrolysis replaced one nitro group in 5-(3-nitroisoxazol-5-yl)-3-nitroisoxazole with the hydroxyl group, it proceeded to degrade one ring to give 3-nitro-5-acetylisoxazole, Eq. (5-32a).[125]

A thermal ring contraction of a 2-alkyl-3-phenyl-4-cyano-5-nitroisoxazolidine (**5-24d**) or (**5-24f**) in methanol to a 1-alkyl-3-cyano-4-phenylazetidin-2-one (**5-138**) (14%) as a 1:1 mixture of *cis* and *trans* isomers, Eq. (5-32b), was enhanced by the presence of a base. The reaction depended on alcoholic or polar aprotic solvents and failed to occur in benzene or hexane. A similar ring contraction to the *trans*-β-lactam was brought about by irradiation at 254 nm. A rationale was presented requiring ring opening to a nitroacyl intermediate followed by ring closure to the azetidinone.[176]

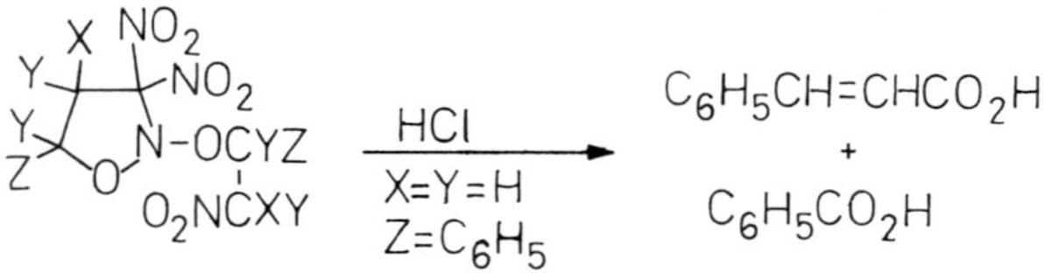

5-58c $\xrightarrow[X=Y=H]{\text{HCl}}$ $C_6H_5CH=CHCO_2H$ + $C_6H_5CO_2H$ $\quad$ (5-29)
$\quad Z=C_6H_5$

5-58c $\xrightarrow[2.H_3O^+]{1.\text{KOR}}$ $HC(NO_2)_2CH_2CH(OH)Z$ + $[ZCH=CHNO_2]$ $\quad$ (5-30)
$X=Y=H, Z=C_6H_5$

$[ZCH=CHNO_2] \xrightarrow{\text{KOR}} ZCHCH_2NO_2$
$Z=\text{alkyl}\qquad\qquad\qquad\qquad\;\;\overset{|}{\text{OR}}$ $\quad$ (5-31)

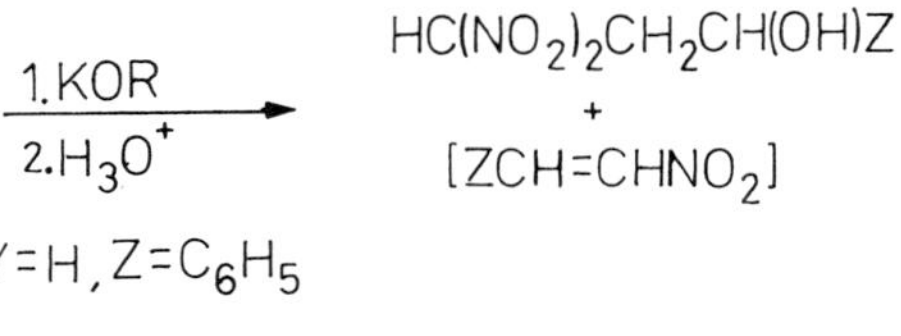

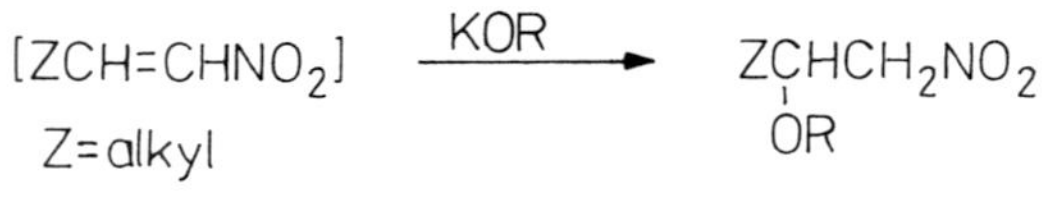

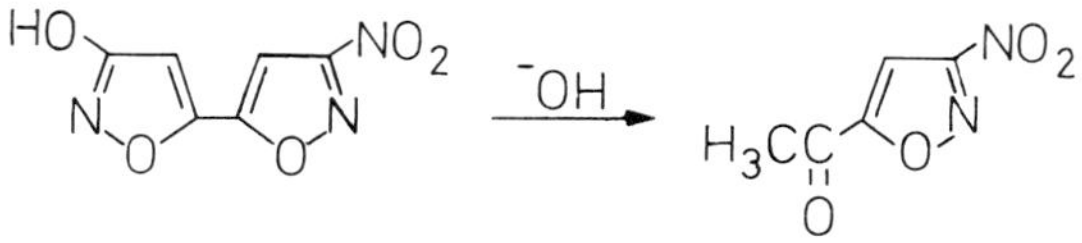

[Structure 5-32a]

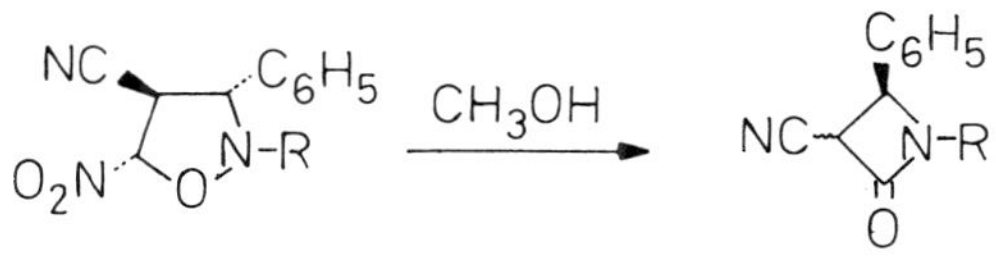

[Structures for 5-24d R = CH₃, 5-24f R = C(CH₃)₃, and 5-138]

(5-32b)

Acids have ring-degraded 3a-nitrotetrahydro-2H-isoxazolo[2,3-b]isoxazole (**5-139**). When no other substitution was present, sulfuric acid (20%) at 45° for 3 h afforded 3-pentanone-1,5-diol (70%), Eq. (5-32c).[177,178] On the other hand anhydrous hydrogen chloride in benzene at 10° for 20 min degraded one ring to give 3-(β-hydroxyethyl)isoxazoline (90%), Eq. (5-33). Concentrated hydrochloric acid brought about the same conversion but in a lower yield (72%). Similar treatment afforded the 5-methyl heterocycle (**5-141**) from the 2-methyl derivative of the isoxazolizidine (**5-139**).[178] Comparable reactions in the presence of other substituents are known.[55]

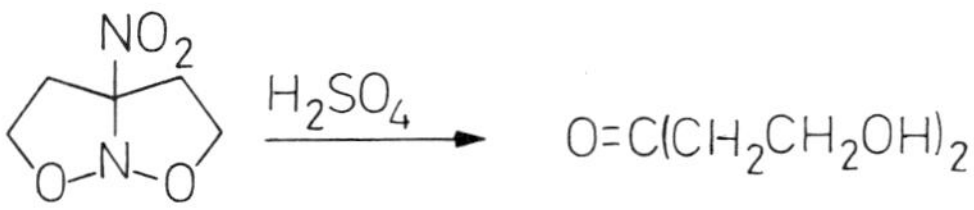

[Structure 5-139 → O=C(CH₂CH₂OH)₂]

(5-32c)

[Structure 5-139 → 5-140]

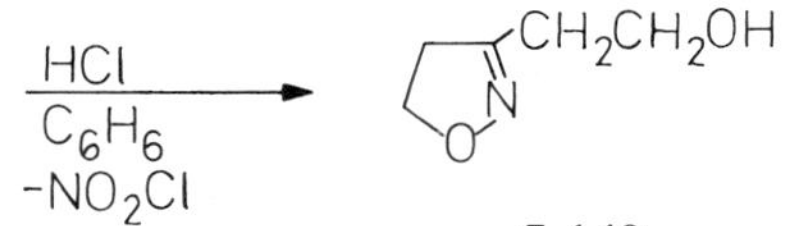

(5-33)

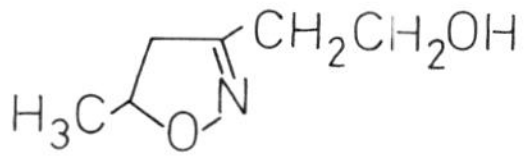

[Structure 5-141]

Properties

Acid-Base

Information on the influence of substituents in isoxazoles is just beginning to be collected. In a spectrophotometric determination of the pK_a values of 20 derivatives of isoxazole and three derivatives of isoxazoline, a pK_a of -3.72 ± 0.06 was given to 3,5-dimethyl-4-nitroisoxazole. The study established higher basicities when other 4-substituents were present: $CH_3 > H > C_6H_5CH_2 > CH_3CO_2CH_2 > I > Br > Cl > NO_2$.[179] A pK_a of -7.90 ± 0.17 for 3-phenyl-4-nitroisoxazole was similarly determined.[180]

Bis(trinitromethyl)mercury formed a complex with 3-nitro-2-isoxazoline-2-oxide (5-83) (X = H) (65%). In the investigation similar complexes were obtained from 18 other molecules showing a wide variety of structures.[181] A brief report described limiting activity coefficients for 11 hydrocarbons in a gas-liquid chromatographic investigation of heterocyclic compounds and afforded a selectivity in the order furans < isoxazoles < oxadiazoles. Thus 5-methyl-4-nitroisoxazole exhibited a much greater selectivity than 2-methyl-3-nitrofuran.[182]

Infrared Spectroscopy

Absorption at 830–870 and 1,240–1,300 cm^{-1} attributed to the N—O and N$^+$—O$^-$ bonds and at 1,310–1,360 and 1,505–1,540 cm^{-1} attributed to v_s and v_{as} values for the C—NO$_2$ group was obtained for 3-nitro-2-isoxazoline-2-oxide (5-83) (X = H) and seven derivatives in an examination of cyclic and acyclic nitronate esters of the *aci* form of di- and trinitromethane. The linear compounds also showed absorption in these four regions. The acyclic esters were also characterized by absorption at 1,610–1,637 cm^{-1} and the cyclic esters at 1,610–1,660 cm^{-1} for the azomethine linkage.[183,184]

Absorption at 1,025–1,065 (O—N—O) and at 1,326–1,335 (v_s) and 1,585–2,590 (v_{as}) cm^{-1} (C—(NO$_2$)$_2$) was found characteristic of 3,3-dinitro-2-methoxyisoxazolidine (5-37) (X = H) and seven derivatives at 1,010–1,040 (O—N—O—) and at 1,355–1,370 (v_s) and 1,540–1,570 (v_{as}) cm^{-1} (NO$_2$) for 3a-nitrotetrahydro-2H-isoxazolo[2,3-b]isoxazole (5-100) and 12 derivatives.[184]

Intensity of the band for the symmetrical CH vibration of the methyl group in 26 five-membered heteroaromatic compounds ($v_{CH} = 2{,}935 \pm 10$ cm^{-1}) including 1-perdeuteromethyl-2-methyl-5-nitroimidazole, 1-perdeuteromethyl-3,5-dimethyl-4-nitropyrazole, and 3,5-dimethyl-4-nitroisoxazole correlated with the calculated (CNDO/2) total charges on the carbon and hydrogen atoms of the methyl group. The ir band intensities and the NMR chemical shifts of the protons (δCH_3) could be united in a single reaction series with analogous data for polysubstituted toluenes and six-membered nitrogen heterocycles within the framework of a common additive scheme.[185]

Other examinations of the ir absorption of heterocycles also included nitroisoxazoles.[186,187]

Ultraviolet Spectroscopy

An absorption maximum in the 320-nm ($\varepsilon \sim 8{,}500$) region was characteristic of 3-nitro-2-isoxazoline-2-oxide (5-83) (X = H), its 4-methyl, 5-methyl, and 4-hydroxy derivatives, and the O-methyl ethers of dinitromethane, dinitro-acetonitrile, and trinitromethane and was assigned to the common segment $RON(O) = CNO_2$.[183,184]

In addition to a maximum absorption in the 220-nm (log ε 4.01) region characteristic of 3,5-dimethylisoxazole, the 4-nitro derivative also showed an absorption maximum in the 260-nm (log ε 4.10) (ethanol) region. The latter band was shared with other 4-nitroisoxazoles but had shifted to the 290-nm (log ε 4.15) region for the 5-(p-methoxystyryl) and the 5-cinnamylidene derivatives of 3-methyl-4-nitroisoxazole.[188] An earlier report also listed uv absorption maxima for nitroisoxazoles.[15]

Nuclear Magnetic Resonance Spectroscopy

Chemical shifts for protons directly attached to the isoxazole ring and in methyl substituents were tabulated for 81 compounds; δ values for isoxazole (**5-142**) (CDCl$_3$), the 4-nitro (**5-1**) (CCl$_4$), 3,5-dimethyl (**5-143**) (H$_2$SO$_4$), 3,5-dimethyl-4-nitro (**5-3**) (CCl$_4$), 3-phenyl (**5-144**) (CCl$_4$), and the 3-phenyl-4-nitro (**5-17**) (CCl$_4$) derivatives are shown.[148,189]

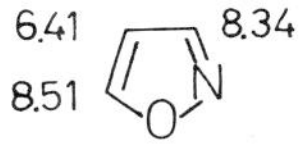

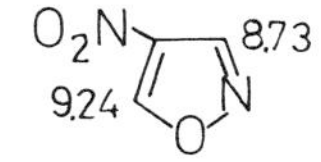

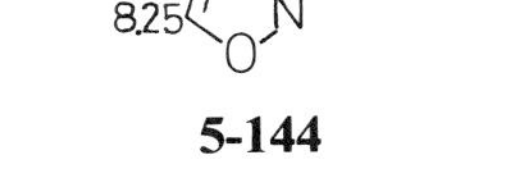

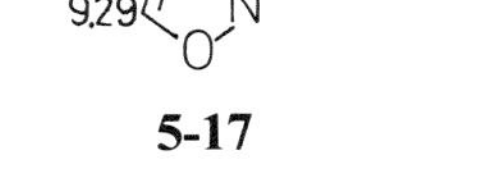

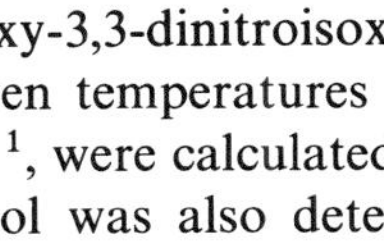

5-145 Y = NO$_2$, Z = CH$_3$
5-37 Y = NO$_2$, Z = H

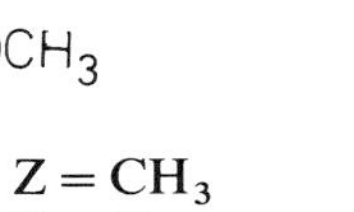

The pmr spectra of 2-methoxy-3,3-dinitroisoxazolidine (**5-37**) (X = H) were calculated and observed at seven temperatures in the range 7.5°–103.5°, and exchange rate constants, $k = \tau^{-1}$, were calculated (0 at 7.5°, 585 at 103.5°). The inversion barrier of 16 kcal/mol was also determined.[190-192] A comparable barrier of 14.6 kcal/mol was determined for 2-methoxy-3,3-dinitro-5,5-

dimethylisoxazolidine (**5-145**) from the change in the shape of the equal doublets of the methyl groups.[192] A barrier of 23–29 kcal/mole for the compound (**5-145**) (invertomers were isolated)[134,135] revealed that the presence of the *gem*-dinitro group lowered the inversion barrier by 8–10 kcal/mole.

Mass Spectroscopy

With the first investigations it was recognized that the mass spectra of alkylisoxazoles, including 3,5-dimethyl-4-nitroisoxazole (**5-3**), were explained in terms of a valence isomeric conversion to an oxazole via an azirine, Eq. (5-34). Although M-RCO and M-R peaks were found in the simpler 5-alkylisoxazoles, initial fragmentation for the nitro compound (**5-3**) occurred by loss of the nitro group.[193]

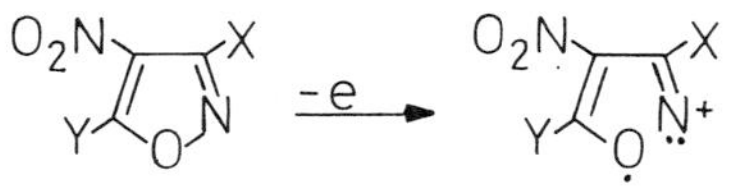

$$\text{5-146} \quad \text{5-147} \tag{5-34}$$

In other work on 3-methyl-4-nitro-5-(substituted)styrylisoxazoles (**5-112a,b,h,i,j,l**), fragmentation to produce a hydroxyl group accounted for $(M-17)^+$ and was explained by loss of a styryl group's α-hydrogen and an oxygen atom from the nitro group. A fragment $(M-46)^+$ correlated with a loss of nitro group, and a subsequent loss of acetonitrile correlated with $(M-87)^+$. An intramolecular rearrangement in which the nitro group was deoxygenated and the styryl group was converted to a ketone, Eq. (5-35), was postulated to account for the appearance of $(XC_6H_4C \equiv O)^+$. A rupture of the isoxazole C—O bond to initiate a chain of events to account for intense peaks at $(M-135)^+$ and $(M-114)^+$[194] was later seen to be brought about by cleavage of the isoxazole N—O bond as shown in Eq. (5-34).[195] Loss of hydroxyl also accounted for the appearance of an $(M-17)^+$ peak derived from 3-methyl-4-nitro-5-(1,3-diaryl-1-oxo-butyl)isoxazoles (**5-115**).[196]

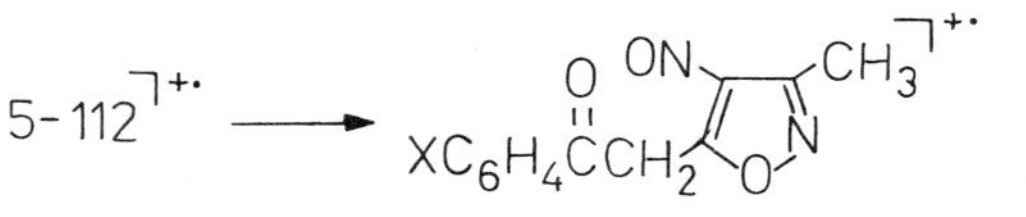

$$\tag{5-35}$$

That the mass spectra of 3a-nitrotetrahydro-2*H*-isoxazolo[2,3-*b*]isoxazoles (**5-54**) and (**5-55**) failed to show an M^+ peak was attributed to the exceptional ease with which the nitro group was eliminated with the generation of the

(M-46)$^+$ peak. Further fragmentation afforded the loss of hydroxyl and acyl groups. Since only a 2-substituted derivative may lose both formyl and another acyl group, Eq. (5-36), the mass spectra provided a differentiation between 1- and 2-substituted derivatives. A differentiation between the two isomers depended on the presence of the fragment (**5-149**) (m/e 85) derived only from the 2-phenyl isomer (**5-55b**) via the cation (**5-148b**), and the tropylium ion (m/e 91) derived only from the 1-phenyl isomer (**5-55a**) via the cation (**5-148a**).[197] The isomeric disubstituted derivatives of 3a-nitrotetrahydro-2H-isoxazolo[2,3-b]-isoxazoles were also differentiated by fragmentation patterns.[198]

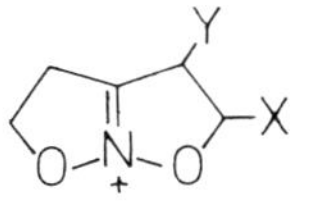

5-148a,b
a X = H, Y = C$_6$H$_5$
b X = C$_6$H$_5$, Y = H

5-148a,b $\xrightarrow{-OH}$

5-148a,b $\xrightarrow{-HCO}$

5-148a,b $\xrightarrow{-XCO}$ (m/e 85)

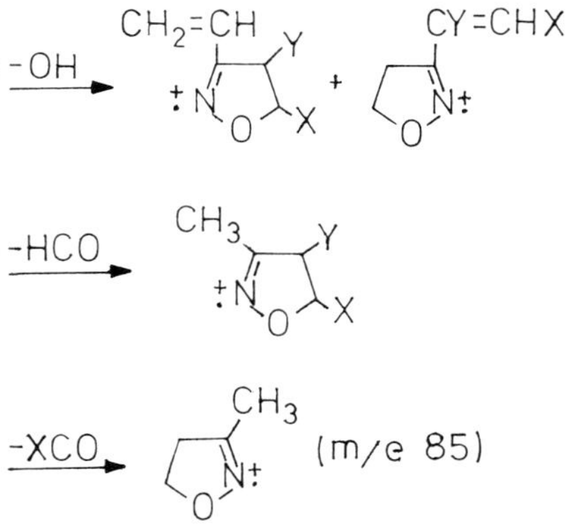

5-149   (5-36)

Other Physical-Chemical Properties

An X-ray structural analysis of 3a-nitrotetrahydro-2H-isoxazolo[2,3-b]isoxazole (**5-100**) has been determined.[49]

Dipole moments for isoxazole and its 3-phenyl-4-nitro and 3,5-dimethyl-4-nitro derivatives were found to be μ(exp) 2.82, 1.10, and 1.39D, and μ(calc) 2.50, 1.55, and 1.36D.[199]

The heats of combustion for 3-nitro-2-isoxazoline (**5-34**) and 3-nitro-2-isoxazoline-2-oxide (**5-83**) (X = H) were found to be -414 ± 0.3 and -406.6 ± 0.5 kcal/mol. Corresponding standard enthalpies of formation were calculated to be 9.3 ± 0.5 and 5 ± 2 kcal/mol (gas phase). The dissociation energy of the N$^+$—O$^-$ bond was then evaluated as 64 ± 3 kcal/mol. Thermolysis of the N-oxide (**5-83**) gave oxygen and the compound (**5-34**).[200]

The heat of combustion for the nitrotetrahydroisoxazoloisoxazole (**5-100**) was found to be -714.8 ± 0.5 kcal mol, and the heat of formation to be -29.9 ± 0.5 kcal/mol; the enthalpy of formation of compound (**5-100**) from the N-oxide (**5-83**) and ethylene was evaluated as 30.2 ± 1.0 kcal.[201]

Various 3-alkyl-5-substituted-4-nitroisoxazoles were examined as corrosion inhibitors for fuels, lubricants, etc.[85]

Biological Properties

Very little is known about the biological properties of nitro derivatives of isoxazoles, isoxazolines, and isoxazolidines. The fungicidal activity of *hymexazol* (**5-150**) was lowered by replacement of hydrogen at the 4-position with a methyl or a nitro group.[202] A variety of 5-isoxazolylureas, including 3- and 4-nitro derivatives, were described as plant-growth regulators.[203] Smooth muscle-relaxing and vasodilating preparations contained 4-nitroisoxazole.[9] Certain styryl derivatives of nitro isoxazoles were useful as β-sympatholytic and antihypertensive agents.[204]

Several 3a-nitrotetrahydro-2H-isoxazolo[2,3b]isoxazoles have exhibited mutagenic activity,[205] and the molecule type has been referred to as biologically active.[197,200] The compounds incorrectly named in *Chemical Abstracts* as 8-nitroisoxazolidines are actually 8-nitroisoxazolizidines (a trivial name for 3a-nitrotetrahydro-2H-isoxazolo[2,3-b]isoxazoles).[205]

5-150   5-151

OXAZOLES

The oxazole ring is resistant to nitration; however, 2-dimethylamino-4-phenyloxazole gave 2-dimethylamino-4-(p-nitrophenyl)-5-nitrooxazole (**5-151**) (97%) on treatment with a mixture of concentrated nitric and sulfuric acids at $-4°$ to $10°$ for 2 h.[206] Recently 18 derivatives of 5-nitrooxazole were prepared (yields from 2% to 52%) by the nitrolysis of the corresponding 5-bromo- or 5-iodooxazole with dinitrogen tetroxide in methylene chloride.[207]

Ring opening on treating ethyl 5-ethoxyoxazole-4-carboxylate (**5-152**) with benzoyl nitrate (from benzoyl chloride and silver nitrate) in chloroform at $-15°$ for 5 h then stored for 6 days gave ethyl α-benzoyloxy-α-benzamidomalonate (**5-154a**) (24%); the 2-methyl derivative (**5-153**) gave the analogous product (**5-154b**) (34%). When perbenzoic acid replaced benzoyl nitrate, the products were obtained in lower yield.[208]

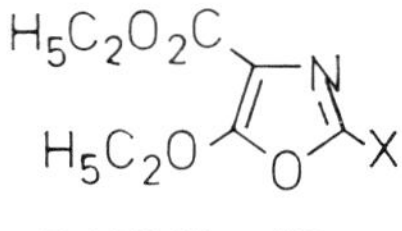

H₅C₂O₂C, H₅C₂O — (ring) — X

5-152 X = H
5-153 X = CH₃

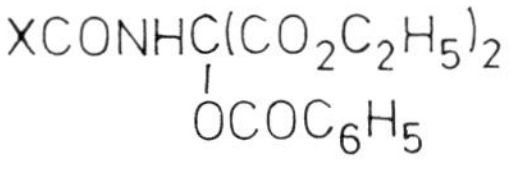

XCONHC(CO₂C₂H₅)₂
|
OCOC₆H₅

5-154a X = H
5-154b X = CH₃

OXADIAZOLES

Except for mesoionic heterocycles, nitro derivatives of 1,2,3- and 1,2,4-oxadiazoles are unknown. A few examples of nitrofurazans and nitrofuroxans have been reported. One review covered the literature of 1960,[209] another to 1983.[210]

Preparations

Dinitrogen trioxide converted propylene to a methylnitrofuroxan (**5-155a**) or (**b**) and cinnamaldehyde to a phenylnitrofuroxan (**5-156a**) or (**b**).[211] These compounds have been confirmed,[212-215] but other structures have been considered to be nitrofuroxans with less certainty.[216,217]

Trifluoroperoxyacetic acid [from hydrogen peroxide (85%) and trifluoroacetic acid] oxidized (45°, 6 h) the 4-amino derivative to 4-nitro-3-phenylfuroxan (**5-156b**) (50%); however, the isomeric 3-amino derivative gave 3-nitro-4-phenylfuroxan (**5-156a**) (5%) and 3-nitro-4-phenylfurazan (**5-157**) (36%). This conversion of a furoxan to a furazan in the presence of a peroxide has not been explained. Trifluoroperacetic acid also converted 3-amino-4-phenylfurazan to the 3-nitro derivative (**5-157**) (60%).[214,215] These results substantiated an earlier differentiation between the isomers (**5-156a**) and (**b**).[218]

In another report trifluoroperoxyacetic acid [from trifluoroacetic anhydride and hydrogen peroxide (95%)] oxidized (40°, 6.5 h) 4-amino-3-phenylfuroxan to the 4-nitro derivative (**5-156b**) (30%) and oxidized (15°–20°, 40 min) the 3-amino

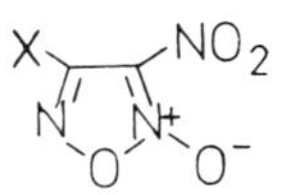

5-155a X = CH₃
5-156a X = C₆H₅

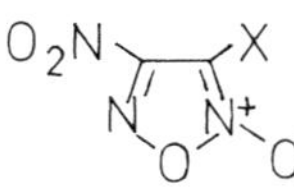

5-155b X = CH₃
5-156b X = C₆H₅

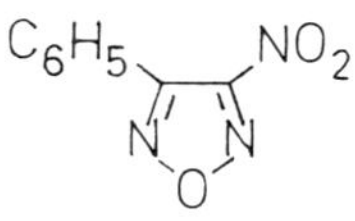

5-157

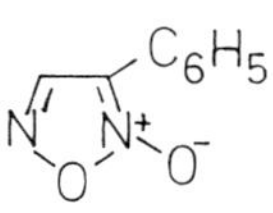

5-158

to a complex mixture of the 3-nitro compound (**5-156a**) (7.2%), the nitrofurazan (**5-157**) (7.4%), 3-phenylfuroxan (**5-158**) (20%), and 4,4'-diphenyl-3,3'-azofuroxan (**5-159**) (2.6%).[219]

The same peracid oxidized 4,4'-diamino-3,3'-bifurazanyl to the corresponding dinitro compound (**5-160a**) (35%); a similar oxidation afforded the picryl derivative (**5-160b**) (79%).[220] Caro's acid (H_2SO_5) and a mixture of ammonium peroxydisulfate [$(NH_4)_2S_2O_8$] with hydrogen peroxide oxidized 3,4-diaminofurazan to give predominantly 3-nitro-4-aminofurazan (**5-161**) and 4,4'-diamino-3,3'-azoxyfurazan (**5-162a**). At pH 6 peroxydisulfate oxidation afforded the azo compound (**5-162b**).[221]

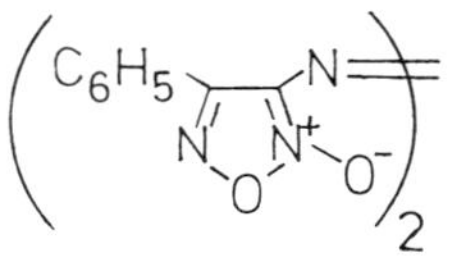

5-159

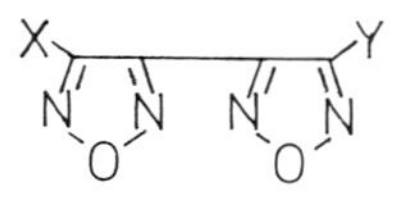

5-160a X = Y = NO_2
5-160b X = NO_2, Y = picrylamino

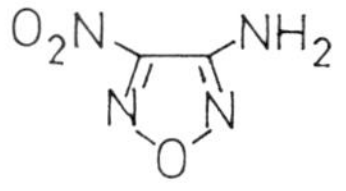

5-161

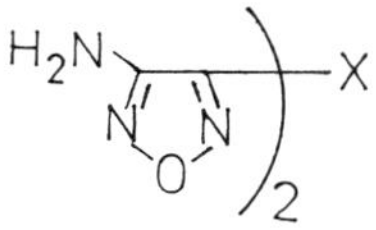

5-162a X = N_2O
5-162b X = N_2

Nitric acid (50°–90°) followed by a mixture of nitric and sulfuric acids (20°) converted 3-phenylfurazan to 3-*p*-nitrophenylfurazan and did not nitrate the furazan ring.[222]

Unsymetrically substituted furoxans can be thermally interchangeable. A thermal reaction has been demonstrated for nitrofuroxans in the isomerization of 3-nitro-4-phenylfuroxan (**5-156a**) in toluene at 110° for 2 h to 3-phenyl-4-nitrofuroxan (**5-156b**) (75%).[219]

Both a mixture of concentrated nitric and sulfuric acids (−10°) and potassium nitrate in concentrated sulfuric acid (−10°) converted mesoionic 3-phenyl-1,2,3-oxadiazolium-5-olate to the 4-nitro derivative (**5-163**) (25%).[223-225] Nitration in the 3-phenyl group was also reported.[224] For a comparison, bromine in aqueous ethanol in the presence of sodium carbonate converted 3-[*p*-(1-pyrazolyl)]phenyl-1,2,3-oxadiazolium-5-olate to the 4-bromo derivative; further bromination brought about substitution in the pyrazole ring.[225]

Reactions

Nitrofuroxans readily underwent nucleophilic displacement. Ethoxide, phenoxide, mercaptide, and thiophenoxide anions and pyrrolidine reacted at 25° with

3-methyl-4-nitrofuroxan (**5-155b**) to give the 4-substituted furoxans (**5-164**): ethoxy (85%), phenoxy (60%), ethylthio (85%), phenylthio (53%), and pyrrolidino (75%).[212,226] Sodium thiophenoxide converted 4-phenyl-3-nitrofuroxan (**5-156a**) and 4-nitro-3-phenylfuroxan (**5-155b**) to the respective sulfides (**5-165a,b**),[214] and sodium methoxide similarly converted each nitrofuroxan (**5-155a,b**) to the expected methoxyfuroxan (**5-166a,b**).[215]

Alkyl phosphite deoxygenated the furoxans (**5-164**) to furazans (45%–95%). Furoxan sulfides (**5-164**) and furazan sulfides were oxidized to sulfones (80%–90%) by hydrogen peroxide (30%) in acetic acid.[212]

5-163a X = C_6H_5
5-163b X = $C_6H_4NO_2$

5-164 X = OC_2H_5, OC_6H_5, SC_2H_5, SC_6H_5 $N(CH_2)_4$

5-165a X = SC_6H_5, Y = C_6H_5
5-165b X = C_6H_5, Y = SC_6H_5
5-166a X = OCH_3, Y = C_6H_5
5-166b X = C_6H_5, Y = OCH_3

Chemical and Physical Properties

By NMR spectroscopy it was shown that 3-methyl-4-substituted-furazans (substituents were H, CH_3, C_2H_5, n-C_3H_5, NH_2, NH_3^+, OCH_3, I, CO_2H, and NO_2) were weak bases with pK_a in the region -5.2 (4-CO_2H) to -2.15 (4-NH_2); the pK_a for the 4-nitro compound could not be determined, since the spectroscopic changes between the protonated and unprotonated forms were slight.[227] From the uv spectra for 3-substituted-4-aminofurazans (substituents were NH_2, CH_3, OCH_3, N_3, CO_2H, and NO_2), pK_a values were in the range -4.46 (3-NO_2) to -1.94 (3-NH_2).[228]

Absorption in the ir characteristic of the nitro group showed at 1,540 (v_{as}) and 1,350 (v_s) cm^{-1} for 3-nitro-4-phenylfurazan (**5-157a**). The corresponding values for 3-nitro-4-phenylfuroxan (**5-156a**) were 1,545 and 1,350 cm^{-1} and for 4-nitro-3-phenylfuroxan (**5-156b**) were 1,570 and 1,370 cm^{-1}.[215,219] Bands associated with the furazan ring in (**5-157a**) were seen at 1,575 and 878 cm^{-1} and with the furoxan ring in (**5-156a**) at 1650, 1,525, 1,455, 1,420, 1,290, 1,265, 1,010, and 850 cm^{-1}, and (**5-156b**) at 1,620, 1,520, 1,470, 1,295, and 1,280 cm^{-1}.[220] Two bands, 1,642 and 1,510 cm^{-1}, were reported for vibrations of the furoxan ring in compound (**5-155b**).[212]

Resonance signals for ^{13}C nmr of 3-nitro-4-phenylfurazan (**5-157a**) were observed at δ 150.8 (C-phenyl), 160.1 (C-nitro), and at 122.8, 130.3, and 133.8 (phenyl); 3-nitro-4-phenylfuroxan (**5-156a**) at δ 153.1 (C-phenyl), 128.4 (C-nitro), and at 125.0, 130.2, 130.3, and 133.4 (phenyl); 4-nitro-3-phenylfuroxan (**5-156b**) at δ 110.2 (C-phenyl), 159.9 (C-NO$_2$), and at 120.4, 130.0, 130.3, and 133.0 (phenyl).[150] The presence of the adjacent N$^+$-O$^-$ group in the furoxan accounted for a shift (31.7 ppm) in the ^{13}C-NO$_2$ signal for the furazan at 160.1–128.4 ppm for the furoxan (**5-156a**), and for a shift (40.6 ppm) in the ^{13}C-phenyl signal for the furazan at 150.8–110.2 ppm for the furoxan (**5-156b**).[219]

The furazan (**5-157a**) showed a mass spectrum with m/e at 191 (M$^+$) (44), 161 (M-NO)$^+$ (0.9), 145 (M-NO$_2$)$^+$ (7), 115 (M-C$_6$H$_5$)$^+$ (100), 103 (38), and 101 (2.5). The furoxans (**5-156a,b**) also showed m/e for M, M-NO, M-NO$_2$, and M-C$_6$H$_5$ ions. The most intense m/e for each of the two furoxans came at M-88. In addition, m/e 177 (M-O) (30) was observed for 3-nitro-4-phenylfuroxan (**5-156a**) but not for the isomer (**5-156b**).[219] This initial information shows that ^{13}C nmr and mass spectroscopy can establish structure assignments to isomeric furoxans.

The structure of 3-nitro-4-phenylfuroxan (**5-156a**) was also confirmed by an X-ray crystallographic analysis.[215]

Thermal decomposition kinetics and energy parameters were correlated for aminonitrofurazan and the isomeric aminonitrofuroxans in an experimental and theoretical study.[229]

Biological Properties

Nitroderivatives of furazan and fuoxan showed antibacterial, antifungal, and antiprotozoal properties.[212,214]

Nitroarylderivatives (see **5-163b**) of mesoionic 1,2,3-oxadiazolium-5-olate and 4-nitro-1,2,3-oxodiazolium-5-olate have been reported to have antiparasitic properties.[224,225]

REFERENCES

1. Barnes, R. A., In R. C. Elderfield, ed., *Heterocyclic Compounds*, Vol. 5, Wiley, New York, 1957, p. 452.
2. Quilico, A., In A. Weissberger, ed., *The Chemistry of Heterocyclic Compounds*. Wiley, New York, 1962, pp. 1–234.
3. Behr, L. C., *Ibid.*, pp. 235–328.
4. Kochetkov, N. C., Sokolov, S. D., *Adv. Heterocycl. Chem.*, **1963**, *2*, 365.
5. Wakefield, J., Wright, D. J., *Adv. Heterocycl. Chem.*, **1979**, *25*, 148.
6. Takeuchi, Y., Furusaki, F., *Adv. Heterocycl. Chem.*, **1977**, *21*, 207.
7. Swinbourne, F. J., Hunt, J. H., Klinkert, G., *Adv. Heterocycl. Chem.*, **1978**, *23*, 103.
8. Kochetkov, N. K., Khomutova, E. D., *Zh. Obshchei Khim.*, **1959**, *29*, 535; *Chem. Abstr.*, **1960**, *54*, 498c.
9. Musante, C., *Gazz. Chim. Ital.*, **1946**, *76*, 131.
10. Quilico, A., Musante, C., *Gazz. Chim. Ital.*, **1941**, *71*, 327.
11. Morgan, G. T., Burgess, H., *J. Chem. Soc.*, **1921**, *119*, 697.

12. Sokolov, S. D., Egorova, T. N., Kuryatov, N. S., *Khim. Getrotsikl. Soedin.*, **1973**, 1329; Eng. 1202.
13. Kusumi, T., Nakanishi, K., *U.S. 4,288,445; Chem. Abstr.*, **1981**, *95*, 209675e.
14. Caradonna, C., Stein, M. L., *Ann. Chim. (Rome)*, **1964**, *54*, 539.
15. Bertini, V., De Munno, A., Tafuri, D., Pino, P., *Gazz. Chim. Ital.*, **1964**, *94*, 915.
16. Langella, M. R., Finzi, P. V., *Chim. Ind. (Milan)*, **1965**, *47*, 996.
17. Kochetkov, N. K., Khomutova, E. D., *Zh. Obshch Khim.*, **1958**, *28*, 359; *Chem. Abstr.*, **1958**, *52*, 13710b.
18. Lynch, B. M., Shiu, L., *Can. J. Chem.*, **1965**, *43*, 2117.
19. Sokolov, S. D., Yudintseva, I. M., *Zh. Organ. Khim.*, **1968**, *4*, 2057; *Chem. Abstr.*, **1969**, *70*, 28857d.
20. Musante, C., *Farm. Sci. Tec. (Pavia)*, **1951**, *6*, 32; *Chem. Abstr.*, **1951**, *45*, 5879.
21. Sokolov, S. D., Ergova, T. N., Yudintseva, I. M., *Khim. Geterotsikl Soedin.*, **1974**, 597; Eng. 516.
22. Katritzky, A. R., Konya, M., Tarhan, H. O., Burton, A. G., *J. Chem. Soc. Perkin 2*, **1975**, 1627.
23. Sychkova, L. D., Shabarov, Y. S., *Zh. Org. Khim.*, **1976**, *12*, 2630; Eng. 2538.
24. Sychkov, L. D., Kalinkeina, O. L., Shabarov, Y. S., *Zh. Org. Khim.*, **1981**, *17*, 1435; *Chem. Abstr.*, **1981**, *95*, 202847d.
25. Sokolov, S. D., Yudintseva, I. M., Petrovskii, P. V., Kalyuzhnaya, V. G., *Zh. Org. Khim.*, **1971**, *7*, 1979; Eng. 2051.
26. Fusco, R., Zumin, S., *Gass. Chim. Ital.*, **1946**, *76*, 223.
27. Gaudiano, R., Ricca, A., Quilico, A., *Atti Accad. Nazl. Lincei, Rend. Classe Sci. Fis. Mat. Enat.*, **1959**, *26*, 154.
28. Desimoni, G., Minoli, G., *Tetrahedron*, **1968**, *24*, 4907.
29. Abushanab, E., Lee, D. Y., Goodman, L., *J. Heterocycl. Chem.*, **1973**, *10*, 181.
30. Kirchner, E., *Monatsh. Chem.*, **1971**, *102*, 159.
31. Worrall, D. E., *J. Am. Chem. Soc.*, **1938**, *60*, 1198.
32. Ibid. **1937**, *59*, 933.
33. Worral, D. E., Levin, E., *J. Am. Chem. Soc.*, **1939**, *61*, 104.
34. Sims, J., Houk, K. N., *J. Am. Chem. Soc.*, **1973**, *95*, 5798.
35. Padwa, A., Fisera, L., Koehler, K. F., Rodriguez, A., Wong, G. S. K., *J. Org. Chem.*, **1984**, *49*, 276.
36. Joucla, M., Grée, D., Hamelin, J., *Tetrahedron*, **1973**, *29*, 2315.
37. Kashutina, M. V., Ioffe, S. L., Shitkin, V. M., Cherskaya, N. O., Korenevskii, V. A., Tartakovskii, V. A., *Zh. Obshch. Khim.*, **1973**, *43*, 1715; Eng. 1699.
38. Shvekhgeimer, G. A., Sobtsova, N. I., Baranski, A., *Rocz. Chem.*, **1972**, *46*, 1543; *Chem. Abstr.*, **1973**, *78*, 84476f.
39. Tartakovskii, V. A., Chlenov, I. E., Morozova, N. S., Novokov, S. S., *Izv. Akad. Nauk SSSR, Ser. Khim.*, **1966**, 370; *Chem. Abstr.*, **1966**, *64*, 17567e.
40. Tartakovskii, V. A., Chlenov, I. E., Lagodzinskaya, G. V., Novikov, S. S., *Dakl. Akad. Nauk SSSR*, **1965**, *161*, 136; *Chem. Abstr.*, **1965**, *62*, 14646f.
41. Tartakovskii, V. A., Duk'yanov, O. A., Shlykova, N. I., Novikov, S. S., *Zh. Org. Khim.*, **1968**, *4*, 231; Eng. 223.
42. Ioffe, S. L., Kashutina, M. V., Shitkin, V. M., Levin, A. A., Tartakovskii, V. A., *Zh. Org. Khim.*, **1973**, *9*, 896; Eng. 922.
43. Ioffe, S. L., Makarenkova, L. M., Shitkin, V. M., Kashutina, M. V., Tartakovskii, V. A., *Izv. Akad. Nauk SSSR, Ser. Khim.*, **1973**, 203; Eng. 212.
44. Erashko, V. I., Shevelev, S. A., Fainzil'berg, A. A., *Izv. Akad. Nauk SSSR, Ser. Khim.*, **1968**, 2117; Eng. 2007.
45. Shevelev, S. A., Erashko, V. I., Fainzil'berg, A. A., *Izv. Akad. Nauk SSSR, Ser. Khim.*, **1968**, 447; Eng. 447.
46. Shevelev, S. A., Erashko, V. I., Fainzil'berg, A. A., *Izv. Akad. Nauk SSSR, Ser. Khim.*, **1968**, 2113; Eng. 2003.
47. Chlenov, I. E., Kashutina, M. V., Ioffe, S. L., Novikov, S. S., Tartakovskii, V. A., *Izv. Akad. Nauk SSSR, Ser. Khim.*, **1969**, 2085; Eng. 1948.
48. Ioffe, S. L., Makarenkova, L. M., Kashutina, M. V., Tartakovskii, V. A., Rozhdestvenskaya, N. N., Kovalenko, L. I., Isagulyants, V. G., *Zh. Org. Khim.*, **1973**, *9*, 905; Eng. 931.
49. Ginzburg, S. L., Neigauz, M. G., Novakovskaya, L. A., Novikov, S. S., Tartakovskii, V. A., Chlenov, I. E., Akopyan, Z. A., Gusev, A. I., Struchkov, Y. T., *Zh. Strukt. Khim.*, **1969**, *10*, 877; Eng. 763.
50. Tartakovskii, V. A., Chlenov, I. E., Smagin, S. S., Novikov, S. S., *Izv. Akad. Nauk SSSR., Ser. Khim.*, **1964**, 583; Eng. 549.

51. Tartakovskii, V. A., Onishchenko, A. A., Smirnyadin, V. A., Novikov, S. S., *Zh. Org. Khim.*, **1966**, *2*, 2225; Eng. 2183.
52. Shitkin, V. M., Korenevskii, V. A., Osipov, V. G., Kashutina, M. V., Ioffe, S. L., Chlenov, I. E., Tartakovskii, V. A., *Zh. Org. Khim.*, **1972**, *8*, 864; Eng. 872.
53. Tartakovskii, V. A., Onishchenko, A. A., Novikov, S. S., *Izv. Akad. Nauk SSSR, Ser. Khim.*, **1967**, 177; *Chem. Abstr.*, **1967**, *66*, 104460d.
54. Shvekhgeimer, G. A., Arslanov, E. V., Baranowski, A., *Rocz. Chem.*, **1972**, *46*, 1249; *Chem. Abstr.*, **1972**, *77*, 152038w.
55. Tartakovskii, V. A., Onishchenko, A. A., Novikov, S. S., *Zh. Org. Khim.*, **1967**, *3*, 588; Eng. 564.
56. Shitkin, V. M., Chlenov, I. E., Tartakovskii, V. A., *Izv. Akad. Nauk. SSSR, Ser. Khim.*, **1977**, 211; Eng. 187.
57. Altukhov, K. V., Perekalin, V. V., *Zh. Org. Khim.*, **1967**, *3*, 2003; Eng. 1953.
58. *Ibid.*, **1966**, *2*, 1902; Eng. 1870.
59. Altukhov, K. V., Tartakovskii, V. A., Perekalin, V. V., Novikov, S. S., *Izv. Akad. Nauk SSSR, Ser. Khim.*, **1967**, 197; *Chem. Abstr.*, **1967**, *66*, 115635r.
60. Buevich, V. A., Altukhov, K. V., Perekalin, V. V., *Zh. Org. Khim.*, **1971**, *7*, 1380; Eng. 1425.
61. *Ibid.*, **1970**, *6*, 187; Eng. 184.
62. Buevich, V. A., Altukhov, K. V., Perekalin, V. V., *Zh. Org. Khim.*, **1970**, *6*, 658; Eng. 661.
63. Ratsino, E. V., Altukhov, K. V., Perekalin, V. V., *Zh. Org. Khim.*, **1972**, *8*, 523; Eng. 526.
64. Ratsino, E. V., Altukhov, K. V., *Zh. Org. Khim.*, **1972**, *8*, 2281; Eng. 2327.
65. Andreeva, L. M., Altukhov, K. V., Perekalin, V. V., *Zh. Org. Khim.*, **1969**, *5*, 220; Eng. 212.
66. Shvekhgeimer, G. A., Sobotsova, N. I., Baranski, A., *Rocz. Chem.*, **1972**, *46*, 1741; *Chem. Abstr.*, **1973**, *78*, 72280z.
67. *Ibid.*, p. 1735; *Chem. Abstr.*, **1973**, *78*, 72285e.
68. Tartakovskii, V. A., Shvekhgeimer, G. A., Sobotsova, N. I., Novikov, S. S., *Zhv. Obshch. Khim.*, **1967**, *37*, 1163; *Chem. Abstr.*, **1968**, *68*, 22000f.
69. Andreeva, L. M., Altukhov, K. V., Perekalin, V. V., *Zh. Org. Khim.*, **1969**, *5*, 1313; Eng. 1281.
70. *Ibid.*, **1972**, *8*, 1419; Eng. 1442.
71. Grünanger, P., *Gazz. Chim. Ital.*, **1954**, *84*, 359.
72. Khisamutdinov, G. K., Trusova, T. V., *Zh. Org. Khim.*, **1982**, *18*, 457; *Chem. Abstr.*, **1982**, *96*, 162575a.
73. Monforte, P., Lo Vecchio, G., *Ann. Chim. (Rome)*, **1956**, *46*, 84.
74. Shvekhgeimer, G. A., Baranski, A., *Tezisy Vses. Soveshch. Khim. Nitrosoedinenii, 5th*, **1974**, 64; *Chem. Abstr.*, **1977**, *87*, 23125e.
75. Shvekhgeimer, G. A., Baranski, A., Grzegozek, M., *Synthesis*, **1976**, 612.
76. Baranski, A., Shvekhgeimer, G. A., *Pol. J. Chem.*, **1982**, *56*, 459; *Chem. Abstr.*, **1984**, *100*, 103223n.
77. Baranski, A., *Pol. J. Chem.*, **1982**, *56*, 1585; *Curr. Abstr. Chem.*, **1984**, *93*, 358831.
78. Mühlstädt, M., Schulze, B., *J. Prakt. Chem.*, **1971**, *313*, 745.
79. Keana, J. F. W., Little, G. M., *Heterocycles*, **1983**, *20*, 1291.
80. Bravo, P., Gaudiano, G., *Gazz. Chim. Ital.*, **1966**, *96*, 454.
81. Piaz, V. D., Pinzauti, S., Lacrimini, P., *Synthesis*, **1975**, 664.
82. Rajappa, S., Advani, B. G., Sreenivasan, R., *Synthesis*, **1974**, 656.
83. Verbruggen, R., Viehe, H. G., *Chimia*, **1975**, *29*, 351.
84. Brittellé, D. R., Boswell, G. A., Jr., *J. Org. Chem.*, **1981**, *46*, 316.
85. Love, R. F., Duranleau, R. G., *U.S. 4,172,079; Chem. Abstr.*, **1980**, *92*, 76487j.
86. Golod, E. L., Novatskii, G. N., Bagal, L. I., *Zh. Org. Khim.*, **1973**, *9*, 1111; Eng. 1139.
87. Tartakovskii, V. A., Onishchenko, A. A., Chlenov, I. E., Novikov, S. S., *Dokl. Akad. Nauk SSSR*, **1966**, *!67*, 844; *Chem. Abstr.*, **1966**, *65*, 3853a.
88. Tartakovskii, V. A., Gribov, B. G., Novikov, S. S., *Izv. Akad. Nauk SSSR, Ser., Khim.*, **1965**, 1074; *Chem. Abstr.*, **1965**, *63*, 8218a.
89. Tartakovskii, V. A., Onischenko, A. A., Novikov, S. S., *Zh. Org. Khim.*, **1967**, *3*, 1079; Eng. 1040.
90. Tartakovskii, V. A., Gribov, B. G., Savost'yanova, I. A., Novikov, S. S., *Izv. Akad. Nauk. SSSR, Ser. Khim.*, **1965**, 1644; *Chem. Abstr.*, **1966**, *64*, 2080b.
91. Fridman, A. L., Gabitov, F. A., Surkov, V. D., Zalesov, V. S., *Khim. Geterotsikl. Soedin.*, **1974**, 571; Eng. 497.
92. Metelkina, E. L., Sopova, A. S., Perekalin, V. V., Ionin, B. I., *Zh. Org. Khim.*, **1974**, *10*, 209; Eng. 213.
93. Hauff, J.-P., Tuaillon, J., Perrot, R., *Helv. Chim. Acta*, **1978**, *61*, 1207.
94. Wieland, H., *J. Liebigs Ann. Chem.*, **1903**, *328*, 224.

95. Wieland, H., Block, S., *J. Liebigs Ann. Chem.*, **1905**, *340*, 65.
96. Wade, P. A., *J. Org. Chem.*, **1978**, *43*, 3665.
97. Rossi, S., Duranti, E., *Tetrahedron Lett.*, **1973**, 485.
98. Claisen, L., *Ber.*, **1903**, *36*, 3665.
99. Moureu, C., Brachin, M., *C. R.*, **1903**, *137*, 795.
100. Moureu, C., Brachin, M., *Bull. Soc. Chim. Fr.*, **1903**, *31*, 343.
101. Eiter, K., Joop, N., *Naturwissenschaften*, **1972**, *59*, 468.
102. Mechkov, T. D., Sulimov, I. G., Usik, N. V., Miladenov, I., Perekalin, V. V., *Zh. Org. Khim.*, **1980**, *16*, 1328; Eng. 1148.
103. Demina, L. A., Khisamutdinov, G. K., Tkachev, S. V., Fainzil'berg, A. A., *Zh. Org. Khim.*, **1979**, *15*, 735; Eng. 654.
104. Khisamutdinov, G. K., Demina, L. A., Severina, N. T., Vandakurova, E. V., *Zh. Org. Khim.*, **1977**, *13*, 230; Eng. 213.
105. Khisamutdinov, G. K., Demina, L. A., Severina, N. T., Vandakurova, E. V., *USSR 536,179; Chem. Abstr.*, **1977**, *86*, 121319v.
106. Baeva, L. N., Demina, L. A., Trusova, T. V., Furin, G. G., Khisamutdinov, G. K., *Zh. Org. Khim.*, **1979**, *15*, 2408; Eng. 2179.
107. Deceuninck, J. A., Buffel, D. K., Hoornaert, G. J., *Tetrahedron Lett.*, **1980**, *21*, 3613.
108. Deswarte, S., Souchay, P., *C. R., Ser. C*, **1966**, *262*, 81.
109. Grundmann, C., Bansal, R. K., Osmanski, P. S., *Justus Liebigs Ann. Chem.*, **1973**, 898.
110. Matsumura, E., Ariga, M., Tohda, Y., Oka, H., *Fukusokan Kagaku Toronkai Koen Yoshishu, 12th*, **1979**, 146; *Chem. Abstr.*, **1980**, *93*, 46356u.
111. Gabitov, F. A., Kremleva, O. B., Fridman, A. L., *Khim. Geterotsikl. Soedin.*, **1978**, 324; Eng. 261.
112. Fridman, A. L., Gabitov, F. A., Nikolaeva, A. D., *Zh. Org. Khim.*, **1971**, *7*, 1309; Eng. 1353.
113. Fridman, A. L., Gabitov, F. A., *USSR 348,570; Chem. Abstr.*, **1973**, *78*, 29752t.
114. Onishchenko, A. A., Chlenov, I. E., Makarenkova, L. M., Tartakovskii, V. A., *Izv. Akad. Nauk SSSR, Ser. Khim.*, **1971**, 1560; Eng. 1460.
115. Nagarajan, K., Rajagopala, P., *Tetrahedron Lett.*, **1966**, 5525.
116. Melnikov, V. V., Tselinskii, I. V., Melnikov, A. A., Terpigorev, A. N., Trubitsin, A. E., *Zh. Org. Khim.*, **1984**, *20*, 658; *Curr. Abstr. Chem.*, **1984**, *94*, 359966.
117. Fridman, A. L., Gabitov, F. A., *Zh. Org. Khim.*, **1968**, *4*, 2259; Eng. 2179.
118. Fridman, A. L., Gabitov, F. A., *USSR 227,327; Chem. Abstr.*, **1969**, *70*, 96787f.
119. Khisamutdinov, G. K., Demina, L. A., Fainzil'berg, A. A., *Zh. Org. Khim.*, **1979**, 2437; Eng. 2205.
120. Khisamutdinov, G. K., Demina, L. A., Cherkasova, G. E., *USSR 829,629; Chem. Abstr.*, **1981**, *95*, 169167g.
121. Barco, A., Benetti, S., Pollini, G. P., Baraldi, P. G., *Synthesis*, **1977**, 837.
122. Shechter, H., Conrad, F., *J. Am. Chem. Soc.*, **1954**, *76*, 2716.
123. Malyuta, N. D., Khisamutdinov, G. K., Demina, L. A., Ivanov, V. V., *Zh. Org. Khim.*, **1979**, *15*, 2427; Eng. 2197.
124. Quilico, A., Fusco, R., Rosnati, V., *Gazz. Chim. Ital.*, **1946**, *76*, 30, 87.
125. Duranti, E., Urbanati, S., *Fac. Farm.*, **1972**, *45*, 65; *Chem. Abstr.*, **1975**, *82*, 72851f.
126. Vorontsova, L. G., Shitkin, V. M., Chizhov, O. S., Petrova, I. M., Chlenov, I. E., Tartakovskii, V. A., *Izv. Akad. Nauk. SSSR, Ser. Khim.*, **1976**, 810; Eng. 790.
127. Chlenov, I. E., Petrova, I. M., Shitkin, V. M., Tartakovskii, V. A., *Izv. Akad. Nauk SSSR, Ser. Khim.*, **1976**, 1405; Eng. 1347.
128. Chlenov, I. E., Petrova, I. M., Tartakovskii, V. A., *Izv. Akad. Nauk SSSR, Ser. Khim.*, **1973**, 2644; Eng. 2589.
129. Chlenov, I. E., Petrova, I. M., Shitkin, Y. M., Tartakovskii, V. A., *Izv. Akad. Nauk SSSR, Ser. Khim.*, **1975**, 1365; Eng. 1258.
130. Chlenov, I. E., Pal'tseva, G. D., Petrova, I. M., Shitkin, V. M., Tartakovskii, V. A., *Izv. Akad. Nauk SSSR, Ser. Khim.*, **1978**, 2649; Eng. 2371.
131. Chlenov, I. E., Petrova, I. M., Khasapov, B. N., Shitkin, V. M., Morozova, N. S., Tartakovskii, V. A., *Izv. Akad. Nauk SSSR, Ser. Khim.*, **1979**, 2613; Eng. 2427.
132. Kornblum, N., Davies, T. M., Earl, G. W., Holy, N. L., Manthey, J. W., Musser, M. T., Swiger, R. T., *J. Am. Chem. Soc.*, **1968**, *90*, 6221.
133. Kornblum, N., Boyd, S. D., *Ibid.*, **1970**, *92*, 5784.
134. Grée, R., Carrié, R., *Tetrahedron Lett.*, **1972**, 2987.
135. *Ibid.*, **1973**, 453.

136. Al'ber, S. I., Lagodzinskaya, G. V., Manelis, G. B., Fel'dman, E. B., *Izv. Akad. Nauk SSSR, Ser. Khim.*, **1975**, 1451; *Chem. Abstr.*, **1975**, *83*, 95875q.
137. Leonard, N. J., Hay, A., *J. Am. Chem. Soc.*, **1956**, *78*, 1984.
138. Ginzburg, S. L., Neigauz, M. G., Novakovskaya, L. A., Novikov, S. S., Tartakovskii, V. A., Chlenov, I. E., Akopyan, Z. A., Gusev, A. I., Struchkov, Y. T., *Zh. Strukt. Khim.*, **1969**, *10*, 877; *Chem. Abstr.*, **1970**, *72*, 36724q.
139. Chlenov, I. E., Petrova, I. M., Khasapov, B. N., Karpenko, N. F., Stepenyants, A. U., Chizhob, O. S., Tartakovskii, V. A., *Izv. Akad. Nauk SSSR, Ser. Khim.*, **1978**, 2551; Eng. 2278.
140. Chlenov, I. E., Petrova, I. M., Khasapov, B. N., Tartakovskii, V. A., *Izv. Akad. Nauk SSSR, Ser. Khim.*, **1975**, 2131; Eng. 2020.
141. Chlenov, I. E., Petrova, I. M., Tartakovskii, V. A., *Izv. Akad. Nauk SSSR, Ser. Khim.*, **1980**, 209; *Chem. Abstr.*, **1980**, *93*, 7383f.
142. Tartakovskii, V. A., Onishchenko, A. A., Novikov, S. S., *USSR 181,120; Chem. Abstr.*, **1966**, *65*, 8917g.
143. Khisamutdinov, G. K., Demina, L. A., Cherkasova, G. E., Klimenko, V. G., *Zh. Org. Khim.*, **1979**, *15*, 2436; Eng. 2204.
144. Khisamutdinov, G. K., Demina, L. A., Fainzil'berg, A. A., *Izv. Akad. Nauk SSSR, Ser. Khim.*, **1981**, 473; *Chem. Abstr.*, **1981**, *95*, 7125q.
145. Kochetkov, N. K., Sokolov, S. D., Luboshnikova, V. M., *Zh. Obshch. Khim.*, **1962**, *32*, 1778; *Chem. Abstr.*, **1963**, *58*, 3409c.
146. Chimichi, S., De Sio, F., Donati, D., Fina, G., Pepino, R., Sarti-Fantoni, P., *Heterocycles*, **1983**, *20*, 263.
147. Pepino, R., Ricci, A., Taddei, M., Tedeschi, P., *J. Organomet. Chem.*, **1982**, *231*, 91.
148. Sokolov, S. D., Setkina, V. N., *Khim. Geterotsikl. Soedin.*, **1969**, 786; Eng. 580.
149. Kandlikar, S., Sethuram, B., Rao, T. N., *Curr. Sci.*, **1974**, *43*, 209.
150. Quilico, A., Musante, C., *Gazz. Chim. Ital.*, **1942**, *72*, 399.
151. Jagannadham, V., Sethuram, B., Rao, T. N., *Curr. Sci.*, **1977**, *46*, 704.
152. Jagannadham, V., Kandlikar, S., Sethuram, B., Rao, T. N., *Natl. Acad. Sci. Lett. (India)*, **1978**, *1*, 207.
153. Krishnamurthy, A., Rao, K. S. R. K. M., Rao, N. V. S., *Indian J. Appl. Chem.*, **1972**, *35*, 90.
154. Reddi, K. M., Rao, C. J., Murthy, A. K., *Indian J. Chem.*, **1981**, *20B*, 607.
155. Rao, C. J., Reddy, K. M., Murthy, A. K., *Indian J. Chem.*, **1981**, *20B*, 997.
156. Reddi, K. M., Rao, C. J., Murthy, A. K., *Bull. Chem. Soc. Jpn.*, **1981**, *54*, 3617.
157. Rao, C. J., Murthy, A. K., *Indian J. Chem.*, **1978**, *16B*, 636; **1981**, *20B*, 282, 335.
158. Rao, K. S. R. K. M., Devi, Y. U., *Proc. Indian Acad. Sci.*, **1976**, *84A*, 79.
159. Jagannadham, V., Sethuram, B., Rao, T. N., *Indian J. Chem.*, **1979**, *17B*, 598.
160. Nesi, R., Chimichi, S., De Sio, F., Pepino, R., Tedeschi, P., *Tetrahedron Lett.*, **1982**, *42*, 4397.
161. Rajanarender, E., Murthy, A. K., *Indian J. Chem.*, **1981**, *20B*, 608.
162. Rajanarender, E., Rao, C. J., Murthy, A. K., *Indian J. Chem.*, **1981**, *20B*, 610.
163. Donati, D., Fiorenza, M., Moschi, E., Sarti-Fantoni, P., *J. Heterocycl. Chem.*, **1977**, *14*, 1951.
164. Donate, D., Fiorenza, M., Sarti-Fantoni, P., *J. Heterocycl. Chem.*, **1979**, *16*, 253.
165. Alberola, A., González, A. M., Guerra, D., Pulido, F. J., *J. Heterocycl. Chem.*, **1982**, *19*, 1073.
166. Souchay, P., Deswarte, S., *C.R. Seances Acad. Sci. Ser.*, **1965**, *260* (group 8), 6379.
167. Arti-Fantoni, P., Donati, D., Fiorenza, M., Moschi, E., Dal Piaz, V., *J. Heterocycl. Chem.*, **1980**, *17*, 621.
168. Sarti-Fantoni, P., Donati, D., De Sio, F., Moneti, G., *J. Heterocycl. Chem.*, **1980**, *17*, 1643.
169. Claisen, L., *Chem. Ber.*, **1909**, *42*, 59.
170. Bell, F., *J. Chem. Soc.*, **1941**, 285.
171. Van der Plas, H. C., *Ring Transformations of Heterocycles*, Vol. 1. Academic Press, London, **1973**, pp. 293–295.
172. Musante, C., Stener, A., *Gazz. Chim. Ital.*, **1959**, *89*, 1579.
173. Hill, H. S., Hale, W. J., *Am. Chem. J.*, **1903**, *29*, 253.
174. Musante, C., *Gazz. Chim. Ital.*, **1943**, *73*, 355.
175. *Ibid.*, **1942**, *72*, 537.
176. Padwa, A., Koehler, K. F., Rodriguez, A., *J. Org. Chem.*, **1984**, *49*, 282.
177. Tartakovskii, N. A., Onishchenko, A. A., Chlenov, I. E., Novikov, S. S., *Dokl. Akad. Nauk SSSR*, **1965**, *164*, 1081; Eng. 994.
178. Tartakovskii, V. A., Onishchenko, A. A., Lazodzinskaya, G. V., Novikov, S. S., *Zh. Org. Khim.*, **1967**, *3*, 765; Eng. 730.
179. Sokolov, S. D., Kazitsyna, L. A., Guseva, I. K., *Zh. Org. Khim.*, **1966**, *2*, 731; Eng. 733.

180. Sokolov, S. D., Tikhomirova, G. B., *Khim. Geterotsikl. Soedin.*, **1976**, 1031; Eng. 851.
181. Fridman, A. L., Ivshina, T. N., Ivshin, V. P., Tartakovskii, V. A., Novikov, S. S., *Izv. Akad. Nauk SSSR, Ser. Khim.*, **1971**, 2279; Eng. 2151.
182. Pul'tsin, M. N., Gaile, A. A., Proskuryakov, A., *Zh. Fiz. Khim.*, **1975**, *49*, 1052; Eng. 618.
183. Ivanov, A. I., Chlenov, I. E., Tartakovskii, V. A., Slovetskii, V. I., Novikov, S. S., *Izv. Akad. Nauk SSSR, Ser. Khim.*, **1965**, 1491; Eng. 1458.
184. Ivanov, A. I., Slovetskii, V. I., Tartakovskii, V. A., Novikov, S. S., *Khim. Geterotsikl. Soedin.*, **1966**, *2*, 197; Eng. 138.
185. Zatsepina, N. N., Tupitsyn, I. F., Belyashova, A. I., Kane, A. A., Kolodina, N. S., Sudakova, G. N., *Khim. Geterotsikl. Soedin.*, **1977**, 1110; Eng. 894.
186. Katritzky, A. R., Boulton, A. J., *Spetrochim. Acta*, **1961**, *17*, 238.
187. Sokolov, S. D., Ashkinadze, L. D., *Zh. Vses. Khim. Obshchestvaim. D. I. Mandeleeva*, **1963**, *8*, 119; *Chem. Abstr.*, **1963**, *58*, 13315e.
188. Murty, A. K., Rao, K. S. R. K. M., Rao, N. V. S., *Indian J. Chem.*, **1973**, *11*, 1074.
189. Sokolov, S. D., Yudintseva, I. M., Petrovskii, P. V., *Zh. Org. Khim.*, **1970**, *6*, 2584; Eng. 2594.
190. Al'ber, S. I., Lagodzinskaya, G. V., Manelis, G. B., Fel'dman, E. B., *Izv. Akad. Nauk SSSR, Ser. Khim.*, **1975**, 1451; Eng. 1346.
191. *Ibid., Mat. Probl. Khim.*, **1975**, *2*, 217; *Chem. Abstr.*, **1976**, *84*, 97482d.
192. Lagodzinskaya, G. V., *Zh. Strukt. Khim.*, **1970**, *11*, 30; Eng. 25.
193. Bowie, J. H., Kallury, R. K. M. R., Cooks, R. G., *Aust. J. Chem.*, **1969**, *22*, 563.
194. Martinez, R., Cortés, E., *J. Heterocycl. Chem.*, **1980**, *17*, 585.
195. Kallury, R. K. M. R., Hemalatha, J., *Org. Mass Spectrom.*, **1980**, *15*, 659.
196. Rao, C. J., Murthy, A. K., *Indian J. Chem.*, **1981**, *20B*, 335.
197. Rozynov, Y. V., Puchkov, V. A., Vul'fson, N. S., Tartakovskii, V. A., Onishchenko, A. A., Novikov, S. S., *Dokl. Akad. Nauk SSSR*, **1966**, *169*, 123; Eng. 672.
198. Rozynov, B. V., Vul'fson, N. S., Puchkov, V. A., Tartakovskii, V. A., Onishchenko, A. A., Novikov, S. S., *Khim. Geterotsikl. Soedin.*, **1969**, *5*, 36; Eng. 28.
199. Mazheika, I. B., Yankobska, I. S., Sokolov, S. D., Yudintseva, I. M., *Khim. Geterotsikl. Soedin.*, **1972**, 460; Eng. 419.
200. Miroshnichenko, E. A., Lebedev, Y. A., *Khim. Geterotsikl. Soedin.*, **1969**, *5*, 963; Eng. 717.
201. Miroshnichenko, E. A., Tartakovskii, V. A., Lebedev, Y. A., *Khim. Geterotsikl. Soedin.*, **1968**, *4*, 351; Eng. 259.
202. Takahi, Y., *Sankyo Kenkyusho Nempo*, **1973**, *25*, 8; *Chem. Abstr.*, **1974**, *81*, 10361e.
203. *Japan Kokai Tokkyo Koho*, 79 70,266; *Chem. Abstr.*, **1979**, *91*, 135604z.
204. Schlecker, R., Frickel, F. F., Franke, A., Lenke, D., Gries, J., *Eur. Pat. Appl. EP 34,754*; *Chem. Abstr.*, **1982**, *96*, 104219b.
205. Gumanov, L. L., *Spetsifichnost Khim. Mutageneza, Mater. Vses. Simp.*, **1967**, 65; *Chem. Abstr.*, **1969**, *71*, 58047c.
206. Gomper, R., Christmann, O., *Chem. Ber.*, **1959**, *92*, 1928.
207. Hammar, W. J., Rustad, M. A., *J. Heterocycl. Chem.*, **1981**, *18*, 885.
208. Grifantini, M., Stein, M. L., Temperilli, A., *Ann. Chim. (Rome)*, **1966**, *56*, 946.
209. Boyer, J. H., In *Heterocyclic Compounds*, R. C. Elderfield, Ed., Vol. 7. Wiley, New York, 1961, pp. 462–508.
210. Khmel'nitskii, L. I., Novikov, S. S., Godovikova, T. I., *Chemistry of Furoxans: Reactions and Use*. Nauka, Moscow, 1983, 311 ff.; *Chem. Abstr.*, **1983**, *99*, 70695y.
211. Ref. 140, p. 478.
212. Gasco, A., Mortarini, V., Rua, G., Serafino, A., *J. Heterocycl. Chem.*, **1973**, *10*, 587.
213. Cameron, A. F., Freer, A. A., *Acta Crystallogr.*, **1974**, *B30*, 354.
214. Calvino, R., Mortarini, V., Gasco, A., Sanfilippo, A., Ricciardi, M. L., *Eur. J. Med. Chem. Chim. Ther.*, **1980**, *15*, 485.
215. Calvino, R., Gasco, A., Serafino, A., Viterbo, D., *J. Chem. Soc., Perkin 2*, **1981**, 1240.
216. Brown, J. F., Jr., *J. Am. Chem. Soc.*, **1955**, *77*, 6341.
217. Ungnade, H. E., Kissinger, L. W., *Tetrahedron*, **1963**, *19*, Suppl. 1, 143.
218. Bianco, M. A., Gasco, A., Mortarini, V., Serafino, A., Menziani, E., *Formaco Ed. Sci.*, **1973**, *28*, 701.
219. Makhova, N. N., Ovchinnikov, I. V., Khasanov, B. N., Khmel'nitskii, L. I., *Izv. Akad. Nauk SSSR, Ser. Khim.*, **1982**, 646; Eng. 573.
220. Coburn, M. D., *J. Heterocycl. Chem.*, **1968**, *5*, 83.
221. Solodyuk, G. D., Boldyrev, M. D., Gidaspov, B. V., Nikolaev, V. D., *Zh. Org. Khim.*, **1981**, *17*, 861; *Chem. Abstr.*, **1981**, *95*, 80839e.

222. Zelenov, M. P., Frolova, G. M., Melnikova, S. F., Tselinski, I. V., *Khim. Geterotsikl. Soedin.*, **1982**, 27; *Chem. Abstr.*, **1982**, *96*, 162603h.
223. Baker, W., Ollis, W. D., Poole, V. D., *J. Chem. Soc.*, **1950**, 1542.
224. Pala, G., Mantegani, A., Coppi, G., Genova, R., *Chim. Ther.*, **1969**, *4*, 31.
225. Havanur, S. B., Badami, B. V., Puranik, G. S., *Can. J. Chem.*, **1983**, *61*, 154.
226. Andrianov, V. G., Eremeev, A. V., *Zh. Org. Khim.*, **1984**, *20*, 150; *Chem. Abstr.*, **1984**, *100*, 174739w.
227. Tselinskii, I. V., Mel'nikova, S. F., Vergizov, S. N., Frolova, G. M., *Khim. Geterotsikl. Soedin.*, **1981**, 35; Eng. 27.
228. Tselinskii, I. V., Mel'nikova, S. F., Vergizov, S. N., *Khim. Geterotsikl. Soedin.*, **1981**, 321; Eng. 228.
229. Zverev, V. V., Saifullin, I. S., Sharnin, G. P., *Izv. Akad. Nauk SSSR, Ser. Khim.*, **1978**, 313; Eng. 269.

BIBLIOGRAPHY

Alberola, A., Gonzalez, A. M., Laguna, M. A., Pulldo, F. J., SYNTHESIS OF 4-FUNCTIONALIZED 2-ISOXAZOLINES BY REDUCTION OF THE ISOXAZOLE RING WITH COMPLEX METAL HYDRIDES, *Synthesis*, **1983**, *5*, 413.
Alberola, A., Gonzalez, N. A. M., Pulido, F. J., THE DIFFERENTIAL REACTIVITY OF 5- AND 3-METHYL GROUPS IN THE REACTION OF 4-SUBSTITUTED 2,3,5-TRIMETHYL-ISOXAZOLIUM SALTS WITH AROMATIC ALDEHYDES, *Heterocycles*, **1983**, *20*, 1035.
Baranski, A., SYNTHESIS AND PROPERTIES OF AZOLES AND THEIR DERIVATIVES. PART V. REGIOCHEMISTRY IN (2+3) CYCLOADDITION REACTIONS OF BENZONITRILE N-OXIDE WITH β-SUBSTITUTED NITROETHYLENES, *Pol. J. Chem.*, **1982**, *56*, 257.
Baranski, A., Shvekhgeimer, G. A., SYNTHESIS AND PROPERTIES OF AZOLES AND THEIR DERIVATIVES. PART VI. (2+3) CYCLOADDITION REACTIONS OF NITROETHYLENE WITH AROMATIC NITRILE N-OXIDES AND SOME CONVERSIONS OF 3-ARYL-5-NITRO-4,5-DIHYDRO-1,2-OXAZOLES, *Pol. J. Chem.*, **1982**, *56*, 459.
Chimichi, S., De Sio, F., Donati, D., Pepino, R., Rabbati, L., Sarti-Fantoni, P., THE PREPARATION OF ARYLPROPIOLIC ACIDS VIA STYRYLISOXAZOLES, *J. Heterocycl. Chem.*, **1983**, *20*, 105.
Gaponic, P. N., Karavai, V. P., Chernavina, N. I., SYNTHESIS OF N-ALLYLTETRAZOLES, *Vestn. Belorus. Un-ta, Ser. 2*, **1983**, 23; *Chem. Abstr.* **1984**, *100*, 174727r.
Keana, J. F. W., Little, G. M., Little, J. F. W., SYNTHESIS AND ELABORATION OF 3-SUBSTITUTED 4-NITROISOXAZOLES, *Heterocycles*, **1983**, *20*, 1291.
Nalyuta, N. G., Khisamutdinov, G. K., Demina, L. A., SYNTHESIS AND PROPERTIES OF ISOXAZOLE AND ISOXAZOLINE 4-ARYLAZO DERIVATIVES, *Zh. Org. Khim.*, **1984**, *20*, 2020; *Curr. Abstr. Chem.*, **1985**, *96*, 368094.

Padwa, A., Fisera, L., Koehler, K. F., Rodriquez, A., Wong, G. S. K., REGIOSELECTIVITY ASSOCIATED WITH THE 1,3-DIPOLAR CYCLOADDITION OF NITRONES WITH ELECTRON-DEFFICIENT DIPOLAROPHILES, *J. Org. Chem.*, **1984**, *49*, 276.

Spear, R. J., Redman, L. D., Bentley, J. R., SENSITIZATION OF HIGH-DENSITY SILVER AZIDE TO STAB INITIATION, *Rep.-Mater. Res. Lab. (Aust.)*, **1983**, MRL-R-881; *Chem. Abstr.*, **1984**, *100*, 9480x.

SUBJECT INDEX

1-Acetamidomethyl-3-nitro-1,2,4-triazolin-5-one, 281
4-Acetyl-3-nitro-1-(4-nitrophenyl)pyrazolin-5-one, 207, 208
Acidity
    nitroimidazoles, 151–154
    nitroisoxazoles, 335
    nitropyrazoles, 242–245, 247, 252, 253
    nitropyrroles, 40, 49, 54, 62, 63
    5-nitrotetrazole, 291
    nitrotriazoles, 291, 292
Acylation
    nitroimidazoles, 109, 122
    nitropyrazoles, 228
    nitropyrroles, 58
Adjacent lone pair effect, 122
Alkylation and arylation,
    nitroimidazoles, 108–122
    nitropyrazoles, 223–228
    nitropyrroles, 59
    nitrotriazoles, 282
1-Alkyl-5-nitroimidazole-2-carboxylic acid, 83
4-Alkyl-4-nitropyrazolenine-1,2-dioxides, 203
Alkyl 5-substituted-3-nitro-1,2,4-triazol-1-ylacetates, 277, 278
Aminoazoles, oxidation to nitroazoles, 40, 47, 94, 95, 203, 275, 341, 342
Aminonitrofurazan, 342, 343
1-(1-Amino-5-nitroimidazol-2-yl)-2-imidazolidinone, 94
Amyl 4-nitroimidazole-1-carboxylate, 123
Antipyrine, 199
Antistatic filaments, 163
1-Aryl-2,3-dimethyl-4-nitropyrazol-5-ones, 253, 254
1-Aryl-4-nitropyrazol-5-ones, 253, 254
4-Azaindole
    3-nitro, 33, 34, 71
5-Azaindole
    1-benzyl-3-nitro-6-methoxy-7-cyano, 34
    3-nitro, 34, 68, 69
7-Azaindole
    2-methyl-3-nitro, 33
    4-methyl-3-nitro, 33
    4-methyl-2-nitro-3-p-nitrophenyl, 34
    3-nitro, 33
    3-nitro-2-p-nitrophenyl, 34
    3-nitro-7-oxide, 34
    3-nitro-2-phenyl, 33, 34
    3-nitro-2-pyridyl, 33
Azathioprine, 164
Azomycin, see 2-nitroimidazole

Benzimidazole
    1-alkyl-2-nitro, 94, 140
    2-nitro, 91, 140
9-Benzoyltetrahydrocarbazole, adduct with nitric acid, 28
Biladiene, 57, 63
Biological properties
    nitroimidazoles, 165, 296
    nitroisoxazoles, 340
    nitrooxadiazoles, 344
    nitropyrazoles, 257
    nitropyrroles, 70–72
    nitrotriazoles, 295–296

3-Chloro-4-nitro-3,5,5-trimethylpyrazoline-1,2-dioxide, 200
Condensation reactions in nitroimidazole derivatives, 122–124
Corrosion of steel, 164
Curtius rearrangement, 141
4-Cyano-3,3-dimethyl-5-nitro-3H-pyrazole, 204

Decarboxylation, 48, 49, 194
1,5-Dehydro-2,6-dimethyl-3,7-dinitro-1,2,3-triazolo[2,1-a][1,2,3]triazole, 276
Deoxygenation of nitro groups, 46
Deoxymethoxatin, 31, 42, 43
Detonation velocity, 163
3-Diazo-2,5-diphenyl-4-nitro-3H-pyrrole, 22
3,5-Dichloro-4-nitropyrrole-2-carboxylic acid, 60
1,2-Diethyl-4-nitropyrazolidine, 239
2,3-Dihydroimidazo[2,1-b]oxazole
    2-chloro-5-nitro, 104
    2-chloro-6-nitro, 104
    2-methoxy-5-nitro, 104
    2-methoxy-6-nitro, 104
    2-methyl-5-nitro, 104

353

2-methyl-6-nitro, 104
5-nitro, 104, 162
2-Dimethylamino-5-nitro-4-(4-nitrophenyl)-
  oxazole, 340
Dimethyl 3,4-dibromo-2,5-dicarboxylate,
  reaction with nitric acid, 10
Dimethyl 2,5-dinitropyrrole-3,4-dicarboxylate
  15–16
2,3-Dimethyl-4-nitro-1-(4-nitrophenyl)-
  Δ^3-pyrazolin-5-one, 198, 237, 239
2,5-Dimethyl-4-nitro-1-(4-nitrophenyl)-
  Δ^4-pyrazolin-3-one, 197
3,3-Dimethyl-5-nitro-4-phenyl-3H-pyrazole,
  204
2H-1,5-Dimethyl-4-nitro-2-phenylpyrazol-
  3-one, 235, 236
1,2-Dimethyl-4-nitro-Δ^4-pyrazolin-3-one, 211
2,5-Dimethyl-4-nitropyrazolin-3-one, 212
2,4-Dimethyl-5-nitropyrrole-3-carboxylic acid,
  40, 68, 69
3,5-Dimethyl-4-nitropyrrole-2-carboxylic acid,
  68, 69
Dimethyl 2-nitropyrrole-3,4-dicarboxylate,
  15, 16
1,4-Dimethyl-3-nitro-1,2,4-triazolin-5-one, 280
Dimetridazole, 165
4,4'-Dinitrobifurazan-3,3'-yl, 342
Dinitrocarbene, 308, 309, 321
3,6-Dinitro-2,5-diphenyl-2H,5H-pyrazolo-
  [4,3-c]pyrazole 222, 223
Dinitrogen trioxide 341
3,4-Dinitro-5-methyl-3-phenyl-5-(4-nitro-
  phenyl)- Δ^1-pyrazoline 320
3,5-Dinitro-1-methyl-pyrazole-2-oxide
  199, 200
1-(2,4-Dinitrophenyl)-3-methyl-4-nitro-4-n-
  propyl- Δ^2-pyrazolin-5-one 197
3,3-Dinitro- Δ^1-pyrazoline 322
3,4-Dinitro- Δ^1-pyrazoline, alkyl and aryl
  derivatives 254
Di(4-nitropyrazol-1-yl)methane 188
Dinitropyrrocoll 56, 63
3,5-Dinitropyrrole-2-carboxylic acid 48
4,5-Dinitropyrrole-2-carboxylic acid 21
1,4-Di-(3-oxobutyl)-3-nitro-1,2,4-triazolin-
  5-one 281
5,7-Dioxo-6-isopropyl-3-nitro-4H,6H-pyrazolo
  [1,5-a]-s-triazine 222
3,5-Diphenyl-3,4-dinitro-pyrazoline 204
1,5-Diphenyl-3-ethyl-4-nitropyrazolidine 206
1,5-Diphenyl-3-ethyl-4-nitro- Δ^2-pyrazoline
  206
1,5-Diphenyl-3-ethyl-4-nitro- Δ^3-pyrazoline
  206
1,5-Diphenyl-3-isopropyl-4-nitropyrazolidine
  206
3,5-Diphenyl-4-methyl-4-nitropyrazolenine-
  1,2-dioxide 203, 237, 238

1,2-Diphenyl-5-methyl-4-nitro- Δ^4-pyrazolin-
  3-one 211
1,2-Diphenyl-4-nitro- Δ^4-pyrazolin-3-one 211
2,5-Diphenyl-4-nitro-3-pyrryl 3,5-dimethyl-4-
  isoxazolyl ketone 7
2,5-Diphenyl-4-nitro-3-pyrryl 3,5-dimethyl-4-
  isoxazolyl ketone 7
1,2-Diphenyl-3,4,5-trioxopyrazolidine 199
trans-4,5-Dipiperidino-3-nitro-2-pyrrolines 51
Dipole moments
  nitroisoxazoles 339
  nitropyrazoles 256
  nitropyrroles 67, 70
  nitrotriazoles 294
Dipyrafene 199
Dislite 318, 333
Distamycin A 7, 24, 40, 41, 71, 77

Electron spin resonance spectroscopy,
  nitroimidazoles 163
5-Ethoxy-6-nitro-2,3-dihydro-1H-
  imidazolo[1,2-b]pyrazole 210, 211
Ethyl 4-acetamido-1-methyl-5-nitropyrrole-2-
  carboxylate 5
Ethyl 4-acetamido-1-methyl-5-nitrosopyrrole-
  2-carboxylate 5
Ethyl 7-amino-3-nitropyrazolo[1,5-a]-
  pyrimidine-6-carboxylate 221, 222
Ethyl 4-aryl-5-nitro- Δ^2-isoxazoline-3-
  carboxylate-2-oxide 317
Ethyl 3-cyclohexyl-4-nitroisoxazole-5-
  carboxylate 319
Ethyl 1,2-dimethyl-4-formyl-5-nitropyrrole-3-
  carboxylate 8
Ethyl 1,2-dimethyl-3-nitropyrrole-5-
  carboxylate 67
Ethyl 1,5-dimethyl-4-nitropyrrole-2-
  carboxylate 7, 15
Ethyl 3,5-dimethyl-4-nitropyrrole-2-
  carboxylate 12, 16
Ethyl 3,5-dinitro-1-methylpyrrole-2-
  carboxylate 15
Ethyl 4,4-dinitropyrazolidin-1-ylacetate
  207, 245
Ethyl 4-formyl-3-nitropyrrole-2-carboxylate
  8, 49
Ethyl 4-formyl-5-nitropyrrole-2-carboxylate
  8, 49
Ethyl 5-formyl-3-nitropyrrole-2-carboxylate
  8, 48
Ethyl 5-formyl-4-nitropyrrole-2-carboxylate 8
Ethyl 3-iodo-5-methyl-4-nitropyrrole-2-
  carboxylate 10, 44
1-Ethyl-5-methyl-4-nitro-3-carboxylic acid
  195, 196
Ethyl 1-methyl-5-nitroimidazole-2-
  carboximidate 124, 125

Subject Index

Ethyl 1-methyl-2-nitroimidazole-5-carboxylate 91
Ethyl 4-[2-(1-methyl-5-nitroimidazol-2-yl)-2-thioethoxy]benzoate 136, 162
Ethyl 1-methyl-3-nitro-5-phenyl-4-(2-pyridyl)-pyrrole-2-carboxylate 46
1-Ethyl-3-methyl-4-nitropyrazole-5-carboxylic acid 195, 196
Ethyl 1-methyl-4-nitropyrrole-2-carboxylate 6, 15, 58
Ethyl 1-methyl-5-nitropyrrole-2-carboxylate 6, 15
Ethyl 5-methyl-4-nitropyrrole-2-carboxylate 8, 11, 48
Ethyl nitrate 96
Ethyl 4(5)-nitroimidazole-5(4)-carbamate 141
Ethyl 4-nitroimidazole-1-carboxylate 123
Ethyl 1-nitro-2-methylthiopyrrolo[2,1-a]-isoquinoline-3-carboxylate 36, 37
Ethyl 4-nitro-5-phenylisoxazole-3-carboxylate 314, 315
Ethyl 5-nitro-3-phenyl-$\triangle^2$-isoxazoline-5-alkanecarboxylate 313, 314, 323
Ethyl 4-nitro-1-phenylpyrazole-3-carboxylate 207, 208
Ethyl 4-nitropyrrole-2-carboxylate 8, 24, 58
Ethyl 5-nitropyrrole-2-carboxylate 8
Ethyl 4-nitropyrazol-1-ylacetate 207, 245
Ethyl 3(5)-4-nitropyrazole-5(3)-carboxylate 205
Ethyl 4-nitropyrazole-3(5)-carboxylate 228
1-Ethyl-4-nitropyrazolidine 207
Ethyl 4-nitro-(2,3,5-tri-O-acetyl-β-D-ribo-furanosyl)pyrazole-3- and 5-carboxylates 228
Ethyl 3-nitro-1,2,4-trimethylpyrrole-5-carboxylate 67
Eulite 318

Flunidazole 119
Formycin 234
2-(2-Furyl)imidazo[1,2-a]pyridine 3,5'-dinitro 98

N-Glycosylation 121

Habraken reaction 201, 272
Halogenation
    nitroimidazoles 107
    nitropyrazoles 228
    nitropyrroles 59, 60
4-Halo-1-methyl-3-nitropyrazole-5-carboxylic acid 194
ω-Halo(pseudohalo)alkylnitroimidazoles 143
Hunsdiecker reaction 108
1-β-Hydroxyethyl-5-nitroimidazole-2-carboxylic acid 166
4-(2-Hydroxyalkyl)-1-methyl-3-nitro-1,2,4-triazolin-5-ones 280

1-Hydroxymethyl-4-methyl-3-nitro-1,2,4-triazolin-5-one 281
1-Hydroxymethyl-3-nitro-1,2,4-triazolin-5-one 281
3-Hydroxy-5-phenyl-4-nitro-3-pyrrolin-2-one 23

Imidazo[1,2-a]benzimidazole
    2,9-dimethyl-3-nitro 102
    9-methyl-3-nitro-2-phenyl 102, 147
Imidazo[2,1-b]benzothiazole
    2-p-nitrophenyl-3-nitro 96
Imidazo[1,2-a]imidazole
    5,6-dihydro-7-methyl-7H-3-nitro 96, 159
    1H-1,6-dimethyl-2(3),5-dinitro 96
9H-Imidazo[1,2-a]indole
    3,x-dinitro 107
    3-nitro 107
Imidazo[2,1-a]isoquinoline
    5,6-dihydro-3-nitro 106
Imidazole
    1-(2-alkoxyalkyl)-5-nitro 89
    4-alkoxy-5-nitro 133
    4(5)-(4-alkoxyphenyl)-5(4)-nitro 84
    2-alkyl-4(5)-bromo-5(4)-nitro 85, 107
    1-alkyl-2-cyano-5-nitro 126
    1-alkyl-2-dialkylamino-5-nitro 88
    1-alkyl-2,4-dinitro-5-halo 131, 132
    1-alkyl-2-formyl-5-nitro 125, 126
    1-alkyl-5-formyl-2-nitro 127
    1-alkyl-2-methyl-4-nitro 108
    1-alkyl-2-methyl-5-nitro 108
    2-alkyl-1-methyl-5-nitro 112
    5-alkyl-1-methyl-2-nitro 91
    1-alkyl-2-nitro 140
    2-alkyl-4(5)-nitro 81
    5-amino-1-ethyl-2-methyl-4-nitro 149
    2-amino-1-methyl-5-nitro 104, 124, 130, 184
    1-amino-2-nitro 114
    4(5)-amino-5(4)-nitro 141
    2-(2-amino-1,3,4-thiadiazol-5-yl)-1-methyl-5-nitro 89
    2-(2-amino-1,3,4-thiadiazol-5-yl)-4(5)-nitro 89
    2-aroyl-1-methyl-5-nitro 109
    2-aryl-5-nitro 89
    4(5)-aryl-5(4)-nitro 118
    2-azido-4(5)methyl-5(4)-nitro 91
    4(5)-azido-5(4)-nitro 141
    2-benzoyl-1-methyl-5-nitro 141, 149
    1-benzyl-5-bromo-4-nitro 108
    5-bromo-1-(2-bromoethyl)-2-methyl-4-nitro 116, 117
    5-bromo-1,2-dialkyl-4-nitro 108, 131, 132
    2-bromo-1,5-dimethyl-4-nitro 107
    4-bromo-1,2-dimethyl-5-nitro 129, 131, 134

5-bromo-1,2-dimethyl-4-nitro 129, 131, 134
1-(2-bromoethyl)-2-methyl-5-nitro 162
2-bromo-1-methyl-4-nitro 86, 154
2-bromo-1-methyl-5-nitro 86, 130
2-bromo-4(5)-methyl-5(4)-nitro 86, 107
4(5)-bromo-2-methyl-5(4)-nitro 86
5-bromo-1-methyl-4-nitro 86, 130, 154
4(5)-bromo-5(4)-nitro 85, 138, 139
1-(ω-carboxyalkyl)-2-methyl-4-nitro-5-amino 124
4-chloro-1,2-dialkyl-5-nitro 86
2-chloro-1-methyl-4-nitro 112, 133
2-chloro-1-methyl-5-nitro 112, 133
4-chloro-1-methyl-2-nitro 139
4-chloro-1-methyl-5-nitro 86, 129, 133
5-chloro-1-methyl-4-nitro 86, 131, 161, 164
2-chloro-4(5)-nitro 138
4(5)-chloro-5(4)-nitro 129, 138
4(5)-(4-chlorophenyl)-5(4)-nitro 83
5-cyano-1,2-dimethyl-4-nitro 129
1-(2-cyanoethyl)-2-nitro 162
4-cyano-1-methyl-5-nitro 129
5-cyano-1-methyl-4-nitro 129
4(5)-dialkylaminomethyl-2-nitro 109
1,2-dialkyl-5-nitro 94
2,4(5)-dibromo-5(4)-nitro 107, 108
2-dichloroacetamido-1-methyl-5-nitro 112
1,5-diethyl-2,4-dinitro 114
2,4(5)-diiodo-5(4)-nitro 87
2-(β-dimethylaminovinyl)-1-methyl-5-nitro 128, 145
2,4-dimethyl-1-ethyl-5-nitro 82
1,2-dimethyl-4-iodo-5-nitro 86
1,2-dimethyl-5-iodo-4-nitro 86
1,2-dimethyl-4-isopropyl-5-nitro 82, 148
1,2-dimethyl-4-nitro 82, 123
1,2-dimethyl-5-nitro 82, 123, 128, 144, 155, 161–163, 165
1,4-dimethyl-2-nitro 123
1,4-dimethyl-5-nitro 82, 107, 123
1,5-dimethyl-2-nitro 91, 123
1,5-dimethyl-4-nitro 82, 107, 122, 123
2,4(5)-dimethyl-5(4)-nitro 82, 119
4,5-dimethyl-2-nitro 91
1,4-dinitro 80
2,4(5)-dinitro 81, 89, 93, 104, 112, 138, 152, 155, 158, 182
4,5-dinitro 80, 81, 93, 122, 138, 140, 156, 157
2,4-dinitro-1-ethyl 113, 114
2,5-dinitro-1-ethyl 113, 144
4,5-dinitro-2-(4-fluorophenyl) 93
2,4(5)-dinitro-5(4)-iodo 93
1,4-dinitro-2-methyl 81
2,4-dinitro-1-methyl 112, 139, 154, 155, 158, 161

2,5-dinitro-1-methyl 112, 154, 155, 161
4,5-dinitro-2-methyl 93, 156
4-dinitromethyl-1-methyl-5-nitro 124
2,4(5)-dinitro-5(4)-trinitromethyl 152
2,4(5)-diphenyl-5(4)-nitro 92, 147
4,5-diphenyl-2-nitro 91
2,4(5)-diphenyl-5(4)-nitroso 92
1-ethyl-2-methyl-5-nitro 162
2-ethyl-4(5)-methyl-5(4)-nitro 82
2-(4-fluorophenyl)-4(5)-nitro 83, 93, 119
2-formyl-1-methyl-5-nitro 123, 127, 129, 145
5-formyl-1-methyl-2-nitro 127, 145, 148, 151
5-halo-1,2-dialkyl-4-nitro 86, 107, 131, 132
2-halomethyl-1-methyl-5-nitro 141
2-halo-4(5)-nitro 112
5-(2-hydroxyanilino)-1-methyl-4-nitro 133, 140
1-(2-hydroxyethyl)-2-hydroxymethyl-5-nitro 109, 166
1-(2-hydroxyethyl)-2-methyl-4-nitro 162
1-(2-hydroxyethyl)-2-methyl-5-nitro 139, 146, 148, 150, 151, 153, 162–166, 182–184
1-(2-hydroxy-3-methoxy)propyl-2-nitro 164–166, 180
2-hydroxymethyl-1-methyl-5-nitro 108
5-hydroxymethyl-1-methyl-2-nitro 148
2-hydroxymethyl-4(5)-nitro 82
4(5)-hydroxymethyl-5(4)-nitro 82, 83
2-(imidazol-1-yl)-1-methyl-5-nitro 115
2-(imidazol-2-yl)-1-methyl-5-nitro 90
2-(imidazol-2-yl)-4(5)-nitro 90
2-iodo-1-methyl-4-nitro 86
4-iodo-1-methyl-5-nitro 85, 111, 133
4(5)-iodo-2-methyl-5(4)-nitro 86, 117
5-iodo-1-methyl-4-nitro 85, 111, 133, 155
2-iodo-4(5)-nitro 85, 155
4(5)-iodo-5(4)-nitro 85, 111, 129
2-isopropyl-1-methyl-5-nitro 144
2-(1-methylimidazol-2-yl)-4(5)-nitro 90
1-methyl-2-(1-methylimidazol-2-yl)-5-nitro 90
1-methyl-2-(1-methyl-4-nitroimidazol-2-yl)-5-nitro 90
1-methyl-2-(1-methyl-5-nitroimidazol-2-yl)-5-nitro 90
1-methyl-2-methylsulfonyl-5-nitro 130, 131, 132, 149
1-methyl-2-morpholino-5-nitro 159
1-methyl-2-nitro 91, 92, 153, 158, 159, 161, 244
1-methyl-4-nitro 81, 110, 151, 154–156, 158, 159, 161, 244
1-methyl-5-nitro 81, 109, 110, 140, 153–155, 158, 159, 161, 244

Subject Index

2-methyl-4(5)-nitro 45, 46, 81, 82, 93, 116–120, 122, 149, 154, 156, 159, 162–164
4(5)-methyl-2-nitro 91, 159
4(5)-methyl-5(4)-nitro 79, 81, 119, 159
2-(1-methyl-5-nitroimidazol-2-yl)-4(5)-nitro 90
2-methyl-4-nitro-1-p-nitrophenyl 152
2-methyl-5-nitro-1-p-nitrophenyl 152
4(5)-methyl-5(4)-nitro-2-(4-nitrophenyl) 83, 84
1-methyl-4-nitro-5-piperidino 130
1-methyl-5-nitro-2-piperidino 130
1-methyl-4-nitro-5-styryl 158
1-methyl-5-nitro-4-styryl 158
1-methyl-4-nitro-2-thioalkyl 88
1-methyl-5-nitro-2-thioalkyl 79, 88, 166
1-methyl-5-nitro-2-(4-thiomethylphenoxy)methyl 166
1-methyl-5-nitro-4-trinitromethyl 124, 157
4-methyl-2-nitro-5-trinitromethyl 89
1-nitro 80
2-nitro 45, 46, 79, 81, 91, 117, 118, 121, 138, 149, 152, 157–159, 165, 181, 244
4(5)-nitro 45, 46, 79, 80, 81, 93, 106, 110, 115, 118–123, 147, 149, 151, 152, 154–158, 161, 162, 244, 246
4(5)-nitro-2-(3-nitro-4-fluorophenyl) 84, 93
4(5)-nitro-2-(4(5)-nitroimidazol-2-yl) 89
4(5)-nitro-2-(4-nitrophenyl) 83
2-nitro-1-phenyl 92
4(5)-nitro-2-phenyl 83
5-nitro-1-phenyl 89
4(5)-nitro-5(4)-phenylazo 141
4(5)-nitro-1-(1-phenyl-3-methyl-4-nitropyrazol-5-yl) 115
5-nitro-1-phenyl-2-thiomethyl 79
4(5)-nitro-5(4)-pyrrolidinomethyl 155
4-nitro-1,2,5-trialkyl 82, 123
4(5)-nitro-2-(3,4,5-trichlorophenyl) 84
4(5)-nitro-2-trideuteriomethyl 89
5-nitro-1,2,4-triiodo 87, 93
5-nitro-1,2,4-trimethyl 123
2-nitro-1-trimethylsilyl 121
2-nitro-1-trityl 92
2,4,5-trinitro 93, 152
Imidazo[2,1-d]-1,4-oxazepine 106
Imidazo[1,2-b]pyridazine
    6-chloro-2-methyl-3-nitro 148
    2-methyl-3-nitro-6-phenyl-1-oxide 101
Imidazo[1,2-a]pyridines
    2-carboethoxy-6-chloro-3-nitro 97
    2-carboethoxy-6-methyl-3,5-dinitro 97
    2-carboethoxy-6-methyl-3-nitro 97, 162
    2-carboethoxy-8-methyl-3-nitro 97
    5-carboethoxy-6-methyl-3-nitro 97
    2-carboethoxy-3-nitro 97, 140
    5-carbomethoxy-3-nitro 97
    2-(3-carboxyacryloyl)-3-nitro 98
    2-[α-chloro-β-(3,5-dinitro-2-furyl)vinyl]-3-nitro 99
    2-chloro-3-nitro 97, 133
    6-chloro-3-nitro 97
    2-dimethylamino-3-nitro 133
    5-ethoxy-3,6-dinitro 97, 162
    5-ethoxy-3,8-dinitro 97
    5-ethoxy-8-nitro 97, 162
    6-methyl-3-nitro 97
    8-methyl-3-nitro 97
    3-nitro 97, 140
    3-nitro,5,5'-dehydrodimer 99
    3-nitro-2-phenyl 97
Imidazo[1,5a]pyridines
    3-bromo-1-nitro 100
    1,3-dinitro 99
    1-methyl-3-nitro 99
    3-methyl-1-nitro 99
    1-nitro 99
    1-nitro-3-phenyl 99
    3-nitro-1-phenyl 99
Imidazo[1,2a]pyrimidine
    2-maleyl-3-nitro 144
    8-methyl-3-nitro-5,6,7,8-tetrahydro 101
    3-nitro-2-(5-nitro-2-furyl) 144
Imidazo[4,5-d]pyrimidine 102
    see Purine
Imidazo[1,2-a]pyrimidin-5-one
    6-carboethoxy-4,5-dihydro-1-methyl-2-nitro 104, 124
    7-carbomethoxy-4,5-dihydro-1-methyl-2-nitro 104, 124
Imidazo[5,1-b]thiazoles
    7-nitro-3-(4-pyridyl) 103
    7-nitro-2-trifluoroacetyl-3-(4-pyridyl) 103
    7-nitro-3-trifluoromethyl 103
Imidazo[2,1-b]thiazole
    2,3-dihydro-5-nitro 96
    5-nitro-6-aryl 95
    5-nitro-6-chloro 95, 149
    5-nitroso-6-aryl 96
Imidazo[1,2-f]xanthine
    6,8-dimethyl-1,2-disubstituted-3-nitro 103
Imidazoylnitroimidazoles 85
Indazole
    3-alkylamino-5-nitro 233, 234
    3-alkylamino-6-nitro 233, 234
    3-Chlorodinitro 233
    2,4-dinitro 216–218
    2,5-dinitro 216–218, 233
    2,6-dinitro 216–218, 233
    2,7-dinitro 216–218
    3,4-dinitro 216–218
    3,5-dinitro 216–218, 233
    3,6-dinitro 216–218, 233
    3,7-dinitro 216–218
    2-nitro 216–218

3-nitro 216–218
4-nitro 216–218
5-nitro 216–218
6-nitro 216–218
7-nitro 216–218
2,3,4,6-tetranitro 216–218
2,3,5,6-tetranitro 216–218
2,3,6-trinitro 216–218
2,4,6-trinitro 216–218
2,5,6-trinitro 216–218
3,4,6-trinitro 216–218
3,5,6-trinitro 216–218
3,5,7-trinitro 216–218
Indole
   3-acetyl-1-methyl-5-nitro-2-phenyl 29, 30
   3-acetyl-1-methyl-6-nitro-2-phenyl 29, 30
   2-anilino-3-nitro 25
   1,2-dimethyl-3,5-dinitro 28
   1,2-dimethyl-3,6-dinitro 28
   1,2-dimethyl-3,4,6-trinitro 28
   1,2-dimethyl-3,5,6-trinitro 28
   3,4-dinitro-2-methyl 27
   3,5-dinitro-2-methyl 27
   3,6-dinitro-2-methyl 27
   3,4-dinitro-1-methyl-2-phenyl 29, 30
   3,5-dinitro-1-methyl-2-phenyl 29, 30
   3,6-dinitro-1-methyl-2-phenyl 29, 30
   3-formyl-1-methyl-1-phenyl-5-nitro 29
   2-methyl-3,4,6-trinitro 27
   3-nitro-1-methyl-2-phenyl 29, 30
   2-phenyl-3,6-dinitro 28, 29
   2-phenyl-3-nitro 28, 29, 31
Indolizine
   2-aryl 35, 36
   1,3-dinitro-2-phenyl 35
   2-methyl-1-nitro 35
   2-methyl-3-nitro 35
   1-nitro 27
   1-nitro-2-methyl-6-cyano 27
   1-nitro-2-*p*-nitrophenyl 35, 71
   3-nitro-2-*p*-nitrophenyl 35, 71
Infrared spectroscopy
   nitroimidazoles 156
   nitroisoxazoles 336
   nitrooxadiazoles 343
   nitrpyrazoles 251
   nitropyrroles 67
   nitrotetrazoles 293
   nitrotriazoles 293
Ipronidazole 166
1-Isopropyl-3-methyl-4-nitropyrazol-
   5-ylthioacetic acid 231
Isoxazole
   5-acetamido-3-methyl-4-nitro 304
   5-acetamido-4-nitro-3-phenyl 304, 323
   5-acetyl-3-nitro 334, 335
   3-acyl-5-alkyl-4-nitro 316
   3-acyl-4-nitro 315, 316
   5-alkyl-3-methyl-4-nitro 329, 330

3-alkyl-4-nitro 314
5-alkyl-3-nitro 314, 315, 318, 319
3-alkyl-4-nitro-5-substituted 340
3-amino-5-methyl-4-nitro 304
5-amino-4-nitro 319, 320, 332
5-amino-4-nitro-3-phenyl 314, 315, 332
3-anilino-5-(2,4-dinitrophenyl)-4-nitro 304
3-anilino-4-nitro-5-phenyl 304
3-aryl-5-dialkylamino-4-nitro 314, 316
3-benzoyl-4-nitro-5-phenyl 314, 315
3-*tert*-butyl-4-nitro 302
5-*tert*-butyl-4-nitro 302
3-*tert*-butyl-4-nitro-5-pyrollidino 314, 316
2,5-diaryl-4-nitro 322
4,5-diaryl-4-nitro-3-phenyl 320
3-(4-chlorophenyl)-4-nitro 303
3,5-diaryl-4-nitro 318, 319, 323
5-(1,3-diaryl-1-oxobutyl)-3-methyl-4-nitro
   330, 338
3-dichloromethyl-5-methyl-4-nitro 302
5-dichloromethyl-3-methyl-4-nitro 302
3,5-dimethyl-4-nitro 215, 301, 303,
   328–330, 333, 336–339
3,5-dinitro 316, 317, 322, 323
3,5-dinitro-4-methyl 317, 322, 323
3-(2,4-dinitrophenyl)-4-nitro 302
3,5-diphenyl-4-nitro 303, 314, 315, 323
5-formyl-3-methyl-4-nitro 330
5-hydroxy-3-methyl-4-nitro 331
3-hydroxy-5-methyl-4-nitro 340
3-hydroxy-4-nitro 320
5-(isoxazol-5-yl)-4-nitro 303
5-methyl-3-(5-methyl-4-nitroisoxazol-3-yl)-
   4-nitro 215, 318, 333
3-methyl-4-nitro 301, 323
5-methyl-4-nitro 214, 332, 336
5-methyl-4-nitro-3-phenyl 314, 315
3-methyl-4-nitro-5-styryl 328, 329, 331, 332
   dimers 331
3-methyl-4-nitro-5-substituted 330, 337, 338
5-methyl-4-nitro-3-succinimido 304
3-nitro 318, 319, 323, 324
4-nitro 301, 319, 332 337, 340
3-nitro-5-(3-hydroxyisoxazol-5-yl) 328, 334
3-nitro-5-(3-nitroisoxazol-5-yl) 323, 324,
   328, 334
4-nitro-5-(4-nitroisoxazol-5-yl) 303
4-nitro-3-(4-nitrophenyl) 302
4-nitro-5-(4-nitrophenyl) 302
3-nitro-5-(3-nitro-4-phenylisoxazol-5-yl)-
   4-phenyl 314, 315
3-nitro-5-phenyl 314, 315, 318, 319, 328
4-nitro-3-phenyl 303, 323, 336, 337, 339
4-nitro-5-phenyl 302
5-nitro-3-phenyl 313, 314, 316
3,4,5-triphenyl 313
Isoxazolidine
   3-benzoyl-2,5-diphenyl-4-nitro 306
   2-*tert*-butyl-4-cyano-5-nitro-3-phenyl 305

2-*tert*-butyl-4-cyano-5-nitro-3-phenyl 305, 334
2-*tert*-butyl-5-cyano-4-nitro-3-phenyl 305
cis- and *trans*-2-*tert*-butyl-4-nitro-3-phenyl 305
2-*tert*-butyl-5-nitro-3-phenyl 305
5-chloro-2-methyl-4-nitro-3-phenyl 305, 324
cis- and *trans*-4-cyano-2-methyl-5-nitro-3-phenyl 305, 334
5-cyano-2-methyl-4-nitro-3-phenyl 305
5,5-dimethyl-3,3-dinitro-2-methoxy 337, 338
3,3-dinitro-2-alkoxy-5-substituted 306, 307
3,3-dinitro-2-isopropoxy-5-methyl 308, 333
3,3-dinitro-2-methoxy 325, 326, 336, 337
3,3-dinitro-5-phenyl-1-(α-phenyl-β-nitroethoxy) 310, 334
3,3-dinitro-2-(x-nitroalkoxy)-4-substituted-5-substituted 310–313, 334
3,3-dinitro-5-phenyl-2-trimethylsilyloxy 308
cis- and *trans*-2-methyl-4-nitro-3-phenyl 305, 324
3-nitro-5-phenyl-2-trimethylsilyloxy 306
3-nitro-2-trimethylsilyloxy 306
4-nitro-2,3,5-triphenyl 306
$\triangle^2$-Isoxazoline
5-alkyl-5-aryl-4-nitro-3-phenyl 320
4-alkyl-3-nitro-2-oxide 317
3-aryl-5-aryl(alkyl)alkyl-4-nitro 318, 319
4-arylazo-3,5-diphenyl-4-nitro 323
4-aryl-3-cyano-5-nitro-2-oxide 317
3-aryl-5,6-dialkyl-4-nitro 318, 319
5-aryl-4-(3-methyl-4-nitroisoxazol-1-yl)-3-phenyl 321
3-aryl-5-nitro 313
5-*n*-butyl-3,5-dimethyl-4-nitro 329
3-cyano-4-ethyl-5-nitro-2-oxide 317
3,5-di-(4-methoxytetrafluorophenyl)-4-nitro 323
4,5-dimethyl-5-nitro-3-phenyl 313, 314
5,5-dimethyl-4-nitro-3-phenyl 313, 314, 328
3,5-dinitro-4-methyl-2-oxide 317, 322, 323
3,5-dinitro-2-oxide 317, 322, 323
3,5-diperfluorophenyl-4-nitro 323
3,5-diphenyl-4-nitro 313, 319, 323
4-ethoxy-3-nitro-2-oxide 317
4,4-fluorenylidino-3-nitro-2-oxide 320, 321
5,5-fluorenylidino-3-nitro-2-oxide 320, 321
4-hydroxy-3-nitro-2-oxide 327, 336
5-imino-4-nitro 319, 320
4-methyl-3-nitro-2-oxide 317, 336
5-methyl-3-nitro-2-oxide 308, 309, 336
5-methyl-4-nitro-3-phenyl 313, 314
5-methyl-5-nitro-3-phenyl 313, 314, 323
5-methyl-5-nitro-3-phenyl 314, 323
3-nitro 306–308, 318, 327, 339
4-nitro-5-nitrophenyl-3-phenyl 313
3-nitro-2-oxide 320–322, 336, 339
4-nitro-5,5-pentamethylene-3-phenyl 328
3-nitro-5-phenyl 306, 307, 327
5-nitro-3-phenyl 314, 323
3-nitro-4-phenyl-2-oxide 308, 309
3-nitro-5-phenyl-2-oxide 308, 309, 320, 321
3-nitro-2-oxide 308–310
3-nitro-5-substituted 327, 328
3-nitro-4,5-tetramethylene-2-oxide 308, 309
3-nitro-5-trimethylsilylalkyl 306, 307, 328

Lithioimidazoles, nitration of 92

Mannich bases 109
Mannich reactions 226, 329
Mass spectroscopy
 nitroimidazoles 159–162
 nitroisoxazoles 338–344
 nitropyrazoles 254–256
 nitropyrroles 67
Meisenheimer adducts 49, 58
Metal salts of nitropyrazoles 241
Methoxatin 71
Methyl 3,5-dinitropyrrole-2-carboxylate 8
Methyl 3,5-dinitropyrrole-2,4-dicarboxylate 20
1-Methyl-4-hydroxy-3-nitro-1*H*-pyrazolo[3,4-*b*]pyridine-5-carboxylic acid 219
Methyl 1-methyl-2-nitroimidazole-2-carboxylate 109
Methyl 1-methyl-3-nitropyrazole-4-carboxylate 202, 203
Methyl 1-methyl-4-nitropyrrole-2-carboxylate 41
Methyl 1-methyl-3-nitro-1,2,4-triazole-5-carboxylate 277, 278
Methyl 2-methoxy-3-nitroisoxazolidine-5-carboxylate 306
Methylnitrofuroxan 341–343
1-Methyl-5-nitroimidazole-2-hydroxamoyl chloride 124, 125
1-Methyl-4-nitroimidazole-5-sulfonamide 155
1-Methyl-5-nitroimidazole-4-sulfonamide 155
2-Methyl-5-nitroimidazol-1-ylacetic acid 166
β-(1-methyl-2-nitroimidazol-5-yl)acrolein 127
1-Methyl-5-nitroimidazol-2-ylnitrile oxide 125
Methyl 3-nitroisoxazole-5-carboxylate 318, 219
3-Methyl-4-nitroisoxazole-5-carboxylic acid 331
Methyl 3-nitro-$\triangle^2$-isoxazoline-5-carboxylate 306, 307, 323, 327
Methyl 3-nitro-$\triangle^2$-isoxazoline-2-oxide-5-carboxylate 309, 310

3-Methyl-4-nitro-1-*p*-nitrophenyl-$\triangle^3$-
pyrazoline-5-one 198, 247
1-Methyl-3-nitro-4-(3-oxobutyl)-1,2,4-
triazolin-5-one 281
4-Methyl-3-nitro-4-(3-oxobutyl)-1,2,4-
triazolin-5-one 281
2-Methyl-4-nitro-3-phenyl-$\triangle^4$-isoxazoline 305
3-Methyl-4-nitro-1-phenyl-$\triangle^2$-pyrazolin-5-
one 234
3-Methyl-4-nitro-1-phenyl-$\triangle^3$-pyrazolin-5-
one 197
1-Methyl-5-nitropyrazole-4-carboxamide 202
3(5)-Methyl-4-nitro-5(3)-pyrazolecarboxamide
239
3(5)-Methyl-4-nitropyrazole-5(3)-
carboxhydrazides 235
1-Methyl-5-nitropyrazole-4-carboxylic acid
253
3(5)-Methyl-4-nitropyrazole-5(3)-carboxylic
acid 195, 196, 248, 240
3(5)-Methyl-4-nitropyrazole-3-diazonium salt
240
3-Methyl-4-nitropyrazole-3,5-dicarboxylic acid
240, 241
1-Methyl-5-nitropyrazole-2-oxide 199, 200,
203, 230
3(5)-Methyl-4-nitro-5(3)-pyrazoleoximino-
carboxamide 235
3-Methyl-4-nitro-$\triangle^3$-pyrazolin-5-one 197
5-Methyl-4-nitro-$\triangle^4$-pyrazolin-3-one 211, 212
3(5)-Methyl-4-nitro-5(3)-(pyrazol-1-yl)-
pyrazoles 233
Methyl 4-nitropyrrole-2-carboxylate 60, 66
Methyl 5-nitropyrrole-2-carboxylate 66
1-Methyl-4-nitropyrrole-2-carboxylic acid 7,
40, 41, 47
2-Methyl-5-nitropyrrole-3-carboxylic acid 48
1-Methyl-3-nitropyrrolidin-2-one 22
1-Methyl-5-nitrosaminopyrazole-4-carboxylic
acid 202
Methyl 3-nitro-4-succinimido-$\triangle^2$-isoxazoline-
5-carboxylate-2-oxide 317
Methyl 3a-nitro-3,3a,4,5-tetrahydro-2*H*-
isoxazolo[2,3-*b*]isoxazole-2-carboxylate
309, 310
Methyl 3-nitro-1-(2,3,5-tri-O-acetyl-β-D-
ribofuranosyl)-5-carboxylate 287
1-Methyl-3-nitro-1,2,4-triazole 278
Methyl 3-nitro-1,2,4-triazole-1-ylacetate 278
Methyl 3-nitro-2-trimethylsilyloxyisoxazo-
lidine-5-carboxylate 306
Metronidazole
see 1-(2-hydroxyethyl)-2-methyl-5-
nitroimidazole
Michael reactions 206, 226, 278, 281, 285, 329
Misonidazole
see 1-(3-methoxy-2-hydroxy)propyl-2-
nitroimidazole

MO calculations
nitroimidazoles 97, 123, 295, 336
nitroisoxazoles 336
nitropyrazoles 187, 197, 220, 257, 295, 336
nitropyrroles 39, 68, 295
nitrotetrazoles 295
nitrotriazoles 294, 295

Nimorazole 165
Nitration/nitrolysis reagent
acetyl nitrate 1–21, 27–39, 84, 88–90, 93,
96–103, 191–193, 216–220, 269, 271,
276, 301–304
alkyl nitrate 3, 22, 27, 31, 92, 96
amyl nitrite 21, 22, 91, 197
benzoyl nitrate 101, 340
dinitrogen pentoxide 2, 270
dinitrogen tetroxide 2, 89, 92, 340
inorganic nitrates in sulfuric acid,
see mixed acid
mixed acid 15–21, 27–39, 80–94, 95–103,
186–200, 217–223, 269, 270, 276,
301–304, 340, 342
nitric acid 6, 10, 12, 14–21, 27–39, 81–88,
93, 96, 100, 102, 189, 196–198, 218,
270, 271, 342
nitric acid in acetic acid,
see acetyl nitrate
nitric acid in acetic anhydride,
see acetyl nitrate
nitric acid in chloroform 98
nitric acid in methanol 98
nitric acid in oleum 6, 83, 93, 194, 304
nitric acid in phosphoric acid 98
nitric acid in propionic anhydride 33
nitric acid in sulfuric acid,
see mixed acid
nitric acid in trifluoroacetic acid 33, 37,
88, 97, 190
nitronium tetrafluoroborate 34, 89, 189,
271, 301
nitrous acid/inorganic nitrites 21, 28,
96, 102
Nitration/nitrolysis substrate
azaindoles 33–35
imidazo[1,2-*a*]benzimidazoles 102
imidazo[1,2*a*]imidazoles 96
imidazoles 80–93
imidazo[1,2-*b*]pyridazines 101
imidazo[1,2-*a*]pyridines 97
imidazo[1,5-*a*]pyridines 99
imidazo[1,2-*a*]pyrimidines 101
imidazo[4,5-*d*]pyrimidines 102
imidazo[1,2-*b*]thiazoles 95
imido[1,2-*f*]xanthines 103
indazoles 216
indoles 27–32
indolizines 35–39

isoxazoles 301–304
oxazoles 340
pyrazoles 186–200
4H-pyrazolo[1,5-a]benzimidazoles 222
2H-5H-pyrazolo[4,3-c]pyrazoles 222
pyrazolo[1,5-a]pyridines 218
pyrazolo[3,4-b]pyridines 219
pyrazolo[4,3-b]pyridines 219
pyrazolo[1,5-a]pyrimidines 220
pyrazolo[3,4-d]pyrimidines 219
pyrazolo[1,5-c]quinazolines 222
pyrazolo[1,5-a]-s-triazines 222
pyrroles 1–21
5H-pyrrolo[2,1-a]isoindoles 39
pyrrolo[2,1-b]thiazoles 39
thiazolo[2,3,4-c,d]pyrrolizines 39
triazoles 268–271
1,2,3-triazolo[1,5-a]pyridines 276
3-Nitro-2-azacyclopentadienone hydrazone 71
8-Nitrocaffeine 79, 91, 102
Nitrodecarboxylation 194
6-Nitro-2,3-dihydro-1H-imidazolo[1,2-b]-
  pyrazol-5-one 210, 211
2-Nitro-5,6-dihydro-4H-imidazo[3,2-b]1,2,4-
  triazole 290, 291
3-Nitro-4,5-dihydro-6H-1,2-oxazine-2-oxide
  310
Nitro 5,6-dihydrooxazolo[3,2-b]-1,2,4-triazoles
  278, 279
5(4)-Nitrohistamine 82
5(4)-Nitrohistidine 82
Nitroimidazolealdehyde derivatives 125–129
ω-2-Nitroimidazole-5-alkyl carbamates 142, 143
ω-2-Nitroimidazole-5-alkyl carbonates 142, 143
ω-2-Nitroimidazole-5-alkyl carboxylates 142,
  143
ω-5-Nitroimidazole-1-alkyl phosphates 142
Nitroimidazole carboxamides 142
4(5)-Nitroimidazole-5(4)-carboxylic acid 108,
  122
4(5)-Nitroimidazole-5(4)-carboxylic acid azide
  141
2-Nitroimidazole-4,5-dicarboxylic acid 94
4(5)-Nitroimidazole-5(4)sulfonamide 150
4(5)-Nitroimidazole-5(4)sulfonic acid 140, 182
Nitroimidazolidines 80
Nitroimidazolines 80
Nitroimidazolyl disulfides 135–137
1-(5-Nitroimidazol-2-yl)-2-imidazolidinone 94
Nitroimidazolyl mercaptans 135–137
Nitroimidazolyl nucleosides 150
Nitroimidazolyl sulfides 135, 137, 139, 145
3-Nitroimidazo[1,2-b]-1,2,4-triazine derivatives
  163
Nitroisoxazoylureas 340
3-Nitro-1-(3-oxobutyl)-1,2,4-triazolin-5-one 285
3-Nitro-4-phenylfuroxan 341, 343, 344
4-Nitro-3-phenylfuroxan 341–344

4-Nitro-5-phenylisoxazole-3-carboxylic acid
  332
4-Nitro-3-phenyl-1,2,3-oxadiazolium-5-olate
  342, 344
2-Nitro-1-phenyl-1-trinitromethylcyclohexane
  311
4-Nitro-4′-picrylaminobifurazan-3,3′-yl 342
3-Nitro-1-picryl-1,2,4-triazol-5-one 279
Nitropolyzonamine 72
Nitroporphyrin 77
Nitropropanes 203
4-Nitro-3(5)-n-propylpyrazole-5(3)-carboxylic
  acid 195, 196, 235
4-Nitropyrazole-3(5)-carboxylic acid 237,
  241
Nitropyrazolidines 203–207, 256
Nitropyrazolines 203–207, 248
4-Nitro-△⁴-pyrazolin-3-one 211
1-Nitropyrazolo[1,5-c]quinazoline 222
4-Nitropyrazolo[1,5-a]-s-triazines 222
1-(4-Nitropyrazolyl)alkane carboxylic acids 258
β-(4-Nitropyrazol-1-yl)propionic acid 226
4-Nitro-3(5)-(pyrazol-1-yl)pyrazoles 233, 235
4-Nitropyrrole-2-carboxylic acid 40, 48, 69
5-Nitropyrrole-2-carboxylic acid 40, 69
2-Nitropyrrole-4-glyoxalic acid 47
Nitropyrrolenines 57
Nitropyrroles, radical reactions 59
Nitropyrrolidines 1
Nitropyrroline 1, 50, 63
5-Nitropyrrolo[2,3-d]pyrimidines 63
Nitrosoazoles, conversion to nitroazoles 21, 22,
  28–30, 84, 93, 197, 200, 203, 218
Nitrosoindoles 28
Nitrosopyrroles 2
cis-6-Nitro-8-substituted-2,9-dioxa-1-
  azabicyclo[4,3,0]nonane 310, 324, 325
3-Nitro-1,2,4-triazole anions 284, 285
3-Nitro-1,2,4-triazole-5-carboxylic acid 272,
  273, 291
Nitrotriazolines 273, 274
3-Nitro-1,2,4-triazolin-5-one 270, 281, 288,
  292, 293
  1,4-dimethyl 270, 292, 293
  1-methyl 270, 280, 281, 283–285
  4-methyl 270, 280, 281
  1-(4-nitrobenzyl) 270
  4-(4-nitrobenzyl) 270
  1-(4-nitrophenyl) 270
  4-(4-nitrophenyl) 270
3-Nitro-1,2,3-triazolo[1,5-a]pyridine 276, 288
3-Nitro-1,2,3-triazol-1-ylacetonitrile 278
Nitro-1,2,4-triazol-1-ylalkanones 278, 285, 286
Nuclear magnetic resonance spectroscopy
  nitroimidazoles 154–156
  nitroisoxazoles 337, 338
  nitrooxadiazoles 343, 344
  nitropyrazoles 248–251

nitropyrroles 64–67
nitrotriazoles 293, 294

Ornidazole 165
Oxidation of aminoazoles
   aminoimidazoles 94, 95
   aminooxadiazoles 341
   aminopyrazoles 203
   aminotriazoles 275
Oxidation by a nitrating reagent 6–13, 17, 20, 21, 27, 34, 47, 48, 82, 83, 98–101, 144, 213, 311
Oxidative dimerization 7
Oxygenation of Nitrenes 46

Pentamine(4-nitroimidazolato)cobalt(III)-perchlorate 81
Photoelectron spectroscopy of nitroimidazoles 162
Photo properties and reactions
   nitroimidazoles 163
   nitropyrazoles 246
   nitropyrroles 60
   nitrostyrylisoxazoles 331
Picrolonic acid 198, 247
1-Picryl-Δ^2-pyrazoline 205, 245
1-Propionyl-2,4-disubstituted-3-nitro-pyrrolidine 33
Propylphenazone 197
Purine
   1,3-dimethyl-8-nitro 102
   3,7-dimethyl-8-nitro 102
   9-methyl-8-nitro 102
   1,3,7-trimethyl-8-nitro 79, 91, 102
Pyrazole
   3-acetamido-1-methyl-4-nitro 251
   3-acetamido-4-nitro 199
   1-acyl-3-anilino-5-dimethylamino-4-nitro 229
   1-acyl-3-chloro-5-methyl-4-nitro 228, 229
   3-alkylamino-5-dialkylamino-4-nitro 209, 210, 237, 252
   3(5)-alkylcarbamoyl-5(3)-methyl-4-nitro 236
   1-alkyl-3,5-dimethyl-4-nitro 215
   5-amino-3-β-aminoethylamino-4-nitro 209
   5-amino-3-benzylamino-1-methyl-4-nitro 209
   5-amino-3-benzylamino-4-nitro 209
   5-amino-1,3-dimethyl-4-nitro 214, 215, 252
   4-amino-3,5-dinitro-1-methyl 231, 232
   3-amino-1-methyl-4-nitro 249, 250, 252
   5-amino-1-methyl-4-nitro 199, 249, 250, 252
   5(3)-amino-3(5)-methyl-4-nitro 199, 221
   5-amino-3-methyl-4-nitro-1-phenyl 214, 215
   5-amino-3-methyl-4-nitro-1-substituted 332, 333
   3(5)-amino-4-nitro 211, 212, 221, 222, 240, 252
   5-amino-4-nitro-1-phenyl 199
   3(5)-(2-aminophenyl)-4-nitro 215, 216, 222
   3(5)-anilino-5(3)-dimethylamino-4-nitro 229
   4-anilino-3,5-dinitro-1-methyl 231, 232
   3-arylamino-5-dialkylamino-4-nitro 209, 210, 237, 252
   1-aryloxyacyl-3,5-dimethyl-4-nitro 228, 229
   3(5)azido-4-nitro 240
   5-benzoyl-1,3-dimethyl-4-nitro 239
   5-benzoyl-1-ethyl-3-methyl-4-nitro 235
   5-benzylamino-3-methyl-4-nitro-1-(4-nitrophenyl) 231
   1-benzyl-3-cyano-5-methyl-4-nitro 225
   1-benzyl-3-cyano-4-nitro-(2,3,5-O-triacetyl-β-D-ribofuranosyl) 225
   1-benzyl-5-cyano-4-nitro-3-(2,3,5-O-triacetyl-β-D-ribofuranosyl) 225
   4-bromo-3,5-dimethyl-1-nitro 202
   4-bromo-3,5-dinitro-1-methyl 194, 195, 231
   4-bromo-3,5-dimethyl-1-(3-methyl-nitro-1-phenylpyrazol-5-yl) 230
   5-bromo-1,3-dimethyl-4-nitro 194, 195, 252
   3(5)-bromo-5(3)-methyl-4-nitro 230
   4-bromo-1-nitro 232
   4-bromo-3(5)-nitro-5(3)-(3-pyridyl) 230
   1-butyl-3(5)-methyl-4-nitro 257
   3(5)-tert-butyl-4-nitro 187, 242, 243, 252
   3(5)-tert-butyl-5(3)-nitro 201, 243, 252
   3-carbamoyl-4-nitro-1-(2,3,5-trihydroxy-β-D-ribofuranosyl) 228, 235
   5-carbamoyl-4-nitro-1-(2,3,5-trihydroxy-β-D-ribofuranosyl) 228, 235
   4-chlorocarbonyl-1-methyl-3-nitro 240
   5-chloro-1-isopropyl-3-methyl-4-nitro 231
   3(5)-chloro-5(3)-methyl-4-nitro 230, 257
   5-chloro-1-methyl-4-nitro 230
   5-chloro-3-methyl-4-nitro-1-(4-nitrophenyl) 231
   5-chloro-3-methyl-4-nitro-1-phenyl 115, 227, 231
   1-chloro-4-nitro 229
   5-chloro-4-nitro 208
   1-(chlorophenyl)-4-nitro 208, 236
   5-cyano-1-ethyl-3-methyl-4-nitro 235
   3-cyano-1,5-dimethyl-4-nitro 225
   5-cyano-1,3-dimethyl-4-nitro 225, 257
   4-cyano-1-methyl-3-nitro 246, 253
   3(5)-cyano-5(3)-methyl-4-nitro 225
   5-cyano-3-methyl-4-nitro-1-phenyl 231
   4-cyano-1-methyl-3-nitroso 246

4-cyano-3(5)-nitro 202, 203, 225, 228
3(5)-cyano-4-nitro-3(5)-(2,3,5-tri-O-acetyl-
 β-D-ribofuranosyl) 234, 235
3(5)-deuteriomethyl-4-nitro 187
3-diacetylamino-1-methyl-4-nitro 251
5(3)-dialkylamino-3(5)-methyl-4-nitro 232
3(5)-dialkylamino-4-nitro 232
3,5-diamino-4-nitro 240
1-(2,4-dichlorophenyl)-4-nitro 208
3(5)-dideuteriomethyl-4-nitro 187
3(5)-(4-diethylaminiophenylazo-4-nitro-
 5(3)-phenyl 225, 226
3,5-diethyl-1-nitro 192
3,5-diethyl-4-nitro 192, 236, 237
1-dimethylcarbamoyl-4-nitro 228, 229
3,5-dimethyl-1-dimethylcarbamoyl-4-
 nitro 228, 229
1,2-dimethyl-5-nitro 257
1,5-dimethyl-3,4-dinitro 191, 235, 252
3,5-dimethyl-1,4-dinitro 192, 244
1,3-dimethyl-5-hydroxy-4-nitro 252
3,5-dimethyl-1-(3-methyl-4-nitro-1-
 phenylpyrazol-5-yl) 230
1,3-dimethyl-4-nitro 187, 235
1,4-dimethyl-3-nitro 187, 193
1,4-dimethyl-5-nitro 187
1,5-dimethyl-4-nitro 187, 235
3,5-dimethyl-1-nitro 192
3,5-dimethyl-4-nitro 187, 192, 196, 209,
 225, 227, 228, 236, 237, 242, 252, 256
3,5-dimethyl-4-nitro-1-perdeuteriomethyl
 336
1,5-dimethyl-4-nitro-3-perfluoroheptyl 188
1-(2,6-dimethyl-3-nitrophenyl)-4-nitro 192
3,5-dimethyl-1-phenylmercury-4-nitro 251
1-(2,6-dimethylphenyl)-4-nitro 192
3,5-dimethyl-4-nitro-1-substituted 333
3,5-dimethyl-4-nitro-1-vinyl 227, 253
1,3-dinitro 232
1,4-dinitro 239
3,5-dinitro 201, 243, 252
3(5),4-dinitro 200, 237, 240, 243, 252
3,5-dinitro-4-ethyl 243, 252
3,5-dinitro-4-halo-1-methyl 194
1,4-dinitro-3-methyl 192, 232, 233
3,4-dinitro-1-methyl 187, 194
3,5-dinitro-1-methyl 200
3(5),4-dinitro-5(3)-methyl 243, 252
1,4-dinitro-3-(4-nitropyrazol-1-yl) 192, 193
3(5),4-dinitro-5(3)-phenyl 243, 252
1,5-di(4-nitrophenyl)-3-methyl-4-nitro
 194, 195
1-(2,4-dinitrophenyl)-3-nitro 227
1-(2,4-dinitrophenyl)-4-nitro 188, 227
1,5-di(4-nitrophenyl)-4-nitro-3-
 perfuoroheptyl 191
1-(2,4-dinitrophenyl)-4-nitro-3-
 ribofuranosyl 193, 194
1,4-dinitro-3-(pyrazol-1-yl) 192, 193

3(5),4-dinitro-5(3)-(3-pyridyl) 191, 236
1,4-dinitro-3-(2,3,5-tri-O-acetyl-β-D-
 ribofuranosyl 233
1,5-diphenyl-3-ethyl-4-nitro 206
1,5-diphenyl-3-methyl-4-nitro 192
1,3-diphenyl-4-nitro 208
3,5-diphenyl-4-nitro 204, 205, 236, 237, 252
1-ethyl-3-nitro 244
1-ethyl-4-nitro 244
1-ethyl-5-nitro 244
3(5)-ethyl-4-nitro 195
4-ethyl-3(5)-nitro 201, 243, 252
4-ethyl-1-methyl-3-nitro 193
4-ethyl-3-nitro-1-phenyl 194, 195
3(5)-fluoro-4-nitro 195, 235
4-formyl-1-methyl-3-nitro 240
3(5)-hydrazino-5(3)-methyl-4-nitro 230,
 240
1-β-hydroxyethyl-4-cyano-3-nitro 225
1-hydroxymethyl-3,5-dimethyl-4-nitro 226
1-hydroxymethyl-3(5)-methyl-4-nitro 226
3(5)-hydroxymethyl-5(3)-methyl-4-nitro
 244
3-hydroxy-5-methyl-4-nitro-1-
 (4-nitrophenyl) 197, 198
5-hydroxy-3-methyl-4-nitro-1-
 (4-nitrophenyl) 197, 198
1-hydroxymethyl-3(5)-nitro 226
3-hydroxy-5-methyl-4-nitro-1-phenyl 197,
 198, 255
5-hydroxy-3-methyl-4-nitro-1-phenyl 197,
 251
3(5)-hydroxy-4-nitro 209, 235
4-hydroxy-3-nitro-1-phenyl-5-phenylazo
 211
3-methoxy-5-methyl-4-nitro-1-
 (4-nitrophenyl) 197, 198
1-methyl-4-iodo-3-nitro 229, 230
3(5)-methyl-5(3)-methylcarbamoyl-4-nitro
 196, 235
3(5)-methyl-4-(1-methyl-5-nitroimidazol-
 2-yl)-5(3)-nitro 201
5-methyl-3-(5-methyl-4-nitropyrazol-3-yl)-
 4-nitro 333, 334
1-methyl-3-nitro 187, 223, 224, 226, 230,
 244, 246, 250, 255, 257
1-methyl-4-nitro 187, 195, 208, 216, 230,
 235, 239, 244, 246, 252, 255
1-methyl-5-nitro 200, 223, 224, 226, 230,
 244, 246, 250, 255, 257
3(5)-methyl-4-nitro 187, 194, 201, 237,
 241–243, 252
3(5)-methyl-5(3)-nitro 201, 243, 252, 255
4-methyl-3(5)-nitro 201, 255
5-methyl-1-nitro 255
1-methyl-3-nitro-4-(2-nitrophenyl) 193,
 194
1-methyl-3-nitro-4-(4-nitrophenyl) 193,
 194

4-methyl-3-nitro-1-(4-nitrophenyl) 193
1-methyl-4-nitro-5-(4-nitrophenyl)-3-
  perfluoroheptyl 191
5-methyl-4-nitro-3-perfluoroamyl 188
5-methyl-4-nitro-3-perfluoroheptyl 188
5-methyl-4-nitro-3-perfluoropropyl 188
1-methyl-3-nitro-4-phenyl 194, 250
1-methyl-3-nitro-5-phenyl 223
1-methyl-5-nitro-3-phenyl 223, 224
3-methyl-4-nitro-1-phenyl 192, 236
5-methyl-3-nitro-1-phenyl 192
5-methyl-4-nitro-1-phenyl 192
3-methyl-4-nitro-1-phenyl-5-(pyrazol-
  3(5)-yl) 227, 231
1-methyl-3-nitroso 246
5(3)-methyl-4-nitro-3(5)-(3-pyridyl) 191
1-methyl-3-nitro-4-(4-tolyl) 188
1-nitro 45, 186, 191, 201, 239, 244, 253, 256
3(5)-nitro 45, 46, 201, 202, 203, 205, 227,
  232, 235, 238, 242, 244, 249, 252, 257,
  295
4-nitro 186, 191, 194, 203, 216, 226, 229,
  235, 238, 240–242, 244, 246, 248, 249,
  252–254, 257
4-nitro-1-(3-nitrophenyl) 188
4-nitro-1-(4-nitrophenyl) 188, 292
3(5)-nitro-5(3)-(4-nitrophenyl) 201, 243
4-nitro-3(5)-(4-nitrophenyl) 243, 252
4-nitro-5(3)-(4-nitrophenyl)-3(5)-
  perfluoroamyl 191
4-nitro-5(3)-(4-nitrophenyl)-3(5)-
  perfluoroheptyl 191
4-nitro-1-(4-nitropyrazol-1-yl) 240
4-nitro-5(3)-(4-nitropyrazol-1-yl) 190
1-(4-nitrophenyl) 188
1-nitro-3(5)-phenyl 192
3-nitro-1-phenyl 203, 227
3(5)-nitro-1-4-phenyl 201
3(5)-nitro-5(3)-phenyl 201, 243, 252
4-nitro-1-phenyl 189, 192, 216
4-nitro-3(5)-phenyl 192, 242, 243, 252
5-nitro-1-phenyl 203
3-nitro-1-picryl 205, 245
1-(4-nitropyrazol-1-ylmethyl) 188
3-nitro-1-(2-pyridyl) 227
3(5)-nitro-5(3)-(3-pyridyl) 213, 214, 230,
  237, 253
4-nitro-3(5)-(3-pyridyl) 191, 230
1-nitro-4-ribofuranosyl 193, 194
3(5)-nitroso 203
4-nitroso 203
4-nitro-1-(4-tolyl) 192
4-nitro-1-tosyl 216
4-nitro-3(5)-trideuteriomethyl 187
1-nitro-3,4,5-trimethyl 192
4-nitro-1,3,5-trimethyl 225, 245, 249, 253,
  258
4-nitro-3-trimethylsilyl 200
Pyrazolo[1,5-a]pyridine

2-alkyl-3-nitro 218, 219
3,4-dinitro 218, 219, 249
2-hydroxy-3-nitro 218, 219
3-nitro 218, 219, 241
3-nitroso 218, 219
1H-Pyrazolo[4,3-b]pyridine
  3,6-dinitro-5-hydroxy 219
  5-methyl-3-nitro 219
Pyrazolo[1,5-a]pyrimidine
  5,7-dimethyl-3-nitro 220–222, 235
  5-methyl-7-ethyl-3-nitro 220–222
  2-nitro 220–222
  3-nitro 220–222
  6-nitro 220–222
  3-nitro-5,7-dihydroxy 220–222
Pyrazolo[3,4-d]pyrimidine
  1-aryl-4-hydroxy-3-nitro 219, 220
  1-cyclohexyl-4-hydroxy-3-nitro 219,
    220, 236
  1-methyl-4-hydroxy-3-nitro 219, 220, 236
Pyrazolo[1,5-a]pyrimidin-7-one
  6-carboethoxy-4-ethyl-3-nitro 220, 221
  6-carboethoxy-3-nitro 220, 221
  4-ethyl-3-nitro 220, 221
Pyrrole
  3-acetamido-1,2-dimethyl-4-nitro 4
  3-acetamido-1,2-dimethyl-5-nitro 4
  3-acetamido-1-methyl-2-nitro 4
  3-acetamido-1-methyl-4-nitro 4
  3-acetamido-1-methyl-5-nitro 4
  5-acetyl-1,2-dimethyl-3-nitro 58, 67
  2-acetyl-3,5-dinitro 7
  2-acetyl-4,5-dinitro 7
  3-acetyl-2,5-diphenyl-4-nitro 22
  3-acetyl-2,5-diphenyl-4-nitroso 22
  2-acetyl-1-(2-hydroxyethyl)-5-nitro 49
  3-acetyl-1-hydroxy-2-methyl-5-
    (5-methylisoxazol-3-yl)-4-nitro 7, 42
  2-acetyl-1-methyl-4-nitro 47, 59
  2-acetyl-1-methyl-5-nitro 59
  4-acetyl-1-methyl-2-nitro 58
  2-acyl-4-methyl-3-nitro 26
  1-acyl-2-nitro 5
  1-acyl-3-nitro 5, 41
  2-acyl-3-nitro 5, 25
  2-acyl-4-nitro 5, 47, 58, 64, 71
  2-acyl-5-nitro 5, 47, 64, 71
  3-acyl-2-nitro 6
  3-acyl-4-nitro 6
  3-acyl-5-nitro 6, 47, 58
  4-amino-2-methyl-1-nitro 70
  1-benzenesulfonyl-3-nitro 5
  3-benzoyl-2,5-diphenyl-4-nitro 22
  3-benzoyl-2,5-diphenyl-4-nitroso 22
  1-benzyl-2,4-dinitro 23
  1-benzyl-3,4-dinitro 23
  1-benzyl-2,5-diphenyl-3-nitro 44
  1-benzyl-2-nitro 3
  1-benzyl-3-nitro 3

Subject Index

1-benzyl-3-nitro-2,4,5-triphenyl 44
2,5-bis(dimethylamino)-3-nitro-4-phenylimino 26
2-bromoacetyl-1-methyl-4-nitro 59
2-bromoacetyl-1-methyl-5-nitro 59
5-bromo-2-bromomethyl-1-methyl-3-nitro 60
5-bromo-1,2-dimethyl-3-nitro 60
3-bromo-2-hydroxymethyl-1-methyl-5-nitro 60
2-bromomethyl-4,5-dibromo-1-methyl-3-nitro 60
4-bromo-2-nitro 59
1-t-butyl-2-dimethylamino-4-nitro 51
1-t-butyl-2,4-dinitro 19
1-t-butyl-3,4-dinitro 19, 51
1-t-butyl-2-nitro 2, 3
1-t-butyl-3-nitro 2, 3
1-t-butyl-2,3,4-trinitro 19
1-t-butyl-2,3,5-trinitro 19
2-chlorocarbonyl-4-nitro 56
2-chloro-4-nitro 60
2-cyano-1-methyl-4-nitro 14
2-cyano-1-methyl-5-nitro 14
2-cyano-4-nitro 14, 26
2-cyano-5-nitro 14, 55
3,5-diacetyl-1,2-dimethyl-4-nitro 58, 67
4,5-dibromo-1,2-dimethyl-3-nitro 60
4,5-dibromo-2-hydroxymethyl-1-methyl-3-nitro 60
2-dichloroacetyl-5-nitro 55
2,3-dichloro-4-nitro 60
2,5-dichloro-4-nitro 60
3,5-dichloro-4-nitro 60
1,2-dimethyl-3,4-dinitro 14, 23, 67
1,2-dimethyl-3,5-dinitro 14, 67
1,2-dimethyl-4,5-dinitro 15
2,5-dimethyl-1,3-dinitro 55, 63, 67, 114
3,5-dimethyl-2-formyl-4-nitro 57
1,2-dimethyl-3-nitro 4, 58, 59, 67
1,2-dimethyl-4-nitro 4, 58
1,2-dimethyl-5-nitro 4
2,4-dinitro 7, 14, 15, 43, 48, 54, 55, 63
2,5-dinitro 14, 15, 54, 60, 61, 63, 66
4,5-dinitro-2-formyl 21
2,4-dinitro-1-isopropyl 19
3,4-dinitro-1-isopropyl 19
1,4-dinitro-2-methyl 24, 43, 44, 49, 55, 63, 64, 70
2,3-dinitro-1-methyl 14, 15
2,4-dinitro-1-methyl 14, 15, 19, 49, 56, 60, 78
2,5-dinitro-1-methyl 14, 15, 49, 50
3,4-dinitro-1-methyl 14, 15, 19, 50
2,4-dinitro-1-(2-nitrophenyl) 19
2,4-dinitro-1-(4-nitrophenyl) 19, 49
3,4-dinitro-1-(2-nitrophenyl) 19
3,4-dinitro-1-(4-nitrophenyl) 19
2,5-diphenyl-1-methyl-3-nitro 26, 67

2,5-diphenyl-1-methyl-4-(2-pyridyl)-3-nitro 46, 67
2,5-diphenyl-3-nitro 21
2,5-diphenyl-4-nitro-3-phenacylsulfonyl 7
2-formyl-1-methyl-3-nitro 6, 41
2-formyl-1-methyl-4-nitro 6, 41
2-formyl-1-methyl-5-nitro 6, 41
2-formyl-3-nitro 65, 67
2-formyl-4-nitro 21, 41, 48, 57, 58, 67, 71
2-formyl-5-nitro 21, 57, 65, 67, 68, 71
3-formyl-2-nitro 49, 65, 67
3-formyl-4-nitro 49, 65, 67, 68
3-formyl-5-nitro 65, 68
1-β-hydroxyethyl-2-nitro 71
2-hydroxymethyl-1-methyl-5-nitro 60
3-isopropylamino-1-methyl-4-nitro 51
1-isopropyl-3-isopropylamino-4-nitro 51
1-isopropyl-3-methylamino-4-nitro 51
1-isopropyl-2-nitro 2, 3
1-isopropyl-3-nitro 2, 3
1-isopropyl-2,3,4-trinitro 19, 64
1-isopropyl-2,3,5-trinitro 19
2-methoxy-1-methyl-4-nitro 50
3-methoxy-1-methyl-4-nitro 50
5-methoxy-1-methyl-2-nitro 49
1-methyl-2-methylamino-5-nitro 113
1-methyl-2-nitro 2, 3, 43, 58–60, 66, 70
1-methyl-3-nitro 2, 3, 43, 47, 58–60, 70
2-methyl-3-nitro 4, 48
2-methyl-4-nitro 4
2-methyl-5-nitro 3, 48, 60
1-methyl-4-nitro-2-piperidino 50
1-methyl-2-nitro-5-piperidino 49, 64
1-methyl-2,3,4,5-tetranitro 19
2-methylthio-3-nitro 24, 47
1-methyl-2,3,4-trinitro 15, 19
1-methyl-2,3,5-trinitro 17–19
1-nitro 2
2-nitro 1, 3, 5, 40, 43–46, 54, 58–60, 62, 63, 66, 68, 69, 70, 246
3-nitro 1, 3, 5, 43–46, 48, 55, 58, 60–62, 66, 68, 70
2-nitro-1-(2-nitrophenyl) 2
2-nitro-1-(4-nitrophenyl) 2
3-nitro-1-(2-nitrophenyl) 2
3-nitro-1-(4-nitrophenyl) 2
2-nitro-1-phenyl 2
3-nitro-1-phenyl 2
3-nitro-4-phenyl 59
1-(4-nitrophenyl)-2,3,4-trinitro 19
4-nitro-2-propionyl 64
5-nitro-2-sulfonamides 71
3-nitro-1,2,4,5-tetraphenyl 43
4-nitro-2,3,5-trichloro 60
2-nitro-3,4,5-triiodo 8
3-nitro-1-triisopropylsilyl 3
3-nitro-1,2,5-trimethyl 4, 67
3-nitro-2,4,5-triphenyl 21
Pyrrole "polymerization" 2

Pyrrolesufonic acids, nitrolysis 19
3-Pyrrolin-2-one
   3-amino-4-nitro-5-phenyl 51, 67
   1-cyclohexyl-3-ethoxy-4-nitro 70
   3-methoxy-4-nitro-5-phenyl 51, 67
Pyrrolizine
   5-nitro 4
   6-nitro 4
   7-nitro 4
Pyrrolo[2,1-c]-[1,2,4]benzotriazine
   3-nitro 38
   1-nitro-3-iodo 38
Pyrrolo[3,2-b]indole, nitro derivatives 77
Pyrrolo[2,1,5-cd]indolizine
   1-nitro 39
5-H-Pyrrolo[2,1-a]isoindole
   2,3-dihydro-2,5-dioxo-3,3-dimethyl, nitration of 39
Pyrrolomycin 60, 70, 78
Pyrrolophenanthrone 77
Pyrrolopyridones 42
Pyrrolo[1,2-a]pyrazines
   1,3-dinitro 38
   1-nitro 38
   3-nitro 38
Pyrrolo[1,2-b]pyridazine
   2,4-dimethyl-5,7-dinitro 38
   2-methyl-4-p-nitrophenyl-5,7-dinitro 38
Pyrrolo[2,3-d]pyrimidin-4-one
   5-nitro 34, 71
Pyrrolo[1,2-a]quinoline
   2,7-dimethyl-3,6-dinitro 36
   2,7-dimethyl-1-nitro 36
   2,7-dimethyl-6-nitro 36
Pyrrolo[1,2-a]quinoxaline
   6,9-dimethoxy-3-nitro 38
   1-nitro 38, 43, 53
   3-nitro 38, 43
   1-nitro-3-sulfonic acid 38, 53
Pyrrolothiazines 42
Pyrrolo[2,1-b]thiazole
   6-methyl, nitration of 39
Pyrrolo[1,2-d]-[1,2,4]triazine
   1,2-dihydro-1-oxo-6,8-dinitro 38
   1,2-dihydro-1-oxo-6-nitro 38
   1,2-dihydro-1-oxo-8-nitro 38
β-Pyrrol-2-ylacrylic acids
   α-aryl 57, 64
   4-nitro 57
   5-nitro 57

1-(8-Quinolinesulfonyl)-3-nitro-1,2,4-triazole 287

Reduction
   nitroimidazoles 146–151
   nitroisoxazoles 323, 324
   nitropyrazoles 235, 240
   nitropyrroles 40–46
   nitrotriazoles 287–289
Release 230, 257
Replacement of ring halogen and pseudohalogen
   nitroimidazoles 129–135
   nitropyrazoles 230–232
   nitrotriazoles 282
Replacement of the nitro group
   nitroimidazoles 137
   nitroisoxazoles 324
   nitrooxadiazoles 342, 343
   nitropyrazoles 232
   nitropyrroles 49–54
   nitrotriazoles 282
Replacements in side chains in nitroimidazoles 141–146
Ring closure and ring transformation to
   nitroimidazo[1,2-a]indoles 107
   nitroimidazo[2,1-a]isosquinolines 106
   nitroimidazoles 93
   nitroimidazo[2,1-d]-1,4-oxazepines 106
   nitroimidazo[2,1-b]oxazoles 104
   nitroimidazo[1,2-a]pyrimidines 103
   nitroimidazo[5,1-b]thiazoles 103
   nitroimidazo[1,5-c]-1,2,4-triazines 106
   nitroisoxazoles 304–322
   nitrooxadiazoles 341
   nitropyrazoles 207–216
   nitropyrazolo[4,3-c]pyrazoles 222
   nitropyrazolo(1,5,a]quinazolines 222
   nitropyrazolo[1,5-a]s-triazines 222
   nitropyrroles 22–27
   nitrotriazoles 273
Ronidazole 166

Sandmeyer reaction 272, 273
Schmidt reaction 281, 290
Styrylnitroisoxazoles 340

1,2,3,4-Tetrahydroimidazo[1,5-c]-1,2,4-triazine
   6-methyl-8-nitro 106
3,3a,4,5-Tetrahydro-2H-isoxazolo[2,3-b]-isoxazole
   cis- and trans-2,5-dimethyl-3,4-diphenyl-3a-nitro 305
   2,5-diphenyl 308, 309
   3-ethoxy-3a-nitro 327
   4-hydroxy-3a-nitro-2-substituted 327
   2-methyl-3a-nitro 308, 309, 338, 339
   3-methyl-3a-nitro 308, 309, 338, 339
   4-methyl-3a-nitro-2-trimethylsilylalkyl 309, 310
   3a-nitro 325, 326, 335, 336, 339, 340
   3a-nitro-2-phenyl 308, 309, 338, 339
   3a-nitro-3-phenyl 308, 309, 338, 339
   3a-nitro-2-trimethylsilylalkyl 309, 310
Tetranitromethane 31, 32, 39, 311–313

Tetrazole
   1-methyl-5-nitro 277, 281, 293, 296
   2-methyl-5-nitro 277, 281, 282, 296
   5-nitro 276, 277, 281, 292, 293, 296
   2-picryl-5-nitro 282, 296
1-Tetrazoyl-4,6-dinitroindazole 206
Thermal properties and reactions
   nitroimidazoles 163, 164
   nitroisoxazoles 339, 340
   nitrooxadiazoles 344
   nitropyrazoles 201, 245
   nitropyrroles 60, 70
   nitrotetrazoles 296
   nitrotriazoles 271, 296
1,3,4-Thiadiazoylnitroimidazoles 85
Thiazolo[2,3,4-c,d]pyrrolizine
   3,4-benzo-5,6-dihydro-5-p-methoxyphenyl-6-nitro-6a-phenyl 39
Tinidazole 164–166, 180
1,2,3-Triazole
   1-alkyl-5-amino-4-nitro 274, 292
   2-(2,4-dinitrophenyl)-4-nitro 269, 293
   2-(2,4-dinitrophenyl)-5-nitro-1-oxide 271
   1,5-diphenyl-4-nitro 273
   4-hydroxy-5-nitro-2-phenyl 275
   5-methyl-2-(2,4-dinitrophenyl)-4-nitro 269
   1-methyl-5-N-methylanilino-4-nitro 274
   4-nitro 273, 274, 286
   4-nitro-1-(4-nitrophenyl) 269
   4-nitro-2-(4-nitrophenyl) 269, 292
   5-nitro-2-(4-nitrophenyl)-1-oxide 271
   4-nitro-2-phenyl 275
   5-nitro-2-phenyl 275
   5-nitro-2-phenyl-1-oxide 271
   4-nitro-2-phenyl-5-(2-phenyl-1,2,3-triazol-4-yloxy) 292, 293
   5-nitro-4-picrylamino 270
   4-nitro-1-β-D-ribofuranosyl 286, 287, 295
   4-nitro-2-β-D-ribofuranosyl 286, 287
   4-nitro-2-(1,2,4-triazol-3-yl) 275, 293
   4-nitro-2-(2,4,6-trinitrophenyl) 269
1,2,4-Triazole
   1-acetamidoalkyl-3,5-dinitro 291
   5-alkoxy-1-methyl-3-nitro 282, 283
   5-alkylamino-1-methyl-3-nitro 282, 283
   1-alkyl-5-amino-3-nitro 290
   1-alkyl-3-nitro 289, 290
   5-alkyl-3-nitro 281, 291
   1-allyl-3,5-dinitro 277, 278
   5-amino-1-methyl-3-nitro 283, 285
   5-amino-1-nitro 271, 272
   5-amino-3-nitro-1-(3-oxobutyl) 289
   5-aryloxy-1-methyl-3-nitro 283
   3-azido-5-nitro 291, 293
   5-azido-1-methyl-3-nitro 288, 291, 292
   1-bromo-3-nitro 289
   3-chloro-1-methyl-5-nitro 286
   3-chloro-5-nitro 270
   5-chloro-3-nitro-1-picryl 279
   5-dialkylamino-1-methyl-3-nitro 286
   1,3-dimethyl-5-nitro 277, 278
   1,5-dimethyl-3-nitro 277, 278
   3,5-dinitro 272, 273, 284, 285, 291, 293, 296
   3,5-dinitro-1-methyl 277, 283–285, 288
   3,5-dinitro-1-(2-nitroethyl) 281, 285
   3,5-dinitro-4-(2-nitro-2-phenylethyl) 281
   3,5-dinitro-4-(2-nitropropyl) 281
   3,5-dinitro-1-(3-oxobutyl) 278, 285, 288, 289–291
   1-(2,4-dinitrophenyl)-3-nitro 294
   5-halo-1-methyl-3-nitro 282, 283, 285, 286
   3-halo-1-nitro 284
   3-halo-5-nitro 272, 273, 284
   5-hydrazino-1-methyl-3-nitro 283, 288
   5-hydroxylamino-1-methyl-3-nitro 288
   5-hydroxy-1-methyl-3-nitro 282, 283
   3-hydroxy-5-nitro 270
   1-(mesitylene-2-sulfonyl)-3-nitro 287, 295
   5-methoxy-1-methyl-3-nitro 280
   5-methoxy-3-nitro 280
   1-methyl-5-(5-methyltetrazolyl)-3-nitro 285
   1-methyl-3-nitro 277, 278, 283, 290, 293, 294
   1-methyl-5-nitro 277, 278, 286, 293, 294
   3-methyl-1-nitro 271
   4-methyl-3-nitro 294
   5-methyl-3-nitro 270, 277, 278, 293, 294
   5-methyl-3-nitro 291
   1-methyl-3-nitro-5-(1,2,4-triazol-1-yl) 284, 285
   2-nitramino 271
   1-nitro 271
   3(5)-nitro 270, 272, 273, 281, 285, 287, 289–291, 293, 294
   3-nitro-1-(2-nitroethyl) 285
   4-nitro-1-(4-nitrophenyl) 273
   5-nitro-3-(4-nitrophenyl) 272
   3-nitro-1-(3-nitro-1,2,4-triazol-1-ylmethyl) 289, 290
   4-nitro-1-phenyl 273
   5-nitro-3-phenyl 272
3,5,5-Triphenyl-3,4-dinitropyrazoline 204
Tubercidin 34

Ullmann reaction 227
Ultraviolet spectroscopy
   nitroimidazoles 157–159
   nitroisoxazoles 336, 337
   nitropyrazoles 252–254
   nitropyrroles 62–64
   nitrotetrazoles 293
   nitrotriazoles 292, 293

Vilsmeier reaction 276
Vinylimidazoles 127, 162

Weedazol 273
Wittig reaction 127

X-ray spectroscopy
    nitroimidazoles 162

nitroisoxazoles 339
nitrooxadiazoles 344
nitropyrazoles 256
5-nitrotetrazole 295
nitrotriazoles 295

RAYMOND H. FOGLER LIBRARY
DATE DUE

BOOKS ARE SUBJECT TO
RECALL AFTER TWO WEEKS

SEP 0 9 1988